Studies in Computational Intelligence

Volume 919

Series Editor

Janusz Kacprzyk, Polish Academy of Sciences, Warsaw, Poland

The series "Studies in Computational Intelligence" (SCI) publishes new developments and advances in the various areas of computational intelligence—quickly and with a high quality. The intent is to cover the theory, applications, and design methods of computational intelligence, as embedded in the fields of engineering, computer science, physics and life sciences, as well as the methodologies behind them. The series contains monographs, lecture notes and edited volumes in computational intelligence spanning the areas of neural networks, connectionist systems, genetic algorithms, evolutionary computation, artificial intelligence, cellular automata, self-organizing systems, soft computing, fuzzy systems, and hybrid intelligent systems. Of particular value to both the contributors and the readership are the short publication timeframe and the world-wide distribution, which enable both wide and rapid dissemination of research output.

Indexed by SCOPUS, DBLP, WTI Frankfurt eG, zbMATH, SCImago.

All books published in the series are submitted for consideration in Web of Science.

More information about this series at http://www.springer.com/series/7092

Yassine Maleh · Mohammad Shojafar ·
Mamoun Alazab · Youssef Baddi
Editors

Machine Intelligence and Big Data Analytics for Cybersecurity Applications

 Springer

Editors
Yassine Maleh🆔
Sultan Moulay Slimane University
Beni Mellal, Morocco

Mamoun Alazab🆔
Charles Darwin University
Darwin, NT, Australia

Mohammad Shojafar🆔
Institute for Communication Systems
University of Surrey
Guildford, UK

Youssef Baddi🆔
Chouaib Doukkali University
El Jadida, Morocco

ISSN 1860-949X ISSN 1860-9503 (electronic)
Studies in Computational Intelligence
ISBN 978-3-030-57026-2 ISBN 978-3-030-57024-8 (eBook)
https://doi.org/10.1007/978-3-030-57024-8

This Springer imprint is published by the registered company Springer Nature Switzerland AG
The registered company address is: Gewerbestrasse 11, 6330 Cham, Switzerland

Preface

As cyber-attacks against critical infrastructure increase and evolve, automated systems to complement human analysis are needed. Moreover, chasing the breaches is like looking for a needle in a haystack. Such organizations are so large, with so much information and data to sort through to obtain actionable information that it seems impossible to know where to start. The analysis of an attack's intelligence is traditionally an iterative, mainly manual process, which involves an unlimited amount of data to try to determine the sophisticated patterns and behaviors of intruders. Besides, most of the detected intrusions provide a limited set of attributes on a single phase of an attack. Accurate and timely knowledge of all stages of an intrusion would allow us to support our cyber-detection and prevention capabilities, enhance our information on cyber-threats, and facilitate the immediate sharing of information on threats, as we share several elements. The book is expected to address the above issues and will aim to present new research in the field of cyber-threat hunting, information on cyber-threats, and analysis of important data.

Therefore, cyber-attacks protection of computer systems is one of the most critical cybersecurity tasks for single users and businesses. Even a single attack can result in compromised data and sufficient losses. Massive losses and frequent attacks dictate the need for accurate and timely detection methods. Current static and dynamic methods do not provide efficient detection, especially when dealing with zero-day attacks. For this reason, big data analytics and machine intelligence-based techniques can be used.

This book brings together researchers in the field of cybersecurity and machine intelligence to advance the missions of anticipating, prohibiting, preventing, preparing, and responding to various cybersecurity issues and challenges. The wide variety of topics it presents offers readers multiple perspectives on a variety of disciplines related to machine intelligence and big data analytics for cybersecurity applications.

Machine intelligence and big data analytics for Cybersecurity Applications comprise a number of state-of-the-art contributions from both scientists and practitioners working in machine intelligence and cybersecurity. It aspires to provide a relevant reference for students, researchers, engineers, and professionals working in this area or those interested in grasping its diverse facets and exploring the latest advances on machine intelligence and big data analytics for cybersecurity applications. More specifically, the book consists of 24 contributions classified into three pivotal sections: Machine intelligence and big data analytics for cybersecurity: Fundamentals and Challenges: Introducing the state-of-the-art and the taxonomy of machine intelligence and big data for cybersecurity. Section 2 Machine intelligence and big data analytics for cyber-threat detection and analysis: Offering the latest architectures and applications of machine intelligence and big data analytics for cyber-threats and malware detection and analysis. Section 3 Machine intelligence and big data analytics for cybersecurity applications: Dealing with the application of machine intelligence techniques for cybersecurity in many fields from IoT health care to cyber-physical systems and vehicle security.

We want to take this opportunity and express our thanks to the authors of this volume and the reviewers for their great efforts by reviewing and providing interesting feedback to the authors of the chapter. The editors would like to thank Dr. Thomas Ditsinger Springer, Editorial Director (Interdisciplinary Applied Sciences) and Prof. Janusz Kacprzyk (Series Editor-in-Chief), and Ms. Jennifer Sweety Johnson (Springer Project Coordinator), for the editorial assistance and support to produce this important scientific work. With this collective effort, this book would not have been possible.

Khouribga, Morocco   Prof. Yassine Maleh
El Jadida, Morocco   Prof. Youssef Baddi
Guildford, UK   Prof. Mohammad Shojafar
Darwin, Australia   Prof. Mamoun Alazab

Contents

About the Editors

Yassine Maleh is an Associate Professor at the National School of Applied Sciences at Sultan Moulay Slimane University, Morocco. He received his Ph.D. degree in Computer Science from Hassan first University, Morocco. He is a cybersecurity and information technology researcher and practitioner with industry and academic experience. He worked for the National Ports Agency in Morocco as an IT manager from 2012 to 2019. He is a Senior Member of IEEE, Member of the International Association of Engineers IAENG and The Machine Intelligence Research Labs. Dr. Maleh has made contributions in the fields of information security and privacy, Internet of things security, wireless and constrained networks security. His research interests include information security and privacy, Internet of things, networks security, information system, and IT governance. He has published over than 50 papers (book chapters, international journals, and conferences/ workshops), four edited books, and one authored book. He is the editor in chief of the International Journal of Smart Security Technologies (IJSST). He serves as an Associate Editor for IEEE Access (2019 Impact Factor 4.098), the International Journal of Digital Crime and Forensics (IJDCF), and the International Journal of Information Security and Privacy (IJISP). He was also a Guest Editor of a special issue on Recent Advances on Cyber Security and Privacy for Cloud-of-Things of the International Journal of Digital Crime and Forensics (IJDCF), Volume 10, Issue 3, July–September 2019. He has served and continues to serve on executive and technical program committees and as a reviewer of numerous international conference and journals such as Elsevier Ad Hoc Networks, IEEE Network Magazine, IEEE Sensor Journal, ICT Express, and Springer Cluster Computing. He was the Publicity Chair of BCCA 2019 and the General Chair of the MLBDACP 19 symposium.

Mohammad Shojafar received his Ph.D. in Information Communication and Telecommunications (advisor Prof. Enzo Baccarelli) from Sapienza University of Rome, Italy, as the second rank university in QS Ranking in Italy and top 100 in the world with an Excellent degree in May 2016. He is Intel Innovator, Senior IEEE member, and Senior Lecturer in the 5GIC/ICS at the University of Surrey, Guildford,

UK. Before joint to 5GIC, he was served as a Senior Member in the Computer Department at the University of Ryerson, Toronto, Canada. He was Senior Researcher (Researcher Grant B) and a Marie Curie Fellow in the SPRITZ Security and Privacy Research group at the University of Padua, Italy. Also, he was a Senior Researcher in the Consorzio Nazionale Interuniversitario per le Telecomunicazioni (CNIT) partner at the University of Rome Tor Vergata contributed to 5g PPP European H2020 "SUPERFLUIDITY" project for 14 months. Dr. Mohammad was principle investigator on PRISENODE project, a 275,000 euro Horizon 2020 Marie Curie project in the areas of network security and fog computing and resource scheduling collaborating between the University of Padua and University of Melbourne. He also was a principal investigator on an Italian SDN security and privacy (60,000 euro) supported by the University of Padua in 2018. He was contributed to some Italian projects in telecommunications like GAUChO—A Green Adaptive Fog Computing and Networking Architecture (400,000 euro), S2C: Secure, Software-defined Cloud (30,000 Euro), and SAMMClouds—Secure and Adaptive Management of Multi-Clouds (30,000 euro) collaborating among Italian universities. His main research interest is in the area of Network and Network Security and Privacy. In this area, he published more than 100+ papers in topmost international peer-reviewed journals and conferences, e.g., IEEE TCC, IEEE TNSM, IEEE TGCN, IEEE TSUSC, IEEE Network, IEEE SMC, IEEE PIMRC, and IEEE ICC/GLOBECOM. He served as a PC member of several prestigious conferences, including IEEE INFOCOM Workshops in 2019, IEEE GLOBECOM, IEEE ICC, IEEE ICCE, IEEE UCC, IEEE SC2, IEEE ScalCom, and IEEE SMC. He was a General Chair in FMEC 2019, INCoS 2019, INCoS 2018, and a Technical Program Chair in IEEE FMEC 2020. He served as an Associate Editor in IEEE Transactions on Consumer Electronics, IET Communication, Springer Cluster Computing, KSII - Transactions on Internet and Information Systems, Tylor & Francis International Journal of Computers and Applications (IJCA), and Ad Hoc & Sensor Wireless Networks Journals.

Mamoun Alazab is the Associate Professor in the College of Engineering, IT and Environment at Charles Darwin University, Australia. He received his Ph.D. degree in Computer Science from the Federation University of Australia, School of Science, Information Technology and Engineering. He is a cybersecurity researcher and practitioner with industry and academic experience. Dr. Alazab's research is multidisciplinary that focuses on cybersecurity and digital forensics of computer systems including current and emerging issues in the cyber environment like cyber-physical systems and the Internet of things, by taking into consideration the unique challenges present in these environments, with a focus on cybercrime detection and prevention. He looks into the intersection use of machine learning as an essential tool for cybersecurity, for example, for detecting attacks, analyzing malicious code or uncovering vulnerabilities in software. He has more than 100 research papers. He is the recipient of short fellowship from Japan Society for the Promotion of Science (JSPS) based on his nomination from the Australian Academy of Science. He delivered many invited and keynote speeches, 27 events in

2019 alone. He convened and chaired more than 50 conferences and workshops. He is the founding chair of the IEEE Northern Territory Subsection: (February 2019–current). He is a Senior Member of the IEEE, Cybersecurity Academic Ambassador for Oman's Information Technology Authority (ITA), Member of the IEEE Computer Society's Technical Committee on Security and Privacy (TCSP) and has worked closely with government and industry on many projects, including IBM, Trend Micro, the Australian Federal Police (AFP), the Australian Communications and Media Authority (ACMA), Westpac, UNODC, and the Attorney General's Department.

Youssef Baddi is full-time Assistant Professor at Chouaïb Doukkali University UCD EL Jadida, Morocco. He received his PhD degree in computer science from ENSIAS School, University Mohammed V Souissi, Rabat. He also holds a Research Master's degree in networking obtained in 2010 from the High National School for Computer Science and Systems Analysis—ENSIAS-Morocco-Rabat. He is a member of Laboratory of Information and Communication Sciences and Technologies STIC Lab, since 2017. He is a guest member of Information Security Research Team (ISeRT) and Innovation on Digital and Enterprise Architectures Team, ENSIAS, Rabat, Morocco. Dr Baddi was awarded as the best PhD student in University Mohammed V Souissi of Rabat in 2013. Dr. Baddi has made contributions in the fields of group communications and protocols, information security and privacy, software-defined network, the Internet of things, mobile and wireless networks security, Mobile IPv6. His research interests include information security and privacy, the Internet of things, networks security, software-defined network, software-defined security, IPv6, and Mobile IP. He has served and continues to serve on executive and technical program committees and as a reviewer of numerous international conferences and journals such as Elsevier Pervasive and Mobile Computing PMC and International Journal of Electronics and Communications AEUE, and Journal of King Saud University—Computer and Information Sciences. He was the General Chair of IWENC 2019 Workshop and the Secretary Member of the ICACIN 2020 Conference.

Machine Intelligence and Big Data Analytics for Cybersecurity: Fundamentals and Challenges

Network Intrusion Detection: Taxonomy and Machine Learning Applications

Anjum Nazir and Rizwan Ahmed Khan

Abstract Information and Communication Technologies (ICT) has revolutionized our lives and transform it into a knowledge centric world. Where information is available just under few clicks. This advancement introduced different challenges and problems. One big challenge of today's world is cybersecurity and privacy issues. With every passing day, number of cyber-attacks are increasing. Legacy security solutions like firewalls, antivirus, intrusion detection and prevention systems etc. are not equipped with right technologies to neutralized advance attacks. Recent developments in machine learning, deep learning have shown great potential to deal with modern attack vectors. In this chapter, we will present: (1) Current state of cyber-attacks. (2) Overview of Intrusion Detection Systems and taxonomy. (3) Recent techniques in machine/deep learning being used to detect and defend against novel intrusion.

Keywords Intrusion detection · Machine learning · Classification

1 Introduction

Internet has completely changed the way we used to live and perform routine tasks. Its exponential growth allows to interconnect and communicate anywhere, anytime and access almost any type of service that was just a dream before. This has become possible due to the advancements in Information and Communication Technologies (ICT), economical access of quality services and easy availability of products and tools. ICT refers to the use of technologies which are responsible for information processing and safe secure transmission and sharing of information. This advancements have opened new challenges and problems for researchers, practitioners and

A. Nazir · R. A. Khan (✉)
Faculty of IT, Barrett Hodgson University, Karachi, Pakistan
e-mail: rizwan17@gmail.com

A. Nazir
e-mail: anjum@geeks-hub.com

Y. Maleh et al. (eds.), *Machine Intelligence and Big Data Analytics for Cybersecurity Applications*, Studies in Computational Intelligence 919,
https://doi.org/10.1007/978-3-030-57024-8_1

3

end users. Security, privacy and trust in public networks is one of the biggest challenge of today that not only impacts industries, government and private organizations but also a common home user as well.

Internet is a public network, which is open and can be used by anyone [1]. Statistics show that there is a deafening increase in the number of cyberattacks performed every year. In computer systems an attack can be defined as an attempt to expose, alter, disable, destroy, steal or gain unauthorized access to or make unauthorized use of an asset [2]. Symantec Internet Security Threat Report (ISTR) 2019 [3] presents an analysis about growth and progression of commonly perpetuated cyberattacks. The summary of ISTR 2019 is presented below.

1. **Web Attacks**: The report shows that overall web attacks on end points is increased by 56% in 2018. In 2018, one in every ten URL was identified as malicious, as compared to previous year in which the ratio was 1 out 16.
2. **Cryptojacking**: Cryptojacking is an emerging threat for web browsers specially for mobile and other smart gadgets. It is a type of malware generally browser-based scripts or plugins that hooks itself and start mining cryptocurrencies. Analysis report shows that there has been at least four times more cryptojacking events were detected.
3. **Email Attacks**: Attackers refocused on using malicious email (or attachments) as a primary infection vector. Microsoft Office users remain the prime target of email-based malware. ISTR report shows that office files are accounting for at least 48% of malicious email attachments, this number has increased by 5% from 2017.
4. **Malware**: Use of malicious "Power Shell" scripts is increased by 1000% in 2018. Like 'Emotet' is a self-propagating malware that is jump up to 16 from 4% in 2017. Cyber crime groups continued to use macros in Office files as their preferred method to propagate malicious payloads.
5. **Ransomware**: Ransomware is also relatively a new type of malware which actually encrypts users data and ask to pay ransom amount to get the decryption key. There is a 12 and 33% growth is observed for enterprise and mobile ransomware.
6. **Mobile Malware**: Information gathered from different sources show that 1 in 36 mobile devices usually have high risk application installed which can be used to launch attacks.
7. **Targeted Attacks**: Number of organized attack groups those use destructive malware has increased by 25%. 65% of groups used spear phishing as the primary infection vector. 96% of groups' primary motivation was to be intelligence gathering. Attacks on supply chain has also increased by 78%.
8. **Internet of Things**: After a massive increase in Internet of Things (IoT) attacks in 2017 (reported upto 600%), attack numbers stabilized in 2018. Routers and connected cameras were the most infected devices and which accounted for 75 and 15% of the attacks respectively.

Attacking physical or virtual infrastructure for malicious purpose is not new. There are many reported incidents which are dated back to World War II (WWII) era [4]. Cyberattack rate has grown exponentially in last few years. In literature we found different reasons and motivations behind the pandemic growth. Taylor [5] discussed several reasons and Brewster et al. pointed out attack motivations taxonomy in [6]. They highlighted several motivations like political, ideological, commercial, emotional, financial, personal, etc, which can be behind a cyberattack.

Main reasons and motivations behind cyberattacks are:

1. Political or social cause: different incidents have been reported where hackers interfere to influence social or a political cause. Bessi and Ferrara [7], Kollanyi et al. [8] and Allcott and Gentzkow [9] discussed and explained how social bots distort 2016 US Presidential Election online discussion. Such hacking activities and groups of hackers are usually sponsored by the state or the competitors of the target organization [10].
2. Easy and control free availability of tools: basic but often neglected reason of increase numbers of cyberattacks is the easy and control free availability of tools and procedures used by hackers. As a result, a user can easily launch an attack without requiring a detail and technical understanding of the underlying technologies and infrastructure. Hansman [11] discussed that attack sophistication has been increased and intruder knowledge or skills which are required to perpetuate an attacks has been reduced over years.
3. Financial gain: Ransomware is the most common type of cyberattack used for obtaining financial gains [12].

Considering the data presented above—traditional security solutions like antivirus, firewalls, Intrusion Detection /prevention Systems (ID/PS) etc. have been questioned for their reliability in detecting and providing safeguard.

Normal endpoint security solutions like antivirus can only block and stop execution of malicious or unwanted programs. They mostly use malware signatures to block them. A virus signature or a signature in general is a continuous sequence or stream of bytes or a pattern that is common for a certain malware sample [13]. Antivirus software usually applies different hooks (kernel hooks) at different locations in the operating system kernel to intercept execution flow of applications. When we run an application, antivirus intercepts and checks file signatures. If the signature is not matched in the signature database it will let it run, otherwise it will stop execution and will take appropriate necessary actions.

Every antivirus software depends upon signature database. Signature database is a repository of signatures of malicious programs. It is also known as virus definition which is pushed by the software vendor several times a day generally through cloud. There are various limiting factors which effect the performance and accuracy of an antivirus solutions discussed below.

- Since it contains signatures of malicious applications only. Therefore it will fail to detect new viruses until the signature is not developed and updated.
- Infinite numbers of signatures cannot be stored in the signature database. Therefore, it is likely possible that antivirus can miss a relatively older infection as well.
- Lastly, as signature database size grows it increases files scanning times as well.

Although latest endpoint security solutions have incorporated many advance techniques like heuristics, Machine Learning (ML) , Indicators of Compromise (IoC) etc. to detect new attacks and compromises.

Similarly conventional firewalls can only allow and deny traffic on the basis of IP Addresses [14] and port numbers [14]. This type of firewall is known as layer 4 or transport layer [15] firewall. These firewalls cannot differentiate between various protocols states. On the other hand stateful firewalls have the capability to understand and distinguish different protocol dialogues and handshaking processes. However, these firewalls still cannot perform deep packet inspection (DPI) [16] to inspect and look inside the packets for any kind of abnormality or intrusions.

With the advent of unified threat management (UTM) [17] and next generation based firewalls (NGFW) [18], firewalls can now look beyond packet headers. They can inspect and filter traffic on the basis of payload. Payload is actual message or data generated by the source machine for its intended recipient. These firewalls are also known as application and user aware firewalls because they can detect applications or protocols streams following through them and allow security administrators to apply policies on the basis of applications or users instead of fixed port numbers and IP Addresses. They also have built-in mechanism to detect intrusions.

Any kind of un-authorized activity on the hosts or in the network is considered as an intrusion. Karen and Mell [19] defines intrusion detection is the process of monitoring the events occurring in a computer system or in networks and analyzing them for the signs of possible incidents, which are violations or imminent threats of violation of computer security policies, acceptable use policies, or standard security practices.

Rest of the chapter is organized as follows. In Sect. 2 detail analysis of intrusion detection systems is presented. In this section IDS taxonomy is presented, which attempts to portray a comprehensive picture of technologies, methodologies, architectures, etc used by well known intrusion detection system. In Sect. 3 recent techniques, approaches and trends being practiced and researched in Network Intrusion Detection System (NIDS) domain from machine learning perspective are presented. In Sect. 3.2 we summarized and highlighted limitations of NIDS datasets. Subsequently, in Sect. 3.3 recent machine learning research conducted in NIDS domain is surveyed. We presented classifiers trends (most common classifiers used in NIDS) in last five years and critically analyzed the published work. Chapter summary is presented in Sect. 4.

2 Overview of Intrusion Detection System

Intrusion Detection System (IDS) plays an integral role to strengthen the security posture of an organization. Historically, intrusion detection systems were categorized as anomaly-based and misuse or signature-based systems [20]. An anomaly is considered as the deviation from the known or established behavior, while signature is a pattern or string that corresponds to a known attack. However, Herve et al. [21], Liao et al. and others [22] classify IDS based on different characteristics. Figure 1 presents IDS taxonomy based on different characteristics and behavior.

1. Detection Methodologies
2. Detection Approach
3. Analysis Target
4. Reaction on Intrusion Event
5. Analysis Timing
6. Architecture.

2.1 Detection Methodologies

The detection methodologies describe the methods followed by detection engine to detect intrusion. Detection engine is the core component of an IDS responsible to detect intrusion. Liao [22] and Scarfone [19] proposed three different intrusion detection methodologies (i) Signature-based (SD), (ii) Anomaly-based (AB) and (iii) Stateful Protocol Analysis (SPA) based.

Signature based IDS uses *Intrusion Signatures Vector (ISV)* to detect intrusions. An ISV is a pattern or string that corresponds to known attack or threat. It builds a database of known attacks and monitors network traffic flowing through it. On a signature match, it generates an alert of malicious activity which can be blocked by an IPS. Snort and Suricata [23] are well-known open-source signature-based intrusion detection systems.

On the other hand, Anomaly-Based (AB) intrusion detection systems analyze network or systems' behavior over a period of time and build an anomaly profile also known as model through training process. The model build after traffic monitoring is considered as the baseline which can be used to detect unkown intrusions through *'deviation measure'*. Any significant difference in the network behavior from the baseline is considered as deviation [24].

The main benefit of anomaly-based IDS is the their potential to detect unknown or novel attacks. However one of the biggest challenge of anomaly based IDS is high False Positive Rate (FPR). Anomaly-based IDS are prune to generate high false positives. When number of alerts generated by an IDS are very high then it becomes difficult for an analyst to investigate them properly and find root cause of the problem.

Stateful protocol analysis-based intrusion detection systems perform deep packet inspection to identify divergence from the standard or predefined protocol definitions

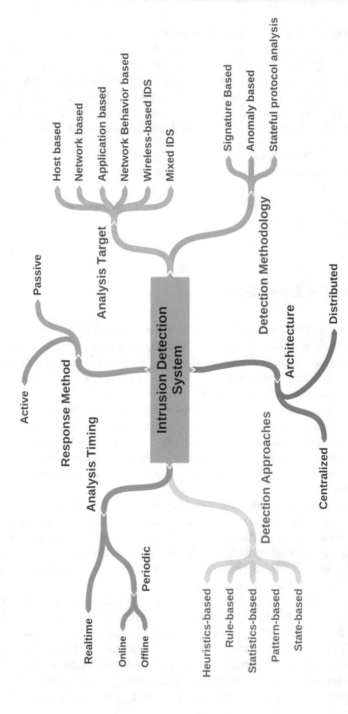

Fig. 1 Taxonomy of intrusion detection system

Table 1 Pros and cons of intrusion detection methodologies

Signature-based	Anomaly-based	Stateful protocol analysis
Pros		
• Simplest and effective detection methodology	• Effective to detect new and unforeseen vulnerabilities and attacks	• Efficient at detecting protocol design level vulnerabilities and flaws
• High detection rate with less false positive	• Facilitate the detection of variant of attacks	• Can distinguish unexpected sequence or protocol dialogues
• Provide more granular contextual analysis of attack(s)		
Cons		
• Ineffective to detect unknown (new) attacks, evasion attacks, and variants of known attacks	• Difficult to build accurate model or profile	• Resource hungry method
• Difficult to maintain signature database up to date	• Requires significant training time	• Limited capabilities to detect OS or API level attacks
	• Generate large false positives	

of normal traffic [19]. These IDS can understand different protocol dialogs and hand-shaking processes [25]. They also have tendency to detect'command injection' at protocol level. Command injection is a sophisticated attack in which attacker tries to inject malicious commands [26]. Comparison of all three detection methodologies are presented in Table 1.

2.2 Detection Approaches

Detection approach is the approach exploited by the detection engine to decipher intrusion from normal traffic. In literature [22] different detection approaches are discussed such as statistics based, pattern based, rule based, state based, heuristics based etc. Each detection approach has its own merits and demerits.

Statistics-based intrusion detection approach uses different statistical methods and techniques like Baye's theorem [27], probability density function, mean, variance, standard deviation etc. to detect abnormal behavior. Statistics based IDS approach is generally used in anomaly based intrusion detection systems discussed in Sect. 2.2. Pattern-based detection techniques focus on *patterns* of known attacks. They apply different pattern matching techniques like string matching, regular expression and tree based pattern recognition to detect known attack. Pattern based detection is usually employed in signature based IDS discussed in Sect. 2.2.

Rule based detection approach has some resemblance with pattern based detection technique. It works on the principle of *'condition matching'*; if-else rules. For instance, if an internal host is trying to establish a connection with an external serverl or domain, then IDS will first check and verify the reputation of the target machine. If the domain name or IP address is blacklisted, the connection attempt will be blocked. Domain Name System based Blackhole List (DNSBL) [28], Real-time Blackhole List (RBL) [29] etc. are few examples of reputation based database services [30, 31] commonly used to check domain/IP reputation.

State-based detection methods exploit the behavior of finite state machine [21]. They continuously monitor and keep tracks of machines' states in terms of sessions, packets transferred/received, number of connections to specific host or IP address etc. Once they establish a state-transition maps or state tables of active connections, then IDS can look for any possible intrusions.

Heuristics based IDS approach applies different problem solving techniques to detect intrusion. They are used to find quality solution within reasonable time frame. For heuristics it is not necessary that it should always give optimal solution. Heuristic based detection approaches are usually inspired from biological behavior of different animals, birds and artificial intelligence [32].

2.3 Analysis Target

Analysis target determines what type of data will be monitored and inspected by the IDS. For example we can categorize IDS into different classes based on what it can monitor, detect and block. Where it should be deployed either on a network segment or at host machine to detect and block attacks. A brief summary of different IDS analysis targets is presented below.

1. Network-based IDS (NIDS)
2. Host-based IDS (HIDS)
3. Application-based IDS (AIDS)
4. Wireless-based IDS (WIDS)
5. Network Behavior Analysis (NBA) based IDS
6. Mixed IDS (MIDS).

2.3.1 Network-Based IDS (NIDS)

Network based intrusion detection systems usually deployed at network transit points where most of the network traffic is pass or exchange [33]. The core principle of network based IDS is to monitor network traffic and looks for possible intrusions by exploiting different methodologies and approaches discussed in Sects. 2.1 and 2.2.

2.3.2 Host-Based IDS (HIDS)

Host based intrusion detection systems actively monitor hosts activities for any potential malicious behaviour [34, 35]. It includes hosts' process tables, network connections (ins and outs), registry entries, filesystem activities, prefetch items etc. and try to analyze their behavior for any signs of abnormality.

2.3.3 Wireless-Based IDS (WIDS)

Wireless-based IDS is similar to network-based IDS (NIDS), but it monitors wireless network traffic, such as wireless LAN (WLAN), wireless (Mobile) Ad-hoc NETworks (MANET), Wireless Sensor Networks (WSN), Wireless Mesh Networks (WMN), Wireless Body Area Networks (WBAN) etc. [36].

2.3.4 Network Behavior Analysis (NBA) Based IDS

Network Behavior Analysis (NBA) based IDS inspects network traffic to recognize attacks with unexpected traffic flows. For example it tries to detect Denial of Service (DoS) attack, certain type of malware, backdoors etc. [37]. NBA based IDS usually have a set of sensors deployed at different network segments and a console for central reporting and monitoring of network alerts.

2.3.5 Application Based IDS

Application based IDS monitors application traffic or flows for any signs of intrusions. Application based IDS solutions generally monitor and inspects few common traffic types like http, dns, smtp, database server traffic etc.

2.3.6 Mixed or Hybrid IDS (MIDS)

Mixed or hybrid IDS can incorporate different family of IDS discussed above. It provides more detail and accurate detection and prevention against attacks [37]. Hybrid IDS solutions actually mitigate the weakness and limitations of one another. Adopting multiple technologies as MIDS can fulfill the goal for a more complete and accurate detection.

2.4 Response Method

IDS can be classified as Passive or Active based on how it responds to an intrusion. Passive IDS can only generates alerts or notifications when it encounters any intrusion event. On the other hand, active IDS have capability to take basic necessary measures based on the type of intrusion. For example, it can terminate live active connections by sending RESET packets, covering holes, shutdown services, and start logging an intruder session.

2.5 Analysis Timing

IDS can also be classified based on how its analysis engine works. Analysis Engine (AE) is the an important component of an IDS. When IDS receives traffic from different streams or sources then it must analyze that traffic in order to detect possible malignancy. AE actually apply different detection techniques and approaches discussed in Sects. 2.1 and 2.2 to detect *true* intrusions. Event analysis can be performed either in (i) Online realtime mode or (ii) Periodic online or offline analysis.

In online realtime mode, AE analyze events on the fly as they hit IDS, detects intrusions and trigger notifications instantaneously. It is suitable for mission critical networks. However it also requires high computational resources to process large traffic volumes to generate useful alerts in timely manner.

On the other hand in Periodic online or complete offline analysis approach, AE does not analyze traffic logs in realtime manner. Rather AE is invoked at periodic intervals for traffic analysis. In Periodic offline mode, AE works on collected historical network traffic. This type of approach does not require high computational resources and often suitable for small size networks. However the biggest drawback of periodic online analysis is that it can miss real intrusion events.

In periodic online analysis, IDS analysis engine becomes online for small duration periodically. For example every hour for minutes. This type of IDS is actually used to gather historical data for weeks or months.

2.6 Architecture

There are two common IDS architectures are used which are (i) centralize and (ii) distributed. In centralized architecture all sensors monitor and collect network traffic and send it to central server. Central Server may constitute a number of components like traffic collector (serializer) which serialize/stream the traffic coming from different sensors (sources), analysis/detection engine, central manager to administer policies, reporting and notification subsystem etc.

In distributed architecture, IDS as a whole or with core components like event detection and notification system is deployed at different zone or network regions. The central manager only receives notification alerts from different sub IDS. This topology/IDS architecture is good when you have offices distributed in different regions.

3 Machine Learning Applications in Intrusion Detection

The data presented in Sect. 1 show that the growth rate of new attacks is unprecedented and exponential in nature. This also reflects that weaknesses of legacy security solutions. Therefore, researchers focused on anomaly based detection approach due to its tendency to detect novel attacks [38, 39]. Although anomaly-based intrusion detection system can detect new attacks but it comes with its own set of limitations. Therefore, to achieve optimal security posture for an organization researchers started to explore Machine Learning (ML)/Deep Learning (DL) approaches to detect new intrusions. Results from several other studies suggest that machine learning has shown great potential to solve some of the very complex problems like cancer detection and prediction [40], genetics and genomics [41], text classification [42], network/data center optimization [43], face recognition [44] and affect analysis [45–47]. Recent studies have also established that machine learning can be used in network intrusion detection systems to detect new unknown attacks [48–50]. In rest of this section we will present machine learning and its classifiers briefly in Sect. 3.1. In Sect. 3.2 we will present well-known datasets developed for NIDS and in Sect. 3.3 we will present work published in machine learning/deep learning in NIDS.

3.1 Brief Overview of Machine Learning and Classification

Computer is an electronic device that can execute millions or billions of instructions per seconds. These are machine-coded instructions which is a result of some algorithm (developed in high-level programming language) used to solve problem. An algorithm is a sequence of unambiguous instructions for solving a problem [51]. For example if you are given a task to sort out a numeric list in ascending or descending order, then you might able to apply more than one algorithm to achieve it. In this case, the input to the algorithm is a numeric list and the output is sorted list of numbers. However, in some scenarios we do not have a clear and well-defined algorithm to solve a problem. For examples, to differentiate a spam email from legitimate emails. In this case, we know that the input will be the email message and the output should be yes (spam) or no (not spam). But we do not have well-defined unambiguous set of instructions that can read hundreds of thousands of different emails and can classify them with higher degree of accuracy. Similarly, there are many other challenging problem for which we do not have a well-defined algorithm e.g. effective face recognition, expressions, identify and classify different objects in an image or a video stream etc.

Machine learning is capable to solve these challenging problems. It is a branch of Artificial Intelligence (AI) that focuses on the study of methods and techniques for programming computers to learn. Mitchell [52] in his classical text defined machine learning as, "if the performance of an algorithm is improved with experience to solve a specific problem over time, then we can say that algorithm is learning from its experience".

Machine learning algorithms are classified based on the type of learning adopted to train the model. The common techniques which are used to train the model are Supervised, Unsupervised and Semi-supervised learning. In supervised learning, training data is provided to the algorithm to create a model. Training data contains a pair of input vector and output (i.e. the class label). When the model is constructed, it can classify unknown examples into a learned class labels. In unsupervised learning training dataset does not include any label. The algorithm tries to establish a pattern in the given dataset without any class label, that is why it is known as unsupervised learning. Semi-supervised learning make use of hybrid approach. Label and unlabeled dataset is feed into the algorithm. Algorithm tries to recognize a pattern to predict the correct class of test dataset.

One fundamental requirement of classical machine learning algorithms is the dataset must be in structured format. It means that the dataset must contain well-defined 'features' or 'classes'. These features are actually input to the classifier and classifier learn and takes decisions on them. Generally features are extracted from raw data, through a process which is known as feature extraction [53]. Feature extraction is a time and memory consuming process due to this it is mostly performed in offline mode. Moreover, feature extraction schemes not always generate strong features, which is basically required to achieve the acceptable accuracy of the classifier.

In some circumstances it is not always possible to perform feature extraction from the raw data. For example in some realtime applications like context recognition in a video, adaptive filters used in channel estimation etc. In addition to this extracting strong features from raw data is also a challenging job. In such situations Deep Learning (DL) comes into picture and plays its role. Deep learning is a subset of machine learning and it does not necessarily require structured or labelled data. Its working is inspired from the working of human brain. All we need to input is the raw data, it has tendency to extract features on the fly and classify them.

There are two core components in any machine learning process (i) dataset and (ii) algorithm or classifier used to build or train model. Dataset is the heart of any ML based system. Without a good and balanced dataset we cannot build reliable and accurate models. It plays a crucial role in deriving the performance of any ML-based system. Secondly, the classifier is the core component or brain of ML-based system, it is responsible for classification. In literature we can find different types of classifiers but broadly we can classify them based on the type of learning utilized i.e. supervised, unsupervised or semi-supervised. In Table 3 we presented the summary of recent papers published in network intrusion detection systems along with the name of the dataset and classifiers used by authors.

3.2 Datasets for Intrusion Detection System (IDS)

IDS datasets are classified into network and host datasets. Network datasets contains normal and attack traffic while host datasets contains host or PC activities over a period of time. Since in this chapter our focus is on network based IDS so we will restrict our discussion to network based datasets only. Network based datasets can be further divided into packet-based and flow-based dataset. Table 2 summarizes basic features and limitations of some of the well-known network-based IDS datasets.

Table 2 Dataset features and limitations

Year	Dataset	Features	Format	Traffic Type	Limitations	Attack classes
1998	DARPA 98-99 [54]	• Created by MIT Lincoln lab i.e. DARPA'98 & '99 • Dataset consists of four type of attacks	Packet-based	Emulated/ synthetic	• Large number of duplicate records • Unbalanced dataset	(i) Denial of Service (DoS) (ii) User to Root (U2R) (iii) Remote to Local (R2L) (iv) Probing Attacks
1999	KDD-Cup99 [55]	• Inherited from DARPA'98 dataset • It consists of 41 features • Comprises of same attack classes as in DARPA'98	Packet-based	Emulated/ synthetic	• It contains redundant records • Low difficulty level of records in the dataset	Same as DARPA 98-99 dataset
2000	NSL-KDD [56]	• Derived from KDD-Cup99 dataset • Remove large number of duplicate record • Improved attacks difficulty level	Packet-based	Emulated/ synthetic	• Attack vector consists of only four type of attacks	Same as KDD-Cup99 dataset
2002	DEFCON-10 [57]	• Traffic captured during a hacking competition • Dataset mostly contain intrusive/offensive traffic • Only useful in alert correlation	Packet-based	• Emulated/ emulated	• Lacks normal background traffic • Not suitable for anomaly based IDS study	(i) Probing Attacks like port scan/ping sweep (ii) Bad packets (iii) Administrative privileges exploitation (iv) FTP by telnet protocol attack [58]

(continued)

Table 2 (continued)

Year	Dataset	Features	Format	Traffic Type	Limitations	Attack classes
2008	Sperotto [59]	• Flow based labeled real traffic • Single node honeypot connected with university campus network	Flow-based	Real/ real	• Amount of traffic captured is very low • Only monitors a single host connected to campus LAN	(i) Attacks on SSH Service: (automated & manual: brute force scan, user-name/password enumeration) (ii) Attacks on HTTP Service: http service compromise (iii) Few attacks on FTP protocol like ftp reconnaissance [59]
2010	MAWI Dataset [60]	• Dataset is contributed by Measurement and Analysis on the WIDE Internet (MAWI) • It consists of labeled real network traffic	Packet-based	Real/ real	• Daily capture is for limited time only (15 min.) • Labeling depends upon classifiers' accuracy which may generate false positive or true negative	(i) Port scan (ii) Network Scan (TCP/ UDP/ICMP), (iii) DoS, etc.
2012	UNB ISCX [57]	• Introduces the concept of traffic profiles for traffic generation • Testbed is created by using 17 Windows XP and 1 Windows 7 machines	Packet-based	Emulated/ synthetic	Traffic capture duration is for limited time Testbed is very simple	(i) Infiltrating the network from the inside (ii) HTTP Denial of Service (DoS) (iii) Distributed Denial of Service (DDoS) using an IRC botnet and (iv) SSH brute force
2013	CTU-13 [61]	It consists of traffic capture of 13 different malware in real network It comprises of normal, botnet and background traffic	Flow-based	Real/real	• Traffic capture duration is short • Creators did not explain the details of background traffic • No documentation is available regarding testbed	Majorly different type of botnet attacks that includes (Menti, Murlo, Neris, NSIS, Rbot, Sogou, Virut)

(continued)

Table 2 (continued)

Year	Dataset	Features	Format	Traffic Type	Limitations	Attack classes
2015	UNSW-NB15 (2015) [62]	Recently proposed by Moustafa et al. to address common issues exist in IDS dataset	Packet-based	Emulated/ synthetic	• Short capture Duration i.e. 31 h of data Class imbalance problem	Dataset includes nine different families of attacks: (i) Fuzzers (ii) Analysis (iii) Backdoors (iv) DoS (v) Exploits (vi) Generic (vii) Reconnaissance (viii) Shellcodes (ix) Worms
2016	UGR'16 [63]	• Used cyclo-stationarity feature in network traffic dataset • Mainly targets anomaly-based IDS detection	Flow-based	Real/real	• Only flows are available to download • Limited attack traffic	(i) Botnet (Neris) (ii) DoS (iii) Port scans (iv) SSH brute force (v) Spam
2017	CICIDS 2017 [64]	• Multiclass dataset built in 2017 • Traffic features are extracted via CICFlowmeter	Packet, flow-based	Emulated/ synthetic	• Class imbalance problem • It contains large number of missing values	(i) Botnet (ii) Web Attacks like Cross-site-scripting/SQL injection (iii) DoS and DDoS attacks (iv) Heartbleed (v) Infiltration (vi) SSH brute force

Traffic type: real, emulated, or synthetic. Real means traffic was captured within a productive network environment. Emulated means that real network traffic was captured within a test bed or emulated network environment. Synthetic means that the network traffic was created synthetically (e.g., through a traffic generator hardware or software)

Following observations are made from Table 2:

- KDD-Cup99 and NSL-KDD datasets are evolved from DARPA98-98 dataset which means that base of both datasets is same.
- Most datasets comprise of packet-based data, however few datasets also include flow-features. Packet and flow are two techniques to capture network traffic. Packet-based dataset often includes complete packet information including payload while flow-based dataset usually contains network flows and connection information only.
- Only few datasets contain real traffic (difficult to build real traffic dataset). Most of the datasets are build using synthetic or emulated traffic.

(continued)

Table 3 Comparison of related work

Year	Author	Dataset	Classifiers	Critical Comments
2014	De la Hoz et. al. [65]	NSL-KDD	GHSOM*, NB, RF[†], DT[‡], AdaBoost	- Proposed scheme's FPR is higher upto 4.22% and overall accuracy is less than A-GHSOM [66] - Details about subsets of features is not provided
2014	Feng et al. [67]	KDD-Cup99	SVM[§], ACO[¶]	- Accuracy / detection rate of the proposed scheme is less than CSOACN and FPR is higher than KDD-Cup99 Winner algorithm [67]
2014	G Kim et.al. [68]	NSL-KDD	SVM, DT	- Authors proposed a hybrid approach based on misuse and anomaly detection models to improve detection performance and speed - Results show that proposed method's training and testing time has improved as compared to other hybrid approaches. However compared to misuse and anomaly detection models alone, its testing and training time is high

Continued on next page

*Growing Hierarchical Self-Organising Maps
[†] Random Forest
[‡] Decision Tree
[§] Support Vector Machine
[¶] Ant Colony Optimization

(continued)

Table 3 (continued)

Year	Author	Dataset	Classifiers	Critical Comments
2015	Eesa et. al. [69]	KDD-Cup99	DT	- Authors claim that AR$^{\|}$ and DR increase with the reduction of features set however they do not provide any details about which features are selected in the subsets. Moreover this paper does not provide comparison with state of the art
2016	A. Hadri et. al [70]	KDD-Cup99	KNN**	- Authors compare PCA and Fuzzy PCA dimension reduction techniques, results obtained from the study show that Fuzzy PCA method performed better.
2016	Praneeth N. et. al. [71]	KDD-Cup99	SVM linear, polynomial and radial basis kernels are compared	- Results show that accuracy of RBF kernel is better than other while polynomial kernel has low detection time
2016	S.Guha et. al. [72]	NSL-KDD, UNSW-NB15	ANN, LR††, DT, NB, SVM	- Results show that propose feature selection approach yields better accuracy. - Authors did not compare their results with other feature selection techniques - Furthermore authors did not present feature set details producing higher accuracy

Continued on next page

$^{\|}$ Accuracy Rate
**K - Nearest Neighbor
††Logistic Regression

Table 3 (continued)

Year	Author	Dataset	Classifiers	Critical Comments
2017	A. Rama & W. Gata [73]	KDD-Cup99	KNN & proposed hybrid classifier based on binary PSO[‡‡] and KNN	- Results show that proposed algorithm has accuracy around 99% while KNN accuracy remain around 97% on KDD-Cup99 dataset - Authors did not provide any details about the final feature set
2017	Chuanlong Y. et. al. [74]	NSL-KDD	J48, NB, RF, MLP[§§], SVM and proposed RNN-IDS[¶¶]	- Paper shows that RNN-IDS results are better than other classifier.
2017	S. Zhao et. al. [75]	KDD-Cup99	Softmax Regression, K-NN,	- Results show that softmax regression algorithm performed better as compared KNN algorithm - Authors did not share detail about the feature set included in the final test - Authors did not compare the results with other feature selection techniques
2017	M. A. Zewairi et. al [76]	UNSW-NB15	*Proposed binomial classifier, DT, LR, NB, ANN, EM clustering etc.	- Experimental results show that proposed DL[***] classifier has better accuracy and FPR however authors did not provide any comparison fo result with other feature selection method

Continued on next page

[‡‡]Particle Swarm Optimization
[§§]Multi Layer Perceptron
[¶¶]Recurrent Neural Network
[***]Deep Learning

(continued)

(continued)

Table 3 (continued)

Year	Author	Dataset	Classifiers	Critical Comments
2017	P. Mishra et. al [77]	UNSW-NB15	MNPD [†††], DT, NB, LR, ANN, DT(RFE[‡‡‡]), DT(RFE+Chi Sq.), RF(RFE), RF(RFE+Chi Sq.) etc.	- Final feature set detail is missing
2017	C. Khammassi & S. Krichen [78]	KDD-Cup99 and UNSW-NB15	DT, LR, NB, ANN, EM	- Comparison with state of the art shows that GA-LR performance is slightly lower than DT [§§§] with full features - Authors did not explain the process of selecting samples from the dataset
2018	M.H. Ali et al. [79]	KDD-Cup 99	Proposed PSO-FLN [¶¶¶] and compared the results with different ELM based techniques	- Authors claim that the proposed model has shown better accuracy - Authors did not compare their results with state of the art

Continued on next page

[†††] Malicious Network Packet Detection
[‡‡‡] Recursive Feature Elimination
[§§§] Decision Tree
[¶¶¶] Fast Learning Network
  Experiential Learning Model

Table 3 (continued)

Year	Author	Dataset	Classifiers	Critical Comments
2018	Muna A.H. et. al. [80]	NSL-KDD, UNSW-NB15	Proposed ADS, F-SVM, CVT, DMM, TANN etc.	- Results show that the proposed algorithm performed better, however one fundamental issue is the use of NSL-KDD and UNSW-NB15 datasets. These datasets are not designed to cater IICS challenges nor they contain IICS specific attacks
2019	Jie Gu et. al. [81]	NSL-KDD	DT-EnSVMData Transformation - Ensemble SVM DT-EnSVM2 EnSVM SVM	- Proposed an intrusion detection framework based on SVM ensemble with feature augmentation. - One fundamental problem identified is the use of old dataset.

Continued on next page

Filter-based SVM
Computer Vision Technique
Dirichlet Mixture Model
Triangle Area Nearest Neighbors
Industrial Internet Control Systems

(continued)

Table 3 (continued)

Year	Author	Dataset	Classifiers	Critical Comments
2020	Zhang et. al. [82]	UNSW-NB15	*MSCNN-LSTM, Lenet-5,MSCN-NMultiscale Convolutional Neural Net-workand HASTHierar-chical spatial-temporal features-based intrusion detec-tion system	- Proposed MSCNN-LSTM model has better accuracy, false alarm rate and false negative rate. - Statistically weak sample formation approach is be-ing followed.

Proposed Multiscale Convolutional Neural Network with Long Short-Term Memory
Classical CNN architecture

3.3 Machine Learning in Intrusion Detection System

This section presents summary of recent work carried out in network intrusion detection systems from the application of machine learning. Notable papers published in last six years are presented in chronological order in Table 3. Figure 2 presents visual representation of most commonly used classifiers in this domain. Few observations from Table 3 and Fig. 2 are presented below.

- Most of the authors worked on KDD-Cup99 dataset. Many authors still use it despite of its many weakness and outdated attack vectors.
- We observed that traditionally researchers focused on classical machine learning algorithms like Decision Tree, Naive Bayes, SVM etc. but recent trend is shifting towards deep learning, ensemble learning etc.
- Only few papers include nature inspired algorithm as a classifier like ACO, PSO, etc. showing potential research gap for future researchers.

4 Summary and Future Directions

In this chapter we initially portrayed overall picture of different attack types which are recently materialized and their motivation factors. We briefly discussed the weaknesses of legacy security solutions like antivirus, firewalls etc. In Sect. 2 we presented a comprehensive taxonomy of network based intrusion detection systems. We discussed several different aspects of IDS architecture, detection methodologies and approaches, response mechanisms etc. In Sect. 3, we presented brief overview of machine learning and its applications in NIDS, then we presented well-known network-based IDS datasets and discussed key findings. In Sect. 3.3 we presented summary of recent research published in IDS domain. We discussed common datasets and classifiers used in the study.

We observed that most authors presented their findings on KDD-Cup99 dataset, which does not reflect the true picture of modern day network traffic/attacks. Dataset is the core component on which classifier build its model. Unfortunately due to large number of novel attacks discovered on routine basis, newer datasets can also get outdated rapidly. Researchers should develop some mechanisms to incorporate new attacks vector in the dataset to keep it up to date.

Furthermore, we suggest that researchers should explore other areas for attack detection, like nature-inspired algorithms, soft computing, evolutionary computing etc, as we found only few papers that utilize these techniques.

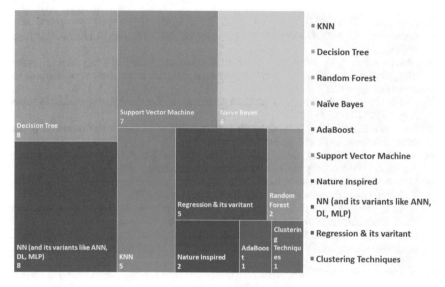

Fig. 2 Graphical overview of classifiers usage statistics in intrusion detection systems

References

1. Venter H, Eloff JH (2003) A taxonomy for information security technologies. Comput Secur 22(4):299–307
2. Cyberattack: cyberattack: computer attack, exploitation, apt. https://en.wikipedia.org/wiki/Cyberattack/. Accessed 18 Dec 2018
3. Symantec (2019) Internet security threat repor, vol 24. Tech. rep., Symentec Corporation
4. Welchman G (1982) The hut six story: breaking the enigma codes. McGraw-Hill Companies, New York
5. Taylor P (2012) Hackers: crime and the digital sublime. Routledge, London
6. Brewster B, Kemp B, Galehbakhtiari S, Akhgar B (2015) Cybercrime: attack motivations and implications for big data and national security. Application of big data for national security. Elsevier, Amsterdam, pp 108–127
7. Bessi A, Ferrara E, Social bots distort the 2016 US presidential election online discussion
8. Howard PN, Kollanyi B, Woolley S, Bots and automation over twitter during the US election. Computational Propaganda Project: Working Paper Series
9. Allcott H, Gentzkow M (2017) Social media and fake news in the 2016 election. J Econ Perspect 31(2):211–36
10. Nazario J (2009) Politically motivated denial of service attacks. In: Perspectives on cyber warfare, The Virtual Battlefield, pp 163–181
11. Hansman S, Hunt R (2005) A taxonomy of network and computer attacks. Comput Secur 24(1):31–43. https://doi.org/10.1016/j.cose.2004.06.011
12. Bhardwaj A (2017) Ransomware: a rising threat of new age digital extortion. In: Online banking security measures and data protection. IGI Global, pp 189–221
13. Kaspersky: antivirus fundamentals: Viruses, signatures, disinfection, https://www.kaspersky.com/blog/signature-virus-disinfection/13233/. Accessed 16 May 2018
14. Forouzan BA (2002) TCP/IP protocol suite, 2nd edn. McGraw-Hill Higher Education, New York

15. Zimmermann H (1980) Osi reference model—the iso model of architecture for open systems interconnection. IEEE Trans Commun 28(4):425–432. https://doi.org/10.1109/TCOM.1980. 1094702

16. Dharmapurikar S, Krishnamurthy P, Sproull T, Lockwood J (2003) Deep packet inspection using parallel bloom filters. In: 11th Symposium on high performance interconnects, Proceedings. IEEE, pp 44–51

17. Dwivedi S, Angeri H, Arora V (2008) Architecture for unified threat management. US Patent App. 11/871,611, 17 Apr 2008

18. Thomason S, Improving network security: next generation firewalls and advanced packet inspection devices. Glob J Comput Sci Technol

19. Scarfone K, Mell P (2007) Guide to intrusion detection and prevention systems (idps), special publication 800–94. Tech. rep, National Institute of Standards and Technology

20. Bace PMR (2001) Intrusion detection systems, technical report special publication 800–31. Tech. rep, National Institute of Standards and Technology (NIST)

21. Debar H, Dacier M, Wespi A (2000) A revised taxonomy for intrusion-detection systems. In: Annales des télécommunications, vol 55. Springer, pp 361–378

22. Liao H-J, Richard Lin C-H, Lin Y-C, Tung K-Y (2013) Review: intrusion detection system: a comprehensive review. J Netw Comput Appl 36(1):16–24. https://doi.org/10.1016/j.jnca.2012. 09.004

23. Park W, Ahn S (2017) Performance comparison and detection analysis in snort and suricata environment. Wirel Pers Commun 94(2):241–252

24. Garcia-Teodoro P, Diaz-Verdejo J, Maciá-Fernández G, Vázquez E (2009) Anomaly-based network intrusion detection: techniques, systems and challenges. Comput Secur 28(1–2):18–28

25. Capone JM, Immaneni P (2010) Protocol and system for firewall and NAT traversal for TCP connections. US Patent 7,646,775

26. Su Z, Wassermann G (2006) The essence of command injection attacks in web applications. ACM Sigplan Not 41:372–382

27. Kabiri P, Ghorbani AA (2005) Research on intrusion detection and response: a survey. IJ Netw Secur 1(2):84–102

28. Ramachandran A, Feamster N, Dagon D et al (2006) Revealing botnet membership using dnsbl counter-intelligence. SRUTI 6:49–54

29. Drako D, Levow Z (2011) Facilitating transmission of email by checking email parameters with a database of well behaved senders. US Patent 7,996,475

30. Perdisci R, Lee W (2010) Method and system for detecting malicious and/or botnet-related domain names. US Patent App. 12/538,612

31. Antonakakis M, Perdisci R, Lee W, Vasiloglou N (2014) Method and system for detecting malicious domain names at an upper dns hierarchy. US Patent 8,631,489

32. Liao H-J, Lin C-HR, Lin Y-C, Tung K-Y (2013) Intrusion detection system: a comprehensive review. J Netw Comput Appl 36(1):16–24

33. Vigna G, Kemmerer RA (1999) Netstat: a network-based intrusion detection system. J Comput Secur 7(1):37–71

34. Chebrolu S, Abraham A, Thomas JP (2005) Feature deduction and ensemble design of intrusion detection systems. Comput Secur 24(4):295–307

35. Deshpande P, Sharma S, Peddoju S, Junaid S (2018) Hids: a host based intrusion detection system for cloud computing environment. Int J Syst Assur Eng Manage 9(3):567–576

36. Can O, Sahingoz OK (2015) A survey of intrusion detection systems in wireless sensor networks. In: 2015 6th International conference on modeling, simulation, and applied optimization (ICMSAO). IEEE, pp 1–6

37. Stavroulakis P, Stamp M (2010) Handbook of information and communication security, 1st edn. Springer Publishing Company, Incorporated

38. Gan X-S, Duanmu J-S, Wang J-F, Cong W (2013) Anomaly intrusion detection based on PLS feature extraction and core vector machine. Knowl-Based Syst 40:1–6

39. Karami A, Guerrero-Zapata M (2015) A fuzzy anomaly detection system based on hybrid pso-kmeans algorithm in content-centric networks. Neurocomputing 149:1253–1269
40. Kourou K, Exarchos TP, Exarchos KP, Karamouzis MV, Fotiadis DI (2015) Machine learning applications in cancer prognosis and prediction. Comput Struct Biotechnol J 13:8–17
41. Libbrecht MW, Noble WS (2015) Machine learning applications in genetics and genomics. Nat Rev Genet 16(6):321
42. Tong S, Koller D (2001) Support vector machine active learning with applications to text classification. J Mach Learn Res 2:45–66
43. Gao J, Machine learning applications for data center optimization
44. Chopra S, Hadsell R, LeCun Y, et al (2005) Learning a similarity metric discriminatively, with application to face verification. In: CVPR, vol 1, pp 539–546
45. Khan RA, Crenn A, Meyer A, Bouakaz S (2019) A novel database of children's spontaneous facial expressions. Image Vis Comput 83:61–69
46. Khan RA, Meyer A, Konik H, Bouakaz S (2012) Human vision inspired framework for facial expressions recognition. In: 2012 19th IEEE international conference on image processing, pp 2593–2596. https://doi.org/10.1109/ICIP.2012.6467429
47. Khan RA, Meyer A, Konik H, Bouakaz S (2019) Saliency-based framework for facial expression recognition. Front Comput Sci 13(1):183–198
48. Sangkatsanee P, Wattanapongsakorn N, Charnsripinyo C (2011) Practical real-time intrusion detection using machine learning approaches. Comput Commun 34(18):2227–2235
49. Winding R, Wright T, Chapple M (2006) System anomaly detection: mining firewall logs. In: Securecomm and workshops. IEEE, pp 1–5
50. Appelt D, Nguyen CD, Briand L (2015) Behind an application firewall, are we safe from sql injection attacks?, In: IEEE 8th international conference on software testing, verification and validation (ICST). IEEE, pp 1–10
51. Levitin A (2012) Introduction to the design & analysis of algorithms. Pearson, Boston
52. Mitchell TM et al (1997) Machine learning
53. Guyon I, Gunn S, Nikravesh M, Zadeh LA (2008) Feature extraction: foundations and applications, vol 207. Springer, Berlin
54. Darpa'98 and darpa'99 datasets. https://www.ll.mit.edu/ideval/docs/index.html. Accessed 28 June 2018
55. Kdd cup 99 dataset. https://kdd.ics.uci.edu/databases/kddcup99/kddcup99.html. Accessed 28 June 2018
56. Tavallaee M, Bagheri E, Lu W, Ghorbani AA (2009) A detailed analysis of the kdd cup 99 data set. In: IEEE symposium on computational intelligence for security and defense applications, CISDA 2009. IEEE, pp 1–6
57. Shiravi A, Shiravi H, Tavallaee M, Ghorbani AA (2012) Toward developing a systematic approach to generate benchmark datasets for intrusion detection. Comput Secur 31(3):357–374
58. Sharafaldin I, Lashkari AH, Ghorbani AA (2018) Toward generating a new intrusion detection dataset and intrusion traffic characterization. In: ICISSP, pp 108–116
59. Sperotto A, Sadre R, Van Vliet F, Pras A (2009) A labeled data set for flow-based intrusion detection. In: International workshop on IP operations and management. Springer, pp 39–50
60. Fontugne R, Borgnat P, Abry P, Fukuda K (2010) Mawilab: combining diverse anomaly detectors for automated anomaly labeling and performance benchmarking. In: Proceedings of the 6th international conference. ACM, p 8
61. Garcia S, Grill M, Stiborek J, Zunino A (2014) An empirical comparison of botnet detection methods. Comput Secur 45:100–123
62. Moustafa N, Slay J (2015) Unsw-nb15: a comprehensive data set for network intrusion detection systems (unsw-nb15 network data set). In Military communications and information systems conference (MilCIS), pp 1–6. https://doi.org/10.1109/MilCIS.2015.7348942
63. Maciá-Fernández G, Camacho J, Magán-Carrión R, García-Teodoro P, Therón R (2018) Ugr '16: a new dataset for the evaluation of cyclostationarity-based network idss. Comput Secur 73:411–424

64. Sharafaldin I, Lashkari AH, Ghorbani AA (2018) A detailed analysis of the cicids2017 data set. In: International conference on information systems security and privacy. Springer, pp 172–188
65. De la Hoz E, de la Hoz E, Ortiz A, Ortega J, Martínez-Álvarez A (2014) Feature selection by multi-objective optimisation: application to network anomaly detection by hierarchical self-organising maps. Knowl-Based Syst 71:322–338
66. Ippoliti D, Zhou X (2012) A-ghsom: an adaptive growing hierarchical self organizing map for network anomaly detection. J Parallel Distrib Comput 72(12):1576–1590
67. Feng W, Zhang Q, Hu G, Huang JX (2014) Mining network data for intrusion detection through combining svms with ant colony networks. Future Gener Comput Syst 37:127–140
68. Kim G, Lee S, Kim S (2014) A novel hybrid intrusion detection method integrating anomaly detection with misuse detection. Expert Syst Appl 41(4):1690–1700
69. Eesa AS, Orman Z, Brifcani AMA (2015) A novel feature-selection approach based on the cuttlefish optimization algorithm for intrusion detection systems. Expert Syst Appl 42(5):2670–2679
70. Hadri A, Chougdali K, Touahni R (2016) Intrusion detection system using pca and fuzzy pca techniques. In: 2016 International conference on advanced communication systems and information security (ACOSIS). IEEE, pp 1–7
71. Nskh P, Varma MN, Naik RR (2016) Principle component analysis based intrusion detection system using support vector machine. In: 2016 IEEE international conference on recent trends in electronics, information & communication technology (RTEICT). IEEE, pp 1344–1350
72. Guha S, Yau SS, Buduru AB (2016) Attack detection in cloud infrastructures using artificial neural network with genetic feature selection. In: IEEE 14th International conference on dependable, autonomic and secure computing, 14th International conference on pervasive intelligence and computing, 2nd International conference on big data intelligence and computing and cyber science and technology congress (DASC/PiCom/DataCom/CyberSciTech). IEEE, pp 414–419
73. Syarif AR, Gata W (2017) Intrusion detection system using hybrid binary pso and k-nearest neighborhood algorithm. In: 2017 11th International conference on information & communication technology and system (ICTS). IEEE, pp 181–186
74. Yin C, Zhu Y, Fei J, He X (2017) A deep learning approach for intrusion detection using recurrent neural networks. IEEE Access 5:21954–21961
75. Zhao S, Li W, Zia T, Zomaya AY (2017) A dimension reduction model and classifier for anomaly-based intrusion detection in internet of things. In: IEEE 15th International conference on dependable, autonomic and secure computing, 15th International conference on pervasive intelligence and computing, 3rd International conference on big data intelligence and computing and cyber science and technology congress (DASC/PiCom/DataCom/CyberSciTech). IEEE, pp 836–843
76. Al-Zewairi M, Almajali S, Awajan A (2017) Experimental evaluation of a multi-layer feed-forward artificial neural network classifier for network intrusion detection system. In: 2017 International conference on new trends in computing sciences (ICTCS). IEEE, pp 167–172
77. Mishra P, Pilli ES, Varadharajan V, Tupakula U (2017) Out-vm monitoring for malicious network packet detection in cloud. In: ISEA asia security and privacy (ISEASP). IEEE, pp 1–10
78. Khammassi C, Krichen S (2017) A ga-lr wrapper approach for feature selection in network intrusion detection. Comput Secur 70:255–277
79. Ali MH, Al Mohammed BAD, Ismail A, Zolkipli MF (2018) A new intrusion detection system based on fast learning network and particle swarm optimization. IEEE Access 6:20255–20261
80. Muna A-H, Moustafa N, Sitnikova E (2018) Identification of malicious activities in industrial internet of things based on deep learning models. J Inf Secur Appl 41:1–11
81. Gu J, Wang L, Wang H, Wang S (2019) A novel approach to intrusion detection using svm ensemble with feature augmentation. Comput Secur 86:53–62
82. Zhang J, Ling Y, Fu X, Yang X, Xiong G, Zhang R (2020) Model of the intrusion detection system based on the integration of spatial-temporal features. Comput Secur 89:101681

Machine Learning and Deep Learning Models for Big Data Issues

Youssef Gahi and Imane El Alaoui

Abstract The growing interest of digital in our daily life makes Big data essential in many fields. Today, more and more companies and communities are turning to big data management to help decision-making. Understanding and better managing big data makes it possible to collect and analyze relevant information to make predictions. However, vulnerabilities exist at all scales of the big data platforms, including at the data level. Despite the tremendous efforts and resources that have been offered by big data tools and providers, big data platforms remain vulnerable to many existing forms of attacks. Therefore, new kinds of solutions should be provided to strengthen Big data security. Predictive models are offering promising solutions for additional security layers. In this paper, we summarize and discuss contributions helping to protect big data environments using Machine learning and Deep learning. We also regroup the most sensitive security aspects that should be addressed to protect valuable data. All the contributions and dimensions were addressed through a set of security use cases, namely, malware detection, intrusion, anomaly, access control, and data ingestion controls. Furthermore, we provide comparison results of different techniques to show their efficiency.

Keywords Machine learning · Deep learning · Big data · Privacy · Cyber-security

1 Introduction

The strength of the data no longer needs to be proved. Big Data has enabled extensive use of data since the 2000s, with the advent of Cloud Computing and the growing interest of digital in our daily life. In the early 2010s, the growth of analytical tools

---

Y. Gahi (✉)
Laboratoire de Recherche en Sciences de l'Ingénieur, Ibn Tofail University, Kénitra, Morocco
e-mail: gahi.youssef@uit.ac.ma

I. El Alaoui
LASTID, Ibn Tofail University, Kénitra, Morocco
e-mail: imane.el.alaoui@uit.ac.ma

Y. Maleh et al. (eds.), *Machine Intelligence and Big Data Analytics for Cybersecurity Applications*, Studies in Computational Intelligence 919,
https://doi.org/10.1007/978-3-030-57024-8_2

29

allowed companies to have access to massive data enabling them to shape specific strategies to predict trends and behaviors. Today, more and more companies and communities are turning to big data management to help decision-making. However, this data strength is often the target of several forms of attacks targeting big data platforms. Despite the tremendous efforts and resources that have been offered by big data tools and providers, vulnerabilities exist at all scales of the big data platforms, including at the data level. These attacks are continuously changing and growing, making traditional protection techniques such as security policies and cryptography techniques less effective.

The biggest challenge faced by the entire security sector is how to detect and deal with coming attacks. During the last few years, new kinds of solutions, based on predictive models, have been proposed to accompany complex and dynamic attack behaviors. Predictive models aim to recognize patterns in attacks and face security weaknesses. Machine learning and deep learning are the kinds of predictive models typically used to enhance security layers. These two models consist of analyzing known attacks, based on stochastical methods, and detect new threats that are not predefined. They have shown a high added value that makes security platforms more resilient. As Big Data platforms are required to manipulate sensitive records and draw strategic business continually, advanced security layers should be implemented to complement existing policies. Machine learning and deep learning are more suitable to bring these required layers.

Many scientific contributions have been oriented to build machine learning and deep learning models for Big Data security and privacy. These models aim to provide advanced security features against continuously changing threats. As security attacks vary considerably in type, complexity, and risk level, the research community has proposed several models to deal with each aspect. In this contribution, we summarize and discuss most of the exciting works, based on Machine learning and Deep learning, helping to protect big data environments against different security and privacy threats. These aspects and its related contributions have been organized under five use cases, including malware detection, intrusion, anomaly, access, and data ingestion controls. For each security use case, we identify the set of security dimensions, criteria interpretations, as well as the recommended models with their detailed results. Furthermore, we provide a comparison of different techniques to show their efficiency.

The rest of the paper is organized as follows; in Sect. 2, we prove the positive impact of Machine learning and Deep learning on strengthening Big Data platforms. In Sect. 3, we project interesting contributions addressing malware issues that are interesting for big data systems. In Sects. 4 and 5, we discuss models that can be used against Big data anomaly and intrusions, respectively. In Sect. 6, we go through research aiming at making access control more resilient for the Big Data environment. In Sect. 7, we show some predictive models that have been designed to make ingested data more reliable. The last section concludes our paper and provides some future directions.

2 Importance of Predictive Analytics for Big Data Security

Predictive analytics is a category of advanced data analytics that is used to make predictions about future outcomes basing on historical data and analytics techniques. It encompasses a variety of technologies such as Machine Learning (ML), and Deep Learning (DL) to predict possible future insights. There are two types of predictive models: classification models that predict class membership, which is a set of categories data belongs to, and regression models that predict a number. Also, there are three types of learning models, supervised learning, unsupervised learning, and semi-supervised learning. In supervised learning, the data used to train the model, called training dataset, is fully labeled. Whereas in semi-supervised learning, the training dataset contains a mixture of labeled and unlabeled data. In unsupervised learning, the data is entirely not labeled, and the model tries to discover a structure by extracting useful information.

Many models can be explored in ML. Here, we quote the most widely used algorithms:

- **Neural Networks** (NN): It was initially inspired by the functioning of the human brain. It relies on the use of "artificial" neurons that perform the learning task. An artificial neuron is defined as a non-linear, parameterized algebraic function with bounded values.
- **Support-Vector Machines** (SVM): are supervised learning models with associated learning algorithms used for classification and regression. It is mainly used to create an input-output mapping model. SVM is a linear learning system using the linear combination of characteristics, which builds classifiers into several classes, such as positive and negative. This classification is mainly based on the sign of this linear combination.
- **Naïve Bayes** (NB): is a family of probabilistic classifiers that rely on Bayes theorem. The class with the highest probability assigned to input data.
- **Decision Trees** (DT): are used for classification and also regression. They are very popular for generating classification and prediction rules. The idea is to split a dataset into several branch-like segments. The decisions with possible consequences, including results of random events, are located at the ends of the branches, called the "leaves" of the tree. The paths from the root to the leaf represent the classification rules.
- **k-Nearest Neighbors** (k-NN): is an unsupervised learning algorithm. Unlike previous learning methods, which learn certain types of models based on training datasets, the k-NN is a lazy learning method. No model is determined from the training dataset, and the learning phase only consists of optimally memorizing examples.
- **Regression**: is one of the most powerful methods in statistics. It allows us to examine and estimate relationships among variables. Standard regression algorithms include Linear Regression and logistic Regression.

As mentioned above, predictive analytics involves Deep Learning models. It is also important to say that DL is a branch of machine learning. DL algorithms consist of multiple consecutive layers of artificial neural networks through which the data is transformed. Each layer contains neurons with activation functions to produce outputs. The layers are interlinked, and each one receives the production of the previous ones as input. The most popular deep learning algorithms are:

- **Convolutional Neural Network** (CNN): most commonly used in analyzing visual imagery in terms of recognition and classification. CNN-based algorithms introduce a convolution layer, the first layer, which consists of applying a filter to input to create a feature map. This filter summarizes the presence of detected features in the input data.
- **Recurrent Neural Networks** (RNNs): they use similar architecture to the traditional NN. The difference is that RNNs introduce the concept of memory. They allow the previous outputs to be used as inputs to the current step while having hidden states.
- **Long Short-Term Memory Networks** (LSTMs): is an RNN-based architecture, capable of learning long-term dependencies. They resolve the vanishing gradient problem of RNN by introducing internal memory, called a cell. The cell allows maintaining a state as long as necessary. This cell is regulated by three control gates: an input gate which decides whether the entry should modify the contents of the cell. This output gate determines whether the contents of the cell should influence the output of the neuron. The forget gate decides whether to reset the contents of the cell to 0 or not.
- **Auto-Encoder** (AE): is a kind of Artificial Neural Network that learns efficient data coding in an unsupervised way. It is typically used for dimensionality reduction. Indeed, AE compresses very-high-dimensional data into a smaller encoded representation using a bunch of layers that are either fully connected layers or convolutional. Furthermore, it tries to generate the data back from the reduced encoded description as close as possible to the original input.

All the previous ML and DL algorithms help to create predictive models to build smart security controls at different levels. They have gained a great interest in Big data security topics due to their efficiency, especially when combined with big data tools. Big data tools and platforms allow real-time and prompt intelligence to launch immediate automated responses to security issues such as intrusions and attacks. In what follows, we review the most recent and sophisticated researches tackling Big data concerns.

3 Predictive Models for Malware Detection

Malware detection is the process of identifying a variety of hostile and intrusive software, including viruses, worms, ransomware, trojan, adware, etc. Malware attacks could be dynamic content, executable code, scripts, and other forms that could spread to other computers and execute on their own. Malware is continually getting smarter, produced in significant numbers, and deployed very fast. Therefore, detecting malware in big data, where large masses of real data are generated, remains a challenging task using traditional ML and DL. Researchers have put in a lot of effort to propose predictive models based on ML and DL to detect today's malware effectively.

Sabar et al. have designed in [1] a hyper-heuristic SVM optimization (HH-SVM) framework for Big Data cybersecurity issues. The proposed framework has shown excellent performance and been tested on two cybersecurity problem instances: Microsoft malware big data classification and anomaly intrusion detection. The experimental results have demonstrated that HH-SVM is a practical methodology for addressing cybersecurity problems and have achieved an accuracy of 85.69%. The framework is the right solution to strengthen Big Data environments as it could be adapted to several contexts.

Chhabra et al. have proposed in [2] an exciting machine learning model for P2P malware analysis and malware reporting. They adopted an approach that relies on the features extraction efficiency to predict authenticity and reliability. The method has been deployed using a cyber-forensic framework for the IoT environment. The authors have relied on Big Data platforms and their related tools to enhance the performance of their approach. A comparative analysis of machine learning models, namely Decision Tree, Ada BOOST, Random Forest, SVM, Linear Model, and Neural Networks, has been detailed. This comparison is based on a set of parameters, such as Recall, Precision, Error Rate, and Specificity. The random forest model has been chosen as the best approach for a network traffic analysis for malware detection (reaching 99.94% of Precision and 99.97% of Specificity). We believe that this contribution does provide a customizable model to build robust malware detection in Big data contexts.

Dovom et al. [3] have presented a model that relies on both feature extraction and fuzzy techniques, which is a significant category of machine learning, to build a robust edge computing malware detection and categorization system. Based on the fuzzy and fast fuzzy pattern tree model, authors have proved that machine learning aided techniques is a suitable solution to deal with malware issues. The developed model has been used for the IoT context to detect malicious activities. The model shows excellent performance with 93.83% accuracy, 89.58% on recall, a precision of 89.70%, and 0.8798 on f-measure, during reasonable run-times. It is worth noting that Fuzzy models are promising techniques to protect Big Data from malware.

Masabo et al. have proposed in [4] novel real-time monitoring, analysis, and malware detection approach for big data using deep learning and SVM. The proposed model mainly relies on the power and scalability of Big data platforms to provide an efficient detection system. The experimental results have shown that deep learning achieves a better accuracy of 97% compared to 95% made through SVM.

Vinayakumar et al. [5] have proposed a novel hybrid method combining visualization and deep learning techniques for malicious software detection. The method relies on Big data capabilities to provide a scalable and hybrid framework that can collect and classify malicious attacks. The authors also considered the real-time aspect by relying on a scalable and hybrid deep learning such as CNN and LSTM. To prove the efficiency of the adopted framework, the authors have conducted a benchmark and compared the classical MLAs and deep learning on the Ember dataset. Results have shown that CNN-LSTM performed well in comparison to all other algorithms by achieving an accuracy of 96.3%.

In Table 1, we provide a comparison and differences between the presented malware detection methods.

4   Predictive Models for Anomaly Detection

The purpose of anomaly detection is to identify things that don't conform to what we are expecting. Anomalies could be stated as rare items, events, trends, or pre-cursors that reduce safety margins. The difficulty of the problem stems from the fact that the underlying distribution is not known beforehand. It is up to the model to learn an appropriate metric to detect anomalies. Exciting Machine learning and Deep learning have been proposed to deal with such kind of issues.

Mulinka and Casas have demonstrated in [7] the effectiveness of their anomaly detection model by comparing the performance of four stream-based machine learning algorithms and their corresponding off-line versions, namely, k-NN, Hoeffding Adaptive Trees, Adaptive Random Forests, and Stochastic Gradient Descent. This model aims to detect security and anomaly issues of continuously evolving data streams. Their experimentation results show that adaptive random forests and stochastic gradient descent models keep high accuracy, 96.12%, and 99.44%, respectively. The model relies on continuous re-training to enhance its efficiency. It represents an attractive solution to address anomaly detection issues related to big data platforms.

Manzoor and Morgan have presented in [8], a real-time anomaly-based intrusion detection system using Apache Storm and Support Vector Machine Classification. The experimental results have shown that the recall and precision of the method reach about 99.5%, and 73%, respectively. The proposed technique is well aligned with Big Data platforms' needs. For this latter objective, Casas et al. have presented in [9] a scalable security and anomaly detection framework, called Big-DAMA. It offers both stream data processing and batch processing capabilities, using Apache Spark. The authors have evaluated many supervised machine learning models such as CART Decision Trees, Random Forest (RF), SVM, Naïve Bayes, Neural Networks, Multi-Layer Perceptron (MLP) using MAWILab. Both MLP and RF models achieve the best performance, around 0.996 average ROC. However, training the MLP model takes a much longer time than RF, which makes RF the appropriate solution for Big-DAMA.

Authors in [8], a real-time anomaly-based intrusion detection system using Apache Storm and Support Vector Machine classification algorithm is designed. The experimental results have shown that the recall and precision of the method reach 99.5% and 73%, respectively.

Table 1 A comparison between Malware detection approaches

Reference	Contribution	Method	Used dataset	Classification	Performance	Big data environment
[1]	Resolving two cybersecurity problem instances: Microsoft malware big data classification and anomaly intrusion detection	Hyper-heuristic SVM optimization	NSL-KDD and BIG 2015	Binary and multi-class	85.69% accuracy	–
[2]	P2P malware analysis	RF	CAIDA	Binary	99.94% precision 99.97% specificity	Hadoop, Hive, Sqoop, and Mahout
[3]	Malicious activities detection in IoT	The fuzzy and fast fuzzy pattern tree model	Vx-Heaven IoT Kaggle Ransomware	Binary	93.83% accuracy 89.58%recall 89.70% precision 0.8798 f-measure	–
[4]	Real-time malware detection	Keras Deep Learning	As described in [6]	Binary	97% accuracy	–
[5]	Real-time detection malware	CNN and LSTM	Malimg and privately collected samples	Multi-class	96.3% accuracy	Spark and Hadoop

Table 2 A comparison of anomaly detection predictive models

Reference	Contribution	Method	Used dataset	Classification	Performance	Big data environment
[7]	Anomaly detection of continuously evolving network data streams	Stochastic gradient descent models	MAWILab	Multi-class	99.44% accuracy	MOA (Massive Online Analysis) (stream)
[8]	Anomaly-based intrusion detection system	SVM	KDD-99	Binary	99.5% recall 73% Precision	Storm
[9]	Real-time network security and anomaly detection	RF	MAWILab	Multi-class	0.996 average ROC	Spark
[10]	Botnet Traffic Analysis and Anomaly Detection	RF	CTU-13	Binary	89% accuracy	Spark and Hadoop (batch mode)
[11]	Based on HTTP real-time anomaly detection	LSTM	Provided by a network security company	Binary	97.4% accuracy	ElasticSearch

Always to predict anomalies, Kozik has designed in [10] a generic system for anomalies detection as well as Botnet Traffic Analysis in big data. The proposed approach analyses the malware activity that is captured through NetFlows using the Random Forest (RF), Spark, and Hadoop. Different configurations of the proposed method have been tested, and the best one reaches an accuracy of 89%. Likewise, a deep learning-based HTTP real-time anomaly detection algorithm module is designed in [11] by Zhang et al. The anomaly detection technique shows good accuracy that reaches up to 97.4% using extended short term memory network.

All the above models present an interesting basis to deal with anomalies in big data platforms. In Table 2, we summarize the adopted techniques showing different criteria and requirements and compare the obtained results.

5   Predictive Models for Intrusion Detection

Intrusion detection is a mechanism that intends to identify abnormal or suspicious activities as well as policy violations. As detecting such kinds of attacks could be challenging to master, predictive models are useful in the early detection of outside and inside intrusions. The data are necessary for training and deducing models from these analyses. The speed of processing and calculation of Big Data technologies, added to this more excellent knowledge, allows more efficient automation of reaction plans. Many researchers are interested in this topic and hardly try to provide a robust technique based on ML and DL.

In this regard, Al-Jarrah et al. have proposed in [12] two novel methods based on Machine learning and feature selection to detect large-scale network intrusions. The authors have adopted RandomForest-Forward Selection Ranking and RandomForest-Backward Elimination Ranking to face intrusion issues. The selected features have improved the detection rate reaching 99.8% and have shown that incorrectly detecting represents 0.001% on the KDD-99 dataset. Acting on the same dataset and with a similar idea, Rathore et al. have provided in [13] a real-time intrusion detection system for high-speed big data environments using Hadoop and machine learning capabilities. The proposed method generates the best results on REP-Tree and J48 ML classifiers in terms of accuracy, up to 99.9% on three files of KDD99 Dataset. In the same context of real-time intrusion detection, Zhang et al. have relied on the Random Forest (RF) classification algorithm and Apache Spark to provide a reliable intrusion detection model [14]. The performance comparison among different models has demonstrated that the proposed method has a shorter detection time and achieves high accuracy, up to 96.6% F1-score.

Mylavarapu et al. [15] have developed a real-time Hybrid Intrusion Detection System using Apache Storm and two neural networks, CC4 instantaneous neural network, and Multi-Layer Perceptron neural network. The accuracy of the detection system is not high, but it is efficient, achieving 89%.

Another cyber intrusion prediction using deep learning and Big data processing capabilities is proposed in [16] by Najada et al. The Authors have first built specific prediction models for each kind of attack separately. Then, they have built prediction models for all attacks together, combining distributed random forest and deep learning techniques. The proposed models can accurately predict the threat and the attack type, as shown by the two performance indexes MSR and RMSE that tend towards zero.

In [17], Vinayakumar et al. have designed a deep neural network model to build an intrusion detection system, called Scale-Hybrid-IDS-AlertNet (SHIA). This contribution aims to identify the best algorithm that can effectively detect and classify cyber-attacks and anomalies in real-time. The authors have conducted a benchmark to choose the optimal parameters and topologies for DNNs, in comparison to classical machine learning classifiers, for both binary and multi-class classification over the KDDCup 99 dataset. The proposed model has been applied to other datasets such as NSL-KDD, UNSW-NB15, Kyoto, WSN-DS, and CICIDS 2017. Experimental results have shown that most of the DNN topologies achieve an accuracy varying between 95 and 99% for KDDCup 99 and NSLKDD datasets, and ranging between 65 and 75% for UNSW-NB15 and WSN-DS. In most cases, the authors proved that DNN outperforms the classical machine learning classifiers such as Decision Tree, AB, Random Forest, Logistic Regression, Naïve Bayes, KNN, and SVM-RBF.

In [18], Faker and Dogdu have combined Big Data and Deep Learning Techniques to improve the performance of intrusion detection systems. Three classifiers have been used to classify traffic datasets, namely, Deep Feed-Forward Neural Network (DNN), Random Forest, and Gradient Boosting Tree (GBT). These techniques have been implemented on Apache Spark to show higher performance. To evaluate the proposed method, two datasets UNSW NB15 and CICIDS2017 have been used. The results show a high accuracy with DNN and GBT for binary classification, 97.01%, and 99.99%, respectively. Better efficiency with DNN and multi-class classification, 99.16%, and 99.56%, respectively. Always in the same context, Hassan et al. [19] have designed a hybrid deep learning model by using CNN and WDLSTM. The experimentation results show that the proposed method achieves satisfactory performance on the UNSW-NB15 dataset (up to 97.17% accuracy).

All intrusion detection techniques presented above are up-and-coming for Big data platforms. They provide suitable predictive models to strengthen intrusion mechanisms. In Table 3, we offer a global comparison of these techniques.

6 Predictive Models for Access Control

In the security sector, access control has been considered a critical factor in protecting access to system resources, either software or hardware, by defining and implementing who has access to what, when, and under what conditions. However, it is hard to predefine every rule so the system could be fully protected. Therefore, predictive models could be used to bring additional security for many sensitive platforms such as Big data. In this regard, researchers in Big data have been focused on enhancing access controls using machine learning and deep learning. There were especially interested in two kinds of access control applications; attacks and threats detection as well as privacy-preserving techniques. In the following, we present some exciting contributions to these two topics.

6.1 Attacks and Threats Detection

The aim is to detect threats and attacks in networks, systems, and applications before they are exploited as false control attacks. These attacks could steal, disable, destroy, alter, and gain unauthorized access to sensitive data.

Hashmani et al. have proposed a cyber-security approach for big data platforms in [20]. This approach combines three classifiers, namely, k-nearest neighbor, support vector machine, and multilayer perceptron, to classify benign and malicious activities. The proposed technique reaches an accuracy of 99.3% in identifying and preventing possible cyber threats.

Table 3 A comparison between intrusion predictive models

Reference	Contribution	Method	Used dataset	Classification	Performance	Big data environment
[12]	Large-scale network intrusions	RF-FSR and RF-BER	KDD-99 (preprocessed by authors)	Binary	99.8% TP 0.001 FP	–
[13]	Real-time intrusion detection (network traffic)	REP-Tree and J48	KDD99 Dataset	Binary	99.9% accuracy	Hadoop and Spark
[14]	Real-time intrusion detection	RF	CICIDS2017open	Binary	96.6% F1-score	Spark and Kafka
[15]	Real-time Intrusion Detection	CC4 instantaneous neural network and Multi-Layer Perceptron neural network	ISCX 2012	Binary	89% accuracy	Storm
[16]	Real-time cyber intrusion prediction	distributed random forest and deep learning	UNB ISCX IDS 2012	Multi-class	MSR and RMSE tend towards zero	–
[17]	Intrusion detection system	DNNs	KDDCup 99, NSL-KDD, UNSW-NB15, Kyoto, WSN-DS and CICIDS 2017	Binary and multi-class	95% to 99% for KDDCup99 and NSLKDD. 65% to 75% for UNSW-NB15 and WSN-DS	Spark and Hadoop
[18]	Intrusion detection	DNN and GBT	UNSW-NB15 and CICIDS2017	Binary and Multi-class	99% average accuracy	Spark
[19]	Intrusion detection of real-time data traffic	CNN and WDLSTM	UNSW-NB15 and ISCX2012	Binary and Multi-class	97.17%accuracy	–

In [21], Jensen et al. have proposed a method to detect attacks in Signaling System No. 7 (SS7). This method is based on the big data analytics platforms Spark, Elasticsearch, and Kibana, as well as some new machine learning algorithms such as k-means clustering algorithm and the Seasonal Hybrid ESD technique. Test results have shown a detection rate of 100% and a false positive rate of 5.6%.

In [22] Subroto and Apriyana have presented a predictive model basing on social media, Big data analytics, and statistical machine learning to predict cyber risks. The prediction is made through several algorithms such as NB, kNN, SVM, DT, and ANN. The comparison using the confusion matrix has shown that ANN is the most accurate among the others with an accuracy of 96.73%.

To enhance access controls against web attacks in different clusters, Chitrakar and Petrović [23] have re-formulated the parallel version of Elkan's k-means with triangle inequality (k-meansTI) algorithm. The model is implemented using the K-means algorithm and relies on Apache Spark to deal with high-dimensional large data sets and a large number of clusters.

In [24], Al Jallad et al. have proposed a solution to detect not only new threats but also collective and contextual security attacks. The solution is based on Networking Chatbot, the deep recurrent neural network LSTM (Long Short Term Memory) on top of Apache Spark. Although the authors claimed that the experiment had shown lower false-positive and higher detection rate traditional learning models, they did not give real simulation results.

In [25], Abeshu and Chilamkurti have introduced a novel distributed deep learning scheme of cyber-attack detection and access controls at the fog level using the NSL-KDD dataset. The experimentation has shown that deep models are superior to shallow models in terms of detection accuracy (99.2% against 95.22%), false alarm rate (0.85% against 6.57%), and Detection rate (99.27% against 97.5%)). Always in the same context, Diro and Chilamkurti have designed in [26] an LSTM network for distributed cyber-attack detection and access controls in fog-to-things communication. The overall accuracy of the model is about 99.91%, which is higher than the shallow model, about 90%. In Table 4, we provide a global comparison between these different techniques based on several criteria, such as the employed algorithms, the used dataset, the classification techniques, and the shown performance.

6.2   Privacy-Preserving Techniques

It is essential to highlight that big data applications continuously collect large amounts of data that could be closely related to our lives. Analyzing these data could reveal hidden patterns and identify secret correlations. Therefore, privacy in terms of big data is an important issue, and its absence makes data and associations easily compromised. For this, researchers have focused on conceiving privacy-preserving systems that allow controlling over how personal information is collected and how it is used.

Chauhan et al. have developed in [29] a novel framework using predictive models to extract knowledgeable patterns from big data in healthcare while preserving the

Table 4 A comparison of predictive detection methods for attacks and threats

Reference	Contribution	Method	Used dataset	Classification	Performance	Big data environment
[20]	Security/cyber threat	k-nearest neighbor, support vector machine and multilayer perceptron	As described in [27]	Binary	99.3% accuracy	–
[21]	Attacks detection in SS7	k-means clustering algorithm and the Seasonal Hybrid ESD algorithm	Authors used the SS7 Attack Simulator [28] to create a dataset	Binary	The detection rate of 100% and a false positive rate of 5.6%	Spark, Elasticsearch and Kibana
[22]	Cyber risks detection	ANN	CVE and cases of cyber risks from Twitter	Binary	96.73% accuracy	–
[23]	Cyber Security Analytics web attacks classification	Reformulated k-meansTI	CSIC	Multi-class	Basing on the processing speed	Spark
[24]	Detection of new threats and also collective and contextual security attacks	LSTM and Networking Chatbot	Flows extracted from MAWI Archives, labels from MAWILAB and aggregated flows from AGURIM	Binary	–	Spark
[25]	Cyber-attack detection in fog-to-things computing	DL	NSL-KDD	Binary	99.2% accuracy 0.85% false alarm 99.27% detection rate	Spark
[26]	Denial of service detection and multi-attack detection in fog-to things computing.	LSTM	ISCX and AWID	Binary and Multi-class	99.91% accuracy	Spark

privacy and security of patients. The authors have proposed a hybrid solution that includes several methods, such as generalization of attributes and K-means clustering.

Another contribution proposed in [30], by Rao and Satyanarayana, deals with privacy-preserving data published based on sensitivity in the context of healthcare. The proposed model is based on nearest similarity-based clustering (NSB) with Bottom-up generalization on top of Hive to achieve (v,l)-anonymity and ensure individual privacy. However, to calculate the sensitivity level, researchers have only considered one kind of index value, which is the mortality rate. Thought, it is an excellent basis to generalize for Big data platforms privacy issues.

In [31], Lv and Zhu have designed two models called k-CRDP and r-CBDP, respectively. These models allow achieving correlated differential privacy in the context of Big data. The r-CBDP uses MIC and neural network-based machine learning to determine dependencies between data, calculates correlated sensitivity, and divides Big data into independent blocks. Then, it implements k-CRDP for blocks to achieve Big data correlated differential privacy.

To provide better protection for trajectory privacy and access control, Pan et al. have proposed in [32] an efficient detection scheme. For this, they have studied many algorithms to generate dummy trajectories to protect privacy. Then, they have found the differences between real trajectories and dummy trajectories from the attacker's point of view, to train a convolutional neural network (CNN) and distinguish the dummy from the real ones. The experiments have demonstrated the efficiency of the proposed model; it can detect 90% of dummy trajectories that are generated according to the current main algorithms (MLN, MN, and ADTGA); meanwhile, its erroneous judgment rate is 10%. The idea is beneficial for communication in big data platforms.

In [33], Andrew et al. have introduced a privacy-preserving high-dimensional data approach that is achieved by using Mondrian Anonymization Techniques and deep neural networks. This approach maintains the balance between data privacy and data utility, as demonstrated by their experimentation.

In [34], Guo et al. have developed a solution to enable IoT big data analytics in a privacy-preserving way using distributed deep learning. For this aim, they have first studied different distributed deep learning techniques that could be suitable for IoT architectures. Then, they have designed a framework with a novel deep learning mechanism to extract patterns and learn knowledge from IoT data in a distributed setting. The simulations have shown that adapted neural networks are better to gain new data while balances the bias and variance by obtaining more than 85% accuracy. In the same vein, Hesamifard et al. [35] have addressed the issue of privacy-preserving classification using convolutional neural networks (CNN). They have introduced new techniques to approximate the activation functions with the low degree polynomials to run CNNs over encrypted data. The experimental results have demonstrated that polynomials are suitable to adopt deep neural networks within the Homomorphic Encryption schemes. When applied to MNIST optical character recognition tasks, the proposed approach achieved 99.25% of accuracy.

In [36], a distributed, secure, and fair deep learning framework, called Deepchain, is proposed by Weng et al. for deep learning privacy-preserving. The goal of the framework is to preserve local gradients' privacy and to guarantee the suitability of the training process. This goal is achieved by employing incentive mechanisms and transactions. DeepChain can perform high training accuracy with up to 97.14% on MNIST data.

Each of these models provides a promising approach to deliver private Big Data platforms. Next, we compare those techniques following several criteria (Table 5).

7 Predictive Models for Reliable Ingestion and Normalization

Ingestion is a composition of steps aiming to collect, clean, and organize data to serve Big data management. The objective of the ingestion phase is having a single storage area for all the raw data that anyone in an organization might need to analyze. However, keeping this process reliable is a real challenge for Big data platforms, especially when they continue adopting manual processes. Therefore, ingestion needs to benefit from emerging analytics and predictive techniques. Many contributions have been redirected in this way.

In [38], Saurav and Schwars have come up with an algorithm to evaluate the correctness of delimiters' choice in tabular data files. This algorithm is based on the logistic-regression classifier to assess the candidate pair, then, the highest score of the candidate pair is chosen as the one most likely to be the correct one.

In [39], researchers have developed an intelligent system for data ingestion and governance based on machine learning and predictive techniques. The system performs the following steps: (i) It receives a set of data requirements from a user, including location information and data policy. (ii) It generates a configuration file automatically. (iii) It initiates retrieval of the new dataset using the configuration file. (iv) It saves the new dataset in a raw zone of the data lake. (v) It identifies and extracts metadata. (vi) It classifies the retrieved dataset. (vii) It saves metadata and classification information. (viii) It retrieves the data policy and converts it to executable code. (ix) It processes the dataset using the executable code and saves it in the specific zone of the data lake. The classification module performs the following tasks: (i) It extracts metadata such as business, technical, and operational metadata. (ii) It classifies dataset using machine learning, supervised, and unsupervised learning algorithms. Also, it extracts metadata that could be used the classify the dataset as either "shared" or "restricted." (iii) It saves metadata into a central repository. (iv) Finally, it exposes metadata to be searched using APIAuthors and suggests that data could be classified as shared, restricted, or sensitive.

In [40], Gong et al. have proposed a project for a normalization method to compress the high-dimensional data and decompress the record whenever necessary. This contribution aims to optimize the storage by using a potential approach, called AutoEncoder, which can support online training. In the same context, Ren et al. [41] have designed a Trust-based Minimum Cost Quality Aware data collection

Table 5 A Comparison of predictive models for privacy

Reference	Contribution	Method	Used dataset	Classification	Performance	Big data environment
[29]	Privacy-preserving of healthcare databases	Generalization of attributes and K-means clustering	Obtained from OTIS (Online Tuberculosis Information System), a data repository of CDC (Centre for Disease Control),	NA	–	–
[30]	Privacy-preserving	NSB with Bottom-up generalization	–	NA	–	Hive and Hadoop
[31]	Privacy-preserving	MIC, neural network and k-CRDP	Air quality data [37]	NA	–	–
[32]	Protection for trajectory privacy	CNN	Microsoft research GeoLift	NA	90% training TP 10% training FP	–
[33]	Privacy-preserving	Mondrian Anonymization Techniques and deep neural networks	Adult dataset download from UCI	NA	minimal information loss	–
[34]	Privacy-preserving distributed learning for big data in IoT	A novel deep learning mechanism	CIFAR-10	NA	85% training accuracy	–
[35]	Privacy-preserving classification	CNN	MNIST	NA	99.25% training accuracy	–
[36]	Deep learning privacy-preserving	Incentive mechanism and transactions	MNIST	NA	97.14% training accuracy	–

scheme for malicious P2P networks basing on the idea of machine learning. For this, the scheme selects a trusted data reporter to collect and normalize data. The experimental comparison among different strategies has demonstrated that the proposed method has a better performance.

Other contributions were rather oriented to tackle fake data detection.

In [42], Miller et al. have used two stream-clustering algorithms, StreamKM++ and DenStream, to detect spam and data disturbers. The recall of the combination of the two algorithms reaches 100% recall and 2.2% false-positive rate. On their side, Van Der Walt et al. [43] have proposed a fascinating Identity Deception Detection Model (IDDM) for social media platforms (SMPs). It employs machine learning to identify appropriate attributes and features of identity-related information on SMPs. To make this happen, they have evaluated several ML algorithms and have found that RF achieves the best accuracy, around 97.49%, to determine if an identity is deceptive or not. In the same context, an attractive model based on a deep neural network (DNN) algorithm, called DeepProfil, has been proposed in [44] by Wanda et al. This algorithm relies on a dynamic CNN algorithm to classify fake profiles. The experimentation has shown high performances, about 94% of Precision, 93.21% recall, and 93.42% F1 Score.

The presented predictive solutions remain very limited for such a significant problem, such as controlling reliability. Still, they could an excellent start to strength ingestion and normalization layer for Big data platforms. Next, we show an overview comparison of the previously presented techniques (Table 6).

8   Conclusion

Predictive analytics could provide additional support in the face of cyber-attacks and other data breaches. This type of analysis would not only identify and alert in the event of an attack but would also prevent them early and analyze them to avoid any danger. Big data platforms are gaining enormous importance, but also inherit the sensitivity of the data and analysis they host. For this, it is crucial to adopt predictive analysis techniques to add advanced security layers to exiting Big data policies. In this paper, we group and discuss most of the exciting works based on Machine learning and Deep learning, presenting promising models to protect big data platforms against different security and privacy attacks. The paper has been organized under five different use cases, including malware detection, intrusion, anomaly, access, and ingestion normalization controls. For each use case, we discuss suitable models and identify the set of security dimensions, criteria interpretations, and obtained results. Furthermore, we provide a comparison of these different models by showing their efficiency.

This contribution is the first step towards a general big data security framework based on predictive analysis.

Table 6 A comparisons of predictive models for ingestion and normalization

Reference	Contribution	Method	Used dataset	Classification	Performance	Big data environment
[38]	Automatic Detection of Delimiters in Tabular Data Files	Logistic regression	Variety of sources	Binary	93% accuracy	–
[39]	Intelligent data ingestion system	ML	–	multiclass	–	–
[40]	Compress the high-dimensional data	AE	Record of events that happened at CERN	NA	0.9497 R2 score (one-layer)	–
[41]	Optimization of data collection in the P2P network	A function based on the idea of ML	Different locations	Binary	improved the QoS by 49.39%	–
[42]	Spam detection on Twitter streams	Modified StreamKM++ and DenStream	Twitter accounts manually labeled	Binary	100% recall 2.2% FP 98% accuracy	–
[43]	Identity deception detection	RF	Collected tweets	Binary	97.49% accuracy	–
[44]	Fake profile detection	Dynamic CNN	OSN dataset	Binary	94% Precision 93.21% Recall 93.42% F1 Score	–

References

1. Sabar NR, Yi X, Song A (2018) A bi-objective hyper-heuristic support vector machines for big data cyber-security. IEEE Access 6:10421–10431. https://doi.org/10.1109/ACCESS.2018.280 1792
2. Chhabra GS, Singh VP, Singh M (2018) Cyber forensics framework for big data analytics in IoT environment using machine learning. Multimed Tools Appl. https://doi.org/10.1007/s11 042-018-6338-1
3. Dovom EM, Azmoodeh A, Dehghantanha A, Newton DE, Parizi RM, Karimipour H (2019) Fuzzy pattern tree for edge malware detection and categorization in IoT. J Syst Architect 97:1–7. https://doi.org/10.1016/j.sysarc.2019.01.017
4. Masabo E, Kaawaase KS, Sansa-Otim J (2018) Big data: deep learning for detecting malware. In: Proceedings of the 2018 international conference on software engineering in Africa, Gothenburg, Sweden, May 2018, pp 20–26. https://doi.org/10.1145/3195528.3195533
5. Vinayakumar R, Alazab M, Soman KP, Poornachandran P, Venkatraman S (2019) Robust intelligent malware detection using deep learning. IEEE Access 7:46717–46738. https://doi.org/10.1109/ACCESS.2019.2906934

6. Marco Ramilli Web Corner, Malware Training Sets: a machine learning dataset for everyone. http://marcoramilli.blogspot.it/2016/12/malware-training-sets-machine-learning. html. Accessed 10 Mar 2020

7. Mulinka P, Casas P (2018) Stream-based machine learning for network security and anomaly detection. In: Proceedings of the 2018 workshop on big data analytics and machine learning for data communication networks, Budapest, Hungary, Aug 2018, pp 1–7. https://doi.org/10. 1145/3229607.3229612

8. Manzoor MA, Morgan Y (2017) Network intrusion detection system using apache storm. Adv Sci Technol Eng Syst J 2(3):812–818

9. Casas P, Soro F, Vanerio J, Settanni G, D'Alconzo A (2017) Network security and anomaly detection with Big-DAMA, a big data analytics framework. In: 2017 IEEE 6th international conference on cloud networking (CloudNet), Sept 2017, pp 1–7. https://doi.org/10.1109/clo udnet.2017.8071525

10. Kozik R (2017) Distributed system for botnet traffic analysis and anomaly detection. In: 2017 IEEE international conference on internet of things (iThings) and IEEE green computing and communications (GreenCom) and IEEE cyber, physical and social computing (CPSCom) and IEEE smart data (SmartData), June 2017, pp 330–335. https://doi.org/10.1109/ithings-gre encom-cpscom-smartdata.2017.55

11. Zhang G, Qiu X, Gao Y (2019) Software defined security architecture with deep learning-based network anomaly detection module. Presented at the 2019 IEEE 11th international conference on communication software and networks, ICCSN 2019, pp 784–788. https://doi.org/10.1109/ iccsn.2019.8905304

12. Al-Jarrah OY, Siddiqui A, Elsalamouny M, Yoo PD, Muhaidat S, Kim K (2014) Machine-learning-based feature selection techniques for large-scale network intrusion detection. In: 2014 IEEE 34th international conference on distributed computing systems workshops (ICDCSW), June 2014, pp 177–181. https://doi.org/10.1109/icdcsw.2014.14

13. Rathore MM, Ahmad A, Paul A (2016) Real time intrusion detection system for ultra-high-speed big data environments. J Supercomput 72(9):3489–3510. https://doi.org/10.1007/s11 227-015-1615-5

14. Zhang H, Dai S, Li Y, Zhang W (2018) Real-time distributed-random-forest-based network intrusion detection system using Apache spark. In: 2018 IEEE 37th international performance computing and communications conference (IPCCC), Nov 2018, pp 1–7. https://doi.org/10. 1109/pccc.2018.8711068

15. Mylavarapu G, Thomas J, Ashwin Kumar TK (2015) Real-time hybrid intrusion detection system using Apache storm. In: 2015 IEEE 17th international conference on high performance computing and communications, 2015 IEEE 7th international symposium on cyberspace safety and security, and 2015 IEEE 12th international conference on embedded software and systems, Aug 2015, pp 1436–1441. https://doi.org/10.1109/hpcc-css-icess.2015.241

16. Najada HA, Mahgoub I, Mohammed I (2018) Cyber intrusion prediction and taxonomy system using deep learning and distributed big data processing. In: 2018 IEEE symposium series on computational intelligence (SSCI), Nov 2018, pp 631–638. https://doi.org/10.1109/ssci.2018. 8628685

17. Vinayakumar R, Alazab M, Soman KP, Poornachandran P, Al-Nemrat A, Venkatraman S (2019) Deep learning approach for intelligent intrusion detection system. IEEE Access 7:41525–41550. https://doi.org/10.1109/ACCESS.2019.2895334

18. Faker O, Dogdu E (2019) Intrusion detection using big data and deep learning techniques. In: Proceedings of the 2019 ACM Southeast conference, Kennesaw, GA, USA, Apr 2019, pp 86–93. https://doi.org/10.1145/3299815.3314439

19. Hassan MM, Gumaei A, Alsanad A, Alrubaian M, Fortino G (2020) A hybrid deep learning model for efficient intrusion detection in big data environment. Inf Sci 513:386–396. https:// doi.org/10.1016/j.ins.2019.10.069

20. Hashmani MA, Jameel SM, Ibrahim AM, Zaffar M, Raza K (2018) An ensemble approach to big data security (cyber security). Int J Adv Comput Sci Appl (IJACSA) 9(9) (2018). https:// doi.org/10.14569/ijacsa.2018.090910

21. Jensen K, Nguyen HT, Do TV, Årnes A (2017) A big data analytics approach to combat telecommunication vulnerabilities. Cluster Comput 20(3):2363–2374. https://doi.org/10.1007/s10586-017-0811-x

22. Subroto A, Apriyana A (2019) Cyber risk prediction through social media big data analytics and statistical machine learning. J Big Data 6(1):50. https://doi.org/10.1186/s40537-019-0216-1

23. Shrestha Chitrakar A, Petrović S (2019) Efficient k-means using triangle inequality on spark for cyber security analytics. In: Proceedings of the ACM international workshop on security and privacy analytics, Richardson, Texas, USA, Mar 2019, pp 37–45. https://doi.org/10.1145/3309182.3309187

24. Al Jallad K, Aljnidi M, Desouki MS (2019) Big data analysis and distributed deep learning for next-generation intrusion detection system optimization. J Big Data 6(1):88. https://doi.org/10.1186/s40537-019-0248-6

25. Abeshu A, Chilamkurti N (2018) Deep learning: the frontier for distributed attack detection in fog-to-things computing. IEEE Commun Mag 56(2):169–175. https://doi.org/10.1109/MCOM.2018.1700332

26. Diro A, Chilamkurti N (2018) Leveraging LSTM networks for attack detection in fog-to-things communications. IEEE Commun Mag 56(9):124–130. https://doi.org/10.1109/MCOM.2018.1701270

27. Ma J, Saul LK, Savage S, Voelker GM (2009) Identifying suspicious URLs: an application of large-scale online learning. In: Proceedings of the 26th annual international conference on machine learning, Montreal, Quebec, Canada, June 2009, pp 681–688. https://doi.org/10.1145/1553374.1553462

28. Jensen K (2020) jss7-attack-simulator. https://github.com/polarking/jss7-attack-simulator. Accessed 11 Mar 2020

29. Chauhan R, Kaur H, Chang V (2020) An optimized integrated framework of big data analytics managing security and privacy in healthcare data. Wirel Pers Commun 1–22. https://doi.org/10.1007/s11277-020-07040-8

30. Rao PS, Satyanarayana S (2018) Privacy preserving data publishing based on sensitivity in context of Big Data using Hive. J Big Data 5(1):1–20. https://doi.org/10.1186/s40537-018-0130-y

31. Lv D, Zhu S (2019) Achieving correlated differential privacy of big data publication. Comput Secur 82:184–195. https://doi.org/10.1016/j.cose.2018.12.017

32. Pan J, Liu Y, Zhang W (2019) Detection of dummy trajectories using convolutional neural networks. Secur Commun Netw 2019. https://doi.org/10.1155/2019/8431074

33. Andrew J, Karthikeyan J, Jebastin J (2019) Privacy preserving big data publication on cloud using Mondrian anonymization techniques and deep neural networks. In: 2019 5th international conference on advanced computing communication systems (ICACCS), Mar 2019, pp 722–727. https://doi.org/10.1109/icaccs.2019.8728384

34. Guo M, Pissinou N, Iyengar SS (2019) Privacy-preserving deep learning for enabling big edge data analytics in internet of things. Presented at the 2019 10th international green and sustainable computing conference, IGSC 2019. https://doi.org/10.1109/igsc48788.2019.8957195

35. Hesamifard E, Takabi H, Ghasemi M (2019) Deep neural networks classification over encrypted data. In: Proceedings of the ninth ACM conference on data and application security and privacy, Richardson, Texas, USA, Mar 2019, pp 97–108. https://doi.org/10.1145/3292006.3300044

36. Weng J, Weng J, Zhang J, Li M, Zhang Y, Luo W (2019) DeepChain: auditable and privacy-preserving deep learning with blockchain-based incentive. IEEE Trans Dependable Secure Comput 1. https://doi.org/10.1109/tdsc.2019.2952332

37. beijingair. http://beijingair.sinaapp.com/. Accessed 11 Mar 2020

38. Saurav S, Schwarz P (2016) A machine-learning approach to automatic detection of delimiters in tabular data files. In: 2016 IEEE 18th international conference on high performance computing and communications; IEEE 14th international conference on smart city; IEEE 2nd international conference on data science and systems (HPCC/SmartCity/DSS), Dec 2016, pp 1501–1503. https://doi.org/10.1109/hpcc-smartcity-dss.2016.0213

39. Okorafor E et al (2020) Intelligent data ingestion system and method for governance and security. US20200019558A1, Jan 16, 2020
40. Gong X, Shang L, Wang Z (2016) Real time data ingestion and anomaly detection for particle physics. Capstone project paper, 2016. https://zw1074.github.io/files/FinalReport_TeamXYZ. pdf. Accessed 13 Mar 2020
41. Ren Y, Zeng Z, Wang T, Zhang S, Zhi G (2020) A trust-based minimum cost and quality aware data collection scheme in P2P network. Peer-to-Peer Netw Appl. https://doi.org/10.1007/s12 083-020-00898-2
42. Miller Z, Dickinson B, Deitrick W, Hu W, Wang AH (2014) Twitter spammer detection using data stream clustering. Inf Sci 260:64–73. https://doi.org/10.1016/j.ins.2013.11.016
43. van der Walt E, Eloff JHP, Grobler J (2018) Cyber-security: identity deception detection on social media platforms. Comput Secur 78:76–89. https://doi.org/10.1016/j.cose.2018.05.015
44. Shama SK, Siva Nandini K, Bhavya Anjali P, Devi Manaswi K (2019) DeepProfile: finding fake profile in online social network using dynamic CNN. Int J Recent Technol Eng (IJRTE) 8:11191–11194

The Fundamentals and Potential for Cybersecurity of Big Data in the Modern World

Reinaldo Padilha França, Ana Carolina Borges Monteiro, Rangel Arthur, and Yuzo Iano

Abstract Information security is essential for any company that uses technology in its daily routine. Cybersecurity refers to the practices employed to ensure the integrity, confidentiality, and availability of information, consisting of a set of tools, risk management approaches, technologies, and methods to protect networks, devices, programs, and data against attacks or non-access authorized. Big Data becomes a barrier for network security to understand the true threat landscape, considering effective solutions that differ from reactive "collect and analyze" methods, improving security at a faster pace. Through Machine Learning it is possible to address unknown risks including insider threats, being an advanced threat analytics technology. Big data analytics, in conjunction with network flows, logs, and system events, can discover irregularities and suspicious activities, can deploying an intrusion detection system, which given the growing sophistication of cyber breaches. Cybersecurity is fundamental pillars of digital experience, so organizations' digital initiatives must consider, from the beginning, the requirements in cyber and privacy, concerning the security and privacy of this data. So, Big data analytics plays a huge role in mitigating cybersecurity breaches caused by the most diverse means, guaranteeing data security and privacy, or supporting policies for secure information sharing in favor of cybersecurity. Therefore, this chapter has the mission and objective of providing an updated review and overview of Big Data, addressing its evolution and fundamental concepts, showing its relationship with Cybersecurity on the rise as well as

R. P. França (✉) · A. C. B. Monteiro · R. Arthur · Y. Iano
School of Electrical and Computer Engineering (FEEC), University of Campinas—UNICAMP,
Av. Albert Einstein, 400, Barão Geraldo, Campinas, SP, Brazil
e-mail: padilha@decom.fee.unicamp.br

A. C. B. Monteiro
e-mail: monteiro@decom.fee.unicamp.br

R. Arthur
e-mail: rangel@ft.unicamp.br

Y. Iano
e-mail: yuzo@decom.fee.unicamp.br

51

Y. Maleh et al. (eds.), *Machine Intelligence and Big Data Analytics for Cybersecurity Applications*, Studies in Computational Intelligence 919,
https://doi.org/10.1007/978-3-030-57024-8_3

approaching its success, with a concise bibliographic background, categorizing and synthesizing the potential of technology.

Keywords Big Data · Cybersecurity · Big data analytics · Malware detection · Prevention · Security · Information security · Machine learning

1 Introduction

Big Data is a nomenclature for the phenomenon, which happens more strongly in the digital environment, allowing the organization to have access to a large amount of information, normally unstructured, which until recently this organization does not have practical practices to access this information. This technology is a massive amount of data that is normally used in data centers. Information security is indispensable for any company that uses technology daily. Preventing disasters, such as loss of important data or even suffering some type of hacker invasion, is a major concern [1, 2].

Moreover, Big Data can partner with the information security industry to detect threats to a company's cloud systems. Thanks to the volume of information collected and attempted invasions, suspicious activities and the spread of viruses can be detected in real-time with precision and responsiveness. The other side of Big Data related to information security is that, as with business strategies, there will be more intelligent protection of critical data. This trend will be a decisive factor of change in the short term. Data analysis will play a key role in security, especially in the early detection of fraud and information theft [2, 3].

The information security sector will have the precise geolocation of these possible threats, will know which individuals participate in these operations and which platforms or means are most used for this sharing of confidential information, in e-mails, cloud systems, social networks, among others [3].

Big Data, based on analytical solutions, will allow organizations to access data faster, both internally and externally, and can correlate information that will help detect possible crimes or threats. Information analysis is fully applicable to security and can help prevent fraud and internal or external threats, shortening response times. All of this facilitates decision-making to improve the information security sector. Through measures such as the categorization and encryption of certain information. It is possible, for example, that only one email recipient has access to the content of the correspondence; automation of certain resources in order to protect the company's database [4].

Improvement in the training of the IT manager, so that he can respond adequately to these threats; definition of stricter criteria for making information available in the cloud. The new technology is evolving to the point of enabling a variety of advanced forecasting capabilities and real-time controls. It will change the nature of conventional security controls, such as anti-malware, data loss prevention, and firewalls. The threat to privacy must be assured since Big Data is expanding the boundaries of information security responsibilities [1, 5].

Analytical applications can be monitored more widely than traditional information security event management systems. The establishment of standards of normality, context information, and external threats will make, based on data analysis, the detection of any anomaly related to information security more efficient. Data analysis allows being extracted relevant information for the devices that make up the Internet in general, as well as offering answers that can be used in information security under any computational environment [6].

Big Data is a radical change in the use and collection of information, in the velocity to analyze and make decisions in real-time. This new way of looking at the world, so to speak, should have an impact on security strategies, which will tackle possible new threats with greater intelligence. Considering that with the advent of the Internet of Things it is the objective to connect several things that generate and return information to and from its users, and that the information must be returned as soon as possible to users so that this becomes relevant, fast and secure data processing methods like Big Data that can handle a large mass of data should be considered [2, 7].

In this sense, Analytics solutions analyze different information from the networks to be able to anticipate cyber threats and act before criminals. Understanding the behavior of the network makes it possible to differentiate irregular activities from normal movements and, thus, change the corporate attitude towards digital security from reactive to proactive. Thus, integrating the Analytics platform for Big Data well positioned in the core of advanced analytics software to provide the market with an additional layer of security and detention. Since it is possible to obtain effective Big Data solutions that differ from the reactive "collect and analyze" methods, and thus aim at behavior analysis and tools to improve security at a faster pace [3, 7].

Machine learning and big data have a crucial relationship with technical processes, including cybersecurity. By themselves, these tools are already major advances in the cyber world. Network security can be the most critical area in companies. If optimized, the volume of data available offers significant opportunities to contextualize more accurate and rapid detection of threats [8].

A machine learning algorithm can work perfectly with a smaller database. But when it is combined with big data, results are maximized. A machine learning model learns much more and faster when it is powered by a large and varied volume of data and information. In this way, machine learning can find, in big data, patterns, and anomalies that can solve problems and even create new insights, allowing technologies and companies to develop. In other words, thanks to the volume of data and the velocity with which it arrives, the actions to be determined by machine learning become more precise and relevant [9].

In turn, machine learning is one of the best ways to bring big data to life. Such a large volume of data is only useful insofar as the data can be effectively analyzed, correlated, and transformed into effective actions. This is, in fact, the main role of machine learning in this case. After all, there is no point in having data volume, variety, and velocity if it is not possible to process them and, above all, add value to them [9, 10].

Therefore, this chapter has the mission and objective of providing an updated review and overview of Big Data, addressing its evolution and fundamental concepts, showing its relationship with Cybersecurity on the rise as well as approaching its success, with a concise bibliographic background, categorizing and synthesizing the potential of technology.

2 Methodology

This survey carries out a bibliographic review of the main research of scientific articles related to the theme of Big Data, addressing its evolution and fundamental concepts, showing its relationship with Cybersecurity, published in the last 5 years on renowned bases.

3 Big Data and Cybersecurity

Information security is essential for any company that uses technology in its daily routine. In the same way as data centers, locally they concentrate servers and equipment for processing big data, this type of architecture works as a "nervous system", storing expressive volumes of information, where it is necessary to prevent disasters, such as loss of important data or until suffering some type of hacker invasion, which are great business concerns, since the preservation of big data, which is a massive volume of data that is normally stored, is of the utmost importance, needs to be as optimized as possible [11].

A data center is composed of several servers working together, which process all digital activities in its software, in services with complete infrastructure, it is common for data to be kept in redundancy, which corresponds to having backup copies being control that needs be considered, with constant backups in the public cloud, that through backup systems, no information is lost, that is, other data centers spread across the globe, giving total security to the integrity of big data [5].

Just as the traditional cybersecurity approach in organizations is proving to be less and less effective in combating more complex and virtually ubiquitous threats, related to "traditional approaches" simply cannot cope with the massive amount of data being created in corporations all the time. What is needed is that real-time predictive technologies accelerate time to detect and combat attacks, which are solutions for analyzing different information from your networks in order to be able to anticipate cyber threats and act before criminals. Since understanding the behavior of the network makes it possible to differentiate irregular activities from normal movement and flow and, therefore, changing the corporate attitude towards digital security from

Fig. 1 Big data 5 Vs illustration

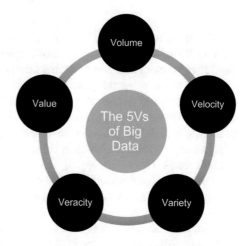

reactive to proactive is necessary, when applying predictive and behavioral analyzes to all available business data, being able to estimate the potential of threats, detecting possible attacks and achieving advanced intelligence [12].

The main aspects of Big Data can be defined by 5 Vs related to Volume, Variety, Velocity, Veracity, and Value, as shown in Fig. 1. The Volume, Variety, and Velocity aspects are related to the large amount of unstructured data that must be analyzed by Big Data solutions at great velocity. Regarding Veracity, it concerns the sources and quality of the data, as they must be reliable. As for Value, it is related to the benefits that Big Data solutions bring to a company, since each business has specific benefits brought by the Big Data analysis that compensate the investment in specific solutions of this technology [13].

The difference between structured and unstructured data is that structured data is data stored in sources that are easy for humans to understand, such as tables, excel spreadsheets, databases, i.e., those that have some standard or format that can be used in reading and extracting data like legacy systems, text files like CSV, txt or XML, among others. Unstructured data is data that does not have a structure defined as a music file, an image, a video, i.e., it does not have a standardized format for reading, it can be word files, internet pages, videos, audios, among others. Semi-structured data is data that is not stored in a database or any other data table, but has some organized internal properties. An example of semi-structured data is HTML code, which does not restrict the amount of desired information to collect in a document, but still imposes hierarchy through semantic elements [14].

Know the evolution of the cybersecurity scenario and understand how Big Data and predictive analysis can be implemented to address threats and risks faced daily, concerning the development of strategies in relation to network threats, associating with existing security systems capable conducting advanced behavior analysis, which is integrated with Big Data Analytics platforms at its core, provides an additional layer of security and detection [15].

Considering the context of government agencies, in addition to multilayer security defenses, they have highly complex infrastructures composed of an extensive amount of application structuring technologies for cloud and mobile, as well as using predictive behavior analysis, replacing their posture for a more proactive defense [16].

Taking into account that network security can be the most critical area in companies, which should always be optimized, reflecting that the volume of data available offers significant opportunities to contextualize more accurate and faster detection of threats. And the identification of threats and solutions for advanced and predictive analysis of Big Data is critical in advancing the cyber order, including regulatory compliance [17].

What with the reduction of gaps and the complexity of digital channels, advanced analytical intelligence solutions, and services have become fundamental technologies for risk managers, data managers, and executives. Since organizations need to take a proactive stance to understand threats before a possible 'attacker' causes any type of damage, which requires constant monitoring of network behavior so that irregular activities can be distinguished from normal activities [18–22].

Applying the predictive and behavioral analysis to all available business data, executed in real-time so that threats are proactively minimized before a significant loss occurs, making it possible to estimate the potential of threats, developing a set of security solutions to deal with the number each increasing sophistication of attacks, detecting possible attacks and achieving advanced intelligence [15, 17].

Big Data becomes a barrier for network security to understand the true threat landscape, considering effective solutions that differ from reactive "collect and analyze" methods, improving security at a faster pace. What impacts on the understanding of the business behavior in each system through surveys of the correlated daily transactions, identifying possible threats, providing organizations, from various segments, with a comprehensive view of risks to obtain the advantage over virtual attackers [23].

4 Machine Learning and Cybersecurity

Machine learning is the basis of artificial intelligence systems, which are methods of analyzing data and information, algorithms, making the systems learn from them and evolve on their own, eliminating or reducing the need for human intervention. It is one of the best ways to bring big data to life, since it is considered such a large volume of data it is only useful insofar as the data can be effectively analyzed, correlated and transformed into effective actions, considering it as the main role of technology, in this case, after all, there is no point in having volume, variety, and velocity of data if it is not possible to process them and, above all, add value to them [24].

In supervised learning, the system receives a previous set of data that contains the correct answer, which consists of labeled data, i.e., the problems and solutions are already defined and associated, leaving the machine to do is to show the right result from the variables, as shown in Fig. 2. In unsupervised learning, the opposite occurs,

SUPERVISED

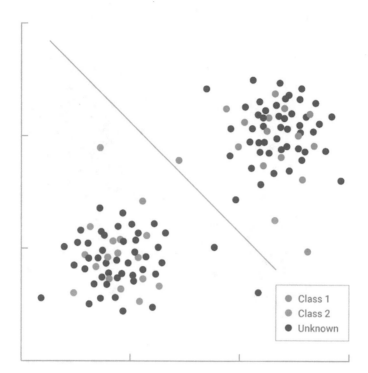

Fig. 2 Supervised learning illustration

it is used against data that do not have historical labels, since there is no specific expected result or correct answer, i.e., the crossing of the data is unpredictable and depends on the variables entered in the system. What makes this type of machine learning, each movement is a discovery, and therefore, it is also much more complex [24–27].

In semi-supervised learning, it is the combination of the two types of data previously, labeled and unlabeled, using both labeled and unlabeled data for training, usually a small amount of labeled data with a large amount of unlabeled data, as the unlabeled data is cheaper and requires less effort to acquire, as shown in Fig. 3. In this sense, there is a small number of responses defined among the uncertainties, which help to direct the discoveries of the machine. And in reinforcement learning it is different from all the previous types, as it does not have any previous data set, it is as if the machine were in an unknown place, where it starts to perform tests to collect impressions and adapt to the environment, and so able to increasingly improve its asset combinations as it analyzes the positive or negative return of the environment [24–27].

UNSUPERVISED

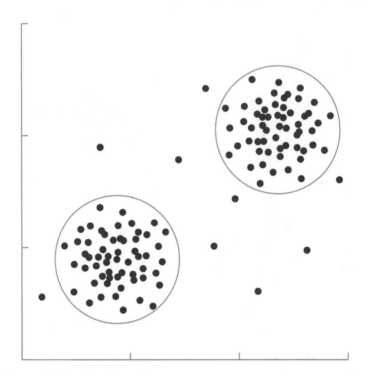

Fig. 3 Unsupervised learning illustration

In other words, it is necessary to find quality and meaning amid so much data, information needs to become productive, in that sense, machine learning and big data work together to create intelligent models that have the ability to make relationships, obtain insights, predict behaviors and even determine actions, through the properties of artificial intelligence. A machine learning algorithm can perfectly work with a smaller database, but when combined with big data, fed by a large and varied volume of data and information, the results are maximized, making the machine learning model learn a lot faster and faster [24–27].

In this way, machine learning can find in big data, provided that dynamic technology, with the help of machine learning algorithms that analyze large volumes of data to determine patterns and anomalies that can solve problems, creating new insights, allowing actions to be determined by machine learning become more accurate and relevant [26].

Machine learning systems are capable of analyzing user behavior, historical demand for a given period and user involvement with a specific event, among many other factors, considering technologies such as User and Entity Behavior Analysis (UEBA) and the design of deep learning algorithms are emerging as two of the most prominent technologies in the field of cybersecurity. Since in the current era it is in the midst of an artificial intelligence security revolution that will make machine learning solutions the new standard, in addition to the known and traditional solutions [27].

Big data and Machine learning have a crucial relationship within technology processes, including cybersecurity, which these tools are major advances for the cyber world, but when used together, they can offer even better results. Taking into account that cyber-attacks are increasingly bold and sophisticated, it is also necessary to have security solutions capable of quickly dealing with known and unknown threats [8, 13].

In a practical everyday context, when it comes to email security, considering that more and more these cyber-attacks use social engineering and spoofing tactics in an attempt to pass through legitimate and harmless emails, managing to override traditional protection filters, Machine Learning and Big Data come together so that the solution can predict and prevent attacks and threats, with advanced algorithms that analyze massive data from legitimate and malicious emails, this analysis capability is essential to prevent phishing and spear-phishing attacks, it is possible to predict and identify risky and dangerous behaviors [28, 29].

Through Machine Learning it is possible to address unknown risks including insider threats, being an advanced threat analytics technology, which is tricky to detect because they are users legitimately logged into corporate systems, which requires advanced analytics. What together with Big Data is possible to identify anomalies in personnel or device behavior creating a model of "normal behavior" for a person, a group of devices on the network or device, and even ones that were not predefined as rules intelligently identifying anomalies [30].

Perform machine learning-based malware detection intelligently analyzing binaries transmitted by email or downloaded, detecting anomalies in the network creating a model of network traffic and intelligently identifying anomalies in traffic, even if not flagged by antivirus, technology through Big Data realizes that something happening that is different than usual for this period of time of day, and understand if it is a benign program or more likely to be a malicious program. In this sense, advanced threat analytics powered by Machine Learning based intrusion detection identifying patterns in network traffic or access control is able to prevent similar to historic intrusions or attacks, increasing business cybersecurity [31].

Through Supervised learning, together with Big Data, it is possible to use it for phishing domain detection, as long as the machine learns from a data set that contains inputs and known outputs. In this respect, the function or model is built allowing to predict what the output variables will be for new, unknown outputs. In which in the context of security, as security tools learn to analyze new behavior and determine if it is "similar to" previously known normal or known suspect behavior [32].

Just like in unsupervised learning also allied with Big Data, the system learns from a dataset that contains only input variables, however, in this context, there is no correct answer, which instead the algorithm or model developed to discover new patterns in the data, as shown in Fig. 3. Using in the context of security, it is possible for these tools to use unsupervised learning to detect and act on abnormal behavior, without classifying it or understanding if it is good or bad, which increases security performance [33].

With respect to data mining, which is the use of analytics techniques, to uncover hidden insights in large volumes of data, this technique can uncover hidden relations between entities, discover classification models which help group entities into useful categories in the same way as discover frequent sequences of events to assist prediction. What effectively applied in the context of security, Data mining techniques are used by security tools concerning tasks like anomaly classification of incidents or network events and prediction of future attacks based on historic data, considering detection in very large data sets, i.e., Big data [34].

Dimension Reduction is the process of converting a data set with a high number of dimensions, or parameters describing the data, to a data set with fewer dimensions, without losing important information. That is, it consists of taking a high-dimensional data set and reducing it to a smaller number of dimensions in a way that represents the original data as much as possible. What applied in a security context related to Security data, consists of logs with a large number of data points about events in IT systems, which is useful and can be used with respect to removing dimensions that are not necessary for answering the question at hand, as long as this criterion helps security tools identify anomalies more accurately, reflecting on a huge set of data like Big Data [35].

So, it is clear that the main advantage of Machine Learning is related to its ability to process and analyze huge volumes of data quickly, making it possible to predict possible "trends" of failure in digital security, allowing the creation and preparation of responses to the possible side effects of these attempts. attacking an organization's cybersecurity.

The main disadvantages of Machine Learning can be seen that the technology can also be employed to improve malware; targeting specific victims and extracting important data, since more and more personal and sensitive data are involved with the digital world; since cybercriminals can look for daily vulnerabilities in digital infrastructures, being able to hijack them through botnets, among others.

5   Big Data Analytics and Cybersecurity

Cybersecurity refers to the practices employed to ensure the integrity, confidentiality, and availability of information, consisting of a set of tools, risk management approaches, technologies, training, and methods to protect networks, devices, programs, and data against attacks or non-access authorized. In practice, ensuring cybersecurity in a company, for example, requires the coordination of efforts across the information system, such as information security, applications, and network; Disaster recovery/business continuity planning; Operational security; End-user education [12, 29].

In short, organizations with good cybersecurity strategies are able to prevent cyberattacks, data breaches, and identity theft; doing risk management. Since the most common cybersecurity approaches adopted are Data Loss Prevention protecting data by focusing on finding, classifying and monitoring information at rest, in use and on the go; Network Security that protects network traffic by controlling incoming and outgoing connections to prevent threats from entering or spreading on the network; Cloud Security providing protection for data used in cloud-based services and applications; Identity and Access Management relying on authentication services to limit and track employee access to protect internal systems against malicious entities; Adoption of intrusion detection systems or intrusion prevention systems acting to identify potentially hostile cyber activities; Encryption to make data unintelligible and is often used during data transfer to prevent theft in transit; antivirus/antimalware solutions, which are applications that scan systems for known threats [29].

One of the most problematic elements of Cybersecurity is the constantly evolving nature of risks, in which the traditional approach has been to focus resources on crucial system components and protect against the biggest known threats, which means leaving components defenseless and not protecting systems against less dangerous risks [36].

Among the biggest challenges, today are hyperconnected environments oriented to APIs (Application Programming Interfaces), which are sets of routines and standards established for the use of system functionality by applications. APIs enhance users' interactive digital experiences and are fundamental to digital transformation. However, they also provide a window into a growing cybersecurity risk, since they present multiple ways to access a company's data and can be used to enable new attacks that exploit mobile and web applications and Internet devices from Things [36, 37].

The categorization and encryption of certain information, related to criteria where only an email recipient has access to the content of digital correspondence; definition of stricter criteria for making information available in the cloud; management of passwords and logins; user confirmation and information storage under a structure that allows categorizing them and establishing their profiles; automation of certain resources, in order to protect the company's database, for example, representing a massive alignment in internal audits facilitating decision-making for the improvement of the information security sector, generated by Big Data [38].

Even more so today, when emerging technologies, mobile devices, the consolidation of cloud computing, the convergence of telecommunications, the advancement of social networks and the concept of Big Data, with the globalization of the internet, is no longer possible treat IT systems in isolation as any business is connected in one way or another to the global digital environment [38].

A data breach can have several devastating consequences for any business, and it can destroy a brand's reputation through the loss of consumer and partner confidence. The loss of critical data, such as source files or intellectual property, can cost the company its competitive advantage and may impact revenues due to non-compliance with data protection regulations. Whether with high profile data breaches or small day-to-day incidents, organizations must adopt and implement a strong cybersecurity approach [39].

Estimates indicate that the amount of information obtained through digital means tends to increase more and more, boosting research in Big Data solutions, and reconciling all this volume of data obtained by Big Data, with information security tends to prevent confidential information to be shared fraudulently, which has produced an increasing trend towards strategic alignment between Big Data and data processing technologies, such as Machine Learning, specifically concerning information security [39, 40].

Big Data can still ally with the information security sector by detecting threats to a company's cloud systems, with respect to the volume of information collected, as long as the use of mobile devices in business environments, which is marked by two main phenomena which are the strong increase in the volume of business data and the need to consider dispersed and diverse information, of all current digital data are not structured, that is, they come from sources that are not in traditional databases, such as videos or images, among other types, that is, meaning that security and digital control go beyond the traditional data center, added to the growing use of Cloud Computing solutions and services to Big Data, has become the main catalysts of evolution in the protection of critical information for organizations [41].

This scenario of an increasing volume of information generated in the virtual environment presents challenges from the point of view of data protection and management, but with Big Data related to information security there will be more intelligent protection of critical data, and storage, however as a consequence, also offers an opportunity to explore some of this data through analytical applications to convert them into useful information for faster and more efficient decision making, related to the processing and increasingly analysis of external data, which in turn provides valuable information for the business [42].

In this sense, data analysis will play a fundamental role in security, it is fully applicable helping to prevent internal or external threats, especially in the early detection of fraud and information theft, attempted intrusions, suspicious activities, and virus spread can be detected in real-time with precision and responsiveness. Provided that through Big Data, from analytical solutions, it will allow organizations to access data faster, both internally and externally, being able to correlate information that will help to detect possible crimes or threats. What allows the information security industry to have the precise geolocation of these possible threats, knowing which individuals participate in these operations and which platforms or means have been used for this sharing of confidential information such as cloud systems, emails, social networks, among others [43].

Thus, to ensure the efficiency of the process and so that data privacy is not compromised, all those that refer to personal identification, record numbers, among other sensitive information, must be masked or removed. In this way, Big Data projects can be customized and have a high-security capacity so that data can be captured and analyzed without any risk. Big data analytics is essentially the process of evaluating large and varied data sets (big data) that are generally not explored by traditional business intelligence and analytics programs [43, 44].

The establishment of standards of normality, context information, and external threats will make, based on data analysis, the detection of any anomaly related to information security more efficient. The information evaluated through Big data analytics includes a mix of unstructured and semi-structured data, such as records from mobile phones, social media content, records from web servers, and data from clickstream on the Internet. Also analyzing includes text from survey responses, machine data captured by sensors connected to the Internet of Things (IoT), even customer emails [44].

Companies are using big data analytics to contend with the continuously evolving, since this technology is a radical change in the use and collection of information, in the velocity to analyze and make decisions in real-time, considering sophisticated cyber threats rising from the increased volumes of data generated daily [45, 46].

The use of big data analytics and machine learning allows a business to perform a thorough analysis of the information collected, so to speak, should have an impact on security strategies, which will tackle possible new threats with greater intelligence, because the results of the analysis through the union of technologies give hints of any potential threats to the integrity of the business, provided that the tools used for big data analysis produce security alerts as per their severity level, which further expanded with more forensic details for fast detection and mitigation of cyber breaches, just like operate in real-time [45, 46].

Analyzing historical data using historical data to predict imminent attacks, by using big data analytics, developing baselines based on statistical information that brings to light what is and what's not normal. Based on data repositories with an architecture that allows information to be managed according to categories and establish profiles and functions supported with tools that allow performing quick analysis and with such a thorough analysis, it is possible to know when there is a variation from the norm using the data collected [47].

This risk assessment combined with a quantitative prediction of susceptibility to cyber-attacks can help the organization come up with counter-attack measures, which will be much more proactive than traditional tools based on signatures or threat detection in the network perimeter, besides helping develop predictive models, analyzing historical data can also help in the creation of statistical models and AI-based algorithms [47].

Many cases of cybersecurity threats are a result of employee-related breaches, also known as inside jobs, through the validation of security controls and monitoring of user access, use of improper software, internal company compliance policies. And with the use of big data analytics, it is possible to significantly reduce the risk of these insider threats, through features for log analysis and integration, file integrity verification, rootkit detection, real-time alert, and active response [17].

Making Monitoring and automating workflows through Big data analytics plays a crucial role in mitigating insider threats to limit access to sensitive information only to those employees that are authorized to access it, whereas only authorized staff will be required to use specific logins and other system applications to view files and change data. Big data analytics, in conjunction with network flows, logs, and system events, can discover irregularities and suspicious activities, can deploying an intrusion detection system, which given the growing sophistication of cyber breaches, intrusion detection systems such as NIDS (network-based intrusion detection systems) that generally monitor packets on a network, analyzing traffic and making decisions, being located at a strategic point in the network topology, on a node configured for this, and have a broad view of the flow, is highly recommended as they are much more powerful when it comes to detecting cybersecurity threats [44].

In this sense, due to the growing number of digital crimes, it is essential that there is a guarantee that data and information manipulated by corporate computers do not leak or be breached. Cybersecurity covers the protection of software, network, hardware, technological infrastructure, and services. In other words, it is restricted to the security of digital data through methods, processes, or techniques for automatic processing. Which is largely depends on the risk management and actionable intelligence that is provided for by big data analysis [38].

Perhaps the main disadvantage of Big Data Analytics is related to data privacy risks. Since the technology acts directly with specific information, about most of the digital and control activities, done with electronic equipment inside a house or even a company. One of the ways to prevent this type of scenario is related to the use of disidentification techniques such as anonymization, encryption, key encryption, pseudo-anonymization, among others, in order to capture user data without harming confidentiality of personal and political information. privacy, ensuring the preservation of the integrity and privacy of users' data.

So, the main advantage of the application of Big Data Analytics for Cybersecurity is in the analysis of data, contributing to the development of increasingly efficient methods and models for the protection of information, focused on "intelligence", meaning having intelligent tools that are linked to actions of Big Data to do the data analysis. Making this model/method through verification and monitoring, allowing the evolution of the ability to make predictions about possible attacks, as well as being able to identify threats and automate functions that protect the databases from possible intruders, i.e., what can include improvements to firewalls, anti-malware, and even perimeter-specific networks, blocking the progression of damage in an IT environment in real-time and in an automated way information security, i.e., what measures should be taken

6 Discussion

Many organizations do not handle information that is directly in the business environment, be structured, especially considering their files in the digital environment. Since a complaint that a particular customer makes at the checkout of a retail store, in general, is unstructured, which in this case, many organizations have the use of Big Data to find out what their customers say about it on social networks. What is there that today there is access to a wealth of information that can be used in order to create better and more efficient ways of interacting with the consumer/customer. However, there are multiple points of contact that create numerous security breaches, or looking at it from another point of view, are opportunities for cybercriminals to take advantage of this information in their illicit activities.

This means that companies have to be extra careful in guaranteeing the protection of their customers' data, since it is not exclusively a problem that is the responsibility of official entities, such as government, in modern times, it is a global issue, of all the organizations, public or private, still evaluating that small and medium institutions also keep important data that can be targets of the offenders. As well as taking into account that as technology evolves, online crimes and fraud also become more sophisticated.

With the use of Big Data, i.e., large amounts of data, it allows improving sales results through digital marketing, greater civic commitment to the government, lower prices due to price transparency, and a better match between products and consumer needs. So, if the threat comes from technology, the solution follows the same path, since Big Data and Machine Learning are used to fight and prevent cybercrime in companies in different sectors, involving data protection and privacy. On the other hand, innovation in hacking tools and techniques and security attacks are increasingly advanced.

It is undeniable that the ability to collect large volumes of data generates competitive advantages for companies, and that this is one of the main ways of having a broad view of your business. Big data, in short, refers to the way data and information are collected, stored, categorized, and updated. In a world that is more and more connected, with so many devices exchanging information all the time, this technology has been much discussed, research and has gained more prominence. After all, information arrives in volume, variety, and velocity never seen before.

The Corporate Information Security Process must exist since the organization's first processing of information. Big Data information security begins with the protection of Small Data information, which is a powerful source of information very useful to improve results and the construction of strategies, known for allowing organizations to access an important range of information intrinsic to the Big Data universe, which leads to real value generation, as it brings together more selected and qualified pieces of data. The key to Small Data is in deciphering specific data that sometimes hide the main information for decision making. They belong to a leaner universe, which encompasses a small and significant proportion of knowledge that makes the real difference in the company's business. While Big Data focuses on quantitative data, Small Data focuses on qualitative information. These are the details related to the customer's perceptions, opinions, and experience. It is as if Small Data were the result of panning in the immensity of Big Data

Thus, the difference lies in the fact that it increases the perception and information about data security, whether in Small Data until it reaches Big Data, that is, the organization acquires a better, and more integrated, knowledge about cyber analytics, which is basically it is possible to analyze and detect potentially vulnerable points, creating attack and defense scenarios, and mitigate impacts. Because there is no way to avoid it, since, in a digital environment, the number of cyber-attacks tends to increase, however, what must be done is to reduce its intensity, i.e., the objective is to hinder the service of cybercriminals.

Information security is for the organization, if it is using Big Data, it will also be considered. However, the concept of security and structural controls and security are the same for everyone, since the Corporate Information Security Process aims to protect information so that the organization achieves its business objectives in terms of information resources.

In this context and as an internal measure, to ensure security and compliance, especially with the expansion of trends such as digital transformation and mobility, as long as this trend provides employees with greater access to the network without having to stay in a cubicle, it also brings new risks to the organization's security, so everyone must understand and practice cyber hygiene.

As demands for mobility and digital transformation have made business networks more accessible, cyber-attacks have also become more frequent and sophisticated, taking advantage of the expanded attack surface, since an unreliable remote connection can leave the network vulnerable, resulting in employees they can unintentionally cause harm due to lack of awareness of cybersecurity. To minimize this risk, especially with connectivity and more interconnected digital resources, organizations need to promote cyber hygiene practices to reduce risks, data leakage, and non-compliance, allowing greater operational flexibility and efficiency.

Access points must be secure, when connecting remotely to the corporate network, cyber hygiene practices recommend the use of a secure access point. Another best practice is to create a secure network for home office business transactions. Update frequently, since installing frequent updates on devices, applications, and operating systems is a step towards achieving strong cyber hygiene. Strong access management, using strong passwords and two-step authentication on all devices and accounts. Passwords must be complex, including numbers and special characters, as well as not reusing passwords in accounts, especially on devices and applications used to access sensitive business information. Safe use of email, since this is the most popular attack vector and is still used by cybercriminals, because of its universal use, email remains the easiest way to distribute malware to unsuspecting users. Use of antimalware, although antimalware software cannot prevent unknown attacks, installing antimalware/antivirus software on all equipment and networks provides protection in the event of a phishing scam or an attempt to exploit a known vulnerability.

With a view of this entire cybersecurity landscape, there must be a prepared response plan and understanding of the details, incident recovery, and measures to minimize cyber-attacks. Since security incidents attributed to people inside the company, that is, active employees, tend to decrease, while those attributed to outside invaders increase.

Thus, to ensure that data privacy is not compromised, the veracity requirement is justified in the context that it is complementary to the reliability requirement, ensuring that the information is accessed only by authorized entities. Regarding the veracity of data in Big Data environments to be analyzed for integrity, ensuring that the information is complete and faithful, that is, that it has not been altered by entities not authorized by its owner, and in the same sense of authenticity, ensuring that the entities involved in a process containing digital information are authentic. The characteristic veracity in Big Data environments, refers to the degree of credibility of the data, and they must have significant reliability to provide value and utility to the results generated from them.

One of the relevant issues is the centralization of data, as it turns this system into a potential target for attacks, which can seriously compromise the organization's reputation due to information leakage, leading to seeking an improvement in the resilience of these systems, materialized by resources such as data mirroring, high availability, resource contingency, among others. Having a proactive attitude means having advanced detection tools; real-time identification of risks, protection, and countermeasures to ensure that most cyber-attacks are identified and their effect mitigated before they cause financial and/or notoriety damage. And this is possible by combining Big Data with predictive analytics.

Thus, in order for this context to be built and constructed, it is necessary to use Big Data analytical models to model and identify threats to model analytical intelligence on cybersecurity threats and incident prevention. Even if Big Data requires processing and storage capacity, qualified and experienced labor is also needed to model analytical applications and code sophisticated algorithms. However, the scarcity of cybersecurity professionals and budget restrictions in organizations ends up reducing the ability to implement sophisticated Big Data solutions. This is another reason why more and more organizations adopt analytical solutions based on cloud services

With increased confidence in cloud models, more and more organizations have come to have critical business functions in the cloud. The adoption of new protection measures for digital business models in the cloud, with the implementation of analytical intelligence programs on threats and information sharing with other organizations to gain knowledge and be more efficient in detecting threats, responding to incidents, and mitigating threats. cyber risks. At the heart of this approach are solutions such as threat analytical intelligence, real-time monitoring, advanced authentication, and open-source software. What the fusion of advanced technologies with cloud architectures allows for faster identification of threats and response to them, understanding of customers and the business ecosystem, and ultimately, cost reduction.

In this new digital context, organizations need to define security solutions that are flexible in order to adapt to technologies and that they manage, anticipate attacks, and simultaneously, equal their sophistication. Internal cybersecurity controls in a Big Data environment such as encryption of sensitive data using HSM (Hardware Secure Module), related to the use of keys for content encryption, whether transactions or data stored on disk, storing keys and applications will access them to perform cryptographic operations; access control by user x cryptographic key (BDAC—Big Data Access Control), the main technology is approximate pattern matching, applying tools based on the set of "big data" technologies, such as clustering, it is a new model-free approach to estimate and control problems, which eliminates separate steps for state estimation and optimal control, process identification, directly synthesizing control actions based on a set of trajectories representative of the data ingestion system automatically mapping and encrypting confidential information, including data dictionary technologies, among others. They aim to increase the maturity of information security and compliance with cybersecurity, information security, Big Data architecture, and Infrastructure, in building solutions and evolution of the Big Data environment to increase performance and information security.

Priority information security controls for a Big Data project must be considered Access Control, since access to the original information or after treatment must be controlled and authorized; Availability, related to the definition of the rigor of availability of information from the Big Data environment; the Authenticity of information, since it must be guaranteed that the information collected for the organization's Big Data has a guaranteed origin; as well as compliance with laws and similar, since there are more and more laws on privacy and treatment of information that is considered for collection in Big Data; as well as the Existence of Information Security Policies and Standards, in principle, it is not necessary to have a specific Policy or Standard for the organization's Big Data, however, it is possible to have a specific regulation that compiles the other controls information security with a focus on Big Data.

Threats are evolving rapidly and entities have to move from a reactive attitude to a proactive approach, and that means knowing and understanding the threat before the attack causes harm, that is, using all available information and applying predictive and analytical tools. Of behavior to discover the potential of a threat, detect the current threat, and collect data about the attack and execute an appropriate response before it becomes meaningful.

Big data solutions can be used in security analysis to capture, filter and analyze millions of events per second, which is why one of the most important issues with big data is the processing and the ability to make useful predictions with everyone this data, since traditional tools cannot deal with so much volume, variety and velocity of information, and in this aspect that machine learning and big data complement each other.

Big Data is related to what happens most strongly in the digital environment, allowing the organization to have access to a large amount of information, normally unstructured, which until recently this organization had no practical conditions to access this information, used to refer to a very large amount of information that organizations are storing, processing and analyzing. Due to its unstructured data mining capacity in search of new knowledge, insights, and technical innovation, Big Data for information security and privacy is closely related to 3Vs = Volume, Velocity, and Variety.

The financial crime most prevalent in organizations is cyber-attack, to protect themselves, companies can use Big Data to separate data, encrypt it, and prevent people who are not allowed to access it from capturing it. The idea of Blockchain is also suitable as a security measure, with data encrypted and separated by blocks in different "locations", hardly an offender will be able to copy the information.

Big Data, in this case, works by seeing movements in the entire database, technology can track information, identifying patterns that are not common, pointing out changes and, even, mapping where changes were made. One of the most used techniques to ensure data security is Machine Learning with the use of learning algorithms that help to identify patterns of fraud, before they occur, causing machines to be taught to read patterns and point out when something different happens, alerting those responsible.

The programmatic media is made through audience data and auctions in real-time, since to bill on top of ads, some sites simulate clicks and audience, and in this sense, the portal is mapped as relevant, but in fact, who gives clicks are robots. Through Big Data analysis of actual accesses to the portals and abnormal performance of the price of an auctioned print, it is possible to identify suspicious portals and prevent fraud.

Another type of cybercrime that affects those who make programmatic media is the use of bots that carry the same ad many times, in the payment system for impressions the advertiser pays for the distribution of the ad and not for the number of clicks. That way, the audience is again not real but the sites end up taking the money and being prioritized in auctions. Another way to prevent fraud in programmatic media is the use of Big Data by classifying profiles according to behavior, excluding fraud robots from the system, ensuring user safety.

The main advantage to avoid fraud is having Big Data Analytics to map abnormal actions, which in addition to ensuring security for users, has ample capacity for collecting data from multiple sources, is customizable and adapts to the requirements of each project, integrating with all online trading platforms, among other benefits.

7 Trends

Next-generation SIEMs can leverage machine learning, Big Data, deep learning and UEBA to go beyond correlation rules, since it is a security service that makes use of data provided by security devices, network infrastructure, systems and applications, and the artificial intelligence to quickly respond to possible vulnerabilities such as Entity behavior analysis, Complex threat identification, Lateral movement detection, Inside threats, and Detection of new types of attacks. The next SIEM will be a solution that analyzes logs of what happens in the company, "looking" at these logs and being able to identify, through analysis of the environment and the usual behavior of the employees or machines of a company, if there is something strange happening, thus it is possible to search for fraud or security events, whether they are server invasion, attacks or breaches [48, 49].

User and Entity Behavior Analysis (UEBA) are solutions based on a concept called baselining, it is a new class of security technology that allows identifying threats coming from this new generation as internal threats, attack targets and financial fraud that it can overcome traditional firewalls and other peripheral systems. They build profiles that model standard behavior for users, hosts and devices (called entities) in an IT environment, is a security technology that can be adapted for various cases, using primarily machine learning techniques, with Big Data, is able to identify anomalous activity, compared to the established baselines, and detect security incidents [49, 50].

The primary advantage of UEBA over traditional security solutions is that it can detect unknown or elusive threats, instead of focusing on equipment or the security of specific events, the technique mainly monitors user behavior, in addition to other "entities" such as endpoints, networks, and applications, as well as digital threats as zero-day attacks and insider threats. That is, UEBA reduces the number of false positives because it adapts and learns current system behavior, it is based on data collection to create a large database of information and make the results more accurate and the detection more effective, rather than relying on predetermined rules which may not be relevant in the current context [49, 50].

It isolates data points by randomly selecting a feature of the data, since in several problems it is necessary to know which points in a data set behave differently from others, as is the case in fraud detection, then randomly selecting a value between the maximum and minimum values of that feature [51–53].

This algorithm is based on random decision trees and basically consists of randomly choosing a variable from the data set and performing a random split that has the advantage over other anomaly detection algorithms to support categorical variables, which represents that this process is repeated until the feature is found to be substantially different from the rest of the data set. The length of the path a point has is how many steps are needed to get from the start node to the end node. In this sense, in the context of security, Isolation Forest is a central technique used by UEBA and other next-gen security tools, is a relatively new technique for detecting anomalies or outliers, to identify data points that are anomalous compared to the surrounding data [53–55].

In the same way that Security Information and Event Management (SIEM) systems are a core component of large security organizations, traditionally, SIEM correlation rules were used to automatically identify and alert on security incidents, related to organizing, capturing, and analyzing log data and alerts from security tools. However, SIEMs provide context on users, devices and events in virtually all IT systems across the company, providing ripe ground for advanced analytics techniques related to Big Data, which currently these systems either integrate with advanced analytics platforms like UEBA, or provide these capabilities as an integral part of their solution [55, 56].

In cybersecurity, as Big Data, and other emerging technologies like Machine Learning, so Deep Learning evolves, their capabilities have become a driving force shaping modern cybersecurity solutions, evokes an air of sophistication when compared to Machine Learning. At the same time security practitioners, fatigued by the barrage of artificial intelligence and machine learning messaging, deep learning is best suited in the image processing and natural language processing fields. In line with cybersecurity, the application of Deep Learning with Big Data has found a useful tool in packet stream and malware binary analysis, which these benefit most from supervised learning, when labeled, that is, legitimate versus malicious data, when they are available.

8 Conclusions

Cybersecurity and trust are fundamental pillars in the digital experience, so organizations' digital initiatives must consider, from the beginning, the requirements and investments in cyber and privacy. Cyber-attacks have increased alertness regarding risks, with respect to the security and privacy of this data.

Big data analytics plays a huge role in mitigating cybersecurity breaches caused by the most diverse means, guaranteeing data security and privacy, or supporting policies for secure information sharing in favor of cybersecurity. It helps by facilitating the timely and efficient submission of any suspicious events, which intelligent machines and other technologies that facilitate the analysis of large amounts of data considerably increasing the predictive potential, to a managed security service for additional analysis. The automation aspect of it enables the system to respond to detected threats, such as malware attacks swiftly, since the technology works more and more autonomously, foreseeing risks and possibilities for optimizing the security employed, based on an increasingly detailed analysis, collected data.

As much as big data is crucial to the success of your business, related to the threat to data privacy is one of the aspects that most arouses concern, it can be ineffective for threat analysis if it is poorly mined and processed. With the growing amount of information collected, Big data analytics solutions are backed by artificial intelligence and machine learning, which digital technologies have more and more resources to guarantee the security of stored and employed data, giving hope to businesses that their data processes can be kept secure in the face of a hacking or cybersecurity breach, in addition to the big data analysis mechanisms themselves can be useful in preventing cyber-attacks.

These systems also enable data analysts to classify and categorize cybersecurity threats while preserving privacy, availability, and data integrity in the context of corporate digitization, without the long delays that could render the data irrelevant to the attack at hand. By employing the power of big data analytics, enhance a cyber threat-detection mechanism and improve data management techniques, referring to a set of strategies to manage processes, tools, and policies necessary to prevent, detect, document, and combat threats to digital data and not digital images of an organization.

References

1. Marz N, Warren J (2015) Big data: principles and best practices of scalable realtime data systems. Manning Publications Co.
2. Zikopoulos P, Eaton C (2011) Understanding big data: analytics for enterprise-class Hadoop and streaming data. McGraw-Hill Osborne Media
3. Bertino E, Ferrari E (2018) Big data security and privacy. In: A comprehensive guide through the Italian database research over the last 25 years. Springer, Cham, pp 425–439
4. Mayer-Schönberger V, Cukier K (2013) Big data: a revolution that will transform how we live, work, and think. Houghton Mifflin Harcourt

5. Erl T, Khattak W, Buhler P (2016) Big data fundamentals: concepts, drivers & techniques. Prentice-Hall Press
6. Kitchin R (2014) The data revolution: big data, open data, data infrastructures and their consequences. Sage
7. Marr B (2016) Big data in practice: how 45 successful companies used big data analytics to deliver extraordinary results. Wiley
8. Zhou L, Pan S, Wang J, Vasilakos AV (2017) Machine learning on big data: opportunities and challenges. Neurocomputing 237:350–361
9. Alpaydin E (2020) Introduction to machine learning. MIT Press
10. Mullainathan S, Spiess J (2017) Machine learning: an applied econometric approach. J Econ Perspect 31(2):87–106
11. Smith RE (2019) Elementary information security. Jones & Bartlett Learning
12. Bodin LD, Gordon LA, Loeb MP, Wang A (2018) Cybersecurity insurance and risk-sharing. J Account Public Policy 37(6):527–544
13. Zomaya AY, Sakr S (eds) (2017) Handbook of big data technologies. Springer, Berlin
14. Golshan B et al (2017) Data integration: after the teenage years. In: Proceedings of the 36th ACM SIGMOD-SIGACT-SIGAI symposium on principles of database systems
15. Apurva A, Ranakoti P, Yadav S, Tomer S, Roy NR (2017) Redefining cybersecurity with big data analytics. In: 2017 international conference on computing and communication technologies for smart nation (IC3TSN). IEEE, pp 199–203
16. Ellis R, Mohan V (eds) (2019) Rewired: cybersecurity governance. Wiley
17. Kao MB (2019) Cybersecurity regulation of insurance companies in the United States. Available at SSRN 3399564
18. França RP, Iano Y, Monteiro ACB, Arthur R (2020) A review on the technological and literary background of multimedia compression. In: Handbook of research on multimedia cyber security. IGI Global, pp 1–20
19. França RP, Iano Y, Monteiro ACB, Arthur R (2020) A proposal of improvement for transmission channels in cloud environments using the CBEDE methodology. In: Modern principles, practices, and algorithms for cloud security. IGI Global, pp 184–202
20. França RP, Iano Y, Monteiro ACB, Arthur R (2020) Improved transmission of data and information in intrusion detection environments using the CBEDE methodology. In: Handbook of research on intrusion detection systems. IGI Global, pp 26–46
21. França RP, Iano Y, Monteiro ACB, Arthur R (2020) Lower memory consumption for data transmission in smart cloud environments with CBEDE methodology. In: Smart systems design, applications, and challenges. IGI Global, pp 216–237
22. Padilha R, Iano Y, Monteiro ACB, Arthur R, Estrela VV (2018) Betterment proposal to multipath fading channels potential to MIMO systems. In: Brazilian technology symposium. Springer, Cham, pp 115–130
23. Lafuente G (2015) The big data security challenge. Netw Secur 2015(1):12–14
24. Monteiro ACB, Iano Y, França RP, Arthur R (2020) Development of a laboratory medical algorithm for simultaneous detection and counting of erythrocytes and leukocytes in digital images of a blood smear. In: Deep learning techniques for biomedical and health informatics. Academic Press, pp 165–186
25. Certo SC (2003) Supervision: concepts and skill-building. McGraw-Hill, New York
26. Wang Z, Li H, Ouyang W, Wang X (2017) Learning deep representations for scene labeling with semantic context guided supervision. arXiv preprint arXiv:1706.02493
27. Jones M (2016) Supervision, learning and transformative practices. In: Social work, critical reflection and the learning organization. Routledge, pp 21–32
28. Raschka S, Mirjalili V (2019) Python machine learning: machine learning and deep learning with python, sci-kit-learn, and TensorFlow 2. Packt Publishing Ltd
29. Shin KS (2019) Cyber attacks and appropriateness of self-defense. Convergence Secur J 19(2):21–28
30. Dunjko V, Briegel HJ (2018) Machine learning & artificial intelligence in the quantum domain: a review of recent progress. Rep Prog Phys 81(7):074001

31. Hardy W, Chen L, Hou S, Ye Y, Li X (2016) DL4MD: a deep learning framework for intelligent malware detection. In: Proceedings of the international conference on data mining (DMIN). The steering committee of the world congress in computer science, computer engineering and applied computing (WorldComp), p 61
32. Zhou ZH (2018) A brief introduction to weakly supervised learning. Natl Sci Rev 5(1):44–53
33. Wang L, Alexander CA (2016) Machine learning in big data. Int J Math Eng Manage Sci 1(2):52–61
34. Ye Y, Li T, Adjeroh D, Iyengar SS (2017) A survey on malware detection using data mining techniques. ACM Comput Surv (CSUR) 50(3):1–40
35. Van Der Aalst W (2016) Data science in action. In: Process mining. Springer, Berlin, pp 3–23
36. Mendel J (2017) Smart grid cyber security challenges: overview and classification. e-mentor 68(1):55–66
37. Baig ZA, Szewczyk P, Valli C, Rabadia P, Hannay P, Chernyshev M, Johnstone M, Kerai P, Ibrahim A, Sansurooah K, Peacock M, Syed N (2017) Future challenges for smart cities: cyber-security and digital forensics. Digit Invest 22:3–13
38. Petrenko SA, Makoveichuk KA (2017) Big data technologies for cybersecurity. In: CEUR workshop, pp 107–111
39. Hubbard DW, Seiersen R (2016) How to measure anything in cybersecurity risk. Wiley
40. Hatfield JM (2018) Social engineering in cybersecurity: the evolution of a concept. Comput Secur 73:102–113
41. Yang C, Huang Q, Li Z, Liu K, Hu F (2017) Big data and cloud computing: innovation opportunities and challenges. Int J Digit Earth 10(1):13–53
42. Manogaran G, Thota C, Vijay Kumar M (2016) MetaCloudDataStorage architecture for big data security in cloud computing. Procedia Comput Sci 87:128–133
43. Maglio PP, Lim CH (2016) Innovation and big data in smart service systems. J Innov Manage 4(1):11–21
44. Ahmed E, Yaqoob I, Hashem IAT, Khan I, Ahmed AIA, Imran M, Vasilakos AV (2017) The role of big data analytics in Internet of Things. Comput Netw 129:459–471
45. Witkowski K (2017) Internet of things, big data, industry 4.0–innovative solutions in logistics and supply chains management. Procedia Eng 182:763–769
46. Reis MS, Gins G (2017) Industrial process monitoring in the big data/industry 4.0 era: from detection, to diagnosis, to prognosis. Processes 5(3):35
47. Asenjo JL, Strohmenger J, Nawalaniec ST, Hegrat BH, Harkulich JA, Korpela JL ... Conti ST (2018) U.S. Patent No. 10,026,049. U.S. Patent and Trademark Office, Washington, DC
48. Al-Duwairi B et al (2020) SIEM-based detection and mitigation of IoT-botnet DDoS attacks. Int J Electr Comput Eng (2088-8708) 10
49. Moreno J et al (2020) Improving incident response in big data ecosystems by using blockchain technologies. Appl Sci 10(2):724
50. Babu S (2020) Detecting anomalies in users–an UEBA approach (2020)
51. Mishra P (2020) Big data digital forensic and cybersecurity. In: Big data analytics and computing for digital forensic investigations, p 183
52. Dey A et al (2020) Adversarial vs behavioural-based defensive AI with joint, continual and active learning: automated evaluation of robustness to deception, poisoning and concept drift. arXiv preprint arXiv:2001.11821
53. Lee T-H, Ullah A, Wang R (2020) Bootstrap aggregating and random forest. In: Macroeconomic forecasting in the era of big data. Springer, Cham, pp 389–429
54. Rutkowski L, Jaworski M, Duda P (2020) Decision trees in data stream mining. In: Stream data mining: algorithms and their probabilistic properties. Springer, Cham, pp 37–50
55. Wang Y, Rawal BS, Duan Q (2020) Develop ten security analytics metrics for big data on the cloud. In: Advances in data sciences, security and applications. Springer, Singapore, pp 445–456
56. Amrollahi M, Dehghantanha A, Parizi RM (2020) A survey on application of big data in fin tech banking security and privacy. In: Handbook of big data privacy. Springer, Cham, pp 319–342

Toward a Knowledge-Based Model to Fight Against Cybercrime Within Big Data Environments: A Set of Key Questions to Introduce the Topic

Mustapha El Hamzaoui and Faycal Bensalah

Abstract It has become universally recognized, by all specialists in the digital world, that cybercrime is a constant threat with serious consequences and includes all forms of digital crime that mostly target data. Big data is a special type of data that has attracted the attention of academics and practitioners over the past two decades. Technically, in the big data field, analysis is a major concern while security is a responsibility which requires qualified skills and high level knowledge. Today, several disciplines (Computer Science, law, etc.) are interested in the inevitable interference between big data and cybercrime what mobilizes various research activities. In addition, the vocation of the mutual relationship between knowledge and data is important because data allows the creation of knowledge while knowledge ensures the protection of data. In this perspective, this chapter aims to propose a knowledge-based approach to support the fight against cybercrime in the big data context. But, we will answer, at the beginning, a large number of comprehension questions to facilitate as best as possible, to those interested in the subject of "big data and cybercrime", the understanding of its different axes.

Keywords Big data · Cybercrime · Cyberspace · Knowledge · Machine learning

1 Big Data Large Context

This first major section is devoted to big data. But, it seems to us necessary to start with the clarification of certain notions relating to classical data, which are still sources of ambiguities and can also emerge to touch the big data field.

M. El Hamzaoui (✉)
LERSEM Laboratory, Commerce and Management School (ENCG-J), Chouaib Doukkali University, El Jadida, Morocco
e-mail: elhamzaoui.m@ucd.ac.ma

F. Bensalah
STIC Laboratory, Faculty of Sciences, Chouaib Doukkali University, El Jadida, Morocco
e-mail: f.bensalah@ucd.ac.ma

Y. Maleh et al. (eds.), *Machine Intelligence and Big Data Analytics for Cybersecurity Applications*, Studies in Computational Intelligence 919,
https://doi.org/10.1007/978-3-030-57024-8_4

1.1 Classical Data: Ambiguities and Misunderstandings

1.1.1 Ambiguities Relating to Definitions and Designations

Unfortunately, many publications and research that dealt with issues related to big data are very late, despite their good scientific value, in reminding the reader of the nature of this type of data and sometimes they may not do that at all. This can sometimes be explained by the fact that the authors, given their long experiences in the field, consider defining the big data nature an axiom and a postulate that requires not to be remembered and to go into details. Given the novelty of big data, this problem may create confusion for readers, especially beginners.

At this early stage, we are content with saying that big data is a particular data which requires processing and manipulations (storage, analysis, use, etc.) almost totally different from that of data from classical databases.

As we have already mentioned previously, the classic data field suffers from a considerable lack of precision in the definition of a certain number of its concepts, which can sometimes push the reader to mix the subjects and to interfere with the meanings of certain terms. Of course, moving quickly to the subject of big data without clarifying some of the ambiguities relating to the subject of classical data may increase the possibility of transferring some confusion on this subject. For example, we can recall the confusion that there is still, in the jargon of classical databases, between the terms "*Data*" and "*information*". Unfortunately, many people still misuse them as two synonyms.

Therefore, it seems to us wise to start first by clarifying at least these two concepts before tackling the notion of Big Data.

1.1.2 Main Differences Between Data and Information

To remove certain ambiguities which essentially affect, on the one hand, the definitions of terms "*data*" and "*information*" and, on the other hand, their uses within organizations (enterprises, administrations, etc.), we will briefly answer, along this subsection, a certain number of comprehension questions about them.

To response these questions we adopt the classical computing principles and mainly our own perception of the subject [1, 2].

Classical Sense of Information

Classical and General Definition of Information Basically, an information could be defined as "*All we can perceive, directly or indirectly, through our five senses, of the things that surround us to increase our level of knowledge and to constitute a sufficient idea on a specific subject for the purpose of achieving a well-defined objective (personal, professional, etc.).*".

In this definition, we have used the term "*All*" instead of the term "*Anything*" just to respect the intangible aspect of the information.

Information Size In general, the information composition is expressed primarily by its size (number of component parts) and depends on its users' objectives.

In the real world, at the first direct contact with a person a minimum of data (name and first name) is enough for us to form the necessary information that allows identifying him and triggering the first discussion with him. But, if we wish after developing a relationship, either in a personal or professional context, the size of this information can grow to contain new data such as phone number, e-mail, office address, etc.

Digitalization of Information

Objective of Information Digitalization Digitization mainly aims to facilitate for humanity, through specific new technologies' approaches and tools, to take advantage of information in active sectors such as marketing, education, and medicine, etc.

Practical Achievement of Digitalization Historically, we can summarize the evolution of the digitalization phenomenon in two main points:

A new representation of information the digitization era was begin when the physicists was trying to link the information concept with certain electricity and light physical properties.

Creation of Computer by taking advantage of the mathematics and electronics progress, scientists were able to open the brilliant history of digitalization thanks to the construction of the first computer. This first computer based on the Von Neumman scheme [3] (processor, memories, etc.) lead to an incessant series of creation and innovation in the computer science field.

Information and Data in the IT Field

Information Meaning in the IT Field In computer science, the term information becomes, as we will see shortly, very precise. The quantification techniques facilitate its expression.

In general, the information's quantitative aspect could be expressed in the following way: «INFORMATION = {Subject + Properties + Values}»

For example, the necessary information to manage enterprise customers could be written, in tabular form, as in Table 1.

In computer science, it is highly desirable to use databases (DB) to store information and link them to each other as needed.

Table 1 Tabular representation of a simple example of the information quantification

Subject	Properties	Values
Customer 1	First name	EL HAMZAOUI
	Second name	Mustapha
	Phone number	(+212)06xxxxxxxx
	etc....	...
Customer 2	...	...
...	...	...

Data Meaning in the IT Field The elementary component of each information is called data and could be defined as "*An elementary information that could be obtained, on a specific subject, in a well-defined environment without any calculation.*"

In computer science, data could be the values of subject properties or derived from the information itself. In general, data could be directly deducted from the surroundings.

New Meaning of Information Considering the Data Concept In computer science, the data-based definition of information is "*In a well-defined environment, the information necessary to define and manage a specific subject (material or abstract) is all the data that can be collected, directly or indirectly, based on methods/languages of analysis and design, on it from the things/persons surrounding it inside this environment.*".

Answering the question about what the data is, Fabio Nelli [4] indicated that the data actually are not information, at least in terms of their form.

In principle, the definition given to term '*data*' by Fabio Nelli, including the definition that we will see in the last main section (Sect. 3.2), align well with our reasoning but there is nevertheless a rare case where the '*data*' could be a '*simple information*'. For example, a single column table in a database means that this data (column header) is the primary key to this table and at the same time constitutes the information necessary to describe well the element represented by this table.

In reality, the column headers of a database table are the names of the data that together constitute the information necessary to properly describe the element of the real world (material or abstract) represented by this table. Practically, the equivalent of this information, in the conception phase, is an entity or an N-N hierarchical relationship between two entities, their attributes will be the data constituting this information.

Always within the context of the information definition in the computer science field, Fabio Nelli added that "*Information is actually the result of processing, which, taking into account a certain dataset, extracts some conclusions that can be used in various ways*". Finally, he called that the process of extracting information from raw data is called data analysis.

This is new information that can be formed from database tables, during its use phase, in order to manage the elements of the real world represented by the constituents of this database or to take decisions which concern them.

Classical Data Carriers: Construct and Obtaining of Data/Information As we have already alluded to it previously, in the context of traditional databases, we limit ourselves, during the "*conception*" and "*realization*" phases of a database, to the representation of data (column headers of the same table) which intervene in the constitution of the information necessary to describe precisely the elements of the real world (material or abstract) are the subject of management and/or decision-making operations. These tables can be at the origin either entities or N-N hierarchical relationships in the conceptual diagram of this database.

In the '*utilization*' phase, we can create new information, based on the contents of the database tables, to manage the elements of the real world (materials or abstracts) represented by these tables or to involve them in decision-making operations.

Data and Information Inside Organization

Contribution of Data/information to the Organization Activities Within organizations, the usefulness of information comes down to support activities, which increases the organization profitability and therefore its overall performance.

Data/Information locations within Organizations In general, Information/data could be in one these two situations:

- *Immobile*: Information/data is either in a permanent (IS databases, digital files, etc.) or in temporary (volatile memories) storage.
- *Mobile*: Information/data circulate between electronic equipment constituting the organization communication platforms (computer network, telecommunications network, etc.) which could be its ICT platforms.

In practice, IS databases are used for storing and managing the organization information/data whereas ICT ensure their communications.

ICT Definition ICTs are a set of electronic equipment made based on industry standards [5] and connected to each other to build a communication platform. ICTs' components operate and communicate based on international standards (OSI, SNMP, http, ftp, etc.) [6–8].

Moreover, ICT are the spine of the organization digital communication and could also implement various security solutions and approaches to secure the data exchanges.

IS Definition IS has numerous definitions. In the context of the classic systemic approach, the IS can be simplified into two main components; namely a database to store data and a Logical Interface (LI) to manage the DB content and to use it properly in the organization operation and management activities.

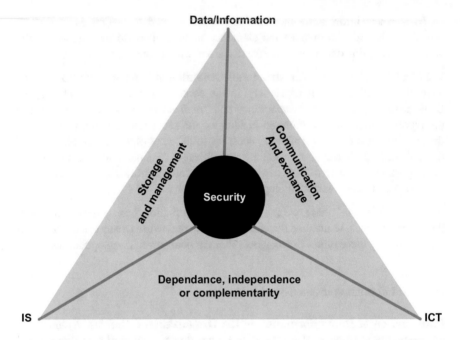

Fig. 1 DIII triangle illustration of the ternary relation between Information/data, IS and ICT

Relations between Data/Information, IS and ICTs within organizations To facilitate the comprehension of this ternary relationship (Data/information, IS, ICT), we have resorted, as it is illustrated on Fig. 1, to the DIII (Data/Information, IS and ICT) triangle [1] which schematized it in a simplified way.

The DIII triangle illustrates, on the one hand, the relationships between the three basic elements Data/information, IS and ICT and, on the other hand, their common management operations such as security.

According to this triangle:

- IS is used to store and manage securely the data/information.
- ICTs are used to assure data/information secured communications.
- Membership, independence and complementarity are the main relations that could link IS to ICT inside organizations.

To close this sub-section, we recall, at this early level, that we must be vigilant when using the fundamental concepts (IS, ICT, data, information, quantification, SQL requests, etc.) of classical computing in specific computing areas such as the big data field.

1.2 Overview of the Big Data Concept

Similar to what we did in the previous subsection, we will take advantage of this subsection to answer to some comprehension questions relative to big data field in order to remove some ambiguities from it, especially for beginners who wish to develop their knowledge in this field.

1.2.1 Big Data Identity

Big Data Nature Katal and his colleagues [9] defined big data as *"large amount of data which requires new technologies and architectures so that it becomes possible to extract value from it by capturing and analysis process. Due to such large size of data it becomes very difficult to perform effective analysis using the existing traditional techniques. Big data due to its various properties like volume, velocity, variety, variability, value and complexity put forward many challenges."*

This definition of big data perfectly clarifies its nature, a new type of data that requires more interest and special studies.

For its part, the Oxford dictionary LEXICO (https://www.lexico.com/definition/big_data) defined big data as *"Extremely large data sets that may be analysed computationally to reveal patterns, trends, and associations, especially relating to human behaviour and interactions."*, and added that *"much IT investment is going towards managing and maintaining big data"* which prove the promising future of this type of data.

Among the many definitions of big data, we have chosen these two examples; one reflects the point of view of academic researchers while the other is general and targets the general public.

Thus, the two definitions agreed on three points in common: Big data is a new type of extremely large data, a data that has several values to exploit, and a data with a very promising future.

For clarity, we add that big data mainly linked to the Internet and the massive exchange of data carried out every day on it. At first, the great merit of the emergence of big data is due to the Internet, where the enormous flow of information has largely exceeded, on the one hand, the expected limit in terms of throughput and quantity, and, on the other hand, the capacities of the means available and implemented on the side of this network and on the side of its users as well. This made the situation very difficult to contain, especially in the early years.

Big data domain is not at all simple because it can take several dimensions depending on the angle of view and the way of interpreting it, which always gives readers the right to continue asking questions of understanding, to analyze and synthesize what has been written and published on this subject.

Main characteristics of Big Data As its name suggests, the first property of big data is the exceptional size or quantity, which is called volume. In addition, big data also has a lot of properties that perfectly distinguish it from other types of data.

Despite the fact that data can generally be subject to common operations and manipulations with the same names (creation, backup, analysis, communication, …, and deletion), the specific properties of big data perfectly distinguish it from the majority of them, especially storage manner, analysis processes, etc.

Big Data is characterized by exceptional capabilities that allow the rapid processing (storage, analysis, management, etc.) of large amounts of data, which allows organization to have a better view of its large amounts of data.

Like any other type of data, big data has its own dimensions that characterize it and also facilitate the accuracy of its study axes such as analysis, communication, security, etc.

The dimensions of big data can be summarized, as it is mentioned in Table 2, in the three famous V (3V: Volume, Variety, Velocity) [10] to which we add a new dimension V (Vigilance).

Historically, according to Zikopoulous [11], perhaps the most well-known version comes from IBM, which suggested that big data could be characterized by any or all of three "V" words to investigate situations, events, and so on: volume, variety, and velocity.

We can note that the time is a determining parameter in the big data field. Thus, its consideration in big data studies, on the one hand, gave rise to the definition of the "*velocity*" dimension and, on the other hand, can simplified the expression of Volume ([data rate per time unit] * [storage time]).

It is true that the big data three V principle, often abbreviated as 3 V, tried to give an abbreviated but complete identity to big data able to distinguish them from any other data type, but unfortunately it did not draw enough attention to Vigilance dimension; a vital component in all active areas which includes as well as possible all the precautionary and watchful activities.

Concerning our own proposed dimension (4th V), we recall that there is no difference today that vigilance, expressed until now in terms of security and preservation of the data content, is one of the first necessities in the field of big data. Thus, if the preservation of data during processing and manipulation becomes a primordial characteristic of big data (no need for systematic data transformations for further

Table 2 Big data four dimensions (4V)

Dimension	Signification
Volume	Collection of large heterogeneous amounts of data from different sources
Variety	Use of data of very different natures without translating them into specific formats
	Storage of different data to respond simultaneously to numerous analyzes of different objectives
Velocity	Simultaneous, fast and sometimes real-time support for too many different analyzes
Vigilance	Non-destructive use of data
	Data security

analysis) then security is also extremely required because of, on the one hand, the large amount of information processed and, on the other hand, the large number of tools and means (hardware and software) implemented. Because of the nature of the data and their informational and economic importance, it is extremely important to add to these two elements the human factor which will always remain, despite the gigantic efforts of automation (optimization, robotic, etc.), one of the most decisive and determinative parameters in several vital areas and sectors.

In short, the vigilance dimension should not be limited to data analysis and security, but it can extend to affect other activities (actions, reactions, manipulations, etc.) that can be carried out, directly or indirectly, with vigilance, in the field of Big Data. The possible interactions *"man"*-*"big data environment"* constitute the major part of this fourth dimension of big data.

Differences between Big Data and Classical data Regardless of its type, a classic database generally belongs to an information system, constitutes its central core, and is also an integral and essential part of its construction project.

Despite their differences (technical, budgetary, purposes, etc.), IS construction projects have many technical points in common. Regarding the structure, we always find in these projects software and hardware architectures for both the IS database and its use interfaces. Despite the differences in names, the IS construction steps are generally united in three phases; namely *'conception'*, *'realization'* and *'use/security'*.

The use of an IS means methodical uses of its interfaced database to meet all the needs of its belonging environment or its existence reasons. Indeed, a database allows, through it interface, the realization of three fundamental tasks on its content (data); namely storage, manipulation (Add, selection, update, and deletion) and control (security).

In general, despite the differences that may exist in the *'conception'* and *'realization'* phases of data carriers (databases and platforms), these latter can more or less resemble each other in the basic principles of the *'use/security'* phase. As illustrated in Fig. 2, independently of the IS use objectives, the data carriers are able to perform, through their interfaces, three main operations on their contents (data); namely storage, manipulation and control (security).

In short, within the framework of traditional IS, the main purpose of using a database is to provide support for the organization functioning and management.

Regarding big data, these are certain types of data that come massively from different sources and grouped in the same framework.

For this reason, we only focus on the *"use/security"* phase where big data undergoes almost the same operations as classical data but in different ways and can also be exposed to the dangers of digital crimes (cybercrime).

We would like to point out that big data is also distinguished from advanced data types such as data warehouse:

In the *"use/security"* phase, the *"non-destructive"* processing (https://inventiv-it. fr/big-data-devez-apprendre/) of big data uses, multi-objective sources of data, for

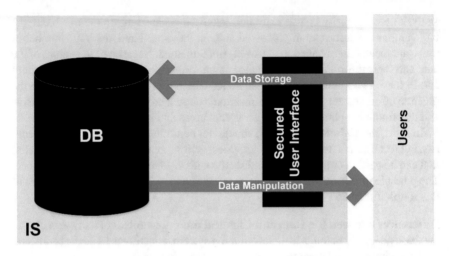

Fig. 2 Possible uses of the content of an IS DB during the IS use/security phase

the analysis of the same batch of data to achieve different objectives. Whereas in the context of data warehouse, which was designed for a specific objective, the data is destroyed, by means of the famous ETL (Extract, Transform and Load) process, to be presented in a very precise format.

Table 3 is the result of a brief comparison between Big Data and traditional data in terms of storage and objectives during the "*Use/Security*" phase:

Relationship between Big Data and Data Science According to EMC Education Services [13] there is enormous value potential in Big Data (innovative insights, improved understanding of problems, and countless opportunities to predict and even to shape the future) that could be discovered and taped by means of data science.

Table 3 A brief comparison between Big Data and traditional data

	Classical data	Big data
Storage	Relational DB, Data warehouse, etc.	Data lakes [12]: Specific carriers to collect and store the endless stream of data
Objectives	The data of an IS DB are useful for	
	• Management and decision making • Operation (production, organization, etc.) • Etc.	• Analysis to deduce past and future behavior of systems • Identification of the premises of a future failure of an industrial installation • Analysis of social networks and all digital crossroads • Etc.

Consequently, Data Science is clearly a primordial means helps man, through specific tools and techniques, to deal with and benefit from Big Data.

1.2.2 Big Data, Smart Intelligent, Machine Learning and Deep Learning

Possible Relationship between Future Prediction, Science and Data For sciences, the development of instruments is necessary to determine the future development of certain phenomena, even for short periods.

For these phenomena, the knowledge of their next evolutions depends mainly on the availability of the necessary and sufficient information which allow the good description of their past and present evolutions in order to also build, as surely as possible, the probable scenarios of their future evolutions.

Therefore, we can say that the prediction of the future, which has long been one of the secular dreams of humanity, is now possible thanks to the great contribution of data to the sciences, especially mathematics (statistics and probability) and computer science. In reality, three computer specialties have benefited from the contribution of data to science, namely Artificial Intelligence (AI), Learning Machine and Deep Learning.

Artificial Intelligence Meaning According to the Cambridge dictionary (https://dictionary.cambridge.org/), artificial intelligence is defined as follows: *"the study of how to produce machines that have some of the qualities that the human mind has, such as the ability to understand language, recognize pictures, solve problems, and learn"*.

To clarify and remove the ambiguity between the AI and the Machine learning, John Paul Mueller and his colleague [14] said that *"AI doesn't equal machine learning, even though the media often confuse the two. Machine learning is definitely different from AI, even though the two are related."*. To explain the relation between these two concepts and what Machine learning does it allow the AI to do, they added that Machine learning is only part of what a system requires to become an AI and helps it to perform:

- Adapt to new circumstances that the original developer didn't envision.
- Detect patterns in all sorts of data sources.
- Create new behaviors based on the recognized patterns.
- Make decisions based on the success or failure of these behaviors.

Finally, they remembered active areas where AI currently has its greatest success, namely logistics, data mining, and medical diagnosis.

Machine learning and Deep learning To briefly answer the question about the difference between these two terms, we refer to an international expert in the field of databases, namely the American software and services' company

Table 4 Oracle comments about AI, machine learning and deep learning

Term	ORACLE comment
AI	Artificial Intelligence as we know it is weak Artificial Intelligence, as opposed to strong AI, which does not yet exist. Today, machines are capable of reproducing human behavior, but without conscience. Later, their capacities could grow to the point of turning into machines endowed with consciousness, sensitivity and spirit
Machine learning	It is able to reproduce a behavior thanks to algorithms, themselves fed by a large amount of data. Faced with many situations, the algorithm learns which decision to adopt and creates a model. The machine can automate tasks depending on the situation
Deep learning	It seeks to understand concepts more precisely, by analyzing data at a high level of abstraction

for professionals (Oracle). Figure 4 lists comments on the three concepts AI, machine learning and deep learning as presented on one of Oracle's French websites (https://www.oracle.com/fr/artificial-intelligence/deep-learning-machine-learning-intelligence-artificielle.html) (Table 4).

In terms of belonging, we can say that the AI contains the Machine Learning, which in turn contains the Deep Learning.

To cloture this last sub-section, we would like to recall that data is the essence of the three disciplines AI, machine learning and deep learning while the accuracy of their returned results increases remarkably with the processed data amount, which qualifies big data to be a natural friend of these three IT disciplines.

2 Cybercrime: Context and Useful Concepts

2.1 Cybercrime: General Context

Cybercrime Definition Historically, cybercrime dates back to the 1980s and 1990s. On one of the publications, on the main FBI website, entitled "*the Morris Worm 30 Years Since First Major Attack on the Internet*" [15], it is written that In1988, a maliciously clever program was unleashed on the Internet from a computer at the Massachusetts Institute of Technology (MIT) and that it is a exactly a cyber worm was soon propagating at remarkable speed and grinding computers to a halt. In reality, the Morris Worm [16] was a malicious program realized on Internet by a student who is called Robert Morris and was classified as one of first digital crimes.

In one of our research studies devoted to cybercrime phenomena [17] where we tried to give a broad definition of cybercrime which takes into account real world crimes can support or become digital crimes when the circumstances allow it, we concluded that: "*Cybercrime is a multidimensional phenomenon (legislation, technical, social, societal,* etc.*) able to target randomly (directly and/or indirectly and at any time), through all illegal means (hacking, destruction, theft, corruption, etc.), cyberspaces composed mainly of information, IS, ICT and any other instrument, platform or electronic/non-electronic device used to store or to communicate information."*

Finally, according to the encyclopedia of crime [18], where the author has chosen to use the term cybercrime in plural, Cybercrimes include illicit uses of information systems, computers, or other types of information technology (IT) devices such as personal digital assistants (PDAs) and cell phones.

Concept of cybersecurity In their research paper focused mainly on the definition of cybersecurity, Dan Craigen and his colleagues [19] recalled that, on the one hand, cybersecurity is a broadly used term, whose definitions are highly variable, often subjective, and at times, uninformative and, on the other hand, the absence of a concise, broadly acceptable definition of this term that captures the multidimensionality of cybersecurity impedes technological and scientific advances.

Dan Craigen' research team newly defined the cybersecurity as *"the organization and collection of resources, processes, and structures used to protect cyberspace and cyberspace-enabled systems from occurrences that misalign de jure from* de facto *property rights."*

During its research for a complete definition of the *"cybersecurity"* term, based on an in-depth literature review and multiple discussions with diverse skills (practitioners, academics, and graduate students), Dan Craigen' research team heavily insisted on the concept of *'Action'*. Because of the cybersecurity term expresses a general framework which could be treated as a specific discipline, in which *"Action"* is a fundamental pillar, it is extremely logical to divert attention to this general framework interferes with other key terms of the cybercrime field.

For us, cybersecurity is now a vital discipline with an own framework. Cybersecurity discipline constantly needs other disciplines, very influential and effective in the context of the fight against cybercrime (cyberattacks), to construct, in a thoughtful and rational way, in the one hand, the cyberspace platform and, on the other hand, a secured space that effectively meets the various conditions of cyberspace protection. For example, cybersecurity badly needs IT security and legislation to secure cyberspaces; these are currently the two major dimensions of cyberspace security.

In general, cybersecurity needs any discipline that can offer it tools and/or approaches to be able to further improve cyberspace security. This certainly leads to the definition of new security dimensions, which will leave nothing to chance in the fight against digital crime.

In short, cybersecurity is a discipline that focuses on the necessary actions (conception, organization, etc.) that must be undertaken in order to be able to secure, based on the Tools and Approaches (TA) provided by other disciplines intervening effectively in the fight against cybercrime, cyberspaces and any other environment that could be threatened by the risks of this phenomenon.

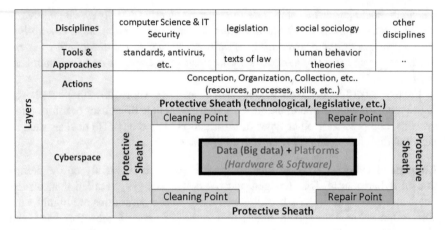

Fig. 3 Layer-based structure of the CGF

For us, the term "*Action*" has two security meanings. In the first one, it summarizes the necessary actions to achieve the following three fundamental objectives: Protection from attacks, cleaning the attacks effects and repairing the attacks' damages. In the second one, it signifies the efforts devoted for building the platforms of the cyberspace itself.

Therefore, now it's really time to talk about the Cybersecurity General Framework (CGF) that we could illustrate as follows:

The *CGF* (Fig. 3) consists of four layers:

- "*Disciplines*" *Layer (DL)*: These are the disciplines that could have impacts on the fight against cybercrime and could also provide support (approaches, tools, etc.).
- "*Tools & Approaches*" *Layer (TAL)*: These are the concrete contributions (technological platforms, standards, law texts, etc.) of different disciplines to support the fight against cybercrime. They can be directly integrated into the cyberspace protective environment.
- "*Actions*" *Layer (AL)*: The set of actions allow using effectively the elements of the TAL to construct the cyberspace itself and its protective environment.
- "*Cyberspace*" *Layer (CL)*: This is exactly the main part of the CGF concerned by the activities and efforts of the cybersecurity discipline. In other words, it is the core of the *CGF* and it is made up of hardware and software platforms and some security components such as protective sheath, cleaning points, and repair points.

For J. Kremling and his colleagues, the cyberspace environment consists of four different layers [20]. From the top down, the important layers are: (1) personal layer (people-people who create websites, tweet, blog, and buy goods online), (2) information layer (the creation and distribution of information and interaction between users),

(3) logic layer (where the platform nature of the Internet is defined and created), and (4) physical layer (physical devices).

Practically, for a good security of the cyberspace environment, cybersecurity discipline requires that the elements of the TAL must be well exploited to build the following components:

- *Protective Sheath (PS)*: Set of tools (firewall, Security servers, etc.) and technological (policy-based management, demilitarized zones, etc.) and legislative (law texts, digital police, etc.) approaches implemented for cyberspace security. Depending on the number of disciplines involved, the main protective sheath is able to contain several protective sub-sheaths (technological, legislative, etc.).
- *Cleaning Points (CP)*: A portion of the cyberspace environment that deals with, on the one hand, cleaning up this environment from any effect or trace of digital attacks (viruses, spam, malware, etc. ...) and, on the other hand, the call for the execution of the necessary maintenance actions (revision, updating, etc.) on the elements of the TAL such as the law texts relating to the fight against cybercrime.
- *Repair Points (RP)*: The cyberspace component responsible for, on the one hand, repairing the technical and technological damage caused by digital attacks and, on the other hand, communicating, at the right time, to the right destination, necessary reports on the other types of damage (economic, political, etc.) in order to trigger the necessary actions which must be undertaken to correct the produced situations.

At the conclusion of this subsection, we will review the relationship ICT-CGF to emphasize that ICT are attached to the CGF, especially to the physical layer of the cyberspace environment. Thus, ICT are technological tools provided by the '*computer science*' discipline, intervene directly in the construction of the 'cyberspace' environments, and can also appear as constructive components of its protective environment, especially at PS and also at the two points CP and RP. Similarly for the IS, it is a tool accompanied by standards and methods/languages and provided by the '*computer science*' discipline. IS (especially its BD component) can appear in the cyberspace layer as a main storage space for all types of data; including security data and big data.

2.2 Fight Against Cybercrime

The main stakeholders in the fight against cybercrime According to the literature relative to cybercrime and the very high number of publications and approaches developed in this context, all stakeholders in the field of the fight against cybercrime, whether organizations, individuals (academics, practitioners), authorities, etc., mainly belong to the IT sector or to the legislation sector. Unfortunately, social psychology is a very promising field which has not taken enough interest in this field. Social psychology facilitated the study of the behavior of criminals and the causality of their illegal acts. In addition, it can also be used as part of academic training to raise awareness, train and educate future users of the digital world.

IT approaches to fight cybercrime Almost all computer security problems are caused by unseen security vulnerabilities in the hardware and software tools used for storing, handling and communicating data. Microsoft team defined the security vulnerability concept as [21]: *"A weakness in a product that could allow an attacker to compromise the integrity, availability, or confidentiality of that product."*.

In the big data context, the component that can manage the security of cyberspace, including the data it contains, is cybersecurity.

According to Kremlig [22] cybersecurity is concerned with three main issues: confidentiality of the data, integrity of the data, and availability of the data, which are the main targets of cybercriminals who try to steal confidential data, manipulate data, or make data unavailable.

Other Contributions of Computer Security to the fight against digital crime Computer security is concerned with the security of, on the one hand, mobiles data/information which are in exchange and, on the other hand, immobile data/information stored on electronic media.

The efforts devoted, by the International Standardization Organization-ISO (https://www.iso.org/home.html), on the subject of network security have focused more on the mobile information security and led to the definition of a general framework of network security architecture. This security framework specified a set of security services with their appropriate security mechanisms.

Thus, ISO 7498-2 standard [23, 24] specified for each security service its own appropriated mechanisms; these include fourteen security services and thirteen security mechanisms.

Table 5 presents examples of security services with their possible implementation mechanisms [24].

Nowadays, among the most used security services we can mainly find availability, integrity and confidentiality. Given their importance, these three elements are considered three most crucial components of security and together compose the famous CIA triad (Confidentiality, Integrity and Availability) [25].

- *Confidentiality*: Refers to protecting sensitive information from being accessed by unauthorized users and being accessed only by authorized ones.
- *Integrity*: Ensures the authenticity of information and means also both no information altering and the information source genuineness.
- *Availability*: Ensures that information and resources are accessible by authorized users and are available to them.

Concerning the immobile information security, it can take advantage of the basic principles of some network security services such as security services interested in authentication and access control.

Generally, in an immobile data carrier environment, the management of security of immobile data consists of two main levels:

Table 5 Some security services and their corresponding mechanisms

		Services				
		Data origin authentication	Access control service	Connection confidentiality	Connectionless integrity	Non repudiation, origin
Mechanisms	Encipherment	Y	–	Y	Y	–
	Digital Signature	Y	–	–	Y	Y
	Access Control	–	Y	–	–	–
	Data Integrity	–	–	–	Y	Y
	Routing Control	–	–	Y	–	–

Y The mechanism is considered to be appropriate (either on its own or in combination with other mechanisms), – The mechanism is considered inappropriate

- *Access Level*: Users (human being, other system, etc.) of this space must have an identity generally determined by a login and a password. This operation is called authentication in IT dictionary authentication.
- *Action Level*: Having the right to access to the immobile data carrier environment does not mean having the absolute right to act over its own resources. The authenticated user must at first know his privileges in this space specifying him the set of actions and tasks that he has the right to perform. In IT dictionary, it is the authorization operation generally verifies the authenticated person privileges/permissions to access resources in a secure environment.

Nowadays, the security management of almost all data carriers' environments is still based on the AA (Authentication and Authorization) principle to which computer security specialists add sometimes, depending on the circumstances of the space itself (nature, NT platforms, users' specificities, etc.), specific security approaches such as policy-based management, authentications servers (Kerberos, RADIUS, DIAMETER (AAA), etc.), and artificial intelligence techniques, etc.

Legislative support for fighting cybercrime Historically, cybercrime is an important area of research that has attracted the attention of many researchers since a long time and pushed, international and regional organizations, to the development of conventions, agreements and guidelines [26]. Therefore, legislative and scientific production was truly impressive [27–29].

However, the challenge of the legislation relative to cybercrime is still significant because the obstacles and constraints (geographic, human, cultural, etc.) are still numerous and complex [30].

On a global scale, the success of the fight against cybercrime begins first with the unification of efforts and the harmonization of the various local legislations in order to federate them in a coherent legislative arsenal which will be able to face the digital crime regardless of any compulsion.

In his interesting publication [31] entitled *"The History of Global Harmonization on Cybercrime Legislation—The Road to Geneva"*, which was published on the website of cybercrime laws (https://www.cybercrimelaw.net/Cybercrimelaw.html), Stein Schjolberg presents a summary of the history of the global harmonizing of computer crime and cybercrime legislation, from the very first efforts in the late 1970ties to the initiatives in Geneva in 2008. He noted that the long history of global harmonization of cybercrime legislation was initiated by Donn B. Parker's research on computer crime and security since the early 1970s and could then evolve through various works and scientific events.

Moreover, the International Telecommunication Union (ITU) in Geneva [32], the most active United Nations agency to achieve harmonization of global cybersecurity and cybercrime legislation, has developed a guide to assist developing countries to understand the legal aspects of cybercrime and to contribute to the harmonization of legal frameworks.

IT approaches and legislative texts are not sufficient to fight cybercrime Given the insufficiency of technical approaches and legislative efforts, it seems necessary to open up to other horizons which can bring more and support to the efforts of the fight against cybercrime. As an example, on the one hand, the interest in the human factor and especially in the study of human behavior and, on the other hand, the exploitation of the results obtained in studies and research devoted to the subject of big data, are capable of providing more support and solutions to fight digital crime in general.

Thus, the advantages drawn from the big data field (exceptional capacities for storing and analyzing huge amounts of information) could strongly support the fight against cybercrime. Therefore, the data collected on digital crimes can therefore be used to decipher and unravel the mysteries of criminals in the digital world: identities, geographic locations, attack strategy, used techniques, etc.

In sum, to support the global fight against cybercrime, we must not continue to limit ourselves to computer security and legislation in the fight against digital crime, but we must also open up seriously to promising disciplines and encourage them in a contractual framework where everyone wins.

3 Big Data Versus Cybercrime: A Knowledge War

3.1 Overview on Our Starting Idea

After a bunch of explanations (ideas, definitions, concepts, etc.) relating to *'big data'* and *'cybercrime'* fields and by inference to the big data certainty property,[1] like any

---

[1]It is one of the important characteristics of big data which expresses that during the analysis of this type of data, the increase in the volume of processed data increases the level of certainty of the extracted information.

researcher, we cannot finish this chapter without having the natural feeling of the birth of a new creative and/or innovative idea.

Thus, the idea that automatically came to us is to think about an approach (theoretical model, practical model, new theory, etc.), even raw, to support the fight against cybercrime in the context of big data.

In reality, after deciphering the main axes of the big data and cybercrime field, we felt that this is the time to reflect and propose the first conception of an approach through which we can support efforts devoted to the fight against cybercrime. Preciously, these include the presentation of an organizational methodology to facilitate the management of these efforts and to focus more on the knowledge when interacting with all disciplines involved in the context of big data-cybercrime relationship.

For that, three good reasons really led us to choose the knowledge as fundamental component in the treatment of the 'big data-cybercrime relationship' subject. The first one is the richness of the subject itself, the second is the exceptional importance of knowledge for all scientific studies and research, and the last one is the fact that knowledge is firmly linked to information, which is in turn inferred from the data.

The main idea of our third reason is clearly explained by Fabio [4] and also aligns perfectly with our perception of the 'big data-cybercrime relationship' subject:

According to Fabio, data are the events recorded in the world and also anything that can be measured or categorized can be then converted into data. At this level, we note a small reservation on the use of the term 'Anything' used, as the case of several authors, in the definition of the data because quite simply this negates the immaterial nature of the data.

Then, Fabio added that to convert data to information and to give it the information properties, it must be studied and analyzed, both to understand the nature of the events and very often also to make predictions or at least to make informed decisions.

Finally, he concluded that knowledge seems to exist when information is converted into a set of rules that helps persons better understand certain mechanisms and therefore make predictions on the evolution of some events.

In accordance with the Fabio's explanations and clarifications, we have opted, as it is illustrated on Fig. 4, for a graphical representation to facilitate the understanding of the sequential relation which links, on the one hand, data to information and, on the other hand, information to knowledge.

3.2 Theoretical Framework of Our Model

In reality, the explanations developed in the previous subsections of this chapter have considerably facilitated the understanding of a large part of the theoretical framework of our approach by presenting sufficient explanations on some of its components.

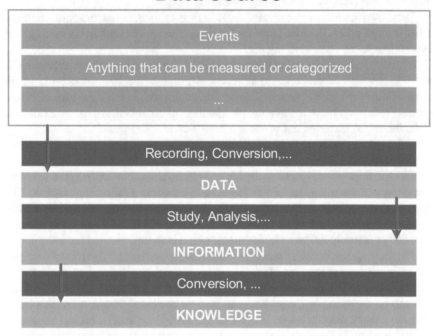

Fig. 4 Illustration of the creation of knowledge from data according to Fabio's perception

3.2.1 Knowledge: Types and Sources

We would like to remind first of all that knowledge can be defined, in a simple way, as 'the understanding of phenomena'.

Since our approach will be based, as we have already mentioned, on the notion of knowledge, it seems wise to start with the distinction between the term knowledge, in its large sense, and its other specific types (Know-why, Know-how and Know-what) used in different fields.

Influenced by engineering, as he showed in the introductory example of one of his publications [33], Raghu Garud (https://www.bbs.unibo.eu/faculty/garud/) first presented, in a raw way, the three specific types of knowledge; which we have listed in Table 6 as following:

Table 6 Raw definitions of some specific types of knowledge

Type of knowledge	Meaning
Know-why	An understanding of the principles underlying phenomena
Know-what	An appreciation of the kinds of phenomena worth pursuing
Know-how	An understanding of the generative processes that constitute phenomena

Table 7 Meaning and creation means of the three specific types of knowledge

Term	Meaning	Means of creation[a]
Know-why	An understanding of the principles underlying the construction of each component and the interactions between them	Learning-by-doing: a process whereby knowledge about how to perform a task accumulates with experience over time
Know-how	An understanding of procedures required to manufacture each component and an understanding of how the components should be put together to perform as a system	Learning-by-studying: It involves controlled experimentation and simulation to understand the principles and theories underlying the functioning of a technological system
Know-what	An understanding of the specific system configurations that different customers groups may want and the different uses they may put these systems to	Learning-by-using: For technological systems, such learning is important because customers invariably use technological systems in ways different from how they were designed or produced

[a]In general, knowledge can be created, directly or indirectly, using all tools (natural or artificial) that allow learning about the environments around us in order to understand and then study them

It is extremely important to remember that the same author also considered that these three specific types of knowledge components of the knowledge itself.

By developing, on the basis of in-depth research carried out by other researchers, the meanings and aspects of each of the three specific types of knowledge, Raghu Garud was able to link them all to their different means of creation. The main conclusions are presented in Table 7.

In general, it is a complete clarification of the different significations of the term knowledge that we intend to adopt exactly in the construction of our theoretical model to fight against digital attacks in the big data context.

3.2.2 Other Components

Practically, the presentation of the general frameworks of cybercrime and big data led us to construct an important idea about both what we must possess and perform, in the context of big data, to succeed the fight against cybercrime. It means, on the one hand, the tools and approach of TAL (Fig. 3) and, on the other hand, the actions of the AL (Fig. 3).

We take this opportunity to highlight two aspects of the subject of big data-cybercrime relationship that characterize the TAL and AL elements (Fig. 3), namely material and immaterial aspects.

The material aspect is materialized by the set of tools and platforms provided, at the TAL layer, by the DL disciplines to build cyberspaces and their protective environments. The immaterial aspect is expressed by necessary actions and knowledge types to act correctly and effectively in the context of the Big data-Cybercrime relationship to design, build, storage, protect, communicate, etc. These include the AL elements and the TAL immaterial approaches.

From a knowledge point of view, the fight against cybercrime, in the context of big data, generally requires, on the one hand, the knowledge, in its broad sense, and other types of high-level knowledge, and, on the other hand, well-qualified skills that master this knowledge and are also able to use it correctly and effectively.

Before proceeding to the detail of our knowledge-based model for the fight against cybercrime in the big data context, it is important to delimit the General Framework of the Big data-Cybercrime Relationship (GFBCR). These include an interference space of three large environments:

- *CGF environment*: It is mainly composed (Fig. 3) of the disciplines (DL) to which cybersecurity uses to fight against cyberattacks, the tools and approaches (TAL) provided by these disciplines, cyberspace (CL), and the actions (AL) to maintain for constructing the cyberspace and to ensure its security basing on the TAL tools and approaches.
- *Cyberspace environment*: It is basically composed of personal layer, information layer, logic layer, and physical layer. Plus of course the productive environment which ensures the complete safety of these four layers.
- *Cybercrime environment*: It consists mainly of black knowledge developed by the digital world criminals, and private or public platforms used partially or entirely by them to launch attacks.

According to our knowledge-based model, knowledge always finds its place in the GFBCR and positions itself there as an effective weapon to counter digital attacks.

3.3   Illustration and Interpretation

It is now very clear that the GFBCR environment is the core of our approach and is composed of three fundamental pillars, namely cybercrime environment, cyberspace environment and CGF. Figure 5 illustrates, from a knowledge point of view, the GFBCR environment by suitably connecting to each of its components the corresponding knowledge layer, which means that each GFBCR layer has a corresponding knowledge-layer (k-layer). Therefore, we derived from the GFBCR environment a knowledge-GFBCR (K-GFBCR) environment.

Concerning the geographical location of the layers of the K-GFBCR environment, the K-cyberspace layer (with the majority of its sub-layers) and both knowledge and specific knowledge layers are all under the direct control of the unit owner of the cyberspace. The K-Disciplines and K-cybercrime environments are considered external components.

Despite the collaborative frameworks that attach cyberspace to disciplines, the disciplines are often considered independent and autonomous units. In reality, two reasons pushed us to consider the k-disciplines as internal elements of the K-GFC: the first one is the fact that the cyberspace has the right to contractual openings on

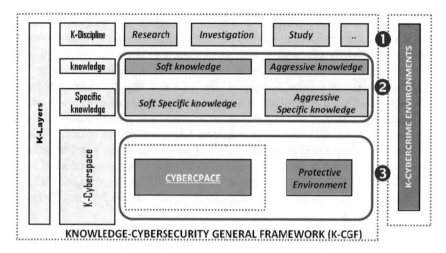

Fig. 5 The K-GFBCR environment

external collaborators while the second one is the fact that knowledge (especially large knowledge) does not belong to nobody, which allows cyberspace to benefit from it for free like others.

From a knowledge point of view, our knowledge-based model states that the k-CGF is made up of three k-layers corresponding to the CGF basic layers while the k-cybercrime environments create and develop black knowledge which will be used, through the points 2 and 3, of the CGF-cybercrime environment contact area, to attack the k-CGF components, especially k-cyberspace data (data, big data, knowledge, etc.).

From the top down, the k-CGF three knowledge-layers are:

- *Knowledge-Discipline layer (K-DL)*: All knowledge developed by research activities undertaken, in private or public frameworks, by the disciplines of the upper layer of the CGF (Fig. 3), in order to support the fight against cybercrime through both development and innovation of approaches that could be used in the construction of cyberspaces and their protective environments. For example, this knowledge can also result from specific scientific activities (surveys, studies, research, etc.) specifically targeting the possible environments of digital crime to create, on the one hand, aggressive knowledge to directly fight cybercrime and, on the other hand, soft knowledge to enrich the humanity knowledge level about this phenomenon.
- *Knowledge Main Layer (KML)*: This layer brings together all the knowledge (ideas, approaches, theories, etc.) provided by the various activities of the DL layer, namely K-DL which must be transferred after to the lower layers to support the construction the cyberspace environment with its security environment. The KML is subdivided into two sub-layers:

– *Knowledge Sub-Layer (KSL)*: This sub-layer captures non-specific knowledge and also knowledge designated to general public like the knowledge communicated within the educational' frameworks. For organizational reasons, we have divided the knowledge at this sub-layer into two main categories; aggressive knowledge come from Disciplines' activities that directly target cybercrime to counter its black knowledge, and soft knowledge developed by other disciplines' activities. In practice, this knowledge can be acquired through education and training in the context of disciplines. The specific types of knowledge are certainly guaranteed in the sciences in general and in the exact sciences in particular, but they are not guaranteed in all areas of learning and in all disciplines as well. It is therefore logical to put this sub-layer of global knowledge to then feed the other types of knowledge or directly transfer it to the cyberspace environment.

– *Specific Knowledge Sub-Layer (SKSL)*: This sub-layer is devoted to specific knowledge generally relating to the engineering within the framework of the disciplines of exact sciences. This type of knowledge can be created from the KML global knowledge or directly from the activities of the DL layer (K-DL) through specific professional education and training. Specific knowledge is also divided, in turn, into two broad categories depending on the adopted information source and the purpose of its future use. It is, on the one hand, a specific aggressive knowledge created by activities, that deal with the cybercrime subject, to counter its counterpart in the k-cybercrime environments (black knowledge) and, on the other hand, a specific soft knowledge created by other activities.

• *Knowledge-Cyberspace Layer (K-CL)*: This layer receives, on the one hand, knowledge including specific knowledge, from the upper layers of the K-CGF, to build cyberspace and its protective environment and, on the other hand, data from its users to put them in its databases, which can facilitate the creation of one of the most important knowledge that greatly attract the cybercriminals interest.

Concerning the confrontation between the K-CGF aggressive knowledge and the K-cybercrime black knowledge, it can take place, along the opposite interfaces of their environments, at points 1, 2 and 3:

• *Point (1)*: Using all possible means, the K-DL could interact (surveys, studies, research, etc.), through this point, with the cybercrime environments, especially the K-Ceybercrime environments, to extract knowledge and to decipher theirs secrets. In case of vulnerability of the K-DL environment, this point can become digital attack paths and/or spy holes which allow criminals to learn about the K-DL hidden secrets and thus create knowledge about the disciplines' knowledge environments (identity, knowledge, platforms, tools, etc.).

• *Points (2)*: As in the case of the first point, this second point can unfortunately become a vulnerability point (attacks, spying, etc.) of the K-CFG, which gives criminals the opportunity to use it in illegal acts or also in the development of the immunity from their environment. The knowledge of the KML layer can become an impulse for the development of black knowledge or can simply become a part of it. In the opposite way, this point could also become a local point to access to black knowledge of the K-cybercrime environments. This means that through this point, the K-CFG can enrich all its knowledge, including specific knowledge. For example, the skills of the GFC can directly use, through the CFG platforms, their know-how to discover and decipher black knowledge and cybercrime environments too.

- *Points (3)*: This is a very sensitive point which represents a direct confrontation area between, on the one hand, the black knowledge of cybercrime and, on the other hand, all the knowledge that contains the different components of k-cyberspace, especially big data-based knowledge. From a knowledge point of view, attacking a k-cyberspace means all irresponsible and illegal acts on its data and platform to create a black knowledge, which facilitates the discovery of data, and continues with bargaining and a whole long series of different forms of fraud, and ends with the absolute destruction of the cyberspace data and software architectures.

In conclusion, the diligence of researchers in finding urgent and direct solutions to the fight against cybercrime must not leave them confined to the IT and legislative dimension of the subject, which can limit and slow down the field of reactions. Rather, they should open up to new horizons to seek new solutions by communicating with the rest of the sciences. In view of the great revolution it may have brought about in the IT field, big data remains highly qualified, through participative work with its related disciplines (AI, Machine learning and Deep Learning), to facilitate opening up to the rest of science and to work on the same side in the fight against digital crime. Our knowledge-based model of fighting cybercrime in the context of big data is only a simple proof that the adoption of new elements in this subject can stimulate the desired addition.

References

1. El Hamzaoui M, Bensalah F (2019) A theoretical model to illustrate the possible relation link the IS and the ICT to the organization digital information/data. In: ACM (eds) NISS19: proceedings of the 2nd international conference on networking, information systems & security, Rabat, Mar 2019, Article No 30, pp 1–10. https://doi.org/10.1145/3320326.3320362
2. El Hamzaoui M, Bensalah F, Bahnasse A (2019) CRUSCom model: a new theoretical model to trace the evolution line of information within the enterprise environment. In: ACM (eds) BDIoT'19: proceedings of the 4th international conference on big data and Internet of Things, Oct 2019, Article No 59, pp 1–8. https://doi.org/10.1145/3372938.3372997
3. Ogban F, Arikpo I, Eteng I (2007) Von neumann architecture and modern computers. Glob J Math Sci 6(2):97
4. Fabio N (eds) (2018) Python data analytics: with Pandas, NumPy, and Matplotlib, 2nd edn. Apress, Berkeley, CA. https://doi.org/10.1007/978-1-4842-3913-1
5. Liotard I (2013) Normes et brevets dans les TIC: une coexistence nécessaire mais sous tension. Innovation, brevets et normes: complémentarités et conflits. Available via HAL https://hal.arc hives-ouvertes.fr/hal-00873156. Accessed 1 Apr 2020

 6. Pras A, Schönwälder J, Stiller B (2007) Peer-to-peer technologies in network and service management. J Netw Syst Manage 15(3):285–288. https://doi.org/10.1007/s10922-007-9072-y
 7. Zhang W, Chen S (2011) Design and implementation of SNMP-based web network management system. Adv Mater Res 341–342:705–709. https://doi.org/10.4028/www.scientific.net/AMR.341-342.705
 8. Kralicek E (2016) Network layer architecture. In: The accidental SysAdmin handbook. Apress, Berkeley, CA, pp 43–60
 9. Katal A, Wazid M, Goudar RH (2013) Big data: issues, challenges, tools and good practices. In: 2013 sixth international conference on contemporary computing (IC3), Noida, India, 8–10 Aug 2013. https://doi.org/10.1109/ic3.2013.6612229
10. O'Leary DE (2013) Artificial intelligence and big data. IEEE Intell Syst 28(2):96–99. https://doi.org/10.1109/MIS.2013.39 (AI Innovation in Industry, IEEE Computer Society)
11. Zikopoulous P et al (eds) (2013) Harness the power of big data. McGraw-Hill, New York
12. Inmon B (2016) Data lake architecture: designing the data lake and avoiding the garbage dump, 1st edn. Technics Publications
13. EMC Education Services (eds) (2015) Data science & big data analytics: discovering, analyzing, visualizing and presenting data. Wiley
14. Mueller J-P, Massaron L (2016) Machine learning for dummies. Wiley
15. FBI Stories (2018) The Morris Worm: 30 years since first major attack on the internet. https://www.fbi.gov/news/stories/morris-worm-30-years-since-first-major-attack-on-internet-110218. Accessed 25 Mar 2020
16. Orman H (2003) The Morris worm: a fifteen-year perspective. IEEE Secur Priv 1(5):35–43. https://doi.org/10.1109/msecp.2003.1236233
17. El Hamzaoui M, Bensalah F (2019) Cybercrime in Morocco: a study of the behaviors of Moroccan young people face to the digital crime. Int J Adv Comput Sci Appl (IJACSA) 10(4):457–465
18. McQuade S-C (2009) Ecyclopedia of cybercrime. Greenwood Press, London
19. Craigen D, Diakun-Thibault N, Purse R (2014) Defining cybersecurity. Technol Innov Manage Rev 4:13–21. https://doi.org/10.22215/timreview/835
20. Clark D (2010) Characterizing cyberspace: Past, present, and future. Mit Csail, Version 1.2. https://projects.csail.mit.edu/ecir/wiki/images/7/77/Clark_Characterizing_cyberspace_1-2r.pdf. Accessed 20 Apr 2020
21. Microsoft. (2015). Definition of a security vulnerability. https://msdn.microsoft.com/enus/library/cc751383.aspx. Accessed 10 Apr 2020
22. Kremling J, Sharp Parker A-M (2018) Cyberspace, cybersecurity, and cybercrime. SAGE Publications, California
23. Kou W (eds) (1997) Networking security and standards. Springer Science Business Media, New York. https://doi.org/10.1007/978-1-4615-6153-8
24. Verschuren J, Govaerts R, Vandewalle J (1993) ISO-OSI security architecture. In: Preneel B, Govaerts R, Vandewalle J (eds) Computer security and industrial cryptography. Lecture notes in computer science, vol 741. Springer, Berlin, pp 179–192
25. Samonas S, Coss D (2014) The CIA strikes back: redefining confidentiality, integrity and availability in security. J Inf Syst Secur (JISSec) 10(3):21–45
26. Schjølberg S (2017) The history of cybercrime (1976–2016). https://www.researchgate.net/publication/313662110_The_History_of_Cybercrime_1976-2016. Accessed 17 Apr 2020
27. Curtis G (2011) The law of cybercrimes and their investigations. CRC Press, Boca Raton
28. Hill B, Marion N-E (2016) Introduction to cybercrime: computer crimes, laws, and policing in the 21st century. Praeger, Santa Barbara
29. Wang Q (2016) A comparative study of cybercrime in criminal law: China, US, England, Singapore and the Council of Europe. Dissertation for obtaining the degree of doctor, Erasmus University Rotterdam
30. Young S-M (2004) Verdugo in cyberspace: boundaries of fourth amendment rights for foreign nationals in cybercrime cases. Mich Telecommun Technol Law Rev 10:139–175

31. Schjolberg S (2008) The history of global harmonization on cybercrime legislation—the road to Geneva. https://cybercrimelaw.net/documents/cybercrime_history.pdf. Accessed 22 Apr 2020
32. Ajayi E (2016) Challenges to enforcement of cyber-crimes laws and policy. J Internet Inf Syst 6(1):1–12
33. Garud R (1997) On the distinction between know-how, know-why and know-what in technological systems. In: Walsh J, Huff A (eds) Advances in strategic management. JAI Press, Greenwich, CT, pp 81–101

Machine Intelligence and Big Data Analytics for Cyber-Threat Detection and Analysis

Improving Cyber-Threat Detection by Moving the Boundary Around the Normal Samples

Giuseppina Andresini, Annalisa Appice, Francesco Paolo Caforio, and Donato Malerba

Abstract Recent research trends definitely recognise deep learning as an important approach in cybersecurity. Deep learning allows us to learn accurate threat detection models in various scenarios. However, it often suffers from training data over-fitting. In this paper, we propose a supervised machine learning method for cyber-threat detection, which modifies the training set to reduce data over-fitting when training a deep neural network. This is done by re-positioning the decision boundary that separates the normal training samples and the threats. Particularly, it re-assigns the normal training samples that are close to the boundary to the opposite class and trains a competitive deep neural network from the modified training set. In this way, it learns a classification model that can detect unseen threats, which behave similarly to normal samples. The experiments, performed by considering three benchmark datasets, prove the effectiveness of the proposed method. They provide encouraging results, also compared to several prominent competitors.

G. Andresini (✉) · A. Appice · F. Paolo Caforio · D. Malerba
Dipartimento di Informatica, Università degli Studi di Bari Aldo Moro via Orabona,
4 - 70126 Bari, Italy
e-mail: giuseppina.andresini@uniba.it

A. Appice
e-mail: annalisa.appice@uniba.it

F. Paolo Caforio
e-mail: f.caforio3@studenti.uniba.it

D. Malerba
e-mail: donato.malerba@uniba.it

A. Appice · D. Malerba
Consorzio Interuniversitario Nazionale per l'Informatica—CINI, Bari, Italy

Y. Maleh et al. (eds.), *Machine Intelligence and Big Data Analytics for Cybersecurity
Applications*, Studies in Computational Intelligence 919,
https://doi.org/10.1007/978-3-030-57024-8_5

1 Introduction

Computer networks and information technology have become ubiquitous in our life. Nowadays, government, financial, military and enterprise infrastructures base the majority of their services on interconnected devices. As a consequence of the ubiquity of network services, the number of cyber-threats is growing at an alarming rate every year. This makes the ability to effectively defend against threats one of the major challenges of both public and private organisations [32].

Traditional methods to detect cyber-threats are based on signatures that match patterns of known threats to identify malicious behaviour. Although these methods can actually detect known threats, they fail to detect unseen threats. On the contrary, machine learning methods can learn the best parameters of a detection model to automatically predict the behaviour of unseen samples. With the emergence of machine learning techniques in various applications, learning-based approaches for detecting cyber-threats have been further improved. At present, they outperform traditional signature-based methods in many studies.

The problem of detecting unseen threats has been explored in depth in the recent cybersecurity literature [2, 30, 36, 62] by using conventional machine learning methods [20, 23]. However, with the recent boom of deep learning, the use of deep neural networks has dramatically improved the state of the art [9, 53, 68]. Particularly, deep learning allows computational models that are composed of multiple processing layers, to learn representations of data with multiple levels of abstraction. From this point of view, deep learning methods are different from conventional machine learning methods, because of their ability to detect optimal features in raw data through consecutive non-linear transformations, with each transformation reaching a higher level of abstraction and complexity. Moreover, the non-linear activation layers of deep neural networks may facilitate the discovery of effective models that keep their effectiveness also under drifting conditions [66].

Although recent research trends in cybersecurity are definitely recognizing deep learning as an important approach in cyber-threat detection, several deep learning methods may still suffer when some hackers deliberately cover their threats by slowly changing their behaviour patterns. Adversarial machine learning [25], also investigated in the area of deep learning, aims at overcoming this issue by allowing the design of machine learning methods that are robust to variants of threats. A common approach investigated in adversarial learning concerns the generation of adversarial samples that look like the original ones, in order to improve the generality of the learned models and their capacity to handle unseen samples correctly [36, 37, 53, 71].

In this study, we account for the recent achievements of deep learning in cybersecurity. In fact, we consider threat detection models trained with deep neural networks. In any case, the novel contribution of this study is complementary to deep learning as it consists in the definition of a new data transformation approach that modifies the training data before training the network. Specifically, our proposal consists of transforming the training data, processed to train a deep threat detection model, by changing the class of a few selected training samples before training the network.

To this aim, we formulate a new machine learning method, named THEODORA (THreat dEtection by moving bOunDaries around nORmal sAmples), that learns a cyber-threat detection model from a training set, which is modified by re-positioning the decision boundary that separates the normal samples and the threats. So, it starts by detecting the decision boundary and proceeds by forcibly assigning the normal training samples, which are close to the detected boundary, to the threat class. Finally, it learns the classification model by training a deep neural network on the modified training set.

The rationale behind the formulated method is that the normal training samples, which are selected to be assigned to the opposite class in the training stage, represent the training samples that behave more closely to threats. Assuming that new threats are often slight changes of existing ones, it is possible that they will look like the normal samples, which are the closest to the boundary. This is based on the idea of exploiting the concept of decision uncertainty by changing the class of the normal samples that the decision boundary situates on the normal side with the highest uncertainty [39, 44, 51]. So, handling the normal samples closer to the boundary as uncertain normal samples, we process them as threats by allowing the training stage to avoid over-fitting and learning a model that is more robust towards possible unseen threats.

This paper is organised as follows. The related works are presented in Sect. 2. The proposed method and the implementation details are described in Sect. 3. The data scenario, the experimental setup and the relevant results of the empirical study are discussed in Sect. 4. Finally, conclusions are drawn and future developments are sketched in Sect. 5.

2 Related Works

Machine learning has been widely adopted in the last decade to address various tasks of threat detection in several cybersecurity applications, e.g. detection of intrusions in critical infrastructures, malware analysis and spam detection [69]. In particular, both supervised and unsupervised machine learning approaches have been investigated. However, in the last three years, there has been a boom in deep learning approaches in cybersecurity. Today, deep learning has undoubtedly emerged as a means to handle threat detection tasks effectively. As this study combines traditional and deep machine learning solutions, we briefly revise the background of both these fields.

2.1 Traditional Machine Learning

The traditional unsupervised machine learning approaches, commonly investigated in cybersecurity, are mainly based on clustering. The basic idea behind threat detection through clustering is that normal data tend to group themselves in large clusters, while threats tend to group themselves in small clusters. This idea is fulfilled in [47], where k-means is used for clustering network flows. Following this research direction, a clustering approach is also adopted in [55] to extract cluster prototypes that model the signature of malicious mobile applications. Cluster-defined malicious signatures are, subsequently, processed to synthesise new malicious samples and balance the sample collection. Recent studies have also explored the use of soft clustering as an alternative to the traditional hard clustering solution. Soft algorithms are employed to yield the confidence of the clustering assignments. For example, the authors of [49] use soft clustering with the number of clusters automatically determined by an incremental learning scheme. They use clusters to build pattern features of both normal samples and intrusions. Soft clustering is also investigated in [48] to identify the encrypted malware traffic by calculating the distance between malicious applications.

Traditional supervised machine learning approaches mainly experiment K-NN [34, 61], SVM [40], Decision Tree [31, 52], Random Forest [8, 12] and Naive Bayes [34] as classification algorithms. For example, the performance of various classification algorithms has been recently compared in [34] in tasks of Android malware detection [40] and network intrusion detection. These studies generally confirm that the SVM-based approaches outperform the competitors based on Linear SVM, RBF SVM, Random Forest and K-NN. The superiority of SVMs compared to Naive Bayes is also proved in [27]. Finally, recent studies have also yielded new achievements by combining fuzzy learning and SVMs [50].

2.2   Deep Learning

The popularity of Deep Neural Networks (DNNs) has greatly increased in recent years, due to the capacity of these models to exploit the availability of large amounts of data and extract high-level information from raw training data. The superiority of deep learning approaches in cybersecurity has been recently proved in [6, 15, 35, 54]. In particular, the experimental study in [54] has shown that several deep learning architectures can gain accuracy compared to various traditional machine learning methods (comprising SVMs).

Like traditional machine learning approaches, deep learning approaches for threat detection in cybersecurity could be divided into unsupervised and supervised methods. Unsupervised deep learning architectures mainly involve autoencoders, which are commonly used for dimensionality reduction [4]. For example, the authors of [9] use autoencoders for feature construction and anomaly detection in tasks of network intrusion detection.

Supervised deep learning approaches include various architectures like Recurrent Deep Neural Networks—RNNs, Long Short-Term Neural Networks—LSTMs and Convolutional Feed forward Deep Neural Networks—CNNs. RNNs and LSTMs are commonly used to process sequence data by using the output of a layer as the input of the next layer. So, they have been experimented in various intrusion detection systems [6, 29, 33, 43, 73], due to their ability to process flow-based data. CNNs have been used in [63, 72] for malware analysis. Both autoencoders and CNNs have been recently combined in [10] for addressing tasks of intrusion detection. The extensive empirical study described in [10] has also shown that this architecture significantly outperforms various recent state-of-the-art deep learning architectures.

Final considerations concern the adversarial learning paradigm that has been recently investigated in combination with deep learning in various cybersecurity tasks. Various approaches have been formulated for defending deep learning models against adversarial samples, i.e. malicious perturbed input that may mislead detection at the testing time [74]. In this direction, a training method, named defensive distillation is presented in [56] to improve the robustness of neural networks in adversarial samples. A game theory-based method is proposed in [75] to modify the training process and train a robust classifier against adversarial attacks. A few studies propose to use Generative Adversarial Networks (GANs) to create synthetic data which are similar to given input data [26]. The authors of [60] propose a deep convolutional generative adversarial network to identify anomalies in unseen data. The effectiveness of the use of GANs in the improvement of the robustness of intrusion detection systems to adversarial perturbations has been recently explored in [5, 36, 45, 76]. Finally, the authors of [77] present a GAN-based intrusion detection method that uses the reconstruction error to classify a threat, based on how far the sample is from its reconstruction.

2.3 Final Remarks

We note that this paper is closely related to the described background as we also investigate the use of machine learning to address a task of threat detection in cybersecurity. As in various existing studies, we adopt the SVM algorithm to train a decision boundary that separates the normal samples from the threats in the training set. However, differently from the described background in this context, we do not use the SVM model for the detection of new threats. In fact, we adopt the SVM-defined decision boundary to modify the training set to account for the behaviour of future threats that may look like the normal samples. To this aim, we re-position the SVM-defined decision boundary to identify the normal samples, which look like threats, and label them in the opposite class.

On the other hand, following the mainstream of research in deep learning, we use a deep neural network, the one defined in [10], to train the final threat detection model. However, the novelty of our proposal is that the neural network training is done on the modified training set instead of on the original data. From this point of view, our proposal has a purpose that is conceptually close to that of adversarial learning, since we intend to train a more robust detection model with a modified training set. In any case, adversarial learning approaches build either new adversarial training samples or a new adversarial representation of training data, while we only modify the class of a selection of existing training samples.

3 The Proposed Method

In this section, we describe the proposed method, **THEODORA**, which performs a supervised learning stage to learn a robust threat detection model. The list of symbols used to describe the method is reported in Table 1. The block diagram of **THEODORA** is shown in Fig. 1. It takes a set $\mathcal{D}$ of training samples $\{(\mathbf{x}_i, y_i)\}_{i=1}^{N}$ with $\mathbf{x}_i \in \mathbf{X}$ and $y_i \in \{normal, threat\}$ as input and learns a threat detection model as a classification function $c: \mathbf{X} \mapsto \{normal, threat\}$.

The learning process is carried out in three phases:

1. It determines the decision boundary between the normal samples and the threats of the training set;
2. It re-positions the decision boundary changing the class assigned to the normal training samples that are the closest to the detected boundary;
3. It learns a classification model through training a supervised deep neural network on the modified training set.

The pseudo code of the three phases is described in Algorithm 1.

Table 1 List of symbols

Symbol	Description
$\mathcal{D}$	Set of training samples $\{(\mathbf{x}_i, y_i)\}_{i=1}^{N}$—training set
$\mathbf{X}$	Independent variable space
$\mathbf{Y}$	Target variable with domain $\{normal, threat\}$
$boundary$	Decision boundary between normal samples and threats
$boundary^{normal}(\mathbf{x})$	Confidence according to a sample $\mathbf{x}$ is estimated with class $normal$
$boundary^{threat}(\mathbf{x})$	Confidence according to a sample $\mathbf{x}$ is estimated with class $threat$
Θ	Threshold for re-positioning the boundary and changing labels

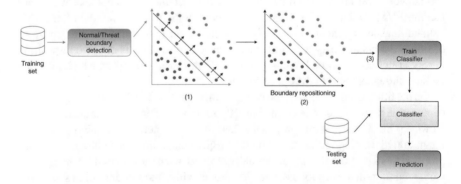

Fig. 1 The block diagram of **THEODORA**: (1) It takes training samples as input and detects the decision boundary between the normal samples and the threats. (2) It re-positions this boundary to change the class assigned to the normal training samples, which are the closest to the boundary. (3) It learns a classifier through training a supervised deep neural network on the modified training set

---

Algorithm 1: Theodora pseudo code

Data:
    $\mathcal{D}$: set of training samples $\{(\mathbf{x}_i, y_i)\}_{i=1}^{N}$ with $y_i \in \{normal, threat\}$
    $\mathbf{X}$: independent variable space
    Y: target variable with domain $\{normal, threat\}$
    Θ: threshold to reposition the boundary

Result:
    c : the learned threat detection model

1 **begin**
        /* Boundary detection   */
2      $boundary^{normal}, boundary^{threat} \leftarrow \text{trainDecisionBoundary}(\mathcal{D})$
3      $\mathcal{N} = \{\mathbf{x}, y \in \mathcal{D} | boundary^{normal}(\mathbf{x}) > boundary^{threat}(\mathbf{x})\}$
4      $\mathcal{T} = \{\mathbf{x}, y \in \mathcal{D} | boundary^{normal}(\mathbf{x}) \leq boundary^{threat}(\mathbf{x})\}$

        /* Boundary re-positioning   */
5   **foreach** $\mathbf{x}, y \in \mathcal{N}$ with $y = normal$ **do**
6   **if** $boundary^{normal}(\mathbf{x}) < \Theta$ **then**
7              $y = threat$

        /* Train threat detection model   */
8      $c \leftarrow \text{trainClassifier}(\mathcal{D})$
9   **return** $model$

3.1 Stage 1—Boundary Detection

We learn a decision boundary function:

$$boundary \colon \mathbf{X} \mapsto \mathbf{R}^{normal} \times \mathbf{R}^{threat},$$

that assigns each training sample to a 2-length vector. This vector represents the confidence according to which a training sample can be assigned to the class "normal" and to the class "threat", respectively.

Function $boundary()$ is determined in a supervised manner by resorting to a statistical classifier trained on $\mathcal{D}$. This learner measures the confidence of assigning a sample to a certain class based on the information enclosed in the independent variable vector $\mathbf{X}$ (line 2, Algorithm 1). Let us consider:

- $boundary^{normal}(\mathbf{x})$ as the confidence according to which $boundary()$ assigns $\mathbf{x}$ to the class "normal",
- $boundary^{threat}(\mathbf{x})$ as the confidence according to which $boundary()$ assigns $\mathbf{x}$ to the class "threat".

Based upon these premises, $boundary()$ partitions the training set $\mathcal{D}$ along the independent variable space $\mathbf{X}$ into two sets, $\mathcal{N}$ and $\mathcal{T}$, respectively, one for each class (lines 3–4, Algorithm 1).

Let us consider a training sample $(\mathbf{x}, y) \in \mathcal{D}$, we define:

$$max_{boundary}(\mathbf{x}) = \max\left(boundary^{normal}(\mathbf{x}), boundary^{threat}(\mathbf{x})\right).$$

We assign $(\mathbf{x}, y)$ to $\mathcal{N}$ if $max_{boundary} = boundary^{normal}$, while we assign $(\mathbf{x}, y)$ to $\mathcal{T}$, otherwise. The highest the $max_{boundary}(\mathbf{x})$, the most confident the assignment decided by the decision boundary on $\mathbf{x}$.

We note that, according to the formulation described, the function $boundary()$ intuitively draws the decision boundary passing through the most uncertain samples, i.e. the training samples that achieve the lowest $max_{boundary}$ in their assignment.

3.2 Stage 2—Boundary Re-positioning

The assumption underlying the idea of learning a decision boundary is that if we use a robust statistical classifier to train $boundary()$, then it should separate, quite correctly, normal training samples from training threats. In theory, all the normal samples should be assigned to $\mathcal{N}$, while all the threats should be assigned to $\mathcal{T}$. Moreover, we expect high confidence in these assignments, that is, the training samples should be assigned far from the decision boundary. In practice, a few training samples may be mis-classified or, even if correctly classified, they are assigned to the correct partition with a low confidence, i.e. they are assigned close to the decision boundary [44].

This scenario suggests that the normal training samples, which are close to the decision boundary, have a more similar behaviour than the other normal samples to the threat behaviour. This makes it plausible to assume that if a new threat is designed in the future by slightly changing a seen one, then this threat may behave similarly to a seen boundary-close normal sample.

Based upon these considerations, we introduce a threshold Θ and select the normal samples assigned to $\mathcal{N}$, which are Θ-close to the decision boundary (i.e. $boundary^{normal}(\mathbf{x}) \leq \Theta$), in order to change their class from "normal" to "threat" (lines 5–7, Algorithm 1). In this way, we modify the training set $\mathcal{D}$ by accounting for the profile of potential new threats, which may look like the normal samples they are close to.

3.3 Stage 3—Classification Model Learning

Let us consider the training set $\mathcal{D}$, as it has been modified according to the boundary re-positioning, and use it to train a supervised deep neural network that learns a robust classification function:

$$c \colon \mathbf{X} \mapsto \{normal, threat\}.$$

This function can be used to classify any new sample (line 8, Algorithm 1). By processing the modified training set, we should avoid possible over-fitting phenomena when training $c()$ and improve the robustness by increasing the ability to recognise new threats.

3.4 Implementation Details

THEODORA has been implemented in Python 3.7 using Scikit-learn 0.22.2.[1] The data are scaled using the Min-Max scaler.[2] The implementation of THEODORA is available online.[3]

Function $boundary()$ is learned as a Support Vector Machine(SVM) [65]. This supervised learner, designed for binary classification [18, 24], maps the input vector into a higher dimensional feature space and determines the optimal hyper-planes that separate the samples which belong to the opposite classes. We select the SVM as a decision boundary learner, since it allows us to estimate the certainty according to which each training sample may be assigned to every class ("normal" or "threat"). We note that our decision is supported by several studies performed in remote sensing [13, 14], medical analysis [19], speech emotion recognition [28], intrusion detection [3, 79] and malware detection [70]. These have repeatedly shown the superiority of the accuracy performance of the boundary decided with SVM compared with the boundary decided with other statistical learners.

In THEODORA, we integrate the Support Vector Classification (SVC),[4] as this is a version of SVM that uses Platt scaling [57] (i.e. a logistic regression on the SVM scores), in order to calibrate the class confidence estimates as probabilities. In this way, for each sample $\mathbf{x}$, the SVC-decision boundary determines $boundary^{normal}(\mathbf{x})$ and $boundary^{threat}(\mathbf{x})$ so that $boundary^{normal}(\mathbf{x}) + boundary^{threat}(\mathbf{x}) = 1$, that is $boundary^{threat}(\mathbf{x}) = 1 - boundary^{normal}(\mathbf{x})$. The used implementation of SVC is based on libsvm a python library for Support Vector Machines (SVMs)[17]. We run the SVC algorithm with the default parameter configuration.

Finally, the classification function $c()$ is learned with the deep neural network architecture recently introduced in [10].[5] It combines an unsupervised stage for multi-channel feature learning with a supervised one, exploiting feature dependencies on cross channels. We note that several traditional and deep learning algorithms have been investigated in the literature to address the threat detection problem (see Sect. 2 for a brief overview). We choose to train the deep neural network architecture

---

[1] https://scikit-learn.org/stable/index.html.

[2] https://scikit-learn.org/stable/modules/generated/sklearn.preprocessing.MinMaxScaler.html.

[3] https://github.com/gsndr/THEODORA.

[4] https://scikit-learn.org/stable/modules/generated/sklearn.svm.SVC.html.

[5] https://github.com/gsndr/MINDFUL.

described in [10], since an extensive empirical study has already proved that this architecture achieves the highest accuracy compared to several recent state-of-the-art systems on various datasets.

4 Empirical Study

THEODORA has been been evaluated to investigate how it can actually gain in threat detection accuracy by re-positioning the decision boundary between training normal samples and threats during the training stage. The datasets processed in the evaluation are described in Sect. 4.1. The metrics measured for the evaluation are introduced in Sect. 4.2, while the results are discussed in Sect. 4.3.

4.1 Dataset Description

A summary of the characteristics of the datasets considered in this evaluation study is presented in Table 2. A detailed description of the dataset is reported in the following:

KDDCUP99[6] was introduced in the KDD Tools Competition organised in 1999. This is a benchmark dataset that is commonly used for the evaluation of intrusion detection systems also in recent studies [21, 41, 76]. It contains network flows simulated in a military network environment and recorded as vectors of 42 attributes (6 binary, 3 categorical and 32 numerical input attributes, as well as 1 class attribute). The original dataset comprised a training set of 4.898.431 samples and a testing set of 311.027 samples. As reported in [64], the testing set collects network flows belonging to 14 attack families, for which no sample is available in the training set. We note that this simulates a *zero-day* threat condition. To keep the cost of the learning stage under control, the original dataset comprises a reduced training set, denoted as 10%KDDCUP99Train, that contains 10% of the training data taken from the original dataset. In this study, we consider 10%KDDCUP99Train for the learning stage, while we use the entire testing set, denoted as KDDCUP99Test, for the evaluation stage.[7] We note that this experimental scenario, with both 10%KDDCUP99Train and KDDCUP99Test, is commonly used in the literature (e.g. [46, 59, 67]). In this dataset, threats represent 22 different network connection attack families grouped into four categories, that is, Denial of service (Dos), User to root (U2R), Remote to local (R2L) and Probe. Furthermore, the entire dataset is imbalanced in both the training and testing set, where the percentage of threats is higher than that of normal flows (80.3 vs 19.7% in the training set and 80.5 vs 19.5% in the testing set).

---

[6]http://kdd.ics.uci.edu//databases//kddcup99//kddcup99.html.

[7]10%KDDCUP99Train and KDDCUP99Test are populated with the data stored in kddcup.data_10_percent.gz and corrected.gz at http://kdd.ics.uci.edu//databases//kddcup99//kddcup99.html.

Table 2 Dataset description

Dataset	Attributes	Total	Normal (%)	Threats (%)
10%KDDCUP99Train	42	494,021	97,278 (19.7%)	396,743 (80.3%)
KDDCUP99Test		311,029	60,593 (19.5%)	250,436 (80.5%)
CICIDS2017Train	79	100,000	80,000 (80%)	20,000 (20%)
CICIDS2017Test		900,000	720,000 (80%)	180,000 (20%)
AAGMTrain	80	100,000	80,000 (80%)	20,000 (20%)
AAGMTest		100,000	80,000 (80%)	20,000 (20%)

For each dataset we report: the number of attributes, the total number of samples collected in the dataset, the number of normal samples (and their percentage on the total size) and the number of threats (and their percentage on the total size)

CICIDS2017[8] was collected by the Canadian Institute for Cybersecurity in 2017. This dataset contains normal flows and the most up-to-date common threats, which resemble the true real-world data (PCAPs). It also comprises the results of the network traffic analysis, performed using CICFlowMeter with the labelled flows based on the timestamp, source and destination IPs, source and destination ports, protocols and attacks. The original dataset was a 5-day log collected from Monday July 3, 2017 to Friday July 7, 2017 [1]. The first day (Monday) contained only benign traffic, while the other days contained various types of attack, in addition to normal network flows. Every network flow sample is spanned over 79 attributes (18 binary and 60 numerical input attributes and 1 class attribute) [1]. In this dataset, threats represent connection traffic attacks that include Brute Force FTP, Brute Force SSH, DoS, Heartbleed, Web Attack, Infiltration, Botnet and DDoS. We note that this dataset is commonly used in the evaluation of anomaly detection approaches with the learning stage performed on the first day [11, 78]. However, a few recent studies have considered these data also in the evaluation of classification approaches, as we do in this paper [10, 38]. In our experimental study, we consider the training and testing sets built according to the strategy described in [38]. So, we build one training set with 100 K samples and one testing set with 900 K samples. Both training and testing samples are randomly selected from the entire 5-day log. For the creation of both the training and testing set, we have used the stratified random sampling to select 80% of normal flows and 20% of threats, as in the original log. This dataset is imbalanced in both the learning stage and the evaluation stage. In fact, the number of normal network flows is significantly higher than the number of threats (80 vs 20%). We note that this resembles the common set-up of an anomaly detection learning task that often occurs in a network.

AAGM[9] was collected by the Canadian Institute for Cybersecurity in 2017. This dataset contains the network traffic captured from Android applications—both malware and benign. Data are obtained by installing Android apps on real smartphones in a semi-automated manner [42]. The dataset is generated from 1900 applications divided into three categories: malware (250 apps) adware (150 apps) and benign

---

[8]https://www.unb.ca/cic/datasets/ids-2017.html.

[9]https://www.unb.ca/cic/datasets/android-adware.html.

(1500 apps). After running the apps on the real Android smartphones (NEXUS 5), the generated traffic has been captured and transformed into samples labelled in two classes (malicious and normal). Specifically, threat samples represent the malicious traffic generated by some popular adware families (e.g airpush, dowgin and kemoge) and malware families (e.g AVpass, FakeAV, FakeFlash/FakePlayer, GGtracker and Penetho). The labelled dataset contains 80 features (3 binary and 76 numerical attributes and 1 class attribute). Attributes are extracted using CICFlowMeter. In this study, we use a subset of the original data, as we build a training set and a testing set with $100K$ samples. In the original dataset, the number of normal apps was higher than the number of malicious apps (80 vs 20%). We preserve this distribution in the training and testing sets prepared for this study.

4.2 Experimental Setting and Evaluation Metrics

In this empirical study, we evaluate:

- how the decision boundary function, that is learned in the first stage of THEO DORA, is able to correctly separate the normal samples from the threats in the training set;
- how we gain accuracy in THEODORA by learning a threat detection model after re-positioning the decision boundary in the training set.

To this aim, we measure the Purity [7] of the decision boundary, as well as the Precision, Recall and F1-score [58] of the threat detection model. The mathematical formulation of the metrics considered in this study is reported in Table 3.

Table 3 Evaluation metrics: Purity, Precision (P), Recall (R) and F1-score (F1)

Metric	Mathematical formulation
Purity	$\dfrac{TP + TN}{TP + TN + FN + FP}$
P	$\dfrac{TP}{TP + FP}$
R	$\dfrac{TP}{TP + FN}$
F1	$2 \cdot \dfrac{P \cdot R}{P + R}$

These metrics are computed by accounting for the number of true positive—TP (number of threats correctly detected), the number of true negative—TN (number of normal samples correctly detected), number of false positive—FP (number of normal samples incorrectly detected as threats) and number of false negative— FN (number of threat samples incorrectly detected as normal samples)

4.2.1 Purity

Purity is a supervised measure that is traditionally adopted in the clustering evaluation. It estimates the accuracy of clustering by measuring the number of correctly assigned samples to each cluster in the total number of samples.

In this empirical study, we analyse the Purity of the decision boundaries, which are learned with a supervised learner—SVM[10] (that is the solution proposed in THEODORA). We compare the Purity of the decision boundaries learned with the SVM to the Purity of the decision boundaries learned with an unsupervised learner—FCM (Fuzzy K-means) [16, 22]. This unsupervised algorithm is selected since it is a good soft clustering approach that, similarly to SVM, can return the confidence according to whether a sample is assigned to a cluster, so that $boundary^{normal}(\mathbf{x}) + boundary^{threat}(\mathbf{x}) = 1$. This experiment is done to confirm that the selected supervised approach can delineate the boundary better than an unsupervised approach.

4.2.2 Precision, Recall and F1-Score

Precision, Recall and F1-score are classification metrics that are commonly used in the cybersecurity literature. They measure the performance of the threat detection models learned on the training sets when they are used to predict the class of the unseen testing samples. In particular, Precision measures the ability of a model to identify threats correctly. It is the ratio of the threats correctly labelled by the model to all the threats predicted by the model. Recall determines the ability to find all the threats, that is, the ratio of the threats correctly labelled by the model to all the samples, which are actually threats. We note that a gain in Recall is the expected outcome of this study, since it indicates that the classification model has actually improved the ability to detect new threats. However, we are interested in increasing Recall, without significantly decreasing Precision. To evaluate this condition, we consider the F1-score that is the harmonic mean of Precision and Recall. The higher the F1-Score, the better the balance between the Precision and Recall achieved by the classification. On the contrary, when one measure is improved at the expense of the other, the F1-Score reported by the model is low.

In this empirical study, we evaluate the Precision, Recall and F1-score of the deep classification models, which are finally learned by THEODORA after the boundary re-positioning of the training data. Since this classification model is learned with the deep learning architecture of MINDFUL [10], we compare the accuracy of THEODORA to that of MINDFUL (trained with the original training set without re-positioning the decision boundary in the training set). This experiment is to quantify

---

[10]In principle, any traditional supervised algorithm, that is able to estimate the classification certainty, can be used in place of SVM. We consider SVM as several studies [27, 34, 40] have repeatedly proved that it outperforms competitors based on Linear SVM, RBF SVM, Random Forest, K-NN and Naive Bayes in various cybersecurity applications.

the gain in accuracy which is actually due to the decision boundary re-positioning. We note that the experimental study in [10] has already proved that the architecture of MINDFUL has outperformed the most prominent, recent competitors. So, proving that THEODORA gains accuracy compared to MINDFUL contributes to assessing that this study has actually over-taken the recent state-of-the-art literature in cyber-threat detection.

We also consider the accuracy of the SVM learned in the first stage of THEODORA as a competitor in this study. This is to prove that the proposed training stage, completed with a deep learning architecture, learns a classification model that is, in any case, more accurate than the decision boundary learned with a traditional learner.

Finally, we evaluate the accuracy performance achieved with several GANs-based competitors, which have been selected from the recent state-of-the-art literature. These results have been collected in the KDDCUP99Test [5, 60, 76, 77]. We pay special attention to the GAN-based competitors as, similarly to THEODORA, they pursue the goal of improving the robustness of the learned models to unseen threats.

4.3   Results

The results are presented as follows. We start by analysing the decision boundaries learned by THEODORA on each dataset (see Sect. 4.3.1). We proceed to study the result of the decision boundary re-positioning by varying the threshold Θ. Particularly, we investigate how re-positioning the decision boundary in the training stage can aid in gaining the ability of detecting new threats in the testing stage (see Sect. 4.3.2). Finally, we compare the accuracy of the threat detection model learned by THEODORA to that of the prominent competitors (see Sect. 4.3.3).

4.3.1   Decision Boundary Detection

Table 4 reports the Purity of the training sets as they are partitioned according to the decision boundaries learned with both SVM and FCM. The results confirm that SVM can actually exploit the supervision when drawing the decision boundary of a training set. In fact, in all the datasets studied, it draws a decision boundary that significantly reduces the number of training samples that are wrongly assigned to the opposite training partition (the Purity of SVM is significantly higher than the Purity of FCM). This aids SVM to identify better the normal samples that are put on the correct side of the boundary even if they are close to the boundary. These are, in principle, the normal samples that will be correctly fitted by the classification model, but that may look like unseen threats.

To confirm these considerations, we analyse the plot of how the training samples are separated by the decision boundary learned with both SVM (Fig. 2a–c) and FCM (Fig. 2d–f), respectively. In general, SVM is can put normal samples and threats on the opposite sides of the decision boundary by diminishing the number of training

Table 4 Decision boundary analysis: Purity of SVC and FCM computed on KDDCUP99Train, UNSW-NB15Train and CICIDS2017Train

Algorithm	Dataset		
	KDDCUP99Train	CICIDS2017Train	AAGMTrain
SVC	99.40	89.10	92.40
FCM	39.40	27.76	53.33

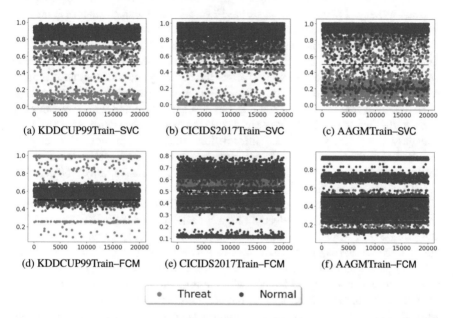

(a) KDDCUP99Train–SVC (b) CICIDS2017Train–SVC (c) AAGMTrain–SVC

(d) KDDCUP99Train–FCM (e) CICIDS2017Train–FCM (f) AAGMTrain–FCM

• Threat   • Normal

Fig. 2 Decision boundary analysis: training data separated into two partitions by the decision boundary learned with SVM (Fig. 2a–c) and FCM (Fig. 2d–f), respectively. The black line draws the boundary detected to separate the normal training samples from the threats. The samples are enumerated on axis X, the confidence of the assignment of a sample to the normal partition ($boundary^{normal}(\mathbf{x})$ is plot on axis Y). A $boundary^{normal}(\mathbf{x})$ greater than 0.5 means that the decision boundary has assigned the sample to the normal partition. A $boundary^{normal}(\mathbf{x})$ lower than 0.5—that is equivalent to $boundary^{threat}(\mathbf{x}) = 1 - boundary^{normal}(\mathbf{x}$ greater than 0.5)—means that the boundary has assigned the sample to the threat partition

samples assigned to the wrong side. In addition, we note that in FCM a high number of normal samples, which are assigned to the normal side, are put close to the boundary. This phenomenon is reduced in SVM. This means that FCM brings a very high number of normal samples to the attention of the boundary re-positioning, with the risk of drastically modifying the training set (once the class of a high number of normal samples has been changed). This risk is lower with SVM that puts a lower number of normal samples close to the boundary for the class change.

4.3.2 Decision Boundary Re-positioning

We investigate how the testing accuracy of the threat detection model depends on the number of normal training samples, which are re-assigned to the opposite class during the training stage. To this aim, we analyse the sensitivity of the threat detection accuracy of THEODORA to Θ. This is the threshold that controls the number of training samples whose label is changed by re-positioning the decision boundary in the training stage. For each dataset, we base the decision of which values of Θ must be experimented on the visual exploration of the normal training samples distribution.

Figure 3a–c show the normal training samples of KDDCUP99Train, CICIDS2017 Train and AAGMTrain that SVM puts on the normal side of the learned decision boundary. These samples are plotted along the certainty (confidence) according to which the SVM assigns them to the normal class. In all the datasets of this study, these plots highlight that the normal training samples are distributed with lower density close to the boundary, while they are distributed with higher density far from the boundary. Considering the density information, we choose threshold Θ to range in the interval of the confidence values, where the density is the lowest. Therefore, we select Θ ranging between 0.50 and 0.80 in KDDCUP99Train (Fig. 3a), 0.50 and 0.60 in CICDS2017Train (Fig. 3b) and 0.50 and 0.80 in AAGMTrain (Fig. 3c).

Figure 4a–c show the curves of the F1 score of THEODORA, which are determined by varying Θ in the selected range for KDDCUP99Test, CICIDS2017Test and AAGMTest, respectively. These results provide the evidence that changing the labels of the normal samples, which are close to the decision boundary, increases the accuracy performance of the final classification model. In fact, the accuracy of the baseline classification model learned with the original training set (i.e. the training set without re-positioning the decision boundary) is lower than the accuracy achieved with the changed training set (i.e. the training set constructed after re-positioning the boundary).

Particularly, we note that THEODORA always improves the F1 score as threshold Θ increases, that is, as a higher number of normal samples is re-positioned on the

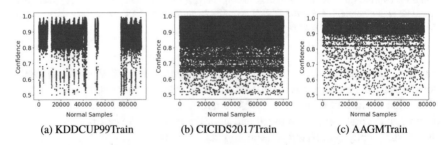

(a) KDDCUP99Train   (b) CICIDS2017Train   (c) AAGMTrain

Fig. 3 Threshold Θ set-up: plot of the normal training samples (axis X) that SVM assigns to the normal side of the decision boundary. The confidence of the assignment is reported on axis Y. Threshold Θ should be set-up to cover the normal samples falling in the area close to the boundary, where the sample density is lower

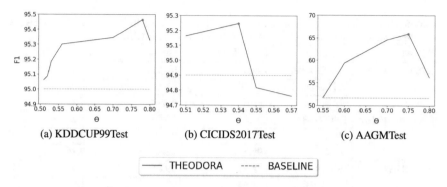

(a) KDDCUP99Test   (b) CICIDS2017Test   (c) AAGMTest

──── THEODORA   ----- BASELINE

Fig. 4 F1 of THEODORA on KDDCUP99Test, CICIDS2017Test and AAGMTest by varying threshold Θ. The dashed line corresponds to the F1 of the classification model (BASELINE) learned from the original training set, without the boundary re-positioning phase. The red point corresponds to the threshold that achieves the highest F1

threat side for completing the training stage. However, after a peak, the F1 score starts decreasing. This happens since the ability to detect correct threats should be linked with the ability to diminish the false alarms (normal samples wrongly detected as threats). If a very high number of normal samples is dealt with as threats during the training stage, then this can excessively bias the trained classification model towards the threat class (by also predicting a high number of normal samples as threats). One limitation of the proposed approach is the automatic identification of Θ to maximise the accuracy improvement. This requires further investigations.

4.3.3 Baseline and Competitor Analysis

For all the datasets in this study, we compare the accuracy performance of THEO DORA to that of its baselines: MINDFUL (i.e. the classification model that is learned with the same deep learning architecture as THEODORA, but processing the original training set with no label changed) and SVM (i.e. the decision boundary learned in the training stage to perform the boundary re-positioning and the label re-assignment). The accuracy performances of these methods, compared in terms of Precision, Recall and F1 score, are reported in Table 5. The results yielded confirm that THEODORA outperforms all its baselines in terms of both Recall and F1 score. However, baselines outperform THEODORA in terms of Precision. So, additional considerations must be formulated to describe the behaviour of Precision.

We recall that the higher the Precision, the higher the percentage of samples, which are classified as threats in the testing set, since they actually present a threatening behaviour. THEODORA, that learns its threat detection model after assigning a few selected normal samples to the opposite class, augments the ability to detect unseen threats at the cost of a higher number of false alarms. On the other hand, we note that higher Precision, that is commonly achieved by the baselines, is always

Table 5 Baseline analysis: Precision, Recall and F1 score of THEODORA, MINDFUL and SVM measured on the testing sets of KDDCUP99, UNSW-NB15 and AAGM

Dataset	Algorithm	Precision	Recall	F1
KDDCUP99Test	THEODORA	99.31	**91.91**	**95.46**
	MINDFUL	99.50	91.10	95.00
	SVM	**99.65**	90.48	94.83
CICIDS2017Test	THEODRA	91.86	**98.90**	**95.25**
	MINDFUL	91.80	98.30	94.90
	SVM	**99.65**	42.78	59.86
AAGMTest	THEODORA	73.37	**59.84**	**65.92**
	MINDFUL	**83.43**	37.37	51.62
	SVM	20.00	17.25	18.51

The accuracy of THEODORA is the best in Fig. 4. The best results are in bold

Table 6 GAN-based competitor analysis: Precision, Recall and F1 score of THEODORA and GAN-based competitors measured on KDDCUP99Test

Method	Precision	Recall	F1
THEODORA	**99.31**	91.91	**95.46**
AnoGAN [60, 77]	87.86	82.97	88.65
Efficient GAN [76]	92.00	**95.82**	93.72
ALAD [77]	94.27	95.77	95.01
MAD-GAN [21]	86.41	94.79	90.00

The accuracy metrics of the competitors are collected from the reference papers. The accuracy of THEODORA is the best in Fig. 4. The best results are in bold

coupled with lower Recall. This means that the number of threats missed by the baselines is higher than the number of threats missed by THEODORA. In any case, the highest F1 score of THEODORA assesses that the proposed method can actually achieve the best balance between Precision and Recall, by definitely gaining overall accuracy compared to the baselines.

Finally, we compare the results achieved by THEODORA on KDDCUP99Test with those of several GAN-based competitors [5, 60, 76, 77]. For all these competitors, we consider Precision, Recall and F1 score, as they are reported in the reference studies. The results reported in Table 6 show that THEODORA also outperforms the considered GAN-based competitors in terms of F1 score. However, the improvement is achieved in terms of Precision, instead of Recall. This behaviour suggests that the idea of re-positioning the training decision boundary should be introduced in a GAN architecture to gain both Precision and Recall, simultaneously.

5   Conclusion

In this study, we address the task of improving the effectiveness of cyber-threat detection models when they are learned with a supervised approach. In particular, we explore the idea of identifying the decision boundary that separates the normal samples from the threats in the training set so that we may re-position this boundary, in order to assign normal training samples that are close to threats to the opposite class. The rationale behind this idea is that training a threat detection model on the modified training set should allow us to learn a more accurate classification model of unseen data. We expect to reach this milestone by accounting for the behaviour of potential unseen threats (which possibly behave similarly to normal samples), instead of learning a threat detection model that over-fits the training data. To this aim, we describe a machine learning method, named **THEODORA**, that uses a supervised approach to identify the decision boundary that separates the normal training samples from the threats. It resorts to a threshold-based approach to re-position the decision boundary and control the number of normal samples to be handled as threats. Finally, it uses a robust deep neural network, recently investigated in the literature, to learn the final threat detection model.

We assess the viability of **THEODORA** by using three benchmark datasets, which contain cyber-data collected in different years and scenarios. The experimental analysis performed allows us to provide the empirical evidence that **THEODORA** can actually gain threat detection accuracy by moving the training set close to the decision boundary. The visual inspection of how the normal data are distributed close to the decision boundary provides useful guidelines to identify a possible range of values. However, the automatic selection of the threshold that maximises the accuracy is still an open problem. Finally, since the idea of learning a classification model from a changed representation of the training set is in some way related to the purposes of adversarial learning, we compare the accuracy of **THEODORA** to that of various adversarial learning models described in the recent literature. We show that **THEODORA** can achieve important results compared with these competitors.

For future work we plan to investigate an automatic approach to identify the threshold used to control the decision boundary in the training stage. In addition, we intend to investigate how the idea of re-positioning the decision boundary can be possibly combined with adversarial techniques (e.g. GANs). Finally, we plan to investigate a fully deep version of the proposed approach, where both the decision boundary and the classification model are learned in an end-to-end fashion.

Acknowledgements We acknowledge the support of the MIUR-Ministero dell'Istruzione dell'Università e della Ricerca through the project "TALIsMan—Tecnologie di Assistenza personALizzata per il Miglioramento della quAlità della vitA" (Grant ID: ARS01_01116) funded by PON RI 2014–2020 and the ATENEO 2017/18 "Modelli e tecniche di data science per la analisi di dati strutturati" funded by the University of Bari "Aldo Moro". The authors wish to thank Lynn Rudd for her help in reading the manuscript.

References

1. Abdulhammed Alani R, Musafer H, Alessa A, Faezipour M, Abuzneid A (2019) Features dimensionality reduction approaches for machine learning based network intrusion detection. Electronics 8:322
2. Abri F, Siami-Namini S, Khanghah MA, Soltani FM, Namin AS (2019) Can machine/deep learning classifiers detect zero-day malware with high accuracy? In: 2019 IEEE international conference on big data (Big Data), pp 3252–3259
3. Al-Qatf M, Lasheng Y, Al-Habib M, Al-Sabahi K (2018) Deep learning approach combining sparse autoencoder with svm for network intrusion detection. IEEE Access 6:52843–52856
4. Aldweesh A, Derhab A, Emam AZ (2020) Deep learning approaches for anomaly-based intrusion detection systems: a survey, taxonomy, and open issues. Knowl-Based Syst 189:105124
5. AlEroud A, Karabatis G (2020) Sdn-gan: generative adversarial deep nns for synthesizing cyber attacks on software defined networks. In: Debruyne C, Panetto H, Guédria W, Bollen P, Ciuciu I, Karabatis G, Meersman R (eds) On the move to meaningful internet systems: OTM 2019 workshops. Springer International Publishing, Cham, pp 211–220
6. Althubiti SA, Jones EM, Roy K (2018) Lstm for anomaly-based network intrusion detection. In: 2018 28th International telecommunication networks and applications conference (ITNAC). IEEE Computer Society, pp 1–3
7. Amigó E, Gonzalo J, Artiles J, Verdejo M (2009) Amigó e, gonzalo j, artiles j et ala comparison of extrinsic clustering evaluation metrics based on formal constraints. Inf Retrieval 12:461–486
8. Andresini G, Appice A, Malerba D (2020) Dealing with class imbalance in android malware detection by cascading clustering and classification. In: Complex pattern mining—new challenges, methods and applications, Studies in Computational Intelligence, vol 880. Springer, pp 173–187. https://doi.org/10.1007/978-3-030-36617-9_11
9. Andresini G, Appice A, Mauro ND, Loglisci C, Malerba D (2019) Exploiting the auto-encoder residual error for intrusion detection. In: 2019 IEEE European symposium on security and privacy workshops, EuroS&P workshops 2019, Stockholm, Sweden, 17–19 June 2019. IEEE, pp 281–290
10. Andresini G, Appice A, Mauro ND, Loglisci C, Malerba D (2020) Multi-channel deep feature learning for intrusion detection. IEEE Access 8:53346–53359
11. Angelo P, Costa Drummond A (2018) Adaptive anomaly-based intrusion detection system using genetic algorithm and profiling. Secur Priv 1(4):e36
12. Appice A, Andresini G, Malerba D (2020) Clustering-aided multi-view classification: a case study on android malware detection. J Intell Inf Systms. https://doi.org/10.1007/s10844-020-00598-6
13. Appice A, Guccione P, Malerba D (2017) A novel spectral-spatial co-training algorithm for the transductive classification of hyperspectral imagery data. Pattern Recognit 63:229–245
14. Appice A, Malerba D (2019) Segmentation-aided classification of hyperspectral data using spatial dependency of spectral bands. ISPRS J Photogrammetry Remote Sens 147:215–231
15. Berman DS, Buczak AL, Chavis JS, Corbett CL (2019) A survey of deep learning methods for cyber security. Information 10(4):1–35
16. Bezdek JC (1981) Pattern recognition with fuzzy objective function algorithms. Kluwer Academic Publishers, USA
17. Chang CC, Lin CJ (2011) Libsvm: a library for support vector machines. ACM Trans Intell Syst Technol 2(3):1–27
18. Cheng F, Yang K, Zhang L (2015) A structural svm based approach for binary classification under class imbalance. Math Probl Eng 2015:1–10
19. Chun M, Wei D, Qing W (2020) Speech analysis for wilson's disease using genetic algorithm and support vector machine. In: Abawajy JH, Choo KKR, Islam R, Xu Z, Atiquzzaman M (eds) International conference on applications and techniques in cyber intelligence ATCI 2019. Springer International Publishing, Cham, pp 1286–1295

20. Comar PM, Liu L, Saha S, Tan P, Nucci A (2013) Combining supervised and unsupervised learning for zero-day malware detection. In: 2013 Proceedings IEEE INFOCOM, pp 2022–2030

21. Dan L, Dacheng C, Baihong J, Lei S, Jonathan G, See-Kiong N (2019) Mad-gan: Multivariate anomaly detection for time series data with generative adversarial networks. In: Artificial neural networks and machine learning, pp 703–716

22. Dunn JC (1973) A fuzzy relative of the isodata process and its use in detecting compact well-separated clusters. J Cybern 3(3):32–57

23. Gandotra E, Bansal D, Sofat S (2016) Zero-day malware detection. In: 2016 Sixth international symposium on embedded computing and system design (ISED), pp 171–175

24. Goh KS, Chang E, Cheng KT (2001) Svm binary classifier ensembles for image classification. In: Proceedings of the tenth international conference on information and knowledge management, CIKM '01. Association for Computing Machinery, New York, NY, USA, pp 395–402

25. Goodfellow I, McDaniel P, Papernot N (2018) Making machine learning robust against adversarial inputs. Commun ACM 61(7):56–66

26. Goodfellow IJ, Pouget-Abadie J, Mirza M, Xu B, Warde-Farley D, Ozair S, Courville AC, Bengio Y (2014) Generative adversarial nets. In: Advances in neural information processing systems 27, Annual conference on neural information processing systems 2014, 8–13 December 2014, Montreal, Quebec, Canada, pp 2672–2680

27. Halimaa A, Sundarakantham K (2019) Machine learning based intrusion detection system. In: 2019 3rd International conference on trends in electronics and informatics (ICOEI), pp 916–920

28. Hao M, Tianhao Y, Fei Y (2019) The svm based on smo optimization for speech emotion recognition. In: 2019 Chinese control conference (CCC), pp 7884–7888

29. Hao Y, Sheng Y, Wang J (2019) Variant gated recurrent units with encoders to preprocess packets for payload-aware intrusion detection. IEEE Access 7:49985–49998

30. Hu Z, Chen P, Zhu M, Liu P (2019) Reinforcement learning for adaptive cyber defense against zero-day attacks. Springer International Publishing, Cham, pp 54–93

31. Ingre B, Yadav A, Soni AK (2018) Decision tree based intrusion detection system for nsl-kdd dataset. In: Satapathy SC, Joshi A (eds) Information and communication technology for intelligent systems (ICTIS 2017), vol 2. Springer International Publishing, Cham, pp 207–218

32. Jang-Jaccard J, Nepal S (2014) A survey of emerging threats in cybersecurity. J Comput Syst Sci 80(5):973–993 Special Issue on Dependable and Secure Computing

33. Jiang F, Fu Y, Gupta BB, Lou F, Rho S, Meng F, Tian Z (2018) Deep learning based multi-channel intelligent attack detection for data security. IEEE Trans Sustain Comput pp 1–1

34. Kedziora M, Gawin P, Szczepanik M, Jozwiak I (2019) Malware detection using machine learning algorithms and reverse engineering of android java code. SSRN Electron J. https://doi.org/10.2139/ssrn.3328497

35. Khan RU, Zhang X, Alazab M, Kumar R (2019) An improved convolutional neural network model for intrusion detection in networks. In: 2019 Cybersecurity and cyberforensics conference (CCC), pp 74–77

36. Kim JY, Bu SJ, Cho SB (2018) Zero-day malware detection using transferred generative adversarial networks based on deep autoencoders. Inf Sci 460–461:83–102

37. Kim JY, Cho SB (2018) Detecting intrusive malware with a hybrid generative deep learning model. In: Yin H, Camacho D, Novais P, Tallón-Ballesteros AJ (eds) Intelligent data engineering and automated learning—IDEAL 2018. Springer International Publishing, Cham, pp 499–507

38. Kim T, Suh SC, Kim H, Kim J, Kim J (2018) An encoding technique for cnn-based network anomaly detection. In: International conference on big data, pp 2960–2965

39. Kremer J, Steenstrup Pedersen K, Igel C (2014) Active learning with support vector machines. WIREs Data Min Knowl Discov 4(4):313–326

40. Krishnaveni S, Vigneshwar P, Kishore S, Jothi B, Sivamohan S (2020) Anomaly-based intrusion detection system using support vector machine. In: Dash SS, Lakshmi C, Das S, Panigrahi BK (eds) Artificial intelligence and evolutionary computations in engineering systems. Springer Singapore, Singapore, pp 723–731

41. Labonne M, Olivereau A, Polve B, Zeghlache D (2019) A cascade-structured meta-specialists approach for neural network-based intrusion detection. In: 16th Annual consumer communications & networking conference, pp 1–6
42. Lashkari AH, Kadir AFA, Gonzalez H, Mbah KF, Ghorbani AA (2017) Towards a network-based framework for android malware detection and characterization. In: PST. IEEE Computer Society, pp 233–234
43. Le T, Kang H, Kim H (2019) The impact of pca-scale improving gru performance for intrusion detection. In: 2019 International conference on platform technology and service (PlatCon), pp 1–6
44. Lewis DD, Gale WA (1994) A sequential algorithm for training text classifiers. In: Croft BW, van Rijsbergen CJ (eds) SIGIR '94. Springer, London, London, pp 3–12
45. Li D, Chen D, Jin B, Shi L, Goh J, Ng SK (2019) Mad-gan: multivariate anomaly detection for time series data with generative adversarial networks. In: Tetko IV, Kůrková V, Karpov P, Theis F (eds) Artificial neural networks and machine learning—ICANN 2019: text and time series. Springer International Publishing, Cham, pp 703–716
46. Li Y, Ma R, Jiao R (2015) A hybrid malicious code detection method based on deep learning. Int J Softw Eng Appl 9:205–216
47. Lin WC, Ke SW, Tsai CF (2015) Cann: an intrusion detection system based on combining cluster centers and nearest neighbors. Knowl-Based Syst 78:13–21
48. Liu J, Tian Z, Zheng R, Liu L (2019) A distance-based method for building an encrypted malware traffic identification framework. IEEE Access 7:100014–100028
49. Liu J, Zhang W, Tang Z, Xie Y, Ma T, Zhang J, Zhang G, Niyoyita JP (2020) Adaptive intrusion detection via ga-gogmm-based pattern learning with fuzzy rough set-based attribute selection. Expert Syst Appl 139:112845
50. Liu W, Ci L, Liu L (2020) A new method of fuzzy support vector machine algorithm for intrusion detection. Appl Sci 10(3):1065
51. Malerba D, Ceci M, Appice A (2009) A relational approach to probabilistic classification in a transductive setting. Eng Appl Artif Intell 22(1):109–116. https://doi.org/10.1016/j.engappai.2008.04.005
52. Malik AJ, Khan FA (2017) A hybrid technique using binary particle swarm optimization and decision tree pruning for network intrusion detection. Cluster Comput pp 1–14
53. Moti Z, Hashemi S, Namavar A (2019) Discovering future malware variants by generating new malware samples using generative adversarial network. In: 2019 9th International conference on computer and knowledge engineering (ICCKE), pp 319–324
54. Naseer S, Saleem Y, Khalid S, Bashir MK, Han J, Iqbal MM, Han K (2018) Enhanced network anomaly detection based on deep neural networks. IEEE Access 6:48231–48246
55. Pang, Y., Chen, Z., Peng, L., Ma, K., Zhao, C., Ji, K.: A signature-based assistant random oversampling method for malware detection. In: 2019 18th IEEE International conference on trust, security and privacy in computing and communications/13th IEEE international conference on big data science and engineering (TrustCom/BigDataSE), pp 256–263
56. Papernot N, McDaniel P, Wu X, Jha S, Swami A (2016) Distillation as a defense to adversarial perturbations against deep neural networks. In: 2016 IEEE symposium on security and privacy (SP), pp 582–597
57. Platt JC (1999) Probabilistic outputs for support vector machines and comparisons to regularized likelihood methods. In: Advances in large margin classifiers. MIT Press, pp 61–74
58. Powers D (2007) Evaluation: from precision, recall and fmeasure to roc, informedness, markedness and correlation. J Mach Learn Technol 2:37–63
59. Qu X, Yang L, Guo K, Ma L, Feng T, Ren S, Sun M (2019) Statistics-enhanced direct batch growth self-organizing mapping for efficient dos attack detection. IEEE Access 7:78434–78441
60. Schlegl T, Seeböck P, Waldstein SM, Schmidt-Erfurth U, Langs G (2017) Unsupervised anomaly detection with generative adversarial networks to guide marker discovery. In: Niethammer M, Styner M, Aylward S, Zhu H, Oguz I, Yap PT, Shen D (eds) Information processing in medical imaging. Springer International Publishing, Cham, pp 146–157

61. Shapoorifard H, Shamsinjead Babaki P (2017) Intrusion detection using a novel hybrid method incorporating an improved knn. Int J Comput Appl 173:5–9. https://doi.org/10.5120/ijca2017914340
62. Stellios I, Kotzanikolaou P, Psarakis M (2019) Advanced persistent threats and zero-day exploits in industrial internet of things. Springer International Publishing, Cham, pp 47–68
63. Stokes JW, Seifert C, Li J, Hejazi N (2019) Detection of prevalent malware families with deep learning. In: MILCOM 2019—2019 IEEE military communications conference (MILCOM), pp 1–8
64. Tavallaee M, Bagheri E, Lu W, Ghorbani AA (2009) A detailed analysis of the kdd cup 99 data set. In: Symposium on computational intelligence for security and defense applications, pp 1–6
65. Vapnik VN (1998) Statistical learning theory. Wiley-Interscience
66. Vigneswaran RK, Vinayakumar R, Soman KP, Poornachandran P (2018) Evaluating shallow and deep neural networks for network intrusion detection systems in cyber security. In: 2018 9th International conference on computing, communication and networking technologies (ICC-CNT), pp 1–6. https://doi.org/10.1109/ICCCNT.2018.8494096
67. Vinayakumar R, Alazab M, Soman KP, Poornachandran P, Al-Nemrat A, Venkatraman S (2019) Deep learning approach for intelligent intrusion detection system. IEEE Access 7:41525–41550
68. Vinayakumar R, Alazab M, Soman KP, Poornachandran P, Venkatraman S (2019) Robust intelligent malware detection using deep learning. IEEE Access 7:46717–46738
69. Virmani C, Choudhary T, Pillai A, Rani M (2020) Applications of machine learning in cyber security. In: Handbook of research on machine and deep learning applications for cyber security
70. Wadkar M, Troia FD, Stamp M (2020) Detecting malware evolution using support vector machines. Expert Syst Appl 143:113022
71. Wang Q, Guo W, Zhang K, Ororbia AG, Xing X, Liu X, Giles CL (2017) Adversary resistant deep neural networks with an application to malware detection. In: Proceedings of the 23rd ACM SIGKDD international conference on knowledge discovery and data mining, KDD '17. Association for Computing Machinery, New York, NY, USA, pp 1145–1153
72. Wang W, Zhu M, Zeng X, Ye X, Sheng Y (2017) Malware traffic classification using convolutional neural network for representation learning. In: 2017 International conference on information networking (ICOIN). IEEE, pp 712–717
73. Yin C, Zhu Y, Fei J, He X (2017) A deep learning approach for intrusion detection using recurrent neural networks. IEEE Access 5:21954–21961
74. Yin Z, Liu W, Chawla S (2019) Adversarial attack, defense, and applications with deep learning frameworks. Springer International Publishing, Berlin, pp 1–25
75. Yin Z, Wang F, Liu W, Chawla S (2018) Sparse feature attacks in adversarial learning. IEEE Trans Knowl Data Eng 30(6):1164–1177
76. Zenati H, Foo CS, Lecouat B, Manek G, Chandrasekhar VR (2018) Efficient gan-based anomaly detection. ArXiv abs/1802.06222
77. Zenati H, Romain M, Foo CS, Lecouat B, Chandrasekhar VR (2018) Adversarially learned anomaly detection. In: 2018 IEEE International conference on data mining (ICDM), pp 727–736
78. Zhang Y, Chen X, Jin L, Wang X, Guo D (2019) Network intrusion detection: Based on deep hierarchical network and original flow data. IEEE Access 7:37004–37016
79. Zhang Z, Pan P (2019) A hybrid intrusion detection method based on improved fuzzy c-means and support vector machine. In: 2019 International conference on communications, information system and computer engineering (CISCE), pp 210–214

Bayesian Networks for Online Cybersecurity Threat Detection

Mauro José Pappaterra and Francesco Flammini

Abstract Cybersecurity threats have surged in the past decades. Experts agree that conventional security measures will soon not be enough to stop the propagation of more sophisticated and harmful cyberattacks. Recently, there has been a growing interest in mastering the complexity of cybersecurity by adopting methods borrowed from Artificial Intelligence (AI) in order to support automation. In this chapter, we concentrate on cybersecurity threat assessment by the translation of Attack Trees (AT) into probabilistic detection models based on Bayesian Networks (BN). We also show how these models can be integrated and dynamically updated as a detection engine in the existing DETECT framework for automated threat detection, hence enabling both offline and online threat assessment. Integration in DETECT is important to allow real-time model execution and evaluation for quantitative threat assessment. Finally, we apply our methodology to a real-world case study, evaluate the resulting model with sample data, perform data sensitivity analyses, then present and discuss the results.

Keywords Bayesian networks · Threat detection · Attack trees · Explainable AI · Risk evaluation · Situation Assesment

1 Introduction

Recent advances in the field of Artificial Intelligence (AI) can be implemented in the development of intelligent cybersecurity frameworks. We are living in times where conventional security measures will soon not be enough to stop the propagation of more sophisticated and potentially more harmful cyberattacks. The implementation

M. J. Pappaterra (✉)
Uppsala University, Uppsala, Sweden
e-mail: mauro.pappaterra@ieee.org

F. Flammini
School of Innovation, design, and engineering, Division of product and realization, Mälardalen University, Västerås, Sweden

Y. Maleh et al. (eds.), *Machine Intelligence and Big Data Analytics for Cybersecurity Applications*, Studies in Computational Intelligence 919,
https://doi.org/10.1007/978-3-030-57024-8_6

129

of AI and Machine Learning (ML) technology as augmentation of cybersecurity is the most promising solution to this increasing problem [1].

The research presented in this chapter tries to answer the following questions:

(1) How can Bayesian Network based probabilistic models be used to detect common cyberthreats scenarios and how uncertainties can be managed?
(2) How can intelligent online cyberthreat detection models be developed based on Bayesian Networks and the DETECT framework? How can detection models be applied in real cyberthreat scenarios?

This research is motivated by the need for an automated, versatile, and easy adapted framework that implements Bayesian Networks and stochastic inference methods for online threat detection. Said framework should accompany the rapid increase of cybersecurity threats.

2 Related Works

Companies, governments, experts and scholars around the world struggle to keep modern cyberthreats in line. A promising solution to the future of cybersecurity is the implementation of different AI techniques such as Bayesian Networks. Different studies on the implementation of BN to mitigate cyberattacks have had generally positive outcome. Based on a literature review of intelligent cybersecurity with Bayesian Networks published by F. Flammini and M. J. Pappaterra it is important to remark that most of the systems and frameworks studied have had positive results when applied in different fields of cybersecurity, and for different purposes. The logical and mathematical underpins of BN are perfect for inferring results when presented with partial observations and uncertainty. Nonetheless, after surveying the related literature, the authors could not find any information about any existent BN based security system that it is widely implemented on a large scale [2]. Moreover, studies suggest that only 3.4% of organizations worldwide implement any AI based automated security implementation on their systems [3].

Previous studies on the implementation of Bayesian Networks in cybersecurity recognized three essential aspects for the construction of stochastic models: modularization of all components in the framework, importance of the credibility of the data use for the population of the CPT tables, and low sensitivity to parameter perturbation. More details on related work can be found on the afore mentioned literature review—a prelude to the work presented in this chapter [2].

A holistic security framework presented by F. Flammini et al., called SENSO-RAIL, explores the application of AI technology, in combination with wireless sensor networks, for monitoring physical infrastructures, in this particular case railway stations. SENSORAIL dwells on the possible application of BNs and wireless sensors for the prevention of events using the DETECT framework [4]. DETECT (*Decision Triggering Event Composer and Tracker*) was developed aimed to an early, real-time threat detection by matching known attacks patterns and signatures. This framework uses soft computing approaches, such as data fusion and cognitive reasoning, as the core of its detection engine. DETECT implements model analysis and sequence of events recognition in order to recognize known threats patterns, and it can be embedded in existing PSIM (Physical Security Information Management) and SIEM (Security Information and Event Management) systems. A specific Event Description Language (EDL) has been developed for threat descriptions to be stored in an appropriate scenario repository used to feed the model-based detection engine [4, 5].

3   Integrating Bayesian Networks in the DETECT Framework

3.1   Introduction to DETECT

The DETECT framework was developed mainly for Critical Infrastructure Protection (CIP). A *critical infrastructure* comprises physical assets and communication services that are critical or high-priority for a private or governmental institution. Other fields of application for the DETECT framework include environment monitoring and control of distributed systems.

The basic idea behind DETECT is that attack scenarios can be inferred from a set of basic events, that can be correlated to build a threat signature, and warn about the threat when it is detected in a specific logical, spatial and temporal combination. To that aim, DETECT includes an attack scenario repository. DETECT aims to early detection, decision support and possible automatic counteraction of threats. The system has been integrated and experimented within existing security management systems for *critical infrastructures.*

For this research, BN are used within the general DETECT framework in order to assess cyberthreats in real time. DETECT processes integrated information possibly enriched with reliability indicators. This feature makes it suitable for online threat detection in presence of uncertainties. The DETECT framework has been demonstrated to be capable to detect complex event-driven scenarios, outlined by heterogeneous events [6–8].

3.2   The Architecture of the DETECT Framework

In order to detects anomalies, possible threats and vulnerabilities, DETECT counts with a complex model-base logic and online detection engine process (Fig. 1). Event occurrences might take place in lapses of time and correlate to other events spatially and temporally. Composite events are reconstructed by DETECT based on its complex engine, that does no regard events separately, but in conjunction with previous and posterior occurrences. DETECT's intrinsic event driven architecture recognizes combinations of events, and how they relate to each other. The architecture of the DETECT framework includes:

Event History: contains a list of all identifiable events that are detected by the system under scrutiny. This database can also be provided by external sources.

Event Adaptor Module: pre-processes the events from the *Event History*. This can also be provided by external sources.

User Interface: provides with an intuitive GUI for the designing and sketching of attack scenarios, control of the detection process, and view of the monitoring status of the system under scrutiny.

XML File Generator: exports the attack scenarios blueprints generated on the Scenario GUI as XML files.

Attack Scenario Repository: indexes all XML files generated for data processing and posterior use.

Model Generator: is responsible for parsing all files from the *Attack Scenario Repository* to EDL, in order to build the correct detection models with the corresponding structures and parameters.

Model Updater: provides with real time update for parameters used in the model.

Detection Model: is one of the main parts of the *Detection Engine,* the central unit of the DETECT architecture. This module is in charge of detecting the potentially harmful events that occur on the monitored system. The engine is designed to implement both a deterministic and a heuristic detection model. Nonetheless, the

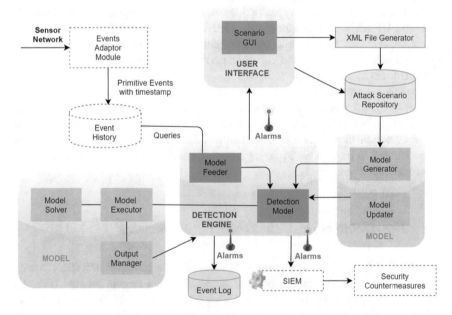

Fig. 1 The architecture of the DETECT framework

deterministic detection model approach is the only one that has been completely implemented with success.

Model Feeder: controls the representation of the input that is taken from querying on system events from the *Event History* database.

Model Executor: prompts the execution of the model, and activation of the *Model Solver*.

Model Solver: is the module in charge of detecting an executing the model. All logical assumptions inferred from the *Model Feeder* are lodged here. The *Model Solver* detects the composite events taking place.

Output Manager: manages the output from the model before it is sent to the *Detection Engine*.

Event Log: saves information about the discovered threats. Metadata saved on the *Even Log* includes detailed information such as time of detection, alarm level severity, all events detected for composite threat detection and more.

The architecture of DETECT allows for alarm hierarchies based on risk levels. Alarms are associated to each event that form part of a set of events that are recognized as composite events by the *Detection Engine*. Alarms are sent from the main engine to the SIEM system for operator decision support and/or immediate application of countermeasures. Warnings and alarms are shown on the *User Interface* for acknowledgement by security operators and CERT (Computer Emergency Response Team); they are also saved to the *Event Log* for late retrieval, investigation and any other forensic activities.

3.3 Bayesian Networks for Online Threat Detection in DETECT

Previous studies have identified the main requirements for security monitoring systems: threats must be represented by using appropriate modelling formalisms; parameters must be contextualized; online detection must be updated in real time; relevant signalizing of the threats must be implemented using a pertinent alert system; finally, threats must be classified and integrated on the systems database [2]. As shown in the previous section, DETECT provides security operators with all those functionalities [4]. Moreover, this security framework has been used before with deterministic models, in combination with both wireless sensors and distributed systems with notorious success [6–8, 9].

On this chapter we show how to implement a BN based detection model, and fully integrate it with the rest of the components of the DETECT framework. An unerring security system should provide with identification of threats, vulnerabilities, and shortcomings in the system, and also be able to determine if a system is under attack. If the latter is detected, the security system should be able to provide with a course of action to countermeasure the attack. This should be achieved by automatic detection of threat scenarios in real time, and correct association and interconnection of different ongoing composite events detected by the system. A cybersecurity monitoring system should provide operators with identification of threats, vulnerabilities and shortcomings in the system, being able to autonomously determine if the monitored system is under attack. If an attack is detected, the security system should be able to provide with a course of action to countermeasure the attack. However, effective decisions and even more autonomous response must be supported by probabilistic analysis in order to assess and justify false positive and false negative detection probabilities. To that aim, we have worked on the idea of creating a DETECT module addressing BN for stochastic inference, along with other soft computing concepts, in order to safely detect possible attacks and provide operators with indicators of threat detection probabilities. In such a probabilistic model, BN inferences can be used offline to support risk assessment, and online to support (early) detection of ongoing attacks. Therefore, BN utility is manifold, including but not limited to the assessment of online threats, to prompt the user to execute the correct course of action to counter measure an ongoing attack; BN models can also be used to mitigate risks by discovering and "measuring" the impact of specific security weaknesses.

For the creation of a BN based detection model for online threats, a simplified implementation will be considered, based on the compositional DETECT architecture (Fig. 2). A modular implementation on a simplified model provides with both encapsulation and isolation, reducing the possible introduction of error by other parts of the system while simplifying the course of action, and allowing for easy extension and further developments in the future.

The attack scenarios will be modelled as Attack Tree (AT) diagrams, that will conform an attack scenario repository. These models, first presented by B. Schneier in 1999 [10], present a visual overall on the security of a system, and have been proven to be easily translated into BNs [11, 12]. Threat models are described in XML to be parsed as machine-readable data.

In online threat assessment, probabilities are updated in real time as events unfold. When cybersecurity relevant events are detected, the *Model Feeder* queries the *Event History* inferring the correlation of the unfolding events inside a defined time window. In our BN assumption, a basic event that is part of a more complex threat scenario can change its state from unknown/estimated (probability less than 1.0) to 'True' (probability equal to 1.0) if it is determined that the event is taking place, and back to unknown/estimated once a new inference update determines otherwise. Examples of events can be firewall alerts, intrusion detections, wrong user login attempts, authentication issues, unauthorized behaviours, access to malicious websites, antivirus alarms, vulnerability scanner alerts, software update alerts, email phishing warnings, etc. When detected, these cybersecurity events will cause the corresponding events on the BN detection model to be set to True. In case an event is only indirectly related to security, with a low correlation to threats, it can still be used to update model parameters, e.g. by increasing expected threat occurrence probability, even though no BN node will be set to 'True'.

The final aim of BN detection model is to enable automatic recognition of threats in order to output an alarm. In order to achieve this, as events unfold, the probabilities on the BN are dynamically updated, and provided that the probabilities reach

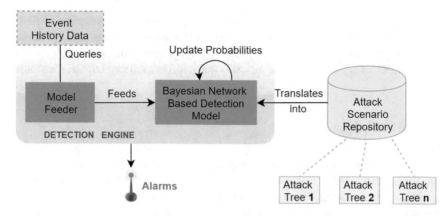

Fig. 2 Model for the application of Bayesian Networks for online threat detection in DETECT

predefined thresholds, warning/alarms/countermeasures are triggered depending on the level of trustworthiness. Therefore, it is essential to associate the appropriate thresholds in the SIEM system. These indicators will be used to inform security managers and operators in security control rooms, also known as Security Operations Centers (SOC), and/or trigger corresponding security countermeasures from the SIEM system. Proper setting of thresholds can be critical and application dependent, but fine-tuning and adequate learning periods can help achieve good trade-offs for best usability and results. Nevertheless, a number of issues need to be addressed. For instance, parameter context should be taken into consideration. It is necessary to discern the right number of events that conform a composite event. To do this it is important to verify that the lapse of time between each event is not too far apart from each other. The model should assure, by following pertinent guidelines, that all events are part of the same composite event that conform the attack scenario. To simplify, events that are outside a relevant window period, will be reset automatically or by security operators/managers; however, if the events take place inside a relevant window period (e.g. 24 h), the detection engine should output the corresponding alarms after making an inference.

On the model proposed in this project, BNs will be obtained from a model-to-model (M2M) transformation from ATs. However, such a transformation only address model structure, while parameters/probabilities should be set directly in BNs. Hence, the conditional probability tables (CPT) of each generated BN will be populated based on available data. In real-world scenarios, these probabilities will be obtained from diverse sources ranging from known detector reliability to expert judgment and historical data or statistics. Since actual data is not essential to validate the methodology, in this chapter we describe and test the BN approach with some realistic assumptions and pseudo-data. Regarding false positive and false negative alarms generated by basic detectors, in BNs those uncertainties can be taken into account together with all other uncertainties related to cybersecurity model structure and parameters. In fact, one strength of BN is that they can easily model the so called "noisy" AND-OR in composite events, implementing some aspects of fuzzy logic. Once the BN detection model for a certain threat has been developed, it should be thoroughly tested for sensitivity analysis, data perturbation, value distortion of probabilities, etc., before it can be included into a fully functional DETECT module.

3.4 Attack Trees

Attack Trees (AT) were first introduced by B. Schneier in 1999. they describe a modelling format based on a tree abstract data type. In an AT, all leaf nodes represent a single event that is the start of a path to the execution of an attack, all middle nodes represent an intermediary step of an attack scenario, and the root node, that is unique to each attack tree, represents the final step necessary for an attack to be deemed as successful (Fig. 3) [10].

The intrinsic flexibility and malleability of ATs make them one of the most implemented models for the representation of attack scenarios [11].

For the model implementation of ATs in conjunction with BN that we intend to present, the following formal definition of an AT is taken into consideration. The definition is based on formal descriptions made by Schneier [10], Mauw and Oostdijk [13], and Gribaudo et al. [12] in the quoted papers, and adapted for the convenience of the proposed model.

The operation functions on the AT are defined graphically in the proposed AT models. These functions define how a node in the tree is to be accomplished during

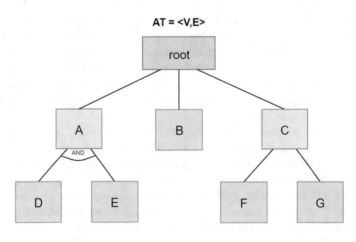

V = {root, A, B, C, D ,E, F, G}
E = {(root,A), (root,b), (root,C), (A,D), (A,E), (C,F), (C,G)}

Fig. 3 Model for the proposed implementation for an Attack Tree. The red node is the root node, all middle nodes are coloured blue, and all leaves nodes are coloured green

the execution of an attack. The operation functions taken into consideration in the presented approach include:

- **AND**: all steps indicated on the child nodes must be accomplished. This is modelled as a line that unites all arcs involved together.
- **OR**: at least one of the steps on the child nodes must be accomplished.
- **XOR**: exclusive **OR**, at most one of the steps on the child nodes must be accomplished.

The default operation function is the inclusive **OR** that should be tacitly understood when no other operation function is directly stated as a label in between the arcs.

3.5 Bayesian Networks

The utility of the use of Bayesian Networks (BN) in the proposed model is based on the possibility to infer the probabilities to different paths in the network by observing all circumventing nodes. The logical underpin of BN is based on the well-known Bayes Theorem (1763), which describes how to implement conditional probability axioms to update probabilities as conditions are proven to be true [14]. A BN graphical model will be represented as directed acyclic graphs (DAGs), in which each node represents a set of variables, and the arcs represent their conditional dependencies or possible state transitions (Fig. 4). Probability values can be assigned as weight to the arcs of the graph.

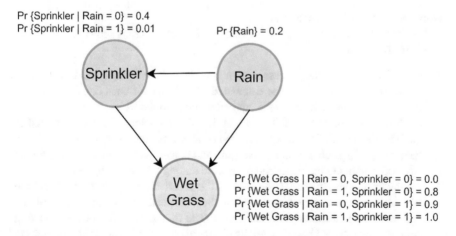

Pr {Sprinkler | Rain = 0} = 0.4
Pr {Sprinkler | Rain = 1} = 0.01

Pr {Rain} = 0.2

Sprinkler ← Rain

Wet Grass

Pr {Wet Grass | Rain = 0, Sprinkler = 0} = 0.0
Pr {Wet Grass | Rain = 1, Sprinkler = 0} = 0.8
Pr {Wet Grass | Rain = 0, Sprinkler = 1} = 0.9
Pr {Wet Grass | Rain = 1, Sprinkler = 1} = 1.0

Fig. 4 Example of a graphical representation of a Bayesian Network with corresponding probability tables using the classic *wet grass* example

Mathematically speaking, a BN is simply a set of joined conditional probability distributions. Every node on the BN is associated with a variable X_i. The relations among nodes, which are graphically represented as edges or arcs, represent the connection with all the parent nodes of the given variable. Every node can be associated with the distribution of probabilities, which are represented by a CPT, whose probabilities are conditionally given by all the parent nodes of X_i. This can be denoted as p (X | parents(X)). Following this simplified explanation, an entire BN, that ends on a target node, can be simply represented by a single joined probability distribution [14]:

$$p(X_1 \ldots X_n) = \prod_1^n p(Xi|parents(Xi)) \tag{1}$$

3.6 Model-to-Model (M2M) Transformation Proposal: From Attack Trees to Bayesian Networks

The M2M transformation proposal that we present in this chapter is based on a combination of the approaches by A. Bobbio et al. implemented for mapping FTs [11], together with the model presented by G. Gribaudo et al. for the analysis of combined ATs using BNs [12]. Both proposed models have been used as a source of inspiration. Nonetheless, we have modified the notation and graphical representation to better suit the implementation for the proposed model with the DETECT framework.

Implemented M2M transformation
The M2N transformation proposal is described in the flow chart shown in Fig. 6. We can identify four main levels:

1. On the first level of the process, each leaf node in the existing AT is translated as a root node in a new BN. The calculated probability for each one of the leaves in the AT is then assigned to the equivalent node in the BN.
2. On level two, for each middle node in the AT, an analogous node is created in the BN. Each one of these nodes is then connected to the matching leaf nodes in the original AT model. Linear, diverging and converging connections within the BN nodes indicate the dependency relation among the nodes in the BN.
3. On the third level, the root node in the AT is translated as a non-root node in the BN. Once again, this non-root node is then connected to the matching middle and leaf nodes in the original AT model. The final probability of this non-root node will be inferred based on all the observed events in the network. As before, linear, diverging and converging connections within the BN nodes indicate the dependency relation among the nodes in the BN.
4. Lastly, on the fourth and last level of the process, each of the node's CPTs must be populated. Operation functions (AND, OR, XOR) assigned in the AT are

translated to the BN models with the implementation of equivalent CPT tables on each node in the BN (Fig. 7).

As the consequence, the resulting BN is closed to an inverted version of the given AT, as depicted in Fig. 5.

Hence the versatility of AT for modelling attack systems can be combined with the stochastic inference powers of BN. Once we have an AT model translated into a BN, we can now model each path or branch using event algebras or UML behavioural diagrams to have a formal or semi-formal representation of each scenario. The objective is to define an algorithm to automate the process of building a BN detection model based on the modelled threats. CPTs can be later be refined in order to take into account any fuzzy correlations and take advantage of the higher modelling power

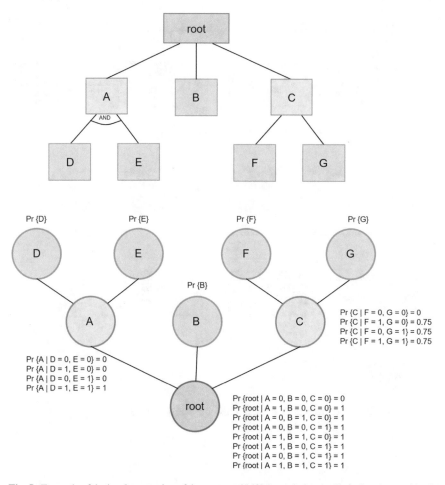

Fig. 5 Example of the implementation of the proposed M2M translation applied, showing an Attack Tree and its corresponding Bayesian Network equivalence

of the BN models. Based on the resulting model, we can start to better define the events/states/conditions need to provide with situation awareness. These factors need to be detected to trigger an early warning during an ongoing attack.

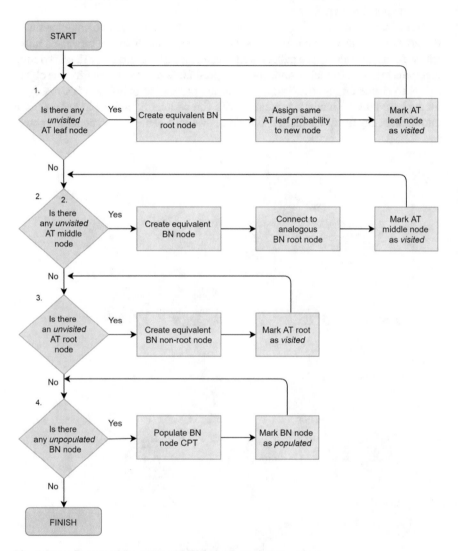

Fig. 6 Flow diagram of the proposed M2M transformation procedure

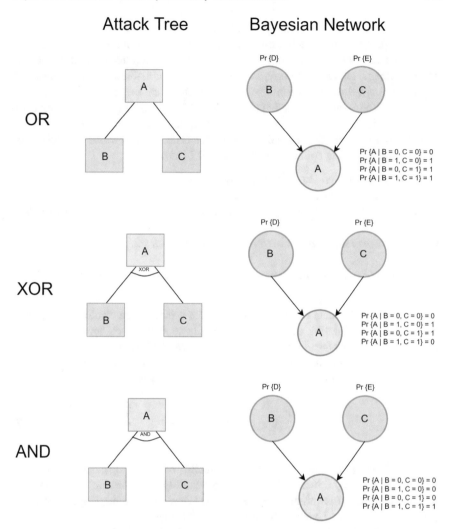

Fig. 7 Example of the implementation of the proposed M2M translation on OR, XOR and AND operation functions. Notice the only visible difference in the resulting BN models is on their CPT tables

3.7 Data Population of the Probability Tables

The next step for the finalization of the model, is to populate the CPTs of the resulting BN.

Initial probability data can be inferred from publicly available datasets or privately generated data. Once the data has been retrieved, it is necessary to translate into feasible redundant probabilistic values. Depending on the scenario, the nature of the system under protection, and many other variables, the results might vary. It is important to remark, that the probabilities will be refined as more precise data is collected. This can be achieved with data collected from real attacks, or other scenarios such as honey pots or data harvested from computer simulations. Using ML techniques, the idea is to progressively tailor the probabilistic inference to each scenario using values that are conjectured from empirical evidence.

As a working example, a study published in 2017 by *Symantec Corporation* titled the *Internet Security Threat Report* (ISTR) was used to estimate the probability rates used for the case study presented in Sect. 4.2. The study revealed the following information from data collected in the year 2017 [15]:

- *Email phishing rate is 1 in 2995 emails.*
- *Email malware rate is 1 in 412 emails.*
- *From more than 1 billion requests analysed every day, 1 in 13 web-requests lead to malware.*
- *76% of websites contain vulnerabilities, out of which 9% are critical vulnerabilities.*
- *Out of 8718 vulnerabilities discovered in 2017, 4262 were zero-day vulnerabilities.*

This and similar data can be customized to a specific organization and updated dynamically by counting the number of emails sent, websites accessed, etc. For instance, if you trust the above statement *"Email phishing rate is 1 in 2995 emails"*, and in your organization you have 1200 emails sent at a certain time, then you can get your custom value for the email phishing probability as:

$$1 - \prod_{1}^{1200}\left(1 - \frac{1}{2995}\right) \tag{2}$$

In this formula $(1 - \frac{1}{2995})$ would be the probability of not having phishing in a single email, whereas the production refers to the probability of not having phishing in any of the 1200 emails (assuming they are not correlated). One minus the production is then the probability of having phishing after 1200 emails.

In other words, it is possible to update in real-time that probability by counting the number of emails received in the organization at any time. The same holds for the other parameters like website access. The SIEM system can be configured to monitor those parameters and provide updates to the BN detection model through DETECT.

3.8 Transformation of Bayesian Networks to Machine-Readable XML Code

In accordance to the proposed architecture that is based on the DETECT framework, it is a requisite to translate the proposed models into a machine-readable format.

Following this, the models can be stored in the *Attack Scenario Repository*, and automatically translated into a BN model which can be appended to the proposed *Detection Engine*.

Our proposal for achieving this is using XML language. Apart from its reliability, malleability and easy parsing; it has been implemented in DETECT [5] and other security frameworks as shown by *A. L. Buczak and E. Guven*'s survey [16].

Transformation procedure

For each resulting BN that is to be attached to the *Attack Scenario Repository* a new *bayes_network* XML element will be created. There are four stages we can identified in the process:

1. On the first stage, for each node on the BN define an XML element *node*, with an attributes *type* indicating if the node is a leaf, middle or root node, an attribute *id* indicating a unique label to identify the node. A description label can be entered as a simple text inside the node element.
2. On the second stage, for each node relation in the BN define an XML element *relation*, with the attributes *parent* and *child*, indicating the nodes involved in the relation using each of the nodes unique *id* attribute, and an attribute *configuration* to explicitly define the configuration of the relation (AND, OR or XOR).
3. On the third stage, for each node that is a leaf node, create and populate the corresponding CPT. Define an XML element *probability* with attribute *node* indicating the leaf node using the unique *id* attribute of the node. Then, for each state on the probability table create an XML element *state* with a *label* attribute indicating an identifier for the state. Finally, enter the probability value as a float between 0.0 and 1.0 inside the XML element.
4. Lastly, for each non-deterministic middle node on the BN, define an XML element *probability*, with the attribute *node* indicating the middle node using its unique *id* attribute. Inside the *probability* element define an XML element *conditional* for each node that conditions the probability, with an attribute *node* indicating the conditional node using its unique *id* attribute. For each state on the probability table create an XML element *state* with a *label* attribute indicating an identifier for the state. Finally, enter the probability value as a float between 0.0 and 1.0 inside the XML element.

Each XML file will be saved in the *Attack Scenario Repository*. An algorithm can be defined to parse each XML file into a working BN that will be implemented in the *Detection Engine* of the proposed model.

The code snippet presented in Appendix 1 reports the example BN showed in Fig. 5 converted from AT to BN and parsed into machine-readable XML code.

4 Case Study: Authentication Violation Scenario

We have derived a model of an attack scenarios in order to test and demonstrate the proposed implementation. The scenario is modelled as an Attack Tree and translated to a Bayesian Network model using the M2M proposed in Sect. 3.6. Since actual data is not essential to validate the methodology, CPTs are populated using realistic assumptions and pseudo-data. The proposed model is then tested for perturbation and relevant results.

4.1 Brief Description of the Scenario and Attack Tree

The following is a scenario for an authentication violation attack. This sort of attack ranked second on OWASP's top 10 vulnerabilities in 2017 [17]. For demonstration purposes the AT presented below displays a simplified version of said scenario (Fig. 8).

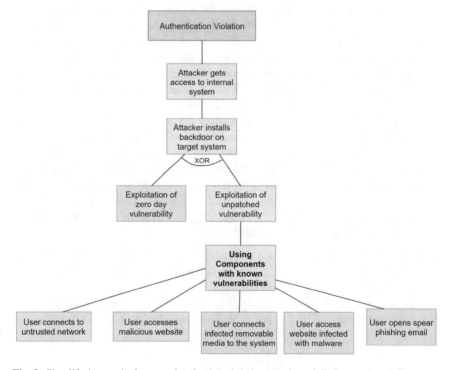

Fig. 8 Simplified scenario for an authentication violation attack modelled as an Attack Tree

4.2 Values for Static Assessment

In order to make a static assessment, it is necessary to assign some probabilistic values to the leave nodes as well as sensor assumptions for the detection of the events. As illustrated in Sect. 3.7, these values can be obtained from reputable sources online or could be tailored to a company or organization with data obtained over a reasonable period of time. For the purpose of this paper, a study published in 2017 by *Symantec Corporation* titled the *Internet Security Threat Report* (ISTR) was used to estimate the probability rates [15]. From this data collected for an entire year—and with some assumptions for the sake of simplicity—we have estimated the daily probabilities for presenting the scenario as shown in Tables 1 and Table 2. Table 1 also displays possible detectors to be used in order to recognize if any part of the scenario has taken place to update the BN parameters accordingly.

Conversion to BN and machine-readable XML code
Following the proposed M2M transformation presented in Sect. 3.6, we convert the given AT into a BN, and populate the CPT with the proposed values. The results are displayed on the BN presented in Fig. 9.

Table 1 Assigned probabilities and sensor assumptions for the proposed scenario

Leaf node	Identifying acronym	Estimated probability	Possible detection sensors
Exploitation of zero-day vulnerability	ZDV	0.03	• Anomaly detection based IDS • User level endpoint monitoring
User connects to untrusted network	UN	0.24	• IDS • SSL certificate missing/rejected • NetFlow analysis • Firewall
User accesses malicious website	MW	0.08	• IDS • SSL certificate missing/rejected • Unexpected flow of data
User connects infected removable media to the system	IM	0.02	• IDS • Antivirus • System event logs
User accesses website infected with malware	IW	0.09	• IDS • Web browser plugin
User opens spear phishing email	SPE	0.03	• Human • User level endpoint monitoring

Table 2 Assigned probabilities for non-deterministic middle nodes

Middle node	Identifying acronym	Estimated probability
Exploitation of unpatched vulnerability	EUV	0.60
Attacker installs backdoor on target system	BD	0.85
Attacker gets access to internal system	AIC	0.90

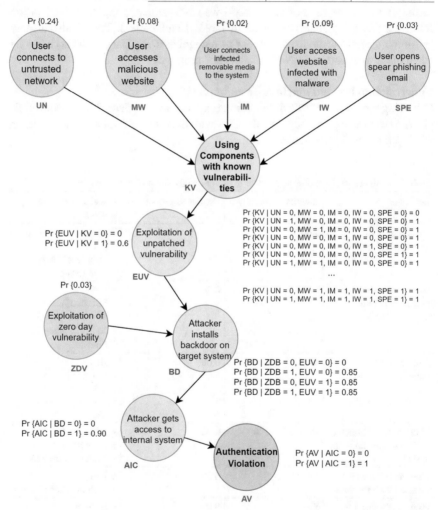

Fig. 9 Resulting BN using the proposed M2M transformation model. Notice that only probabilities for True are represented on each leaf node

The machine-readable XML code derived from this scenario was created as depicted in Sect. 3.8 and can be found in Appendix 2.

BN perturbation test

In order to make an assessment on the sensibility of the BN generated, two different sensitivity tests were performed.

The first perturbance test applies relative variations. To do this we modified all leave nodes one at the time, by increasing/decreasing the estimated probabilities values by 10%, 20%, 25% and up to 50% relative to the originally estimated value (multiplication).

Such that:

$$\text{test value} = \text{estimated value} + (\text{estimated value} * \text{percentage } \%)$$

Relative variation perturbation test results for all leave nodes are presented in Table 3 and Fig. 10, results for middle nodes are presented in Table 4 and Fig. 11.

The second perturbance test applies absolute variations. The approach is similar to the first test, but this time the estimated probability values are increased/decreased by 10%, 20%, 25% and up to 50% (simple addition). Such that:

$$\text{test value} = \text{estimated value} + \text{percentage } \%$$

Table 3 Relative variation perturbation test results for each leaf node on the BN

Node	ZDV		UN		MW	
Perturbance percentage	Value	Results	Value	Results	Value	Results
–50%	1.5	19	12	15.6	4	18.7
–25%	2.25	19.5	18	17.8	6	19.3
–20%	2.4	19.5	19.2	18.2	6.4	19.4
–10%	2.7	19.7	21.6	19	7.2	19.7
0%	3	19.9	24	19.9	8	19.9
10%	3.3	20.1	26.4	20.7	8.8	20.1
20%	3.6	20.2	28.8	21.6	9.6	20.4
25%	3.75	20.3	30	22	10	20.5
50%	4.5	20.8	36	24.1	12	21.1
Node	IM		IW		SPE	
Perturbance percentage	Value	Results	Value	Results	Value	Results
–50%	1	19.6	4.5	18.6	1.5	19.5
–25%	1.5	19.8	6.75	19.2	2.25	19.7
–20%	1.6	19.8	7.2	19.4	2.4	19.7
–10%	1.8	19.8	8.1	19.6	2.7	19.8
0%	2	19.9	9	19.9	3	19.9
10%	2.2	19.9	9.9	20.2	3.3	20
20%	2.4	20	10.8	20.4	3.6	20.1
25%	2.5	20	11.25	20.6	3.75	20.1
50%	3	20.2	13.5	21.2	4.5	20.3

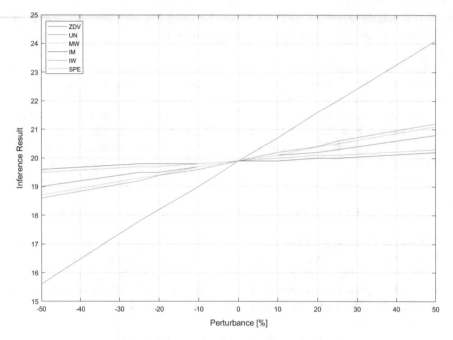

Fig. 10 Relative variation perturbation test results

Table 4 Relative variation perturbation Test results for each non-deterministic middle node on the BN for the proposed scenario

Node	EUV		BD		AIC	
Perturbance percentage	Value	Results	Value	Results	Value	Results
-50%	30	11.1	42.5	9.94	45	9.94
-25%	45	15.5	63.75	14.9	67.5	14.9
-20%	48	16.4	68	15.9	72	15.9
-10%	54	18.1	76.5	17.9	81	17.9
0%	60	19.9	85	19.9	90	19.9
10%	66	21.6	93.5	21.9	99	21.9
20%	72	23.4	100	23.4	100	22.1
25%	75	24.3	100	23.4	100	22.1
50%	90	28.7	100	23.4	100	22.1

Absolute variation perturbation test results for all leave nodes are presented in Table 5 and Fig. 12, results for middle nodes are presented in Table 6 and Fig. 13.

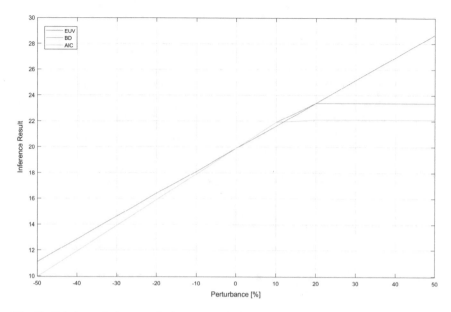

Fig. 11 Relative variation perturbation test results for the proposed scenario

Table 5 Absolute variation perturbation test results for each leaf node on the BN for the proposed scenario

Node	ZDV		UN		MW	
Perturbance percentage	Value	Results	Value	Results	Value	Results
–50%	0	18.1	0	11.4	0	17.5
–25%	0	18.1	0	11.4	0	17.5
–20%	0	18.1	4	12.8	0	17.5
–10%	0	18.1	14	16.3	0	17.5
0%	3	19.9	24	19.9	8	19.9
10%	13	25.7	34	23.4	18	22.8
20%	23	31.6	44	27	28	25.7
25%	28	34.5	49	28.7	33	27.2
50%	53	49.1	74	37.6	58	34.5
Node	IM		IW		SPE	
Perturbance percentage	Value	Results	Value	Results	Value	Results
–50%	0	19.3	0	17.2	0	19.1
–25%	0	19.3	0	17.2	0	19.1
–20%	0	19.3	0	17.2	0	19.1
–10%	0	19.3	0	17.2	0	19.1
0%	2	19.9	9	19.9	3	19.9
10%	12	22.6	19	22.8	13	22.7
20%	22	25.4	29	25.8	23	25.4
25%	27	26.8	34	27.3	28	26.8
50%	52	33.6	59	34.7	53	33.8

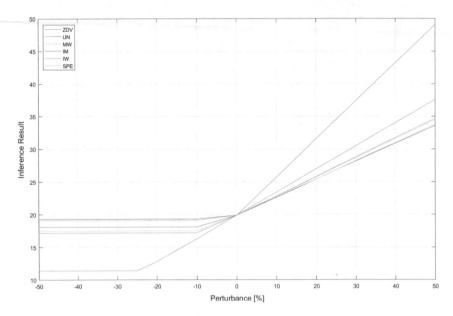

Fig. 12 Absolute variation perturbation test results for the proposed scenario

Table 6 Absolute variation perturbation test results for each leaf node on the BN for the proposed scenario

Node	EUV		BD		AIC	
Perturbance Percentage	Value	Results	Value	Results	Value	Results
–50%	10	5.23	35	8.19	40	8.84
–25%	35	12.6	60	14	65	14.4
–20%	40	14	65	15.2	70	15.5
–10%	50	17	75	17.5	80	17.7
0%	60	19.9	85	19.9	90	19.9
10%	70	22.8	95	22.2	100	22.1
20%	80	25.8	100	23.4	100	22.1
25%	85	27.2	100	23.4	100	22.1
50%	100	31.6	100	23.4	100	22.1

5 Analysis

In order to make an assessment on the sensibility of the BN generated, two different sensitivity tests were performed. The first perturbance test applies relative variations. To do this we modified all leave nodes one at the time, by increasing/decreasing the estimated probabilities values by 10%, 20%, 25% and up to 50% relative to the originally estimated value (multiplication). The second perturbance test applies absolute variations. The approach is similar to the first test, but this time the estimated probability values are increased/decreased by 10%, 20%, 25% and up to 50% (simple

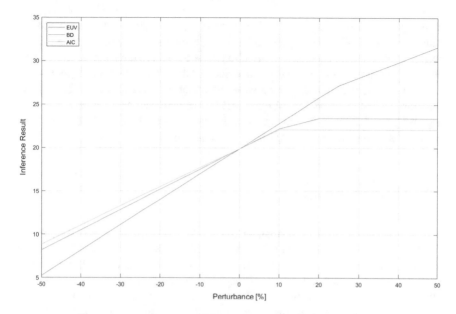

Fig. 13 Absolute variation perturbation test results for the proposed scenario

addition). Results were plotted in a graph where each node was assigned a label to ease identification. In each graph presented, the X axis indicates the perturbation percentage applied, while the Y axis represents the overall BN inference result on the isolated modification of that single probability. Each test result is represented in a different colour on the graph.

Results shown when setting the probability to 100% (or 'True') simulate what happens when the corresponding event is directly recognized by DETECT and hence updated in the BN.

Regarding the criteria for generating the warnings or alarms to the SIEM, they can be based on pre-set thresholds on the root node probability or even a percentage increase with respect to the initial value.

5.1 Relative Variations

For the test on relative variations, the results indicate a perturbance in the range of −25% and +25% in each isolated leaf node, keeping the overall result of the inference within an acceptable margin error threshold in range of positive-negative 10.56% from the outcome on the originally estimated probability of 19.99%. This margin shrinks significantly when the original estimated value reached the 100% value (i.e. 'True'). This indicates that higher estimated probabilities are more prone to perturbation errors, and this can be noticed on the pronounced slope of variables with high probability estimations. For instance, all middle nodes have estimated values of at least 60% and higher, and as a result they are more prone to sensibility perturbance. That means a greater attention must be paid in the estimation and fine-tuning of those values compared to leave nodes. Obviously, there is an upper bound limit on the perturbation range when an estimated value reaches 100%, and this explains the flat lines on the BD and AIC node perturbation results; conversely, negative perturbation values have a lower bound value of 0% (i.e. 'False'). For this proposed scenario, the average probability value is of 32%.

5.2 Absolute Variations

On the absolute variation test results, the impact of perturbation is higher. The second test results on absolute variation perturbation, indicate that in the range of −25% and +25% perturbances are in a range between −8.5% and +73.36% of the originally estimated value for all nodes. Nonetheless, in the range of −10% and +10%, the perturbance range is between −18.09% and +29.14% of the originally estimated value. Since perturbations are higher, upper and lower bounds are reached more often than in the first test. On the range −25% and +25% the average maximum obtain value is 27.11% (+36.23% more from the originally estimated value of 19.9%) and the average minimum obtained value is 15.95% (−19.84% less from the originally estimated value of 19.9%). Nonetheless, on the range −10% and +10%, the average maximum obtain value is 23.01% (15.62% more from the originally estimated value of 19.9%) and the average minimum obtained value is 17.74% (−10.85% of the originally estimated value of 19.9%).

5.3 Overall Analysis

The perturbation tests implementing relative variations demonstrate a low sensibility on the modelled BN. With a relative variation of positive-negative 25% inside an acceptable threshold of ~10% of the originally estimated value. The impact of relative variations depends on the estimated probability of each node. The results in the perturbation test indicate that higher estimated probabilities are more prone to perturbation errors. This can be noticed on the pronounced slope of variables with high probability estimations, indicating that some parameters are more sensitive than others. As a result, greater attention must be paid in the estimation and fine-tuning of nodes with higher estimated probabilities in contract to those of lower probabilities.

Absolute variations are of course, not as dependant on the estimated values as in relative variations but are a good parameter to measure the impact of possible misrepresentation of probabilistic values. In our scenario, perturbation test implementing absolute variations demonstrated a higher sensibility in comparison to relative variations. Nonetheless, on a range of positive-negative 10% perturbance, the average minimum and maximum values obtained are in average inside an acceptable threshold of ~10% of the originally estimated value, revealing an acceptable perturbation sensibility result.

6 Discussion

The future of cybersecurity strongly depends on the application of Artificial Intelligence methods to improve cyber situational awareness and to automate threat detections and countermeasures. Statistical and probability-based models, such as Bayesian Networks, can be very powerful tools. Based on conjectures derived from Bayes models, it is possible to automatize the best course of action for online defence. By retrieving knowledge from previous studies, we can infer the course of action taken by attackers and systematize them into a framework.

The results from the attack scenario we have modelled suggested a low sensitivity to parameter perturbation, with a lower error margin for absolute variations than for relative variations, confirming the feasibility of implementing BN techniques as part of a security framework.

Even thought that for practical purposes we have implemented hypothetical probabilistic values, for future implementations the data feed to the algorithm can be inferred from available datasets or studies. Data can also be harvested using collection techniques such as honey pots and attack simulations. In order to improve the performance of the proposed model, the probabilities could later be refined using Machine Learning techniques.

The implementation of Attack Trees and Bayesian Networks on the proposed model, would allow us to make stochastic inferences in non-deterministic scenarios. The model presented provides with framework integration, a modular architecture scheme, and a formal language for threat modelling. This can be combined with a complete framework such as DETECT, that includes a database and a detection engine. Moreover, BN simplify probabilistic analysis by automating most computations including forward and backward inference. BN also allow modellers to manage the inherent uncertainty in threat scenarios by using aspects of fuzzy logic and stochastic inference. This features and characteristics are desired for a rapid adaptation on an everchanging and expeditious cybersecurity landscape.

7 Conclusion

Following the architecture of the DETECT framework as a mean of inspiration, we have designed a blueprint for the application of Bayesian Networks for online threat detection that implements hybrid-detection analytics and complies with the recommendations from the existent literature.

For a static assessment of our case study attack scenario, we have implemented estimated probabilities for all leave nodes in the BN, by consulting reputable sources and available statistics. The results of the sensitivity analysis revealed an overall low perturbance sensitivity, with insignificant variations. When applying variations relative to the inferred probabilities, perturbations are inside a 10% threshold of acceptance. Therefore, a margin of error of 25% relative to the inferred value is acceptable and will not impact negatively on the overall probability outcome. Special attention must be paid when nodes have higher probabilities as miscalculations will have a bigger impact on the results. Moreover, when the variations applied are absolute, the margin of error lowers to 10% to get similar estimations.

The sensibility analysis of the presented model can help a security analyst to recognized the most influential factors on a global threat, in order to adjust the protection mechanism accordingly. This could save time and money, as well as help reallocate other resources in a more efficient manner.

The proposed model has been proved to be feasible, and it could be further developed into a fully functional implementation, that could be integrated into DETECT. The design presented can be used for off-line risk assessment and on-line risk evaluation/threat recognition. The reliability of the proposed model relies on the power of Bayesian Networks for stochastic assessments in indeterministic scenarios, and it has been proven to be a potentially powerful tool, that will contribute to the improvement of cybersecurity and situational awareness.

Appendix 1

Appendix 1

The following code snippet is the proposed transformation of BN into machine-readable XML code as presented in **Chapter 3.8**. The example code is derived from BN shown in **Fig. 5**.

```xml
<?xml version="1.0" encoding="UTF-8"?>

<bayes_network>
  <!-- Define Nodes-->
  <node type="leaf" id="B">B Description</node>
  <node type="leaf" id="D">D Description</node>
  <node type="leaf" id="E">E Description</node>
  <node type="leaf" id="F">F Description</node>
  <node type="leaf" id="G">G Description</node>
  <node type="middle" id="A">A Description</node>
  <node type="middle" id="C">C Description</node>
  <node type="root" id="root">Root Description</node>

  <!-- Assign Node Relations -->
  <relation parent="D" child="A" configuration="AND"></relation>
  <relation parent="E" child="A" configuration="AND"></relation>
  <relation parent="F" child="C" configuration="OR"></relation>
  <relation parent="G" child="C" configuration="OR"></relation>
  <relation parent="A" child="root" configuration="OR"></relation>
  <relation parent="B" child="root" configuration="OR"></relation>
  <relation parent="C" child="root" configuration="OR"></relation>

  <!-- Populate Leave Nodes CPTs -->
  <probability node = "D">
    <state label='True'>0.5</state>
    <state label='False'>0.5</state>
  </probability>

  <probability node = "E">
    <state label='True'>0.1</state>
    <state label='False'>0.9</state>
  </probability>

  <probability node = "F">
    <state label='True'>0.6</state>
    <state label='False'>0.4</state>
  </probability>

  <probability node = "G">
    <state label='True'>0.8</state>
    <state label='False'>0.2</state>
  </probability>

  <!-- Populate Middle Nodes CPTs -->
  <probability node="C">

    <conditional node= "F" state="True">
      <state label='True'>0.75</state>
      <state label='False'>0.25</state>
    </conditional>
```

```
<conditional node= "F" state="False">
  <state label='True'>1.0</state>
  <state label='False'>0.0</state>
</conditional>

<conditional node= "G" state="True">
  <state label='True'>0.75</state>
  <state label='False'>0.25</state>
</conditional>

<conditional node= "G" state="False">
  <state label='True'>1.0</state>
  <state label='False'>0.0</state>
</conditional>

  </probability>
</bayes_network>
```

Appendix 2

Appendix 2

The following code snippet is the case study BN presented in **Section 4** parsed into machine-readable XML code. The code is derived from BN shown in **Fig. 9**.

```xml
<?xml version="1.0" encoding="UTF-8"?>


<bayes_network>
  <!-- Attack Scenario: Attacker gets access to internal system -->

  <!-- Define Nodes-->
  <node type="leaf" id="un">User Connects to Untrusted Network</node>
  <node type="leaf" id="mw">User Access Malicious Website</node>
  <node type="leaf" id="im">User Connects Infected Removable Media to the System</node>
  <node type="leaf" id="iw">User Access Website Infected with Malware</node>
  <node type="leaf" id="spe">User Opens Spear Pishing Email</node>
  <node type="leaf" id="zdv">Exploitation of Zero-Day Vulnerability</node>

  <node type="middle" id="kv">Using Components with Known Vulnerabilities</node>
  <node type="middle" id="euv">Exploitation of Unpatched Vulnerability</node>
  <node type="middle" id="bd">Attacker Installs Backdoor on Target System</node>
  <node type="middle" id="aic">Attacker Gets Access to Internal System</node>

  <node type="root" id="av">Authentication Violation</node>

  <!-- Assign Node Relations -->
  <relation parent="un" child="kv" configuration="OR"></relation>
  <relation parent="mw" child="kv" configuration="OR"></relation>
  <relation parent="im" child="kv" configuration="OR"></relation>
  <relation parent="iw" child="kv" configuration="OR"></relation>
  <relation parent="spe" child="kv" configuration="OR"></relation>
  <relation parent="kv" child="euv" configuration="OR"></relation>
  <relation parent="euv" child="bd" configuration="OR"></relation>
  <relation parent="zdv" child="bd" configuration="OR"></relation>
  <relation parent="bd" child="aic" configuration="OR"></relation>
  <relation parent="aic" child="av" configuration="OR"></relation>

  <!-- Populate Leave Nodes CPTs -->
  <probability node = "un">
    <state label='True'>0.24</state>
    <state label='False'>0.76</state>
  </probability>
```

```xml
<probability node = "mw">
  <state label='True'>0.08</state>
  <state label='False'>0.92</state>
</probability>

<probability node = "im">
  <state label='True'>0.02</state>
  <state label='False'>0.98</state>
</probability>

<probability node = "iw">
  <state label='True'>0.09</state>
  <state label='False'>0.91</state>
</probability>

<probability node = "spe">
  <state label='True'>0.03</state>
  <state label='False'>0.97</state>
</probability>

<probability node="zdv">
  <state label='True'>0.03</state>
  <state label='False'>0.97</state>
</probability>

<!-- Populate Middle Nodes CPTs -->
<probability node="euv">

  <conditional node= "kv" state="True">
    <state label='True'>0.6</state>
    <state label='False'>0.4</state>
  </conditional>

  <conditional node="kv" state="False">
    <state label='True'>0.0</state>
    <state label='False'>1.0</state>
  </conditional >



</probability>

<probability node="bd">

  <conditional node="euv" state="True">
    <state label='True'>0.85</state>
    <state label='False'>0.15</state>
  </conditional>

  <conditional node="euv" state="False">
    <state label='True'>0.0</state>
    <state label='False'>1.0</state>
  </conditional>

</probability>

<probability node="aic">

  <conditional node="bd" state="True">
    <state label='True'>0.9</state>
    <state label='False'>0.1</state>
  </conditional>

  <conditional node="bd" state="False">
    <state label='True'>0.0</state>
    <state label='False'>1.0</state>
  </conditional>

</probability>

</bayes_network>
```

References

1. IEEE, Syntegrity (2017) Artificial intelligence and machine learning applied to cybersecurity, presented in Washington DC, USA, 6th–8th October 2017, [Online]. Available at https://www. ieee.org/content/dam/ieeeorg/ieee/web/org/about/industry/ieee_confluence_report.pdf?utm_ source=lp-linktext&utm_medium=industry&utm_campaign=confluence-paper. Accessed 20 Mar 2018
2. Pappaterra MJ, Flammini F (2019) A review of intelligent cybersecurity with Bayesian Networks. In: 2019 IEEE international conference on systems, man and cybernetics (SMC), Bari, Italy, pp 445–452
3. Shackleford D (2016) SANS 2016 Security Analytics Survey, SANS Institute. [Online]. Available at https://www.sans.org/reading-room/whitepapers/analyst/2016-securityanalytics-survey-37467. Accessed 3 Mar 2018
4. Flammini F, Gaglione A, Otello F, Pappalardo A, Pragliola C, Tedesco A (2010) Towards wireless sensor networks for railway infrastructure monitoring. Ansaldo STS Italy, Università di Napoli Federico II
5. Flammini F, Gaglione A, Mazzocca N, Pragliola C (2008) DETECT: a novel framework for the detection of attacks to critical infrastructures. In: Proceedings of ESREL'08, safety, reliability and risk analysis: theory, methods and applications. CRC Press, Taylor & Francis Group, London, pp 105–112
6. Gaglione A (2009, November) Threat analysis and detection in critical infrastructure security, Università di Napoli Federico II, Comunità Europea Fondo Sociale Europeo
7. Flammini F, Gaglione A, Mazzocca N, Moscato V, Pragliola C (2009) Online Integration and reasoning for multi-sensor data to enhance infrastructure surveillance. J Inf Assur Secur 4:183–191
8. Flammini F, Gaglione A, Mazzocca N, Moscato V, Pragliola C (2009) Wireless sensor data fusion for critical infrastructure security. In: CISIS, Springer, Berlin Germany, pp 92–99
9. Flammini F, Mazzocca N, Pappalardo A, Vittorini V, Pagliola C (2015) Improving the dependability of distributed surveillance systems using diverse redundant detectors. Dependability problems of complex information systems, Springer International Publishing. https://www.researchgate.net/publication/282269486_Improving_the_Dependability_of_Distributed_Surveillance_Systems_Using_Diverse_Redundant_Detectors
10. Schneier B (1999) Attack trees. Dobb's J 21–22, 24, 26, 28–29. [Online]. Available at https:// www.schneier.com/academic/archives/1999/12/attack_trees.html. Accessed 20 Mar 2018
11. Bobbio A, Portinale L, Minichino M, Ciancamerla E (2001) Improving the analysis of dependable systems by mapping fault trees into Bayesian Networks. In: Reliability engineering and system safety, vol 71, Rome, Italy, pp 249–260
12. Gribaudo M, Iacono M, Marrone S (2015) Exploiting Bayesian Networks for the analysis of combined attack trees. In: Electronic notes in theoretical computer science, vol 310. Elsevier B.V., pp 91–11
13. Mauw S, Oostdijk M (2005) Foundations of attack trees. In: International conference on information security and cryptology ICISC 2005. LNCS 3935. Springer, pp 186–198
14. Charniak E (1991) Bayesian networks without tears: making Bayesian networks more accessible to the probabilistically unsophisticated. AI Mag 12(4):50–63
15. Symantec Corporation (2017) The Internet Security Threat Report (ISTR) 2017. [Online]. Available at https://www.symantec.com/content/dam/symantec/docs/reports/istr-22-2017-en. pdf. Accessed 13 Mar 2018
16. Buczak A, Guven E (2016) A survey of data mining and machine learning methods for cybersecurity intrusion detection. IEEE Commun Surv Tutorials 18(2)
17. OWASP (2017) Top 10—2017. [Online]. Available at https://www.owasp.org/index.php/Top_10_2017-Top_10. Accessed 13 Mar 2018

Spam Emails Detection Based on Distributed Word Embedding with Deep Learning

Sriram Srinivasan, Vinayakumar Ravi, Mamoun Alazab, Simran Ketha, Ala' M. Al-Zoubi, and Soman Kotti Padannayil

Abstract In recent years, a rapid shift from general and random attacks to more sophisticated and advanced ones can be noticed. Unsolicited email or spam is one of the sources of many types of cybercrime techniques that use complicated methods to trick specific victims. Spam detection is one of the leading machine learning-oriented applications in the last decade. In this work, we present a new methodology for detecting spam emails based on deep learning architectures in the context of natural language processing (NLP). Past works on classical machine learning based spam email detection has relied on various feature engineering methods. Identifying a proper feature engineering method is a difficult task and moreover vulnerable in an adversarial environment. Our proposed method leverage the text representation of NLP and map towards spam email detection task. Various email representation methods are utilized to transform emails into email word vectors, as an essential step for machine learning algorithms. Moreover, optimal parameters are identified for many deep learning architectures and email representation by following the hyper-parameter tuning approach. The performance of many classical machine learning

S. Srinivasan (✉) · V. Ravi · S. Ketha · S. Kotti Padannayil
Amrita School of Engineering, Center for Computational Engineering and Networking (CEN), Coimbatore, Amrita Vishwa Vidyapeetham, Coimbatore, India
e-mail: sri27395ram@gmail.com

V. Ravi
e-mail: Vinayakumar.Ravi@cchmc.org; vinayakumarr77@gmail.com

S. Ketha
e-mail: simiketha19@gmail.com

V. Ravi
Division of Biomedical Informatics, Cincinnati Children's Hospital Medical Center, Cincinnati, OH, USA

M. Alazab
Charles Darwin University, Darwin, Australia
e-mail: mamoun.alazab@cdu.edu.au

A. M. Al-Zoubi
King Abdullah II School for Information Technology, The University of Jordan, Amman, Jordan
e-mail: alaah14@gmail.com; a.alzoubi@jisdf.org

© The Editor(s) (if applicable) and The Author(s), under exclusive license to Springer Nature Switzerland AG 2021
Y. Maleh et al. (eds.), *Machine Intelligence and Big Data Analytics for Cybersecurity Applications*, Studies in Computational Intelligence 919,
https://doi.org/10.1007/978-3-030-57024-8_7

classifiers and deep learning architectures with various text representations are evaluated based on publicly available three email corpora. The experimental results show that the deep learning architectures performed better when compared to the standard machine learning classifiers in terms of accuracy, precision, recall, and F1-score. This is essentially due to the fact that the deep learning architectures facilitate to learn hierarchical, abstract and sequential feature representations of emails. Furthermore, word embedding with deep learning has performed well in comparison to the other classical email representation methods. The word embedding simplify to learn the syntactic, semantic and contextual similarity of emails. This endows word embedding with deep learning methods in spam email filtering in the real environment.

Keywords Cybersecurity · Cybercrime · Spam · Intrusion detection · Digital forensic techniques · Content based filters · Machine learning · Deep learning · Natural language processing · Text representation

1 Introduction

Electronic mail (email) has become a preferred medium for communication due to the vast advantages of its inexpensive infrastructure, high efficiency, and quick method of exchanging business and personal information over the internet. Over the past years, email has turned out to be a standout among the most commonly utilized methods for communication. However, at the same time, emails are also widely used for distributing malware, viruses, and phishing links, which are known generally as spam emails. These emails not only affects the performance of the platform but also would bring financial loss to the users if they disclose their private information in response to phishing emails [1–4]. Even though email detection exists for many years, detecting these emails is still a very challenging problem because of the new attack mechanisms emerging every day [5–11].

Spam emails are typically called unsolicited commercial email (UCE) or unsolicited bulk email (UBE) [12]. It is considered as one of the real issues of the present internet and it can be defined as an act of sending useless information or mass information in large amounts to many email accounts. In short, it is a subset of electronic spam including almost identical messages sent to several beneficiaries by email. In cyberspace, spam emails are an effective medium that is widely utilized by cybercriminals to spread malware, viruses, and to steal money from an individual or organization [4]. Overall, spam emails are a major threat to internet safety and personal privacy [1–3]. Therefore, it is necessary to develop security solutions to deal with it.

Automated identification of spam emails has become a challenging task because of the new sophisticated attack mechanisms that emerge now and then. Over the years, several anti-spam solutions have been proposed to block unsolicited email messages over the internet. These systems are based on acts to control email communication [13, 14], email protocols [15, 16], refining the control policies of email protocols, address protection [17], list-based systems [18–20], keyword filtering [21], challenge-response (CR) systems [22], collaborative spam filtering [23], and Honeypots [24]. The self-learning system leverages the machine learning algorithms typically the supervised and unsupervised to learn the behaviors of spam and legitimate emails to distinguish these emails automatically without the human intervention [25–29]. Along with this, the machine learning-based systems can adapt quickly to the highly dynamic nature of spam emails [30].

The process of detecting spammers through classical and standard machine learning algorithms consists of three steps, which are feature construction, feature selection, and classifier modeling. The feature selection step is the task of identifying the most relevant features from a large set of available features. This helps to remove the noisy features and enhance the performance of the classifier by decreasing the computational complexity. Most prominent feature selection techniques that have been proposed in the literature are information gain (IG) [31], document frequency (DF) [32], term frequency variance (TFV) [34], chi-square [32], odds ratio (OR) [33], and term strength (TS) [32]. Feature construction is the next step which converts the features into feature vectors by distinguishing the relationship between the remaining features.

The most well-known feature construction method in spam detection is Bag-of-Words (BoW) [35]. The last step is to build a classification model using some standard machine learning algorithms such as Naive Bayes (NB), Adaptive Boosting (AB), Random forest (RF), Maximum Entropy Modeling (MaxEnt), Decision tree (DT), and Support vector machine (SVM), which are regularly in the various studies [30, 36].

Deep learning (DL) is an immature sub-field of the machine learning domain. It can learn the optimal feature by representing themselves in a long row of input samples. The DL has two fundamental characteristics which distinguish it from other approaches. First, it can learn the hierarchical and complex feature representation efficiently. Secondly, it can memorize the previous information in a large sequence of inputs. In recent days, deep learning has been used as a tool to improve the accuracy rate of various areas like image processing, natural language processing, speech processing, etc. [37].

Recurrent neural networks (RNNs) and convolutional neural network (CNN) are two sub-types of deep learning architectures [37]. CNN is a well-known method that is used often in image recognition, and it exceeds human-level performance easily on other image processing tasks. As for the RNN, it is used in the sequential data modeling problems in which the inputs are passed in variable length. Along with the RNN, there are several recurrent structures usually used in different applications, such as Long Short-Term Memory (LSTM), Identity Recurrent Neural Network (IRNN), Clockwork Recurrent Neural Network (CWRNN), Gated Recurrent Neural Network (GRU) and Bidirectional Recurrent Neural Network (BRNN).

Followed these applications of DL architectures, spam email detection considers one of the most critical applications that desired to be solved. Lennan et al. [38] proposed a CNN with word embedding based spam email detection method and compared its performance with SVM and CNN at the character level. The proposed method obtained better performance when compared to character-based methods with CNN and SVM. Following, Eugene and Caswell [39] used the application of LSTM and CNN for the email prioritization task and compared it with feature engineering based random forest classifier and SVM. Both LSTM and CNN performed better when compared to RF and SVM. The application of CNN and RNN combinations used to disentangle email threads originating from forward and reply behavior [40]. In literature, most of the studies, which are based on classical machine learning, have given more importance towards enhancing the accuracy of spam detection. Even though deep learning is often used in other domains, in the field of Cybersecurity, particularly for spam detection, it is in the early stage. Moreover, the existing works have not given any importance of leveraging the text representation of NLP towards spam email detection. Moreover, the literature shows that spammers regularly change their techniques to break spam filters. Primarily, the behaviors of spam email keep on changing and to accurately identify, new features have to be identified by following proper feature engineering methods. This gives a strong motive to implement a systematic and accurate spam detection system that can learn the optimal features automatically from the email samples. The detailed comparative analysis of classical machine learning and deep learning with various email representation methods is yet to be done.

The main contributions of this work illustrated in the following points:

- This paper presents a cyber threat situational awareness framework named as DeepSpamNet, a scalable and robust content-based spam detection framework [41]. With additional computing support, the proposed framework can be scaled out to process large volumes of email data. The proposed framework stands out when compared with any system of a similar kind due to its scalability and its ability to detect malicious acts from early warning signals in real-time.
- A prototype implementation of DeepSpamNet has been developed to evaluate the performance of systems over large publicly available corpora consisting of a mix of spam, phishing, and legitimate email samples.
- A unique deep learning architecture is proposed to detect spam emails. Due to the absence of feature engineering steps, deep learning permits to rapidly reworks to the diversified nature of spammers. Deep learning architectures are complex and act as a black box. Therefore, it is not easy for an adversary to reverse engineer them without the same set of training samples.
- Finally, the comparison between our proposed method and various standard machine learning classifier with various text representations such as linear and non-linear is done in detail.

The rest of the paper is organized as follows: Sect. 2 discusses the related works for spam detection. Section 3 provides the background details of text representation and classical machine learning and deep learning architectures. The working flow and proposed architecture for the spam detection process are explained in Sect. 4. Section 5 contains results and observations. Conclusion and Future works are available in Sect. 6.

2 Related Work

Spam was first recognized in 1978 and has been described in various ways over the years. Recently, the spam identified in the literature as the process of sending different kinds of entities to various parties and has the characteristics of unwanted, repetitive, and unavoidable messages. In general, spam can exist in several places such as social network [42–45], web [46, 47], and mobile messaging [48]; but, it is widely recognized in emails [49, 50]. Spam emails are considered as one of the most common problems that threaten the privacy and security of users, where over 3.8 billion users utilize the emailing services on the internet; therefore, it is imperative to detect these unwanted emails.

There are many techniques to detect spam emails, from using Blacklist [51] or Real-Time Blackhole List [52] to Content-Based Filters [53]; however, recently the most recognized method in the literature is based on machine learning [25, 54, 55]. The work in [25] for example, used a novel set of features to recognize the regularities in emails in which have malicious content. Another recent work [54], introduced a detection system based on Random Weight Network and Genetic Algorithm to detect spam emails.

Nevertheless, with evolving skills and techniques from spammers to avoid detection, it is hard to keep identifying these spam emails [56]. Currently, malicious spam looks like a genuine email more than ever and detecting these emails with standard machine learning methods have become less effective [57, 58]. One way to discriminate these emails and capture the tiny different details that the spam emails have is by using deep learning. Deep learning is a specific approach used for developing and training neural networks, that are capable of learning important features from data. Several works have been proposed in the literature that shows that deep learning performs better when compared with the standard machine learning procedures.

For instance, Tzortzis and Likas [56] proposed a novel system for spam detection based on Deep Belief Network (DBN). DBN is a type of feedforward neural network (FFN) that has many hidden layers, unlike the standard conventional FFN which has one or two layers. The work evaluated the performance of the proposed approach on three public datasets, with the help of the greedy layer-wise in the unsupervised algorithm to solve the low generalization problem. The proposed approach compared against widely applied classifiers such as Support Vector Machines (SVMs) which is one of the best methods for detecting spam emails. The results show that the DBNs outperform the SVMs with the three datasets.

Sometimes, the spammer can shape their threatening emails into other kinds to trick regular users. The work in [59], tackle the phishing emails problem, which is a type of spam email that aims to gather sensitive information from the user via the internet. The authors, presented a classification approach based on deep learning to detect these kinds of emails. They have also utilized the word2vec technique to represent emails rather than applying the classical rule-based and keyword methods, and then they generated a learning model by using the neural network from the created vector representations. Their deep learning classification approach achieved over 96% in the accuracy rate, which is better than the standard machine learning algorithms results.

The emergence of email problems in the last decade such as fraud, spam, and suspicious patterns in emails affect numerous users all over the internet. To solve such problems, Repke and Krestel [40] introduced a recurrent neural networks (RNNs) approach to untangle emails into threads. They classify each email into two or five zones. The zones not only consist of header and body but also contain signatures and greetings information. Their deep learning approach achieves a better result than the popular existing methods such as hand-crafted rules and traditional machine learning.

In [38], the authors argue that spammers in recent years have developed their techniques to trick and overcomes the standard machine learning based detection engines. Hence, they present an end-to-end spam classification approach based on deep learning using the Convolutional Neural Networks (CNN). CNN can overcome the missing phase of feature engineering that allows spammers to adapt to overcome the traditional methods. The CNN outperformed the baseline linear Support Vector Machines in the accuracy measure.

Tyagi utilized the Stacked Denoising Autoencoder (SDAE) which is a primary type of deep learning to detect spam emails [60]. The proposed approach is compared with other deep learning techniques for spam filtering including, Dense Multi-Layer Perceptron (DenseMLP) and Deep Belief Network (DBN). Secondly, they compare their method with state-of-the-art SVM classifier. The experimental results proved that the SDAE has the upper hand against all other methods.

Another work that adapted deep learning to detect spam emails is [57]. In this paper, they used one type of deep network named Stacked Auto-encoder. The performance of the proposed approach was measured using five benchmark datasets, which are PU1, PU2, PU3, PUA, and Enron-Spam. The performance of the proposed method is compared with various classifiers named Support Vector Machine, Decision Tree, traditional Artificial Neural Network, Boosting, and Naive Bayes. The Stacked Auto-encoder achieved better results than all the other classification models in terms of both F1 measure and accuracy measure.

Barushka and Hajek [61] proposed a novel approach to deal with different types of spam. Their approach consists of N-gram TF-IDF method, deep multi-layer perceptron and balance distribution-based algorithm. The approach evaluated four types of spam datasets, which are SpamAssassin, Enron, Social networking and SMS spam. Further, the use of additional layers can capture the complex features that are hard to distinguish, and by applying the distribution-based algorithm, the proposed approach can overcome the imbalanced datasets. They have compared their approach with several traditional classification models namely NB, SVM, Random Forest, C4.5, Convolutional Neural Network, and Voting. The presented approach shows better results than other state-of-the-art spam filters.

In [62], the authors argue of how phishing emails bother internet users with wastage of time, storage, resource, and money. They introduced a word embedding criteria to represent the text in the supervised classification system to detect phishing emails. The utilized traditional machine learning and ruled based models failed to identify the increasing new forms of threats. Consequently, a deep learning model has been applied to surpass this dilemma. They aim to use the MLP, RNN, and CNN network with the Word2vec method to detect these phishing emails. The Word2vec can capture the semantic and synaptic similarity of legitimate and phishing emails.

This work applies a deep learning approach to spam detection. To evaluate the efficacy of deep learning techniques to spam detection, a comparative study of deep learning models over prevalent classical machine learning algorithms is done. To transform email into numeric vectors, many email representation methods are used. Deep learning models have used sequential email representation methods whereas classical machine learning classifiers have used non-sequential representation methods. Additionally, this module presents a new in-house model called DeepSpamNet which is an amalgamation of CNN and LSTM. DeepSpamNet can be utilized to identify spam emails in daily email flow.

This work differs from the previous works in the following points:

- The proposed work used a deep learning approach for spam email detection. The optimal features are extracted implicitly from raw spam email samples, therefore, the performance of the model can be improved by continuously training with new types of spam emails. This method can stay safe in an adversarial environment since the features are not manually engineered.
- Various NLP text representation methods were mapped to spam email representation to learn the linguistic, structural, and syntactic features automatically.
- The framework is highly scalable on the high commodity hardware server which helps to handle a very large volume of spam emails.
- To find an optimal deep learning architecture and optimal spam email representation, various experiments were carried out with different types of datasets. The performances on these datasets help to identify a more generalizable method.

3 Preliminaries

3.1 Classical Machine Learning Models

Classical machine learning models aim to learn a separating line in an n dimensional space which can be best utilized to differentiate the classes. This algorithm works on features that are extracted using various well-known feature engineering methods. The feature engineering contains feature extraction and feature selection steps. Many classical machine learning algorithms exist and most commonly used are Logistic Regression, Naive Bayes, K nearest neighbor, Decision Tree, AdaBoost, Random Forest and support vector machine (SVM).

3.2   Text Representation

Text representation is an essential topic in the field of natural language processing (NLP). There are various text representation exists and each has its pros and cons. As machine learning algorithms are not capable of dealing with the raw text directly, the text has to be transformed into numbers. Specifically, vectors of numbers. Most commonly used text representation are:

1. Bag-of-Words and Vector space model: Bag-of-words is a collection of words. Every unique word passed as an input will have a position in this bag (vector). Term document matrix (TDM) and term frequence-inverse document frequency (TF-IDF) are vector space model which make use of BoW. TDM records the frequency of the words in the document. The term-document matrix will have each corpus word as rows and documents as columns where the matrix will have the frequencies of the words occurring in that particular document. The most used words are highlighted because of more frequence. Term frequency-inverse document frequency measures how often the word occurs in that document—term frequency. So it tells how often a word occurs in a particular document compared to the entire corpus. The rare words are highlighted to show their relative importance. The vectors of the above can be used to understand the similarities between them. Each term can be a dimension and the documents can be a vector in the vector space model.
2. Vector space model of semantics: The matrices of TDM and TF-IDF spans a high dimension, sparse, and noisy feature in further processing and it would lead to incorrect classification. Thus, SVD and NMF are used for dimensionality reduction purposes. Latent semantic analysis (LSA) using TDM or TF-IDF reduces the number of rows of the matrix but keeps the similarity structure. One of its disadvantages is that it can give negative values. NMF helps to generate only non-negative matrices using TDM and TF-IDF matrices.
3. Embedding: Embedding is basically converting words into a vector in such a way that sequence and word similarities are also preserved. This embedding can be either character or word level. Google Word2vec, continuous bag-of-words (CBOW) or Skip-Gram, FastText, Keras word embedding, neural bag-of-words are few types of embedding models belongs to sequence and semantic category. Word2vec uses a neural network in which all the words are given in one-hot representation. The words are split into pairs in which neighbors of the word are the target. And each word (one-hot vector) is passed through the neural network so that we get the target word. The weights of the hidden layer after backpropagation are the vector representation of the word. CBOW and Skip-Gram are two types of word embedding. CBOW predicts the given word based on the context. That is sum up the vectors of the surrounding words. Given the current word, Skip-Gram can predict the surrounding words.
Keras embedding converts the dense vector into a continuous vector representation. This takes three parameters such as embedding dimensions, vocabulary size, and a maximum length of the vector. Initially, the weights of embedding are

initialized randomly and updated during backpropagation.

FastText works on n-grams of character level where n could range from 1 to the length of the word and is better for morphologically rich languages. It uses the Skip-Gram model and a subword model. The subword model will see the internal structure of the words. It learns an efficient vector representation for rare words and it can learn the vector representation for words that are not present in the dictionary when compared to word embedding and Keras embedding representation. This works well when compared to other embeddings on small datasets. Neural bag-of-words takes an average of the input word vectors and performs classification using logistic regression. Generally, NBOW is a fully connected feed forward network with Classical BoW input. The basic difference between FastText, NBOW, and BoW is that FastText specifically learns word vectors targeted for the classification task and it does not explicitly model the words that are important for the given task.

3.3 Deep Learning

1. Deep neural network (DNN): It is a more advanced version of classical feed-forward networks (FFNs). It generally has an input layer, more than one hidden layer, and an output layer. The hidden layer is also called as fully connected or dense layer because each neuron in the ith layer is connected to all the neurons in the $i + 1$ layer. This layer uses a *ReLU* non-linear activation method to prevent issues such as vanishing and exploding gradient.
2. Recurrent structures (RS): Recurrent neural network (RNN) and long short-term memory (LSTM) are two primary types of recurrent structures. RNN is a variant to a classical neural network model in which the neurons in the hidden layer contains a self-recurrent connection. This aids to preserve the previous time-step information across time-steps. Generally, RNN performs well in learning sequences and has obtained significant performances in various well-known long-standing artificial intelligence tasks. When the number of time-step increases, RNN might end up in a vanishing and exploding gradient issue. Further to handle the vanishing and exploding gradient issue, LSTM is introduced. Unlike RNN, LSTM contains a memory block instead of a simple neuron. This memory block contains a memory cell and several gating methods to control the information across time-steps. LSTM outperformed RNN in several long-standing artificial intelligence problems.
3. Convolutional Neural Network (CNN) and hybrid of CNN and long short-term memory (LSTM): the convolutional neural network (CNN) is a modified version of classical neural network which can outperform human perception in several computer vision problems. CNN is composed of convolution, pooling, and fully connected layers. Convolution operation can be either 1D, 2D, etc. Generally, 1D convolution is used on text and time-series data. CNN contains many convolution operations that help to learn various features and together called a feature map.

The feature map dimension will be very large and sometimes sparse in nature which may lead to overfitting or underfitting issues, therefore, pooling is used to reduce the feature map dimension. Finally, the pooling layer is followed by a dense or fully connected layer to perform classification. Otherwise, the pooling output can be flattened and fed into recurrent structures such as RNN and LSTM to extract sequence information.

4 Methodology

4.1 Proposed Architecture

The working flow of our proposed spam email detection approach is shown in Fig. 1. The approach is known as DeepSpamNet and it consists of three different phases. They are (1) preprocessing (2) features extraction (3) and classification. In the preprocessing phase, the emails are transformed into a feature vector using text representation methods that are mentioned in the previous section. In this study, we have applied different techniques to examine the best text representation technique for our problem as each method has its unique way to represent the text. These methods are TDM, TF-IDF, TDM with SVD, TDM with NMF, TF-IDF with SVD, TF-IDF with NMF, Keras embedding, FastText, NBOW, and word embedding. Unlike the existing methods on feature engineering, the proposed work doesn't rely on any feature engineering. Instead, the features are learned automatically. However, the keywords are extracted by processing the emails so that they can be categorized based on their contents. The information such as source IP and email address are also extracted from emails. The email addresses are also extracted, as well as the source IP addresses of these emails. After processing each set of emails, their statistics are reported with word list and their frequency. This information is also stored in a separate database that can be updated whenever a new dataset is developed. It can be observed that databases that are updated frequently to new data, leads to higher spam detection rate and lower false positive rate. Predicting a spam email as spam is known as true positive while predicting a legitimate email as spam is known as false positive. Loss of legitimate emails due to false flagging is a major concern. Therefore, to deal with this problem and to enhance the performance of the system, the model has to be trained on the latest datasets. This process is illustrated in Fig. 1.

Due to the high dimensionality of the feature representation, the optimal number of features extracted using many deep learning models such as DNN, CNN, RNN, LSTM, and CNN-LSTM. Further, to compare with our approach different text representation methods utilized and combined with various classical machine learning algorithms, including Logistic Regression, GaussianNB, K-nearest neighbor, Decision Tree, AdaBoost, Random Forest, and SVM.

Finally, the optimal features of the deep learning layers are passed into a fully connected layer. It composed of the linear combination of inputs followed by non-

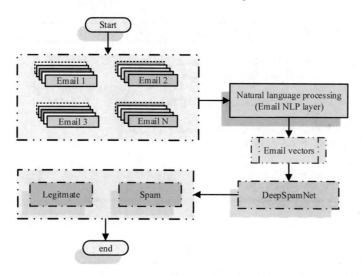

Fig. 1 Spam email detection engine working flow

linear activation methods, *sigmoid*. The loss function that is utilized is the binary cross-entropy as shown in Eq. 1.

$$loss(pd, ed) = -\frac{1}{N} \sum_{i=1}^{N} [ed_i \log pd_i + (1 - ed_i) \log(1 - pd_i)] \qquad (1)$$

where *pd* is a predicted probability vector for all samples in testing dataset, *ed* is expected class label vector, values are either 0 or 1. The pseudo-code for the proposed can be found in Algorithm 1.

Algorithm 1: Email Spam Detection Algorithm

Input: A set of emails $E_1, E_2, .., E_n$.
Output: Labels $y_1, y_2, .., y_n$ (0: Legitimate or 1: Spam).
1 **for** *each email E_i* **do**
2   extractedIPs = extractIpAddresses(E_i)
3   extractedWords = extractWords(E_i)
4   IPReputationDBChecker(extractedIPs)
5   TokenDBChecker(extractedWords)
6   vectorizedEmail = DataPreprocessing(E_i)   `// Email into numerical vectors using text representation method`
7   featureVector f_i = DLModel(vectorizedEmail)   `// Email is passed into DL model in order to obtain optimal feature vector`
8   Compute $z_i = DenseLayer(f_i)$
9   Calculate $y_i = Sigmoid(z_i)$
10 **end for**

Algorithm 2: IP reputation DB checker Algorithm

Input: An email E
Output: Email labelled as either ham or spam.
1 N= Get the number of lines in the input email E.
2 **for** $i=1$ to N **do**
3 $\quad$ $WN=$ Get the total number of words in i^{th} line.
4 $\quad$ **for** $j=1$ to WN **do**
5 $\quad\quad$ **if** j^{th} $word$ $matches$ $with$ any IP $from$ the DB **then**
6 $\quad\quad\quad$ Label email as spam
7 $\quad\quad$ **else**
8 $\quad\quad\quad$ Label email as ham

Algorithm 3: Token DB checker Algorithm

Input: An email E
Output: Email labelled as either ham or spam.
1 N= Get the number of lines in the input email E.
2 **for** $i=1$ to N **do**
3 $\quad$ $WN=$ Get the total number of words in i^{th} line.
4 $\quad$ **for** $j=1$ to WN **do**
5 $\quad\quad$ **if** j^{th} $word$ $matches$ $with$ any $token$ $from$ the DB **then**
6 $\quad\quad\quad$ Label email as spam
7 $\quad\quad$ **else**
8 $\quad\quad\quad$ Label email as ham

The proposed scalable architecture for spam email detection is shown in Fig. 2. This module can be added to the existing framework for cyber threat situational awareness to enhance the malicious detection rate [41]. The architecture consists of three main modules, which are Data collection, Identifying spam email, and Continuous monitoring.

Distributed log collector collects emails from various sources, inside an Ethernet LAN in a passive way. The collected emails are fed into a distributed database. Furthermore, the emails are parsed by the distributed log parser and its output is fed into DeepSpamNet. The DeepSpamNet composed of the following three different security modules to effectively detect spam email activity in real-time.

Fig. 2 Proposed architecture: DeepSpamNet

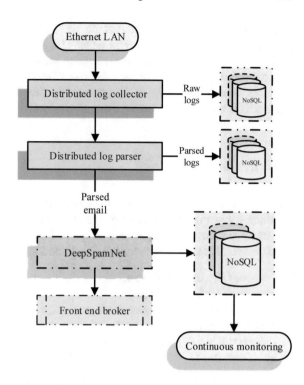

1. IP Reputation System: An IP blocked knowledge base is developed by continuous crawling of public blocklists, blacklists, online malware dumps, and reports related to known botnet IPs and domains from the Internet. The Blocked IP Database contains several blocked IP addresses and the emails originating from these IP addresses will immediately be flagged as spam. Additionally, further analysis has been done to identify If an email body, subject, and signature contains the blacklisted IP so that those emails can be marked as spam. The pseudo-code for the IP reputation DB checker can be found in Algorithm 2.
2. Word Count Database: Using Cybersecurity domain knowledge, a large database is developed which contains the words that are found in spam emails and their counts. Each line of the email is parsed for spam words and any email that has spam words in its lines more than 40% will be flagged as a spam email. This threshold can be further modified to enhance spam email detection performance. The pseudo-code for the Token DB checker Algorithm can be found in Algorithm 3.
3. Deep learning Model: The deep learning module loads the pre-trained module and extracts the features implicitly for the given input email. These features are highly non-linear in nature and they are fit into an n dimensional plane to classify the email as spam or legitimate.

The above mentioned three different modules collectively work together to classify the emails into either spam or legitimate. The preprocessed emails are stored in a distributed database for further use. The deep learning module has a front end broker which displays the analysis of the email data. The framework also has a module that monitors the detected spam email continuously once in 30 s. This helps to detect the spam emails which are generated using Digitally Generated Algorithms (DGA).

4.2 Evaluation Metrics

The main objective of this work is to detect and classify whether an email is either legitimate or spam. The emails are given to the proposed architecture, DeepSpam-Net which outputs either legitimate or spam. To evaluate the performance, various measures such as accuracy, precision, recall, and F1-Score are used based on:

1. Positive (P): legitimate email.
2. Negative (N): spam email.
3. True positive (T_P): legitimate email that is correctly classified as legitimate email.
4. True negative (T_N): spam email that is correctly classified as spam email.
5. False positive (F_P): legitimate email that is incorrectly classified as spam
6. False negative (F_N): spam email that is incorrectly classified as legitimate.

Generally, T_P, T_N, F_P, and F_N are obtained using a confusion matrix. The confusion matrix is represented in the form of a matrix where each row denotes the email samples of a predicted class and each column denotes email samples of an actual class. The various statistical measures considered in this study are defined as follows:

The accuracy measures the proportion of the total number of correct classifications.

$$Accuracy = \frac{T_P + T_N}{T_P + T_N + F_P + F_N} \qquad (2)$$

The recall measures the number of correct classifications penalised by the number of missed entries.

$$Recall = \frac{T_P}{T_P + F_N} \qquad (3)$$

The precision measures the number of correct classifications penalised by the number of incorrect classifications.

$$Precision = \frac{T_P}{T_P + F_P} \qquad (4)$$

The F1-score measures the harmonic mean of precision and recall, which serves as a derived effectiveness measurement.

$$\text{F1-score} = \frac{2 * \text{Recall} * \text{Precision}}{\text{Recall} + \text{Precision}} \tag{5}$$

The performance of the classifiers that are trained on a biased dataset is not reflected accurately by previously mentioned metrics. For such classifiers, the metrics such as geometric mean, true negative rate, false positive rate, true positive rate, and false negative rate are popularly used.

The Geometric Mean (G-Mean) is a performance measure that estimates the balance between classification performances on both the majority and minority classes.

$$\text{G-Mean} = \sqrt{precision \times \text{Recall}} \tag{6}$$

The receiver operating characteristic (ROC) curve is a graph that shows the performance of a model at all classification threshold settings. Generally, ROC is a probability curve and area under curve (AUC) is a degree or measure of separability. To plot a ROC curve, TPR is used on y-axis and FPR is used on x. AUC value 0.5 indicates that TP and FP are equal, and 1 for a perfect classification model.

$$AUC = \int_0^1 \frac{T_P}{T_P + F_N} d \frac{F_P}{T_N + F_P} \tag{7}$$

$$\text{TPR} = \frac{T_P}{T_P + F_P} \tag{8}$$

$$\text{FPR} = \frac{F_P}{F_P + T_N} \tag{9}$$

FPR represents the fraction of all legitimate emails that are predicted as spam emails:

$$\text{FPR} = \frac{F_P}{F_P + T_N} \tag{10}$$

In contrast, FNR represents the fraction of all spam emails that are predicted as legitimate emails:

$$\text{FNR} = \frac{F_N}{F_N + T_P} \tag{11}$$

Note that the lower the FPR and FNR, the better the performance.

5 Experimental Results and Discussions

All deep learning architectures are implemented using TensorFlow [63] with Keras [64] and conventional machine learning algorithms are implemented using Scikit-learn [65]. All the experiments related to deep learning architectures are run on GPU enabled machine. All classical machine learning algorithms are run on the CPU enabled machine. The GPU was NVidia GK110BGL Tesla K40 and CPU had a configuration (32 GB RAM, 2 TB hard disk, Intel(R) Xeon(R) CPU E3-1220 v3 @ 3.10 GHz) running over 1 Gbps Ethernet network.

5.1 Datasets

Many standard email datasets are publicly available and widely used. The quality of the dataset plays an important role in assessing the performance of any spam filter using machine learning and deep learning. In our experiment, there are 3 different datasets are used. They are (1) Lingspam [66] (2) PU [67] (3) Spam Assassin and Enron [68] The detailed description of the dataset is provided in Table 1. Enron is released most recently when compared to Lingspam, PU, and Spam Assassin.

There are many reasons for using these datasets. First of all, the emails that are present in these datasets are mailed out between 2000 and 2010. This interesting scenario characterizes the change in wordings in emails for a period of ten years. Secondly, the Enron dataset is used due to its bias towards spam class. Thirdly, the LingSpam dataset is used due to its domain-specific ham mails which extracted from scholarly linguistic discussions. Lastly, the PU dataset is used since it is not used often.

5.2 Observations and Results

In this work, there are three sets of experiments are done. The experimental use cases are presented as follow:

Table 1 Detailed statistics of email corpus

Dataset	Legitimate	Spam
Lingspam [66]	2412	481
PU [67]	3516	2414
Enron and Apache Spam Assassin train [68]	16,491	24,746
Enron and Apache Spam Assassin test [68]	7048	10,625

1. Experiments with classical text representation and classical machine learning algorithms
2. Experiments with Keras embedding and deep learning architectures
3. Experiments with distributed text representation with deep learning architectures.

Initially, for both the Lingspam and PU datasets, the preprocessing step is followed. All the characters are converted into lower cases and stop words and unnecessary new lines are removed. The dataset is randomly divided into 67% for training and 33% for testing. The default parameters in Scikit-learn as such used for TDM and TF-IDF text representation. To identify the parameters for the TDM and TF-IDF text representation, two trials of experiments are run on the training dataset with 5-fold cross-validation and performing grid search over finite parameter space. For both TDM and TF-IDF, the stop_words parameter is set to English, ngram_range to (1,1), and min_df and max_df to 2 and 1. Additionally, the norm parameter value is set to L2 normalization for TF-IDF. For NMF and SVD, the dimensionality is reduced into 30 and the number of iterations is set to 7. The features that are generated from TDM, TF-IDF, TDM with NMF, TDM with SVD, TF-IDF with NMF, and TF-IDF with SVD converted into dense matrices and passed into many classical machine learning classifiers for classification.

For a comparative study with classical text representation methods and classical machine learning algorithms, the Keras embedding with RNN is used. The maximum number of features is set to 2000, embeddings_initializer to uniform, and the embedding size parameter value is set to 128. The Keras embedding follows RNN which contains 128 units followed by dropout and batch normalization to reduce the overfitting and speed up the training process. Finally, the fully connected layer is used that contains one neuron and *sigmoid* activation function. Three trails of experiments are run to identify the number of features required for Keras embedding with embedding size 64, 128, and 256. All the experiments are run till 100 epochs with 0.01 learning rate. The experiments with Keras embedding size 128 performed better. Due to the size of the dataset is less, the word embedding and the FastText method is not employed.

The distributed text representation methods such as word2vec, FastText, and neural bag of words (NBOW) are employed on the Enron and SpamAssian email corpus. The more representative Word2vec models are the CBOW model and the Skip-Gram model. In this work, Skip-Gram is used. The detailed configuration details of word2vec are reported in Table 2. In Neural Bag-of-Words (NBOW), the word vectors are combined into text vectors. The maximum length of the text vector is set to 500. The semantic information of the text is then represented as a vector of fixed length. The advantage of the NBOW model is that it is simple and fast. The disadvantage is that vector linear addition will inevitably lose a lot of word and word related information and cannot express the semantics of sentences more accurately. The main issue with the Word2vec and NBOW model is that both models are inca-

Table 2 Detailed configuration of FastText

Parameter	Value
Dimension	100
Minimum word count	1
Epochs	100
N-grams	3
Loss function	Softmax
Learning rate	0.01

Table 3 Detailed configuration of Word2vec

Parameter	Value
batch_size	250
embedding_size	300
skip_window	10
num_skips	32
num_sampled	128
learning_rate	0.001
n_epoch	500

pable of handling words that are not present in the dictionary. To handle the unknown words during testing, the FastText representation is employed. Finally, the embedding representation is fed into several deep learning architectures. All the 3 embedding representation methods have parameters and we run three trails of experiments to identify the optimal parameters. The best parameters of word2vec and fastText are given in Tables 2 and 3 respectively.

Finally, the trained models of all experiments are loaded and evaluated the performances using the test dataset. The detailed results for Lingspam, PU, and Enron corpus are reported in Tables 4, 5, and 6 respectively. All tables contain results in terms of Accuracy, Precision, Recall, F1-score, G-Mean, true positive (TP), false positive (FP), false negative (FN), and true negative (TN).

In spam detection, FNR is more important metrics when compared to FPR as FNR represents the group of spam emails that are incorrectly predicted as legitimate emails by the classifier. Figure 3 represents the performance of the classical ML classifiers trained on LingSpam dataset using metrics such as FPR and FNR. From the figure, it can be observed that KNN classifiers with TF-IDF, TF-IDF + NMF, and TF-IDF + SVD representation methods produced the least FNR of 0.021. Adaboost classifiers with TDM and TDM + SVD representation methods also produced FNR of 0.027 which is very close to previously mentioned KNN classifiers. But the FPR of Adaboost classifiers are only 0.007 whereas the FPR of KNN classifiers are 0.054.

Table 4 Detailed test results of classical machine learning algorithms and recurrent neural network with various email representation methods for Lingspam

Model	Accuracy	Precision	Recall	F1-score	TN	FP	FN	TP	G-Mean
Email representation: TDM + NMF									
Logistic regression	0.867	0.947	0.125	0.221	810	1	126	18	0.344
Gaussian NB	0.866	0.531	0.965	0.685	688	123	5	139	0.716
K nearest neighbor	0.860	0.544	0.431	0.481	759	52	82	62	0.484
Decision Tree	0.951	0.801	0.896	0.846	779	32	15	129	0.847
AdaBoost	0.985	0.951	0.951	0.951	804	7	7	137	0.951
Random Forest	0.978	0.942	0.910	0.926	803	8	13	131	0.926
SVM	0.940	0.978	0.618	0.757	809	2	55	89	0.777
Email representation: TDM									
Logistic regression	0.988	0.972	0.951	0.961	807	4	7	137	0.961
Gaussian NB	0.960	0.982	0.750	0.850	809	2	36	108	0.858
K nearest neighbor	0.965	0.894	0.875	0.884	796	15	18	126	0.884
Decision Tree	0.959	0.883	0.840	0.861	795	16	23	121	0.861
AdaBoost	0.990	0.959	0.972	0.966	805	6	4	140	0.965
Random Forest	0.978	1.000	0.854	0.921	811	0	21	123	0.924
SVM	0.986	0.965	0.944	0.954	806	5	8	136	0.954
Email representation: TDM + SVD									
Logistic regression	0.988	0.972	0.951	0.961	807	4	7	137	0.961
GaussianNB	0.960	0.982	0.750	0.850	809	2	36	108	0.858
K nearest neighbor	0.965	0.894	0.875	0.884	796	15	18	126	0.884
Decision Tree	0.958	0.888	0.826	0.856	796	15	25	119	0.856
AdaBoost	0.990	0.959	0.972	0.966	805	6	4	140	0.965
Random Forest	0.975	1.000	0.833	0.909	811	0	24	120	0.913
SVM	0.986	0.965	0.944	0.954	806	5	8	136	0.954
Email representation: TF-IDF + NMF									
Logistic regression	0.982	1.000	0.882	0.937	811	0	17	127	0.939
GaussianNB	0.957	0.981	0.729	0.837	809	2	39	105	0.846
K nearest neighbor	0.951	0.762	0.979	0.857	767	44	3	141	0.864
Decision Tree	0.957	0.832	0.896	0.863	785	26	15	129	0.863
AdaBoost	0.988	0.959	0.965	0.962	805	6	5	139	0.962
Random Forest	0.979	1.000	0.861	0.925	811	0	20	124	0.928
SVM	0.860	1.000	0.069	0.130	811	0	134	10	0.263
Email representation: TF-IDF + SVD									
Logistic regression	0.982	1.000	0.882	0.937	811	0	17	127	0.939
GaussianNB	0.957	0.981	0.729	0.837	809	2	39	105	0.849
K nearest neighbour	0.951	0.762	0.979	0.979	767	44	3	141	0.864
Decision Tree	0.951	0.798	0.903	0.903	778	33	14	130	0.849
AdaBoost	0.988	0.959	0.965	0.962	805	6	5	139	0.962
Random Forest	0.976	1.000	0.840	0.913	811	0	23	121	0.917
SVM	0.860	1.000	0.069	0.130	811	0	134	10	0.263

(continued)

Table 4 (continued)

Model	Accuracy	Precision	Recall	F1-score	TN	FP	FN	TP	G-Mean
Email representation: TF-IDF									
Logistic regression	0.982	1.000	0.882	0.937	811	0	17	127	0.939
GaussianNB	0.957	0.981	0.729	0.837	809	2	39	105	0.846
K nearest neighbour	0.951	0.762	0.979	0.857	767	44	3	141	0.864
Decision Tree	0.956	0.827	0.896	0.860	784	27	15	129	0.861
AdaBoost	0.988	0.959	0.965	0.962	805	6	5	139	0.962
Random Forest	0.975	1.000	0.833	0.909	811	0	24	120	0.913
SVM	0.860	1.000	0.069	0.130	811	0	134	10	0.263
Keras embedding+RNN	0.994	0.993	0.965	0.979	810	1	5	139	0.979

Similar to Fig. 3, Fig. 4 represents the performance of the classical ML classifiers trained on the PU dataset. It can be inferred from the figure that naive Bayes classifier with the TDM + NMF representation method obtained the least FNR of 0.039. However, its FPR is 0.851 which shows that the classifier is biased towards spam class and it is simply predicting most of the emails as spam. The logistic regression classifier with TDM and TDM + SVD representation methods has produced the second least FNR of 0.066 with an FPR of 0.031. RNN with Keras embedding also produces similar results when compared to a logistic regression classifier. It has 0.069 FNR and 0.026 FPR.

Figure 5 represents the performance of deep learning architectures that are trained on Enron and Apache spam assassin datasets. It can be observed from the figure that CNN-LSTM models with word embedding and FastText representations performed very similarly with zero FNR. The LSTM model with Keras embedding representation has performed very similar to the previously mentioned models with FNR of 0.005 and FPR of 0.1.

6   Conclusion

In this work, deep learning models are applied to spam email detection and compared its efficacy with the prevalent classical machine learning classifiers that are commonly used in this domain. Comprehensive analysis of experiments was done on publicly available benchmark corpus and various text representation methods were transferred to email to represent them in numeric vector form. Deep learning models performed better when compared to the classical machine learning classifiers. The performance

Table 5 Detailed test results of classical machine learning algorithms with various email representation methods for PU

Model	Accuracy	Precision	Recall	F1-score	TN	FP	FN	TP	G-Mean
Email representation: TDM + NMF									
Logistic regression	0.749	0.848	0.455	0.593	1107	64	428	358	0.621
GaussianNB	0.475	0.431	0.962	0.596	174	997	30	756	0.802
K nearest neighbour	0.793	0.738	0.749	0.744	962	209	197	589	0.743
Decision Tree	0.843	0.803	0.807	0.805	1015	156	152	634	0.805
AdaBoost	0.852	0.827	0.798	0.812	1040	131	159	627	0.812
Random Forest	0.890	0.897	0.821	0.857	1097	74	141	645	0.858
SVM	0.808	0.831	0.655	0.733	1066	105	271	515	0.738
Email representation: TDM									
Logistic regression	0.955	0.953	0.934	0.943	1135	36	52	734	0.943
GaussianNB	0.914	0.935	0.845	0.888	1125	46	122	664	0.889
K nearest neighbour	0.845	0.800	0.817	0.809	1011	160	144	642	0.808
Decision Tree	0.909	0.873	0.905	0.889	1068	103	75	711	0.889
AdaBoost	0.945	0.937	0.925	0.931	1122	49	59	727	0.931
Random Forest	0.949	0.951	0.920	0.935	1134	37	63	723	0.935
SVM	0.937	0.934	0.907	0.921	1121	50	73	713	0.920
Email representation: TDM + SVD									
Logistic regression	0.955	0.953	0.934	0.943	1135	36	52	734	0.943
GaussianNB	0.914	0.935	0.845	0.888	1125	46	122	664	0.889
K nearest neighbour	0.845	0.800	0.817	0.809	1011	160	144	642	0.808
Decision Tree	0.898	0.857	0.894	0.875	1054	117	83	703	0.875
AdaBoost	0.945	0.937	0.925	0.931	1122	49	59	727	0.931
Random Forest	0.956	0.962	0.926	0.944	1142	29	58	728	0.944
SVM	0.937	0.934	0.907	0.921	1121	50	73	713	0.920
Email representation: TF-IDF									
Logistic regression	0.926	0.934	0.878	0.905	1122	49	96	690	0.956
GaussianNB	0.905	0.950	0.805	0.872	1138	33	153	633	0.874
K nearest neighbour	0.858	0.765	0.933	0.841	946	225	53	733	0.845
Decision Tree	0.891	0.862	0.868	0.865	1062	109	104	682	0.865
AdaBoost	0.940	0.928	0.921	0.925	1115	56	62	724	0.924
Random Forest	0.956	0.972	0.917	0.944	1150	21	65	721	0.944
SVM	0.819	0.941	0.587	0.723	1142	29	325	461	0.743
Email representation: TF-IDF + SVD									
Logistic regression	0.926	0.934	0.878	0.905	1122	49	96	690	0.906
GaussianNB	0.905	0.950	0.805	0.872	1138	33	153	633	0.874
K nearest neighbour	0.858	0.765	0.933	0.841	946	225	53	733	0.844
Decision Tree	0.893	0.866	0.866	0.866	1066	105	105	681	0.866
AdaBoost	0.940	0.928	0.921	0.925	1115	56	62	724	0.924
Random Forest	0.951	0.968	0.910	0.938	1147	24	71	715	0.939
SVM	0.819	0.941	0.587	0.723	1142	29	325	461	0.743

(continued)

Table 5 (continued)

Model	Accuracy	Precision	Recall	F1-score	TN	FP	FN	TP	G-Mean
Email representation: TF-IDF + NMF									
Logistic regression	0.926	0.934	0.878	0.905	1122	49	96	690	0.906
GaussianNB	0.905	0.950	0.805	0.872	1138	33	153	633	0.874
K nearest neighbour	0.858	0.765	0.933	0.841	946	225	53	733	0.844
Decision Tree	0.893	0.856	0.883	0.869	1054	117	92	694	0.869
AdaBoost	0.940	0.928	0.921	0.925	1115	56	62	724	0.924
Random Forest	0.949	0.966	0.905	0.934	1146	25	75	711	0.935
SVM	0.819	0.941	0.587	0.723	1142	29	325	461	0.743
Keras embedding + RNN	0.957	0.961	0.931	0.946	1141	30	54	732	0.946

of deep learning models with word embedding as an email representation method is good when compared to other email representation methods. Word embedding produces dense feature representation which captures the syntactic, contextual, and semantic similarity of words and information concerning the closeness of email samples. Finally, this work proposes DeepSpamNet, a highly scalable framework that uses the CNN-LSTM pipeline to detect spam within the daily email flow.

The proposed model outperforms the existing spam detection approaches that are based on blacklisting and machine learning classifiers. DeepSpamNet overcomes the drawbacks of previously mentioned approaches like the need for a domain-level expert for continuous maintenance of the database due to the ever changing nature of spam emails. Also, the insights about the employed feature engineering methods can help the attacker to bypass the defenses of the system. Dynamic generation of emails makes these classifiers redundant in real life. Since deep learning models are capable of learning abstract features, they can easily adapt to the dynamic nature of the inputs. For future works, we intend to work on real-time email dataset collection and apply the proposed methods on the same. As the email samples are highly imbalanced in a real-world situation, cost-sensitive deep learning architectures can perform better than the cost-insensitive architecture. Developing an optimal cost-sensitive deep learning architecture can be considered as one of the significant directions towards the future works. As the proposed framework is highly scalable in nature, future work can be based on generative adversarial networks (GANs) to generate a large number of email samples with the aim to build stronger and robust classifiers. Thus it can stay safe against an adversary in an adversarial environment.

Table 6 Detailed test results of deep learning architectures with various email representation methods for Enron and Apache Spam Assassin

Email representation + model	Accuracy	Precision	Recall	F1-score	TN	FP	FN	TP	G-Mean
TDM + DNN	0.665	0.990	0.448	0.617	6998	50	5868	4757	0.666
TF-IDF + DNN	0.795	0.773	0.932	0.845	4147	2901	723	9902	0.849
TF-IDF with NMF + DNN	0.829	0.80	0.954	0.871	4521	2527	487	10,138	0.874
TF-IDF with SVD + DNN	0.840	0.812	0.955	0.877	4691	2357	477	10,148	0.881
Keras embedding + RNN	0.940	0.921	0.985	0.952	6149	899	161	10,464	0.952
Keras embedding + LSTM	0.947	0.924	0.995	0.958	6178	870	58	10,567	0.959
Keras embedding + CNN	0.945	0.930	0.983	0.956	6260	788	184	10,441	0.955
Keras embedding + CNN with LSTM	0.948	0.926	0.993	0.958	6199	849	72	10,553	0.959
Word embedding + CNN with LSTM	**0.959**	**0.936**	**1.000**	**0.967**	**6318**	**730**	**0**	**10,625**	**0.967**
FastText + CNN with LSTM	**0.959**	**0.935**	**1.000**	**0.967**	**6315**	**733**	**0**	**10,625**	**0.967**
NBOW + DNN	0.927	0.901	0.986	0.942	5899	1149	145	10,480	0.943

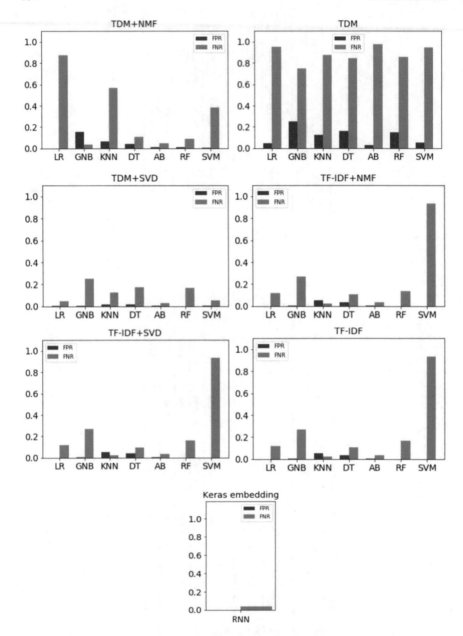

Fig. 3 FPR and FNR bar charts for classical machine learning algorithms with various email representation methods for Lingspam

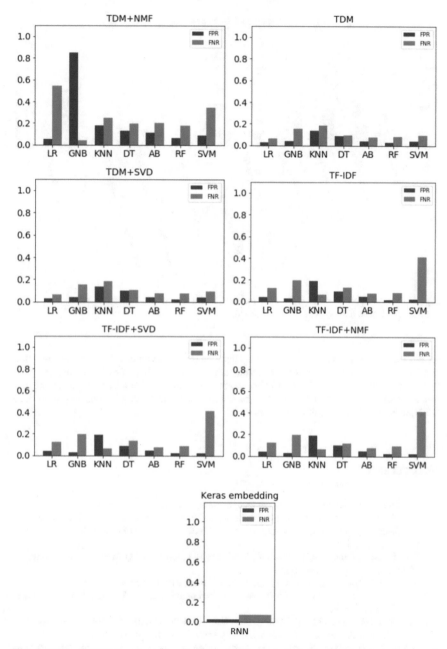

Fig. 4 FPR and FNR bar charts for classical machine learning algorithms with various email representation methods for PU

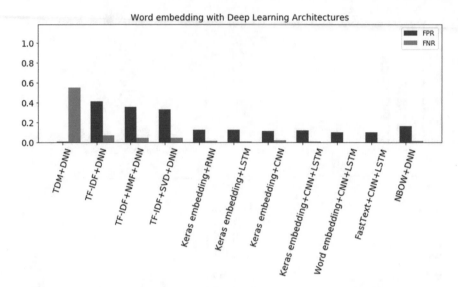

Fig. 5 FPR and FNR bar charts for deep learning architectures with various email representation methods for Enron and Apache Spam Assassin

Acknowledgements This work was in part supported by the Department of Corporate and Information Services, Northern Territory Government of Australia and in part by Paramount Computer Systems and Lakhshya Cyber Security Labs. We are grateful to NVIDIA India, for the GPU hardware support to the research grant. We are also grateful to Computational Engineering and Networking (CEN) department for encouraging the research.

References

1. Alazab M, Broadhurst R (2015) Spam and criminal activity
2. Alazab M, Layton R, Broadhurst R, Bouhours B (2013) Malicious spam emails developments and authorship attribution. In: Cybercrime and trustworthy computing workshop (CTC), 2013 fourth. IEEE, pp 58–68
3. Broadhurst R, Grabosky P, Alazab M, Bouhours B, Chon S (2014) An analysis of the nature of groups engaged in cyber crime
4. Symantec (2014) Internet security threat report. Tech, Rep, Symantec
5. Mamoun A, Roderic B (2014) Spam and criminal activity. SSRN Electron J. https://doi.org/10.2139/ssrn.2467423
6. Broadhurst R, Alazab M (2017) Spam and crime. In: Peter D (ed) Regulation, institutions and networks. The Australian National University Press, Canberra, pp 517–532
7. Alazab M, Broadhurst R (2015) The role of spam in cybercrime: data from the Australian Cybercrime Pilot Observatory. In: Smith R, Cheung R, Lau L (eds) Cybercrime risks and responses: eastern and western perspectives. Palgrave Macmillan, New York, pp 103–120. ISBN 9781349557882
8. Alazab M, Broadhurst R (2016) An analysis of the nature of spam as cybercrime. In: Clark RM, Hakim S (eds) Cyber-physical security protecting critical infrastructure at the state and local level. Springer, Switzerland, pp 251–266. ISBN 9783319328225

9. Karim A, Azam S, Shanmugam B, Kannoorpatti K, Alazab M (2019) A comprehensive survey for intelligent spam email detection. IEEE Access 7:168261–168295. https://doi.org/10.1109/access.2019.2954791
10. Mamoun A (2015) Profiling and classifying the behavior of malicious codes. J Syst Softw 100:91–102. https://doi.org/10.1016/j.jss.2014.10.031 [Q1, 2.559]
11. Sriram S, Vinayakumar R, Sowmya V, Krichen M, Noureddine DB, Shashank A, Soman KP (2020) Deep convolutional neural networks for image spam classification
12. Cranor LF, LaMacchia BA (1998) Spam! Commun ACM 41(8):74–83
13. Nicola L (2004) European Union vs. spam: a legal response. In: CEAS
14. Moustakas E, Ranganathan C, Penny D (2005) A comparative analysis of US and European approaches. In: CEAS, combating spam through legislation
15. Marsono MN (2007) Towards improving e-mail content classification for spam control: architecture, abstraction, and strategies. PhD thesis, University of Victoria
16. Duan Z, Dong Y, Gopalan K (2007) DMTP: controlling spam through message delivery differentiation. Comput Netw 51(10):2616–2630
17. Hershkop S (2006) Behavior-based email analysis with application to spam detection. PhD thesis, Columbia University
18. Sanz EP, Hidalgo JMG, Pérez JCC (2008) Email spam filtering. Adv Comput 74:45–114
19. Heron S (2009) Technologies for spam detection. Netw Secur 2009(1):11–15
20. Resnick P (2001) Internet message format—RFC 2822. Tech. Rep. https://tools.ietf.org/html/rfc2822
21. Cormack GV (2007) Email spam filtering: a systematic review. Found Trends Inf Retriev 1(4):335–455
22. Isacenkova J, Balzarotti D (2011) Measurement and evaluation of a real world deployment of a challenge-response spam filter. In: Proceedings of the 2011 ACM SIGCOMM conference on internet measurement conference (IMC '11), pp 413–426
23. Sophos (2013) Security threat report 2013. Tech. rep, Sophos
24. Andreolini M, Bulgarelli A, Colajanni M, Mazzoni F (2005) HoneySpam: honeypots fighting spam at the source. In: Proceedings of the steps to reducing unwanted traffic on the internet workshop, Cambridge, MA, pp 77–83
25. Tran KN, Alazab M, Broadhurst R (2014) Towards a feature rich model for predicting spam emails containing malicious attachments and urls
26. Cormack G (2007) Email spam filtering: a systematic review. Found Trends Inf Retriev 1(4):335–455
27. Carpinter J, Hunt R (2006) Tightening the net: a review of current and next generation spam filtering tools. Comput Secur 25(8):566–578
28. Kotsiantis S (2007) Supervised machine learning: a review of classification techniques. Informatica 31:249–268
29. Qian F, Pathak A, Hu YC, Mao ZM, Xie Y (2010) A case for unsupervised-learning-based spam filtering. In: ACM SIGMETRICS performance evaluation review, vol 38, no 1. ACM, pp 367–368
30. Bhowmick A, Hazarika SM (2016) Machine learning for e-mail spam filtering: review, techniques and trends. arXiv preprint arXiv:1606.01042
31. Yang Y (1995) Noise reduction in a statistical approach to text categorization. In: Proceedings of the 18th annual international ACM SIGIR conference on research and development in information retrieval. ACM, pp 256–263
32. Yang Y, Pedersen JO (1997) A comparative study on feature selection in text categorization. ICML 97:412–420
33. Koprinska I, Poon J, Clark J, Chan J (2007) Learning to classify e-mail. Inf Sci 177(10):2167–2187
34. Shaw WM Jr (1995) Term-relevance computations and perfect retrieval performance. Inf Process Manag 31(4):491–498
35. Guzella TS, Caminhas WM (2009) A review of machine learning approaches to spam filtering. Expert Syst Appl 36(7):10206–10222

36. Mujtaba G, Shuib L, Raj RG, Majeed N, Al-Garadi MA (2017) Email classification research trends: review and open issues. IEEE Access 5:9044–9064
37. LeCun Y, Bengio Y, Hinton G (2015) Deep learning. Nature 521(7553):436
38. Lennan C, Naber B, Reher J, Weber L (2016) End-to-end spam classification with neural networks
39. Eugene L, Caswell I (2017) Making a manageable email experience with deep learning
40. Repke T, Krestel R (2018) Bringing back structure to free text email conversations with recurrent neural networks. In: European conference on information retrieval. Springer, Cham, pp 114–126
41. Vinayakumar R, Poornachandran P, Soman KP (2018) Scalable framework for cyber threat situational awareness based on domain name systems data analysis. Big data in engineering applications. Springer, Singapore, pp 113–142
42. Ala'M AZ, Faris H (2017). Spam profile detection in social networks based on public features. In: 2017 8th international conference on information and communication systems (ICICS). IEEE, pp 130–135
43. Ala'M AZ, Faris H, Hassonah MA (2018) Evolving support vector machines using whale optimization algorithm for spam profiles detection on online social networks in different lingual contexts. Knowl-Based Syst 153:91–104
44. Madain A, Ala'M AZ, Al-Sayyed R (2017) Online social networks security: threats, attacks, and future directions. Social media shaping e-publishing and academia. Springer, Cham, pp 121–132
45. Al-Zoubi AM, Alqatawna JF, Faris H, Hassonah MA (2019) Spam profiles detection on social networks using computational intelligence methods: the effect of the lingual context. J Inf Sci 0165551519861599
46. Li Y, Nie X, Huang R (2018) Web spam classification method based on deep belief networks. Expert Syst Appl:261–270
47. Asdaghi F, Soleimani A (2019) An effective feature selection method for web spam detection. Knowl-Based Syst 166:198–206
48. Gupta M, Bakliwal A, Agarwal S, Mehndiratta P (2018) A comparative study of spam SMS detection using machine learning classifiers. In: 2018 eleventh international conference on contemporary computing (IC3). IEEE, pp 1–7
49. Hijawi W, Faris H, Alqatawna JF, Aljarah I, Al-Zoubi AM, Habib M (2017) EMFET: e-mail features extraction tool. arXiv preprint arXiv:1711.08521
50. Faris H, Ala'M AZ, Aljarah I (2017) Improving email spam detection using content based feature engineering approach. In: 2017 IEEE Jordan conference on applied electrical engineering and computing technologies (AEECT). IEEE, pp 1–6
51. Cook D, Hartnett J, Manderson K, Scanlan J (2006) Catching spam before it arrives: domain specific dynamic blacklists. In: Proceedings of the 2006 Australasian workshops on grid computing and e-research, vol 54. Australian Computer Society, Inc., pp 193–202
52. Kshirsagar D, Patil A (2013) Blackhole attack detection and prevention by real time monitoring. In: 2013 fourth international conference on computing, communications and networking technologies (ICCCNT). IEEE, pp 1–5
53. Wang B, Pan WF (2005) A survey of content-based anti-spam email filtering. J Chin Inf Process 5
54. Faris H, Ala'M AZ, Heidari AA, Aljarah I, Mafarja M, Hassonah MA, Fujita H (2019) An intelligent system for spam detection and identification of the most relevant features based on evolutionary random weight networks. Inf Fusion 48:67–83
55. Alghoul A, Al Ajrami S, Al Jarousha G, Harb G, Abu-Naser SS (2018) Email classification using artificial neural network
56. Tzortzis G, Likas A (2007) Deep belief networks for spam filtering. In: 19th IEEE international conference on tools with artificial intelligence, 2007. ICTAI 2007, vol 2. IEEE, pp 306–309
57. Mi G, Gao Y, Tan Y (2015) Apply stacked auto-encoder to spam detection. In: International conference in swarm intelligence. Springer, Cham, pp 3–15

58. Yawen W, Fan Y, Yanxi W (2018) Research of email classification based on deep neural network. In: 2018 second international conference of sensor network and computer engineering (ICSNCE 2018). Atlantis Press
59. Hassanpour R, Dogdu E, Choupani R, Goker O, Nazli N (2018) Phishing e-mail detection by using deep learning algorithms. In: Proceedings of the ACMSE 2018 conference. ACM, p 45
60. Tyagi A (2016) Content based spam classification—a deep learning approach. Doctoral dissertation, University of Calgary
61. Barushka A, Hajek P (2018) Spam filtering using integrated distribution-based balancing approach and regularized deep neural networks. Appl Intell 48:3538–3556
62. Coyotes C, Mohan VS, Naveen JR, Vinayakumar R, Soman KP (2018) ARES: automatic rogue email spotter
63. Abadi M, Barham P, Chen J, Chen Z, Davis A, Dean J, Devin M, Ghemawat S, Irving G, Isard M, Kudlur M (2016) Tensorflow: a system for large-scale machine learning. OSDI 16:265–283
64. Chollet F (2015) Keras
65. Pedregosa F, Varoquaux G, Gramfort A, Michel V, Thirion B, Grisel O, Blondel M, Prettenhofer P, Weiss R, Dubourg V, Vanderplas J (2011) Scikit-learn: machine learning in Python. J Mach Learn Res 12:2825–2830
66. Lingspam. Available at: http://www.aueb.gr/users/ion/data/lingspam_public.tar.gz. Accessed 08 May 2018
67. PU. Available at: http://www.aueb.gr/users/ion/data/PU123ACorpora.tar.gz. Accessed 08 May 2018
68. Enron and Apache Spam Assassin. Available at: http://www.cs.bgu.ac.il/¬elhadad/nlp16/spam.zip. Accessed 08 May 2018
69. Sahami M, Dumais S, Heckerman D, Horvitz E (1998) A Bayesian approach to filtering junk e-mail. In: Learning for text categorization: papers from the 1998 workshop, vol 62, pp 98–105
70. Androutsopoulos I, Paliouras G, Karkaletsis V, Sakkis G, Spyropoulos CD, Stamatopoulos P (2000) Learning to filter spam e-mail: a comparison of a Naive Bayesian and a memory-based approach. arXiv preprint cs/0009009
71. Woitaszek M, Shaaban M, Czernikowski R (2003) Identifying junk electronic mail in Microsoft outlook with a support vector machine. In: 2003 symposium on applications and the internet, 2003. Proceedings. IEEE, pp 166–169
72. Amayri O, Bouguila N (2010) A study of spam filtering using support vector machines. Artif Intell Rev 34(1):73–108
73. Yeh CY, Wu CH, Doong SH (2005) Effective spam classification based on meta-heuristics. In: 2005 IEEE international conference on systems, man and cybernetics, vol 4. IEEE, pp 3872–3877
74. Toolan F, Carthy J (2010) Feature selection for spam and phishing detection. In: eCrime researchers summit (eCrime), 2010. IEEE, pp 1–12
75. Wu CH (2009) Behavior-based spam detection using a hybrid method of rule-based techniques and neural networks. Expert Syst Appl 36(3):4321–4330. Soranamageswari M, Meena C (2010) Statistical feature extraction for classification of image spam using artificial neural networks. In: 2010 second international conference on machine learning and computing (ICMLC). IEEE, pp 101–105
76. Fdez-Riverola F, Iglesias EL, Díaz F, Méndez JR, Corchado JM (2007) Applying lazy learning algorithms to tackle concept drift in spam filtering. Expert Syst Appl 33(1):36–48
77. Joulin A, Grave E, Bojanowski P, Douze M, Jégou H, Mikolov T (2016) Fasttext.zip: compressing text classification models. arXiv preprint arXiv:1612.03651
78. Almeida TA, Yamakami A (2010) Content-based spam filtering. In: The 2010 international joint conference on neural networks (IJCNN), Barcelona, pp 1–7
79. Clark J, Koprinska I, Poon J (2003) A neural network based approach to automated e-mail classification. In: IEEE/WIC international conference on web intelligence, 2003. WI 2003. Proceedings. IEEE, pp 702–705
80. Kalchbrenner N, Grefenstette E, Blunsom, P (2014) A convolutional neural network for modelling sentences. arXiv preprint arXiv:1404.2188

AndroShow: A Large Scale Investigation to Identify the Pattern of Obfuscated Android Malware

Md. Omar Faruque Khan Russel, Sheikh Shah Mohammad Motiur Rahman, and Mamoun Alazab

Abstract This paper represents a static analysis based research of android's feature in obfuscated android malware. Android smartphone's security and privacy of personal information remain threatened because of android based device popularity. It has become a challenging and diverse area to research in information security. Though malware researchers can detect already identified malware, they can not detect many obfuscated malware. Because, malware attackers use different obfuscation techniques, as a result many anti malware engines can not detect obfuscated malware applications. Therefore, it is necessary to identify the obfuscated malware pattern made by attackers. A large-scale investigation has been performed in this paper by developing python scripts, named it AndroShow, to extract pattern of permission, app component, filtered intent, API call and system call from an obfuscated malware dataset named Android PRAGuard Dataset. Finally, the patterns in a matrix form have been found and stored in a Comma Separated Values (CSV) file which will be the base of detecting the obfuscated malware in future.

Keywords Android malware · Obfuscated malware · Obfuscated malware pattern · Obfuscated malware pattern identification

Md. O. F. K. Russel (✉) · S. S. M. M. Rahman
Department of Software Engineering, Daffodil International University,
Dhaka, Bangladesh
e-mail: russel35-1170@diu.edu.bd

S. S. M. M. Rahman
e-mail: motiur.swe@diu.edu.bd

M. Alazab
College of Engineering, IT and Environment,
Charles Darwin University, Darwin, Australia
e-mail: alazab.m@ieee.org

© The Editor(s) (if applicable) and The Author(s), under exclusive license to Springer
Nature Switzerland AG 2021
Y. Maleh et al. (eds.), *Machine Intelligence and Big Data Analytics for Cybersecurity Applications*, Studies in Computational Intelligence 919,
https://doi.org/10.1007/978-3-030-57024-8_8

1 Introduction

Android based smartphones become more popular than other platforms because of the usability and large amount of android applications on the market. It offers more applications and administrations than other devices like PCs. As of Feb 2020 reports, worldwide android base OS market share is 38.9% where windows OS market share is 35.29% [1]. For this trendy platform malware attackers focusing on this area. Their main goal is obtaining user's sensitive information illegally for the financial benefits [2]. By spreading malicious applications, spam emails [3] etc. android devices can be affected. Therefore these are the easiest ways to gain access devices. Installing Android Application Package (APK) from unknown and unverified markets or sources makes it a lot easier to access. Third party sources make a simpler way to do this [4]. Malicious activity performed by background services, these are threats to user's sensitive information and privacy. Without the user's attention these dangerous background services run by malicious software. Malicious software performs some common operations on affected devices i.e. get contacts of users, steal text messages from inbox, login details, subscribe premium services without user's attention [5]. In the first quarter of 2019, android based devices sold around 88% worldwide of the smartphone market [6]. Therefore, mobile clients' very own data is in danger. Malware aggressors are abusing cell phones confined zones and taking points of interest of nonattendance of standard security, by spreading convenient explicit malware that get personal data, access the credit card information of users, access the bank account information, and also can remove some access of device functionalities. Presently-a-days, Code obfuscation adjusts the program code to make clones which have a comparative handiness with different byte gathering. The new copy is not detected by antivirus scanners [7]. In March 2019, around 121 thousand new variations of portable malware were seen on Android mobiles in China. Therefore, android malware detection has been a challenging field to research [8]. Malware researchers are trying to find out various solutions and it's rising day by day. Intrusion detection system (IDS) [9], malicious codes behaviour [10], privacy specific [11], security policies [12, 13], privilege escalation [14, 15] typed specific attacks prioritise on those research to obliged information protection. Some visible drawbacks are present to those proposed solutions. Quick changes of intelligent attacks [16], development of artificial intelligence methods are increasing in the cyber security area [17]. There are two major categories of mechanism that exist in android malware detection [18]. One is static analysis [19] and the other one is dynamic analysis [20]. Inspecting source code to determine suspicious patterns is called static analysis. Most antivirus companies use static analysis for malware detection. Behavior based detection [21] also well known as dynamic analysis. The principle contribution of this paper is given below:

- Static analysis has been performed on an obfuscated Android malware dataset.
- Proposed an approach to extract the five features permission, app component, filtered intent, system call, API call from obfuscated malware in android.

- The seven obfuscated techniques such as Trivial Encryption, String Encryption, Reflection Encryption, Class Encryption, Combination of Trivial and String Encryption, Combination of Trivial, String and Reflection Encryption, Combination of Trivial, String, Reflection and Class Encryption has been considered.
- The pattern of five features has been represented in 2D vector matrix which is stored in csv files.
- Identify the usage trends of five features in Obfuscated Android Malware.

This paper is ordered as follows. The Literature Review is represented in Sect. 2. Research methodology is in Sect. 3. Section 4 described the result analysis and discussion. Section 5 finally concludes the paper.

2 Literature Review

2.1 Permission

A massive bit of Android's work in security is its permission structure [22]. Securing protection of Android clients is generally significant in the online world. This significant undertaking is finished by permission. To get to touchy client information (Contacts, SMS), in addition obvious system highlights (Camera, Location) permission must be mentioned by android applications. In view of highlight, the system permits the consent naturally or might incite the client to permit the call. All permission present openly in <uses-permission> labels in the manifest file. Android applications that require ordinary consent (don't mischief to client's protection or gadget activity) system naturally permit these authorizations to application. Application that requires risky authorization (permission that can be unsafe for client's protection or gadget ordinary activity) the client should explicitly permit to acknowledge those authorizations [23]. Permissions empower an application to find a workable perilous API call. Various applications need a couple of approvals to work fittingly and customers must recognize them at install time. Permission gives a progressively top to bottom scenario on the functional characteristics of an application. Malware creators insert dangerous permission in manifest that isn't important to application and furthermore declare considerably more permissions than actually required [24, 25]. Therefore, the importance of the permission pattern has been noticed from state-of-art tabulated in Table 1.

From Table 1, it's been stated that permission pattern analysis has significant effect to develop anti-malware tools or to detect the malware in android devices.

2.1.1 App Component

Android application's one of principal structure is the application component. Each part is an entry point of your application which any system can enter [35]. These

Table 1 Recent works on android permission to detect malware

Ref.	Feature set	Samples	Accuracy (%)	Year
[26]	Permission	7400	91.95	2019
[27]	Permission	1740	98.45	2019
[28]	Permission	2000+	90	2018
[29]	Permission	7553	95.44	2019
[30]	Network traffic, system permissions	1124	94.25	2018
[31]	Permission	399	92.437	2018
[32]	Permission	100	94	2018
[33]	8 features including permission	5560	97	2019
[34]	8 features including permission	5560	97.24	2018

parts are estimated coupled by AndroidManifest.xml that delineates each section of the application and how they link [36]. Some of them rely upon others. Also, Some malware families may have an identical name of parts. For example, a couple of varieties of the DroidKungFu malware use a comparative name for explicit organizations [37] (e.g., com.google.search). There are the accompanying four component of segment utilized in Android application:

- **Activities**. An activity is the segment point for speaking with the user. It addresses a single screen with a UI. For example, an email application may have one development that shows a once-over of new messages, another activity to make an email, and another activity for scrutinizing messages. Despite the way that the activities coordinate to shape a solid customer inclusion in the email application, everybody is liberated from the others. All things considered; a substitute application can start any of these activities if the email application grants it. For example, a camera application can start the development in the email application that makes new mail to empower the customer to share a picture.

- **Services**. A service is a comprehensively helpful segment point for keeping an application coming up short immediately for a wide scope of reasons. n service is a section that continues coming up short immediately to perform long-running exercises. It doesn't give a UI. For example, a service may play music far out while the user is in a substitute application, or it might get data over the system without blocking user association with a connection.

- **Broadcast Receivers**. A broadcast recipient is a section that enables the structure to pass on events to the application outside of standard user stream, allowing the application to respond to structure wide correspondence revelations. Since correspondence authorities are another inside and out portrayed area into the application, the system can pass on imparts even to applications that aren't starting at now running. Along these lines, for example, an application can design an

Table 2 Recent works on android app component to detect malware

Ref.	Feature set	Samples	Accuracy (%)	Year
[36]	8 features including app component	8385	99.7	2017
[38]	4 features including app component	1738	97.87	2012
[39]	7 features including app component	35,331	98.0	2018
[40]	3 features including app component	308	86.36	2014
[41]	7 features including app component	19,000	≈99	2018
[42]	7 features including app component	11,120	94.0	2018

alarm to introduce a notice on illuminating the customer with respect to the best in upcoming events. Likewise, by passing on that alert to a Broadcast Receiver of the application, there is no prerequisite for the application to remain running until the alarm goes off. Regardless of the reality that correspondence beneficiaries don't show a UI, they may make a status bar cautioning to alert the user when a correspondence event happens [38].

- **Content Providers**. In content provider part supplies data from one application to others on requesting. The data may be taken care of in the record system, the database or somewhere else out and out. Through the substance supplier, various applications can request or change the data if the substance supplier grants it. For example, the Android system gives a substance supplier that manages the customer's contact information. Content supplier are moreover supportive for scrutinizing and forming data that is private to your application and not mutual.

In this way, the hugeness of app component pattern has been seen from state-of-art arranged in Table 2.

From Table 2, it's been expressed that app component pattern analysis has remarkable impact to develop anti-malware tools or to recognize the malware in android gadgets.

2.1.2 Filtered Intent

Intent is an informing object you can use to ask for an operation from another application component. Despite the fact that intent makes easier communication between components in a few different ways, there are three basic ways (1) starting an activity (2) starting a service (3) delivering a broadcast. Two types of intent are there (1) Explicit Intents (2) Implicit Intents [43]. Explicit Intents identify the components to start with by containing targeted package names and class names.

Table 3 Recent works on android intent to detect malware

Ref.	Feature set	Samples	Accuracy/findings	Year
[44]	Intent	17,290	97.4%	2016
[45]	Intent	2644	Shows effective solution, need to detect collusion attack	2015
[46]	Intent	–	Resilience to some obfuscation techniques in detection	2014
[47]	Permission, network, intent	7406	95.5%	2017
[48]	Intent	2283	96.6%	2015
[49]	Intent	2000	75%	2014

Normally, Explicit Intents are utilized to interface parts inside a similar application and intended for inter application communications. In contrast to Explicit Intents, Implicit Intents do not name a particular segment, however rather proclaim general activities to perform. At the point when an application makes an Implicit Intent, the Android framework finds the suitable segment to begin by contrasting the substance (i.e., action, category and data) of the Intent to the pronounced Intent Filters. On the off chance that the Intent matches an Intent Filter, the framework begins that segment and conveys it the Implicit Intent item [44]. If multiple Intent filters are matched then the system shows a dialog box to the user to pick up which app to use. An Intent filter is a declaration in an app's manifest.xml file that states the type of intents the component will receive. Suppose, an activity declares an intent filter, means that other apps can directly start the activity with an undoubtable type of intent. Similarly, if an activity does not declare an intent, then it can be activated only by Explicit Intent [43]. Intent used in inter component and inter app communication. Intent filters identify a particular access for a component as well as the application. Intent filters can be used for spying specific intents. Malware is responsive to a particular set of system events. So, Intents that exist in <intent filters> tag can be indicators. For future reference we will call it filtered intent. The importance of filtered intent pattern has been seen from state-of-art arranged in Table 3.

From Table 3, it's been expressed that filtered intent pattern analysis has remarkable impact to develop anti-malware tools or to recognize the malware in android gadgets.

2.1.3 API Call

API stands for Application Programming Interface. In simple terms, APIs simply enable applications to speak with each other. Envision the accompanying situation: You (as in, your application, or your customer, this could be an internet browser)

Table 4 Recent works on android API call to detect malware

Ref.	Feature set	Samples	Accuracy/findings	Year
[16]	API call	20,000	99.0%	2013
[51]	Permission, API call	2510	96.39%	2013
[53]	Permission, API call	28,558	99.0% [online], 84.9% [offline]	2015
[54]	Permission, API call	10,449	90–94%	2017
[55]	API call	8598	97.6% TP, 91.0% TN	2018
[56]	API call, manager class	–	SMSManager, Telephony Manager most used in malware	2015

needs to get to another application's information or usefulness. For instance, maybe you need to get to all Twitter tweets that notice the #malware hashtag. You could email Twitter and request a spreadsheet of every one of these tweets. In any case, at that point you'd need to figure out how to bring that spreadsheet into your application; and, regardless of whether you put them away in a database, as we have been, the information would end up obsolete in all respects rapidly. It is difficult to stay up with the latest. It would be better and easier for Twitter to give you an approach to question their application to get that information, so you can view or utilize it in your own application. It would remain state-of-the-art consequently that way [50]. API includes a principle set of packages and classes. Most apps use a large number of API calls, so it helps us to characterize and differentiate malware from benign apps. Peiravian and Zhu [51] state that benign apps use more APIs than malware apps. The author's in [52] has listed some suspicious API calls used by malware applications. For example, sendTextMessage, getPackageManager, getDeviceId, Runtime.exec. We considered these api calls in this paper analysis.

From Table 4, it's been stated that api call pattern analysis has significant effect to develop anti-malware tools or to detect the malware in android devices.

2.1.4 System Call

It is also known as system command. Android core is the modified version of Linux 2.6 kernel. For adopting mobile operating system devices this modification was done. The Android Kernel explicitly upgrades on power management, shared memory drivers, alert drivers, folios, bit debugger and lumberjack and low memory executioners. System calls connect Android application and kernel. Whenever a client asks for administrations like calling a telephone in client mode through the telephone call application, the demand is sent to the Telephone Directory Service in the application structure. The Dalvik Virtual Machine in Android runtime changes the client ask

Table 5 Recent works on system calls to detect malware

Ref.	Feature set	Samples	Accuracy/findings	Year
[57]	System call	645	Malware app invokes system calls more frequently than benign app	2016
[58]	System call	12,660	93.0%	2015
[59]	System call	1100	92.5%	2015
	Directory path, code based			
[60]	System call	152	>93.0%	2016
[61]	System call	1958	80.3–80.7%	2018
	Broadcast receiver			
	API call			
[62]	System call	Malgenom (DroidDream)	Click event perform malicious tasks	2013
[63]	System call	460	90.0% (polynomial kernel)	2018
			86.0% (RBF Kernel)	

for gone by the Telephone Manager Service to library calls, which results in various framework calls to Android Kernel. While executing the system call, there is a change from client mode to part mode to play out the delicate activities. At the point when the execution of activities asked for by the system call is finished, the control comes back to the client mode [57]. As talked about over, the system calls are the communicator between the client and the bit. This implies all solicitations from the applications will go through the System Call Interface before its execution through the equipment. So, catching and dissecting the system call can give data about the conduct of the application. Seo et al. [52] listed some system calls that are often used in malware applications. For example, chmod, su, mount, sh, killall, reboot, mkdir, ln, ps. We considered these system commands in this paper analysis.

From Table 5, it's been stated that system call pattern analysis has significant effect to develop anti-malware tools or to detect the malware in android devices.

2.2 Obfuscation Techniques

By the term obfuscation, they [64] proposes that any change of the Android executable bytecode (i.e., .dex record) or potentially .xml documents (for instance AndroidManifest.xml or String.xml), that doesn't impact the key functionalities of the application. They partitioned procedures into two sets that they acquired. One set is Trivial Obfuscation Techniques and the other set is Non-Trivial Obfuscation Techniques. There are four key head techniques close by the mix of those systems in full scale; seven [64] strategies as shown by the dataset are considered right in this paper analysis.

2.2.1 Trivial Obfuscation Technique

This methodology simply alter strings in the classes.dex record without changing the bytecode headings. This methodology can alter names everything being equivalent, procedures, classes, fields and source codes of an Android application with unusual letters. Disassembling, reassembling and repacking the classes.dex records are included for these activities.

2.2.2 Non-trivial Obfuscation Strategy

The two strings and byte-codes of the executable are influenced by these procedures. Reflection and Class Encryption are essentially powerful against hostile to malware frameworks that investigate the bytecode sign to distinguish malware. Also, different sorts of strings (e.g., constants) are changed, and this may deal with machines which resort to investigating them in order to perform acknowledgment.

- **Reflection Encryption**. Reflection is fundamentally the property of a class of assessing itself, hence, getting data on its methodologies, fields, and so on. They [64] use the reflection property for summons. Three summons are utilized to restore the main: (i) forName, finds a class with a particular name, (ii) get-Method, gives the point object strategy and (iii) invoke, delayed consequence of the second gather and plays out the right conjuring on the method object. It is used unmistakably in code progression under explicit conditions since abuse of bytecode directions.
- **String Encryption**. This method obfuscates each string that is portrayed inside a class by deriving a calculation subject to XOR assignments. At runtime, the right string is made by passing the encoded string. Disregarding the way that this system doesn't swear by DES or AES calculations, it is critical that it is more confounding than different methods for string encryption that have been proposed in the creation, which got a handle on a Caesar move [65].
- **Class Encryption**. They [64] got this framework as by and large potential and starting strategy from others strategies they got. This muddling procedure absolutely scrambles and shrinks (with GZIP computation) each class and stores its nuances in a data show. During the execution of the scattered application, the jumbled class should be first decoded, decompressed, and some time later stacked in memory. This methodology can unimaginably fabricate the overhead of the application as a great deal of course are fused. Regardless, it makes it extraordinarily hard for a human expert to perform static examination.

The other three obfuscation techniques are a combination of previous techniques. 1. Trivial + String Encryption; 2. Trivial + String + Reflection Encryption and 3. Trivial + String + Reflection + Class Encryption. The third one combination of four techniques is one of the most advanced obfuscation techniques.

3 Methodology

The procedural strategy obtained in this study work has illustrated in Fig. 2 and described in this section.

3.1 Dataset

PRAGuard [64, 66], an obfuscated malware dataset has been investigated here. It has 10479 malware samples. By applying seven different obfuscation techniques on MalGenome [67] and Contagio Minidump [68] datasets, it has been obtained. Each technique has 1497 malware samples.

3.2 Environment

HP 2.30 GHz computing environment, operating system Windows 10, programming language Python 3, python packages matplotlib, csv, pandas, androguard [69] used in this paper work.

3.3 Data Preprocessing

AndroShow inspect all 10,479 malware apk samples whet-her they are valid or not. After checking, 62 APKs were corrupted. So, 10,417 obfuscated malware considered in the final dataset.

3.4 Feature Extraction

AndroShow disassembled the APKs and extracted all features (Fig. 1) from APKs. AndroShow uses customized python scripts with the help of androguard [69]. It takes an 40 min runtime on every obfuscation technique (Fig. 2).

3.5 Vector Matrix (Final Pattern)

2D vector matrix pattern is generated from seven obfuscation techniques. Every obfuscation technique has five features patterns which are separated. The pattern of

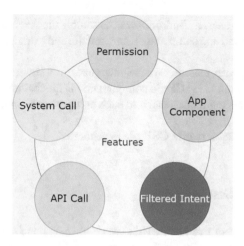

Fig. 1 Features

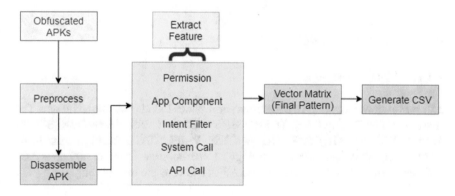

Fig. 2 Overview of proposed approach

the matrix consists of rows and columns. Rows are labeled as 1 or 0 and columns represent the features. If the feature name is related to a column found in APK, it will be 1, 0 for not found. To represent the pattern, a Comma Separated Value (CSV) file is generated.

3.6 Summary

- At the beginning, AndroShow clean the APKs of the dataset to check APK's format is valid or not, letter found that, some APKs format is not valid. So, these APKs removed from consideration.

- After cleaning the dataset, AndroShow disassemble the APKs and extract permission, intent filter and app component from .xml file and extract API call and system call from .dex files.
- After extracting each feature, AndroShow generate a row matrix for each APK and put the matrix in CSV. CSV's column name is the features tag name and row is 1/0. If a features tag name related to each column found in the APK than it's 1 otherwise 0.
- At he end, a 2D vector matrix CSV file is generated.

4 Results and Discussion

Experimented results of five features of seven obfuscation techniques are illustrated in this section. This section has five subsections based on five features.

4.1 Permission Analysis

4.1.1 Trivial Encryption

152 permissions including 18 dangerous permissions found from this technique. Top five usage of INTERNET, READ_PHONE_STATE, ACCESS_NETWORK_STATE, WRITE_EXTERNAL_STORAGE, ACCESS_W-IFI_STATE found in total 1437, 1359, 1185, 1000, 844 APKs accordingly. 12 permissions found in 500 up APKs, 119 permissions found in less than 100 APKs and rests are in between.

4.1.2 String Encryption

Same number of normal permissions and requested permissions found from this technique as Trivial enc. Top five usage of INTERNET, READ_PHONE_STATE, ACCESS_NETWORK_STATE, WRITE_EXTERNA-L_STORAGE, ACCESS_WIFI_STATE found in total 1432, 1354, 1180, 996, 845 APKs accordingly. Same as previous technique 12 permissions found in 500 up APKs, 119 permissions found in less than 100 APKs and rests are in between.

4.1.3 Reflection Encryption

As previous techniques number of permissions including requested permissions are also same in this technique. Top five usage of INTERNET, READ_PHONE_STATE, ACCESS_NETWORK_STATE, WRITE_EXTE-RNAL_STORAGE, ACCESS_WIFI_STATE found in total 1440, 1362, 1188, 1003, 845 APKs respectively. Number of permissions found in upper than 500 APKs and less than 100 APKs are the same as previous techniques.

4.1.4 Class Encryption

This technique also uses 152 permissions including 18 requested permissions. Top five usage of INTERNET, READ_PHONE_STATE, ACCESS_NETWORK_STATE, WRITE_EXTERNAL_STORAGE, ACCESS_W-IFI_STATE found in total 1437, 1359, 1185, 1000, 843 APKs respectively. Number of permissions found in upper than 500 APKs and less than 100 APKs are 12 and 119 accordingly as same as previous techniques.

4.1.5 Trivial + String Encryption

Number of permissions decreased in this technique, as previous techniques used 152 permissions but here is only 115. Though the number of requested permissions are the same. Top five usage of INTERNET, READ_PHONE_STATE, ACCESS_NETWORK_STATE, WRITE_EXTERNA-L_STORAGE, ACCESS_WIFI_STATE found in total 1448, 1368, 1212, 1023, 826 APKs respectively. Number of permissions found in upper than 500 APKs and less than 100 APKs are 13 and 82 accordingly.

4.1.6 Trivial + String + Reflection Encryption

Number of requested permissions is also the same in this technique. Permissions number is decreased to 111 which is second lowest among all techniques. Top five usage of INTERNET, READ_PHONE_STATE, ACCESS_NETWORK_STATE, WRITE_EXTERNA-L_STORAGE, ACCESS_WIFI_STATE found in total 1444, 1358, 1196, 1006, 830 APKs respectively. Number of permissions found in upper than 500 APKs and less than 100 APKs are 13 and 78 accordingly.

4.1.7 Trivial + String + Reflection + Class Encryption

Number of requested permissions is also the same in this technique, is 18. Permissions number is decreased to 107 which is lowest among all techniques. Top five usage of INTERNET, READ_PHONE_STATE, ACCESS_NETWORK_STATE, WRITE_EXTERNA-L_STORAGE, ACCESS_WIFI_STATE found in total 1443, 1358, 1199, 1002, 828 APKs respectively. Number of permissions found in upper than 500 APKs and less than 100 APKs are 13 and 74 accordingly (Fig. 3).

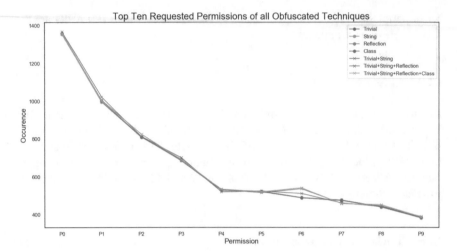

Fig. 3 Top ten requested permissions uses trend in all obfuscation techniques. Where READ_PHONE_STATE, WRITE_EXTERNAL_STORAGE, READ_SMS, SEND_SMS, RECEIVE_SMS, ACCESS_COARSE_LOCATION, READ_CONTACTS, ACCESS_FINE_LOCATION, CALL_PHONE, WRITE_CONTACTS are denoted as P0, P1, P2, P3, P4, P5, P6, P7, P8, P9

4.2 App Component Analysis

4.2.1 Trivial Encryption

AndroShow found that there are 1774 app components used by this technique. Some app components like Receiver, MainA, BaseABroadcastReceiver, BootReceiver, MainActivity found in total 468, 252, 186, 133, 118 APKs accordingly. 22 app components used by more than 100 APKs.

4.2.2 String Encryption

Analysis obtained 2381 app components from this technique. Top five usage of Receiver, UpdateService, AdwoAdBrowserActivity, Dialog, Setting found in total 469, 390, 356, 350, 285 APKs respectively. 28 app components found in more than 100 APKs.

4.2.3 Reflection Encryption

2429 app components found from this technique. Top five usage of Receiver, UpdateService, AdwoAdBrowserActivity, Dialog, Setting found in total 469, 384, 356, 349, 256 APKs correspondingly. More than 100 APKs used 28 app components in this technique.

4.2.4   Class Encryption

Highest number of app components found from this technique, is 2733. Top five usage of Receiver, UpdateService, AdwoAdBrowserActivity, Setting, History found in total 468, 389, 355, 281, 255 APKs on an individual basis (Fig. 4).

4.2.5   Trivial + String Encryption

After analysing this technique, AndroShow found that there are 1579 app components present. Five usage of Receiver, MainA, BaseABroadcastReceiver, BootReceiver, MainActivity found in total 473, 254, 186, 121, 114 APKs correspondingly. 24 app components found in more than 100 APKs.

4.2.6   Trivial + String + Reflection Encryption

AndroShow found that there are 1439 app components present in this technique. Five usage of Receiver, MainA, BaseABroadcastReceiver, MainActivity, BootReceiver found in total 478, 255, 187, 118, 117 APKs correspondingly. 21 app components found in more than 100 APKs.

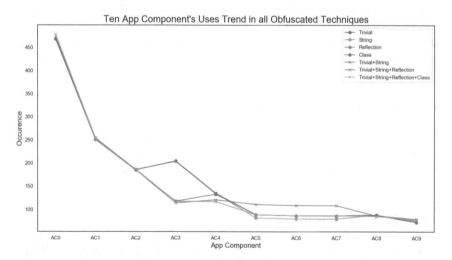

Fig. 4 Ten app components use trends in all obfuscation techniques. Where Receiver, MainA, BaseABroadcastReceiver, MainActivity, BootReceiver, NotificationActivity, OperaUpdaterActivity, AutorunBroadcastReceiver, BaseBroadcastReceiver, SmsReceiver denoted as AC0, AC1, AC2, AC3, AC4, AC5, AC6, AC7, AC8, AC9 on individual basis

4.2.7   Trivial + String + Reflection + Class Encryption

AndroShow found the lowest number of app components used in this technique, is 1365. Five usage of Receiver, MainA, BaseABroadcastReceiver, MainActivity, BootReceiver found in total 478, 253, 187, 117, 117 APKs respectively. 21 app components found in more than 100 APKs.

4.3   Filtered Intent Analysis

4.3.1   Trivial Encryption

Analysis found 90 filtered intents from <intent-filters> of android manifest files of this technique. Top five usage are VIEW, SENDTO, SEND, DIAL, BOOT_COM-PLETED in total 1340, 823, 793, 741, 674 accordingly. More than 50 APKs used 26 intents in this technique.

4.3.2   String Encryption

AndroShow found 34 filtered intents by analyzing this technique. Top five usage of BOOT_COMPLETED, CONTENT_CHANGED, PHONE_STATE, EXTERNAL_ APPLICATIONS_AVAILABLE, EXTERNAL-_APPLICATIONS_UNAVAILABLE found in total 99, 79, 43, 12, 12 APKs correspondingly. It shows that any intents are not used by 100 APKs.

4.3.3   Reflection Encryption

In total 90 filtered intents found from this technique. Highest number of VIEW, SENDTO, SEND, DIAL, BOOT_COMPLETED found in total 1342, 823, 793, 742, 675 APKs accordingly. 25 intents found in more than 50 APKs.

4.3.4   Class Encryption

In this technique, about 68 filtered intents were found. Top five usage of BOOT_ COMPLETED, VIEW, MAIN, SEARCH, PACKAGE_AD-DED found in total 663, 386, 84, 77, 75 APKs respectively. More or equal to 50 APKs used only 13 intents.

4.3.5   Trivial + String Encryption

21 filtered intents found from this technique. BOOT_COMPLETED, CONTENT_
CHNAGED, PHONE_STATE, EXTERNA-L_APPLICATIONS_AVAILABLE,
EXTERNAL_APPLICATIONS_UNAVAILA-BLE are the highest usage found in
total 112, 76, 52, 18, 18 APKs on an individual basis. It shows that only one intent
used by 112 APKs, rest intents usage are below 100.

4.3.6   Trivial + String + Reflection Encryption

AndroShow analyzed this technique and found that only 21 filtered intents used in
this technique. BOOT_COMPLET-ED, CONTENT_CHNAGED, PHONE_STATE,
EXTERNAL_APPLICATION-S_AVAILABLE, EXTERNAL_APPLICATIONS_
UNAVAILABLE are the highest usage found in total 113, 79, 55, 26, 26 APKs
on an individual basis. Like the previous one, one intent used by 113 APKs, rest
intents usage are below 100.

4.3.7   Trivial + String + Reflection + Class Encryption

Lowest number of filtered intents found from this technique is only 3. BOOT_COM-
PLETED, NEW_OUTG-OING_CALL, PHONE_STATE intents used only 1 APK
(Fig. 5).

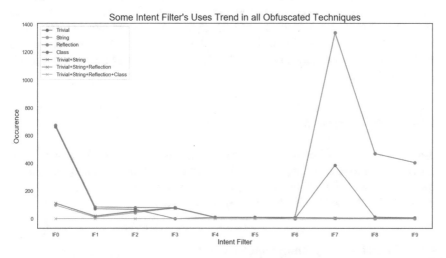

Fig. 5 Ten filtered intents usage trends in all obfuscation techniques. Where BOOT_COMPLETED,
NEW_OUTGOING_CALL, PHONE_STATE, CONTENT_CHANGED, HEART_CODE,
START_AGENT, SMS_SENT, VIEW, SCREEN_OFF, SCREEN_ON denoted as IF0, IF1,
IF2, IF3, IF4, IF5, IF6, IF7, IF8, IF9 on individual basis

4.4 API Call Analysis

4.4.1 Trivial Encryption

AndroShow analyzed all APKs of this technique and found that 23 suspicious API calls were present in the technique. Top five usage of API calls getInputStream, openConnection, getDeviceId, getPackageManager, getSubscriberId found in total 1347, 1328, 1299, 1260, 1049 APKs accordingly. Only two API calls installPackage and Socket found in total 16 and 1 APKs respectively. Rests API call found in more than 100 APKs.

4.4.2 String Encryption

Number of APKs found in this technique is the same as the previous one. Top five usage of suspicious API calls getInputStream, openConnection, getDeviceId, getPackageManager, getSubscriberId found in total 1349, 1335, 1295, 1260, 1050 APKs correspondingly. Four API calls used in less than 100 APKs, are installPackage, mailto, Socket, pdus in total 31, 1, 1, 1 APKs respectively. Rest of all API calls used in more than 100 APKs.

4.4.3 Reflection Encryption

AndroShow found that this technique also has the same number of suspicious API calls like previous techniques. Highest usage of API calls getInputStream, open-Connection, getDeviceId, getPackageManager, getSubscriberId found in total 1356, 1342, 1302, 1260, 1050 APKs on an individual basis. Only two APKs installPackage and Socket used in less number of APKs, 47 and 1 accordingly. More than 100 APKs used 21 API calls.

4.4.4 Class Encryption

Usage of API calls decreased in this technique, only 20 found. Highest number of usage getPackageManager, startActivityForResult, getInputStream, openConnection, pdus found in total 266, 199, 172, 164, 152 APKs accordingly. Analysis found that 12 API calls used in less than 100 APKs (Fig. 6).

4.4.5 Trivial + String Encryption

AndroShow analyzed this combination technique and found that 22 suspicious API calls are in this technique. Top five usage of API calls getInputStream, openConnec-

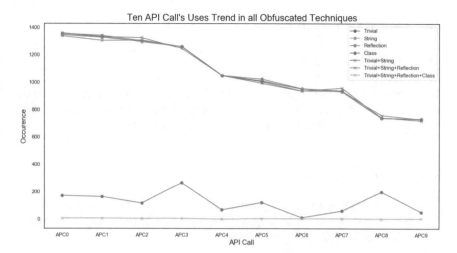

Fig. 6 Top ten API call usage trends in all obfuscation techniques. Where getInputStream, openConnection, getDeviceId, getPackageManager, getSubscriberId, getAssets, getOutputStream, openFileOutput, startActivityForResult, getLine1Number denoted as APC0, APC1, APC2, APC3, APC4, APC5, APC6, APC7, APC8, APC9 on individual basis

tion, getDeviceId, getPackageManager, getSubscriberId found in total 1360, 1333, 1325, 1247, 1049 APKs correspondingly. Only 4 API calls used in less than 100 APKs, are URL, pdus, mailto, Socket in 21, 2, 1, 1 APKs respectively.

4.4.6 Trivial + String + Reflection Encryption

Analysis found that a total 22 suspicious API calls used this technique. Top usage of API calls getInputStream, getDeviceId, openConnection, getPackageManager found in total 1338, 1308, 1305, 1260, 1048 APKs on an individual basis. Like previous technique 4 API calls used in less than 100 APKs.

4.4.7 Trivial + String + Reflection + Class Encryption

Only 19 suspicious API calls found in this technique which is lowest from other techniques. None of the API calls used in 100 APKs. Top five usage getInputStream, openConnection, getPackageManager, getDeviceId, openFileOutput found in total 7, 7, 6, 5, 5 APKs respectively.

4.5   System Call Analysis

4.5.1   Trivial Encryption

AndroShow analyzed the trivial encryption technique and found that 12 system commands are there. Top five usage of system command mkdir, su, sh, ps, rageagainst-thecage found in total 928, 535, 313, 78, 76 APKs accordingly.

4.5.2   String Encryption

Our study found that in string encryption technique, total 13 system calls are used. Top five usage of system call mkdir, getprop, m7, rageagainstthecage, exploid obtained in total 923, 354, 197, 24, 16 APKs respectively.

4.5.3   Reflection Encryption

In total 14 system calls found from our investigation in this technique. Top five usage of system call mkdir, su, getprop, sh, m7 found in total 928, 527, 369, 315, 197 APKs correspondingly.

4.5.4   Class Encryption

AndroShow found less number of system calls in this technique than previous techniques. Total 9 are found. Top five system calls ln, mkdir, su, chown, mount usage found in total 60, 36, 29, 6, 6 APKs accordingly. Number of usage ratio much lower than previous techniques (Fig. 7).

4.5.5   Trivial + String Encryption

Research study shows that total 10 system commands present in this combination technique. Top five usage of system command mkdir, su, rageagainstthecage, explod, sh found in total 908, 15, 15, 14, 12 APKs on individual basis. Results show that all system commands used in less than 20 APKs except one.

4.5.6   Trivial + String + Reflection Encryption

Analysis found that total 8 system calls used in this technique, also it's the second lowest. Usage of all system calls mkdir, ln, ps, su, sh, rageagainstthecage, exploid, getprop found in total 922, 62, 26, 24, 17, 15, 14, 4.

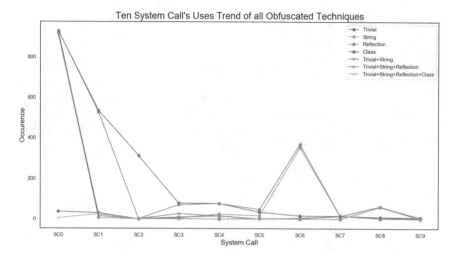

Fig. 7 Top ten system call usage trends in all obfuscation techniques. Where mkdir, su, shh, ps, rageagainstthecage, killall, getprop, exploid, ln, mount denoted as SC0, SC1, SC2, SC3, SC4, SC5, SC6, SC7, SC8, SC9 on individual basis

4.5.7 Trivial + String + Reflection + Class Encryption

Only 4 system calls found in this complex combination technique, which lowest of them all. They are ln, su, mkdir, getprop found in total 59, 24, 4, 2 APKs accordingly.

4.6 Existing Tools and Approaches

In this section, we review the approaches or tools that have been proposed for Android malware detection. Table 6 illustrates some existing tools and methods of Android malware studies.

In [16], authors developed a tool DroidAPIMiner upon on Androguard [69] to extract critical API calls, information of their package level and some dangerous parameters. They also used it for data flow analysis. An innovative detection model, named PermPair have been proposed in [29], that constructs and compares the graphs for dangerous and normal samples by extracting the permission pairs from Android manifest file of an application. Authors in [30], proposed a hybrid Android malware detection model, named NTPDroid, that extracts permissions and network traffic features from the application. By applying FP-growth algorithm to the model, they got enhanced detection rate as compared to use the traffic or permissions alone. A parallel machine learning and information fusion-based Android malware detection model have been introduced in paper [36], named Mlifdect. They first extract eight types of features and then developed a parallel machine learning detection model for speed-

Table 6 Existing tools and approaches

Proposed tools/approach	Feature	Year	Purpose
DroidAPIMiner [16]	API level features	2013	To perform API level feature extraction and data flow analysis
PermPair [29]	Permission pairs	2019	To detect Android malicious application
NTPDroid [30]	Network traffic	2018	To detect Android malware
	System permissions		
Mlifdect [36]	App components	2017	To detect malicious Android application
	Intents, requested		
	Permission, hardware		
	API calls, protected		
	Strings, commands		
	Network		
DroidMat [38]	Permission, activity	2012	To detect Android application benign or malicious
	Service, receiver		
	Intent, API call		
AndroShow (our proposed approach)	Permission, app	2020	To identify pattern of obfuscated Android malware apps
	Component, intent filter,		
	System call, API call		

ing up the process of classification and finally investigate information fusion-based approaches for obtaining detection result. DroidMat [38] is a static feature-based mechanism for detecting the Android malware. It extracts requested permissions, intent, components for tracing API call related to permissions. Next, it applies K-means algorithm to classify the application as benign or malicious. In this paper, we proposed a static analysis-based model, named AndroShow, that extract features and identify the pattern of features. For machine learning based malware detection, malware pattern analysis is very crucial. So, AndroShow will play an important role here as a base for detection of Android malware.

5 Conclusion

5.1 Findings and Contributions

In this study, AndroShow performs a static analysis of obfuscated Android malware applications. Permission, API call, filtered intent, App component and System call features are analyzed. AndroShow demonstrates obfuscated malware using the trend on these features. Several works have been done. Main contribution of this paper analysis is given below:

- Static analysis has been performed on obfuscated Android malware applications.
- Analysis performs on five features—permission, API call, filtered intent, app component, system call.
- Features pattern proposed in 2D matrix. Where column name is the feature tag name and rows are the 0/1 with a family name.
- Most uses are demonstrated in line charts.
- Features extracted from obfuscated malware dataset, PRAGuard. This data-set contains 10,479 obfuscated malware applications with seven different obfuscation techniques.

5.2 Recommendations for Future Works

Future work will be classifying every apk malware family wise. Detection of new malware apps by machine learning based on features patterns can be a good thought.

References

1. Operating system market share worldwide. https://gs.statcounter.com/os-market-share
2. Sen S, Aysan AI, Clark JA (2018) SAFEDroid: using structural features for detecting Android malware. In: Security and privacy in communication networks: SecureComm 2017 international workshops, ATCS and SePrIoT, Niagara Falls, ON, Canada, 22–25 Oct 2017. Proceedings 13. Springer, pp 255–270
3. Alazab M, Broadhurst R (2016) Spam and criminal activity. Trends Issues Crime Criminal Just (Aust Inst Criminol) 52
4. Arp D, Spreitzenbarth M, Hubner M, Gascon H, Rieck K, Siemens CERT (2014) DREBIN: effective and explainable detection of android malware in your pocket. NDSS 14:23–26
5. Saracino A, Sgandurra D, Dini G, Martinelli F (2016) Madam: effective and efficient behavior-based android malware detection and prevention. IEEE Trans Depend Secure Comput
6. Number of smartphones sold to end users worldwide from 2007 to 2020. https://www.statista.com/statistics/263437/global-smartphone-salesto-end-users-since-2007/
7. Huda S, Abawajy J, Alazab M, Abdollalihian M, Islam R, Yearwood J (2016) Hybrids of support vector machine wrapper and filter based framework for malware detection. Future Gener Comput Syst 55:376–390

8. Reina A, Fattori A, Cavallaro L (2013) A system call-centric analysis and stimulation technique to automatically reconstruct android malware behaviors. EuroSec
9. Vinayakumar R, Alazab M, Soman KP, Poornachandran P, Al-Nemrat A, Venkatraman S (2019) Deep learning approach for intelligent intrusion detection system. IEEE Access 7:41525–41550
10. Alazab M (2015) Profiling and classifying the behavior of malicious codes. J Syst Softw 100:91–102
11. Gibler C, Crussell J, Erickson J, Chen H (2012) AndroidLeaks: automatically detecting potential privacy leaks in android applications on a large scale. In: International conference on trust and trustworthy computing. Springer, Berlin, Heidelberg, pp 291–307
12. Backes M, Gerling S, Hammer C, Maffei M, von Styp-Rekowsky P (2014) AppGuard–Fine-grained policy enforcement for untrusted Android applications. Data privacy management and autonomous spontaneous security. Springer, Berlin, Heidelberg, pp 213–231
13. Bugiel S, Davi L, Dmitrienko A, Fischer T, Sadeghi AR, Shastry B (2012) Towards taming privilege-escalation attacks on android. In: NDSS, vol 17, p 19
14. Viswanath H, Mehtre BM (2018) U.S. Patent No. 9,959,406. U.S. Patent and Trademark Office, Washington, DC
15. Zhong X, Zeng F, Cheng Z, Xie N, Qin X, Guo S (2017) Privilege escalation detecting in android applications. In: 2017 3rd international conference on big data computing and communications (BIGCOM). IEEE, pp 39–44
16. Aafer Y, Du W, Yin H (2013) Droidapiminer: mining API-level features for robust malware detection in android. In: International conference on security and privacy in communication systems. Springer, Cham, pp 86–103
17. Demontis A, Melis M, Biggio B, Maiorca D, Arp D, Rieck K, Corona I, Giacinto G, Roli F (2017) Yes, machine learning can be more secure! A case study on Android malware detection. IEEE Trans Depend Secure Comput
18. Egele M, Scholte T, Kirda E, Kruegel C (2012) A survey on automated dynamic malware-analysis techniques and tools. ACM Comput Surv (CSUR) 44(2):6
19. Papadopoulos H, Georgiou N, Eliades C, Konstantinidis A (2017) Android malware detection with unbiased confidence guarantees. Neurocomputing
20. Shabtai A, Moskovitch R, Elovici Y, Glezer C (2009) Detection of malicious code by applying machine learning classifiers on static features: a state-of-the-art survey. Inf Secur Tech Rep 14(1):16–29
21. Burguera I, Zurutuza U, Nadjm-Tehrani S (2011) Crowdroid: behavior-based malware detection system for android. In: Proceedings of the 1st ACM workshop on security and privacy in smartphones and mobile devices. ACM, pp 15–26
22. Fereidooni H, Moonsamy V, Conti M, Batina L (2016) Efficient classification of android malware in the wild using robust static features. Protecting mobile networks and devices: challenges and solutions, p 181
23. Permissions overview. https://developer.android.com/guide/topics/permissions/o-verview
24. Huang, C-Y, Tsai Y-T, Hsu C-H (2013) Performance evaluation on permission-based detection for android malware. Advances in intelligent systems and applications, vol 2. Springer, Berlin, Heidelberg, pp 111–120
25. Felt AP, Chin E, Hanna S, Song D, Wagner D (2011) Android permissions demystified. In: Proceedings of the 18th ACM conference on computer and communications security, pp 627–638
26. Arslan RS, Dogru IA, Barişçi N (2019) Permission-based malware detection system for android using machine learning techniques. Int J Softw Eng Knowl Eng 29(01):43–61
27. Yildiz O, Dogru IA (2019) Permission-based android malware detection system using feature selection with genetic algorithm. Int J Softw Eng Knowl Eng 29(02):245–262
28. Li J, Sun L, Yan Q, Li Z, Srisa-an W, Ye H (2018) Significant permission identification for machine-learning-based android malware detection. IEEE Trans Ind Inform 14(7):3216–3225
29. Arora A, Peddoju SK, Conti M (2019) PermPair: android malware detection using permission pairs. IEEE Trans Inf Forens Secur 15:1968–1982

30. Arora A, Peddoju SK (2018) NTPDroid: a hybrid android malware detector using network traffic and system permissions. In: 2018 17th IEEE international conference on trust, security and privacy in computing and communications/12th IEEE international conference on big data science and engineering (TrustCom/BigDataSE). IEEE, pp 808–813

31. Şahın DO, Kural OE, Akleylek S, Kiliç E (2018) New results on permission based static analysis for Android malware. In 2018 6th international symposium on digital forensic and security (ISDFS). IEEE, pp 1–4

32. Wang C, Xu Q, Lin X, Liu S (2018) Research on data mining of permissions mode for Android malware detection. Cluster Comput 22:13337–13350

33. Motiur Rahman SSM, Saha SK, (2019) StackDroid: evaluation of a multi-level approach for detecting the malware on android using stacked generalization. In: Santosh K, Hegadi R (eds) Recent trends in image processing and pattern recognition. RTIP2R 2018. Communications in computer and information science, vol 1035. Springer, Singapore

34. Rana MS, Rahman SS, Sung AH (2018) Evaluation of tree based machine learning classi-fiers for android malware detection. In: International conference on computational collective intelligence. Springer, Cham, pp 377–385

35. App components. https://developer.android.com/guide/components/fundamentals

36. Wang X, Zhang D, Su X, Li W (2017) Mlifdect: android malware detection based on parallel machine learning and information fusion. Secur Commun Netw 2017

37. Android—application components. https://www.tutorialspoint.com/android/androidapplicationcomponents.htm

38. Wu DJ, Mao CH, Wei TE, Lee HM, Wu KP (2012) Droidmat: android malware detection through manifest and API calls tracing. In: 2012 seventh Asia joint conference on information security. IEEE, pp 62–69

39. Kim T, Kang B, Rho M, Sezer S, Im EG (2018) A multimodal deep learning method for android malware detection using various features. IEEE Trans Inf Forens Secur 14(3):773–788

40. Shen T, Zhongyang Y, Xin Z, Mao B, Huang H (2014) Detect android malware variants using component based topology graph. In: 2014 IEEE 13th international conference on trust, security and privacy in computing and communications. IEEE, pp 406–413

41. Li C, Mills K, Niu D, Zhu R, Zhang H, Kinawi H (2019) Android malware detection based on factorization machine. IEEE Access 7:184008–184019

42. Rana MS, Gudla C, Sung AH (2018) Evaluating machine learning models for android malware detection: a comparison study. In: Proceedings of the 2018 VII international conference on network, communication and computing. ACM, pp 17–21

43. Android developers, intents and intent filters. https://developer.android.com/guide/components/Intents-filters

44. Xu K, Li Y, Deng RH (2016) Iccdetector: ICC-based malware detection on android. IEEE Trans Inf Forens Secur 11(6):1252–1264

45. Elish KO, Yao D, Ryder BG (2015) On the need of precise inter-app ICC classification for detecting Android malware collusions. In: Proceedings of IEEE mobile security technologies (MoST), in conjunction with the IEEE symposium on security and privacy

46. Feng Y, Anand S, Dillig I, Aiken (2014) Apposcopy: semantics-based detection of android malware through static analysis. In: Proceedings of the 22nd ACM SIGSOFT international symposium on foundations of software engineering. ACM, pp 576–587

47. Feizollah A, Anuar NB, Salleh R, Suarez-Tangil G, Furnell S (2017) Androdialysis: analysis of android intent effectiveness in malware detection. Comput Secur 65:121–134

48. Li L, Bartel A, Bissyandé TF, Klein J, Le Traon Y, Arzt S, Rasthofer S, Bodden E, Octeau D, McDaniel P (2015) Iccta: detecting inter-component privacy leaks in android apps. In: Proceedings of 60 ©Daffodil International University the 37th international conference on software engineering, vol 1. IEEE Press, pp 280–291

49. Li L, Bartel A, Klein J, Le Traon Y (2014) Automatically exploiting potential component leaks in android applications. In: 2014 IEEE 13th international conference on trust, security and privacy in computing and communications. IEEE, pp 388–397

50. What exactly IS an API? https://medium.com/@perrysetgo/what-exactly-is-an-API-69f36968a41f
51. Peiravian N, Zhu X (2013) Machine learning for android malware detection using permission and API calls. In: 2013 IEEE 25th international conference on tools with artificial intelligence. IEEE, pp 300–305
52. Seo SH, Gupta A, Sallam AM, Bertino E, Yim K (2014) Detecting mobile malware threats to homeland security through static analysis. J Netw Comput Appl 38:43–53
53. Yang M, Wang S, Ling Z, Liu Y, Ni Z (2017) Detection of malicious behavior in android apps through API calls and permission uses analysis. Concurr Comput: Pract Exp 29(19):e4172
54. Skovoroda A, Gamayunov D (2017) Automated static analysis and classification of Android malware using permission and API calls models. In: 2017 15th annual conference on privacy, security and trust (PST). IEEE, pp 243–24309
55. Shen F, Del Vecchio J, Mohaisen A, Ko SY, Ziarek L (2018) Android malware detection using complex-flows. IEEE Trans Mob Comput
56. Ghani SMA, Abdollah MF, Yusof R, Mas'ud MZ (2015) Recognizing API features for malware detection using static analysis. J Wirel Netw Commun 5(2A):6–12
57. Malik S, Khatter K (2016) System call analysis of android malware families. Indian J Sci Technol 9(21)
58. Dimjašević M, Atzeni S, Ugrina I, Rakamaric Z (2015) Android malware detection based on system calls. Tech. Rep, University of Utah
59. Firdaus A, Anuar NB (2015) Root-exploit malware detection using static analysis and machine learning. In: Proceedings of the fourth international conference on computer science & computational mathematics (ICCSCM 2015). Langkawi, Malaysia, pp 177–183
60. Da C, Hongmei Z, Xiangli Z (2016) Detection of Android malware security on system calls. In: 2016 IEEE advanced information management, communicates, electronic and automation control conference (IMCEC). IEEE, pp 974–978
61. Kedziora M, Gawin P, Szczepanik M, Jozwiak I (2018) Android malware detection using machine learning and reverse engineering. Comput Sci Inf Technol (CS&IT) 95–107
62. Tchakounté F, Dayang P (2013) System calls analysis of malware on android. Int J Sci Technol 2(9):669–674
63. Wahanggara V, Prayudi Y (2015) Malware detection through call system on android smartphone using vector machine method. In: 2015 fourth international conference on cyber security, cyber warfare, and digital forensic (CyberSec). IEEE, pp 62–67
64. Maiorca D, Ariu D, Corona I, Aresu M, Giacinto G (2015) Stealth attacks: an extended insight into the obfuscation effects on android malware. Comput Secur 51:16–31
65. Rastogi V, Chen Y, Jiang X (2013) Droidchameleon: evaluating android anti-malware against transformation attacks. In: Proceedings of the 8th ACM SIGSAC symposium on information, computer and communications security. ACM, pp 329–334
66. Android PRAGuard Dataset. http://pralab.diee.unica.it/en/AndroidPRAGuardD-ataset
67. MalGenome. http://www.malgenomeproject.org/
68. Contagio. http://contagiominidump.blogspot.com/
69. Androguard. https://github.com/androguard/androguard

IntAnti-Phish: An Intelligent Anti-Phishing Framework Using Backpropagation Neural Network

Sheikh Shah Mohammad Motiur Rahman, Lakshman Gope, Takia Islam, and Mamoun Alazab

Abstract Among the cybercriminals, the popularity of phishing has been rapidly growing day by day. Therefore, phishing has become an alarming issue to solve in the field of cybersecurity. Many researchers have already proposed several anti-phishing approaches to detect phishing in terms of email, webpages, images, or links. This study also aimed to propose and implement an intelligent framework to detect phishing URLs (Uniform Resource Locator). It has been observed in this study that Backpropagation Neural Network-based systems need to tune various hyperparameters to obtain the optimized output. With a maximum of two hidden layers along with 400 epochs can reach maximum accuracy of 0.93, the minimum mean squared error of 0.27, and also a minimum error rate of 0.07 which measurements lead this study to generate an optimized model for phishing detection. The detailed process of feature extraction and optimized model generation along with the detection of unknown URLs are considered and proposed during the development of IntAnti-Phish (An Intelligent Anti-Phishing Framework).

Keywords Malicious URLs detection · Anti-phishing framework · Intelligent phishing detection · Neural network · Backpropagation

S. S. M. M. Rahman (✉) · L. Gope · T. Islam
Department of Software Engineering, Daffodil International University, Dhaka, Bangladesh
e-mail: motiur.swe@diu.edu.bd

L. Gope
e-mail: lakshmangope012@gmail.com

T. Islam
e-mail: takiaislam9@gmail.com

M. Alazab
College of Engineering, IT and Environment, Charles Darwin University, Casuarina, Australia
e-mail: alazab.m@ieee.org

© The Editor(s) (if applicable) and The Author(s), under exclusive license to Springer
Nature Switzerland AG 2021
Y. Maleh et al. (eds.), *Machine Intelligence and Big Data Analytics for Cybersecurity Applications*, Studies in Computational Intelligence 919,
https://doi.org/10.1007/978-3-030-57024-8_9

217

1 Introduction

Phishing is a blueprint or scheme based on criminal activities getting attracted to attackers. Exposing users financial information such as credit card information, pin numbers as well as sensitive information like passwords, login credentials, and some personal information for social engineering is possible through phishing. After that, using the exposed information attackers are able to gain financial access and commit fraudulent actions [1, 2]. Using various types of social media (such as emails, private chat messages, blogs, forums as well as on banners), attackers are used to sending a URL—Universal Resource Locator to perform phishing. The malicious URL or link represents itself as an authentic source. According to an estimation, these kinds of attacks have made over 3 billion dollar financial losses annually [3]. Generally, cybercriminals are using three ways to exploit phishing attacks [4]. The first one is making replication of trusted sources web interfaces which is known as web-based phishing. By which, victims will think about the sources as authentic and will provide their sensitive credentials. Secondly, email-based phishing where criminals will send an email with phishing content including web-based techniques as well. Finally, malware-based phishing is also being performed where attackers will inject malicious codes into the victim's system [5].

Web-based phishing has more activities than others such as it is also included during performing the email-based techniques. Furthermore, intelligent anti-phishing frameworks and approaches can be separated into 2 groups such as email-based and web-based. Among the studies and solutions from literature, the maximum is based on malicious links or URLs [6, 7]. However, why is machine learning-based intelligent anti-phishing frameworks? Because at the beginning stages of phishing detection research, there were multiple approaches commonly used including blacklisting, regular expression, and signature matching approaches which failed to detect the new URLs or the variant of existing URLs. Moreover, the database of signatures has to be regularly updated for handling the new patterns of malicious URLs. After that, machine learning techniques and approaches were used to detect the new as well as the variant of malicious URL effectively. However, by the growth of research in machine learning-based research, it is found that deep learning-based architectures performed well in comparison to the conventional machine learning algorithms [8]. Thus, it is considered in this study to make a new framework so that it can be easily identified whether the provided URL is phishing or not by means of a neural network-based approach. The main contributions of this paper can be stated as follows:

- Generate the pattern and detect the new URL by extracting features in real-time.
- Identify the URL whether phishing or not from unlabeled and unknown URLs.
- Representing an intelligent Anti-phishing framework with a detailed procedure to develop anti-phishing approaches or architectures.
- Practical implementation of Backpropagation Neural Network-based anti-phishing frameworks.
- Impact of learning rate (configurable hyperparameter) in neural network-based approaches.

- Evaluation of training accuracy in terms of the changes in learning rate (configurable hyperparameter).

The organization of the rest of the paper is constructed as: Sect. 2 represents the background study. In Sect. 3, the proposed framework and related procedures are described in detail. Experiments, evaluation parameters along with the result discussion are discussed in Sect. 4. Finally, Sect. 5 concludes with future steps.

2 Background

Phishing is a criminal mechanism that employs both social engineering and technical subterfuge to steal customers' personal identity information and financial account data and financial account credentials of consumers [9]. There are many types of phishing and they are Algorithm-Based Phishing, Email Phishing, Link manipulation, Spear phishing, Domain Spoofing, Phishing via HTTPS, SMS, Pop-ups. Phishing attributes are IP address, URL length, '@' symbol, double slash, prefix, suffix, sub-domain, port, https token, request URL, URL anchor, links in tags, age of the domain. Detection of phishing has become a significant concern among researchers.

There are lots of proposed solutions that have been provided by researchers. For example, Cui et al. [10] proposed to detect phishing websites via a hierarchical clustering approach which groups the vectors generated from DOMs together according to their proportional distance. Some studies [11–14] focused on detecting phishing URLs by leveraging the potential characteristics of URLs. Zhang et al. [15] proposed CANTINA, a completely unique HTML content method for identifying phishing websites. Xiang et al. [16] is an upgraded version of CANTINA and proposed CANTINA + . Huang et al. [17] proposed an SVM based technique to detect phishing URLs. Yuancheng et al. [18] proposed a semi-supervised based method for the detection of phishing web pages. Islam et al. [19] proposed filtering phishing emails with the message content and header using a multi-tier classification model. Chen et al. [20] have proposed a hybrid approach that mixes extraction of key phrase, textual, financial data to ascertain the various phishing attacks using supervised classification strategies. Nishanth et al. [21] have proposed a method in which the structured style of the financial data is mined using machine learning algorithms. However, maximum researchers worked with the labeled data and supervised learning rather than unknown data labeling. The general process of any classification process by supervised learning is to work with labeled data performing train: test splitting where this study focused on labeling the unlabeled data. On the other hand, Deep Learning is a subfield of machine learning concerned with algorithms inspired by the structure and function of the brain called artificial neural networks. Behind the deep learning methods, the neural network architectures work. Usually, neural networks with large numbers of layers are known as deep neural networks or deep learning. There are various types of neural networks including Artificial Neural

Table 1 Determination rules of hidden layers

Num hidden layers	When to use
0	For representing the linear decisions
1	A continuous mapping from one finite to another
2	An arbitrary decision with rational activation functions
>2	Computer vision, time series or with complex datasets

Networks (ANN), Convolutional Neural Networks (CNNs), Restricted Boltzmann Machines (RBMs), and so on. The backpropagation neural network architecture is a hierarchical design consisting of fully interconnected layers or rows of processing units which is developed by Rumelhart et al. (1986), which is the most prevalent of the supervised learning models of ANN. Back-propagation is one of the self-learning methods of ANN to give the desired answers [22]. A number of hidden layers in neural networks are usually 1–2 layers and in case of deep learning the number of layers varies but it requires almost more than 150 layers. There are some rules to determine the number of layers that include two or fewer layers for simple data sets and for computer vision, time series, or with complex datasets additional layers can provide better results. The determination rules of hidden layers [23] are tabulated in Table 1.

3   IntAnti-Phish: The Proposed Approach

The methodology of the implemented approach includes three major phases such as: Model Generation Phase, Features Extraction and Pattern Generation Phase, and finally Detection or Test Phase with Output. The phases will be broadly explained in this section which is depicted in Fig. 1.

3.1   Model Generation Phase

In this phase, a model will be trained and stored from existing labeled data. Thus, a publicly available real dataset has been collected as labeled data for training of the model. For generating the model, a neural network approach has been created using backward function rather than forward function because backward function refers to learning mode. Get the network result or output using forward mode which has been compared to the expected result for a known data point and from the result layer to the input layer propagate the error back which is known as backpropagation process. During the training process, parameter adjustment has been performed. While the model has been trained with optimized parameters then it's been saved for detection of new URLs.

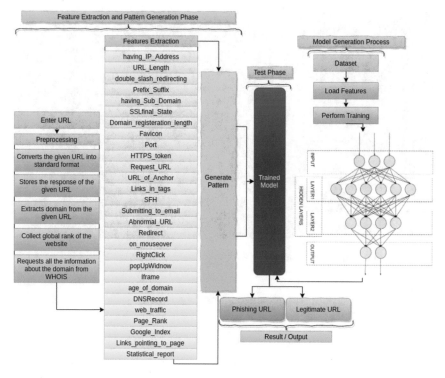

Fig. 1 Architectural framework of proposed approach

3.2 Feature Extraction and Pattern Generation Phase

Feature Extraction and Pattern Generation Phase is the main phase of the proposed and implemented framework. In this phase, it will extract the required features in the required format and will generate the pattern of any given URL to detect whether it is phishing or not. This phase has three sub-phase such as preprocessing, feature extraction, and generating patterns.

In the preprocessing phase, the information about the given URL has been extracted in a standard format. For example, a URL has been given to the system as input and as a preprocessed information of the provided URL has been found as follows:

Standard Format of URL:
https://netcloud.jdevcloud.com/wp-includes/zinnet/2e164eca08165a9365735c1f2d46a3f9
Response Code: 200.
Response HTML Code (Partial from Terminal):
The responses from the provided link have been visualized in Fig. 2. Here, the partial view of the full HTML (Hypertext Markup Language) has been captured.

```
<li><span class="tips-recovery"></span>Use our <a href="/login/lostpassword">login recovery tool</a> to retrieve your Toolbox username and pas
sword. You'll need some ID or the mobile number you supplied as your billing contact.</li>
<li><span class="tips-password"></span>Changing passwords every few months ensures the security of your account. If you'd like to change your
password, visit <strong>My Account</strong> after logging in.</li>
<li><span class="tips-support"></span>Keep track of your support requests - you can find all the open and recent tasks associated with your ac
count under <strong>My Account</strong>.</li>
<li><span class="tips-bills"></span>Save $1.49 each month by choosing to receive bills via email. Visit <strong>Update Contact Details</strong
> in <strong>My Account</strong> and <strong>edit</strong> your preference.</li>
</ul>
</div>
</div>
<div class="footer"></div>
<!--[if lt IE 9]>
        <script src="/bundles/ie8scripts?v=AqnKJnbq_hYIQCcAbhcZXyllYhYEuyNISiN2xRJ_lwo1"></script>

    <![endif]-->
<script src="/bundles/footer-scripts?v=-rfgjfrchiBDOxmRs1nSVKU00Dty_MxXFVrd_OOlXZ41"></script>
<script src="/bundles/iinet?v=OsxAUqisYGvcfBnPl2t-G1IML3bAtFMdy8BgmThVG_Y1"></script>
<script type="text/javascript">
        $(function() {
            $("#show-password")
                .change(function() {
                    var checked = $(this).is(":checked");
                    $('#Password')
                        .clone()
                        .attr('type', checked ? 'text' : 'password')
                        .insertAfter('#Password')
                        .prev()
                        .remove();
                });
        });
</script>
<!-- BANNER MANAGEMENT SYSTEM -->
<script src="https://www.iinet.net.au/_library/banners/toolbox-homepage/script"></script>
<!-- END BANNER MANAGEMENT SYSTEM -->
<script id="page-alerts" type="text/javascript">
```

Fig. 2 Output screenshot of response HTML code

Domain/Subdomain Name: netcloud.jdevcloud.com.
WHOIS Information:
{

    "domain_name": [

        "JDEVCLOUD.COM",
        "jdevcloud.com".

],

    "registrar": "ENOM, INC.",
    "whois_server": "WHOIS.ENOM.COM",
    "referral_url": null,
    "updated_date": "2020–02-17 01:15:06",
    "creation_date": [

        "2015–02-16 16:43:09",
        "2015–02-16 16:43:00"

],
    "expiration_date": [

        "2022–02-16 16:43:09",
        "2022–02-16 16:43:00"

],
    "name_servers": [

        "NS1.GRIDFAST.NET",
        "NS2.GRIDFAST.NET"

],
"status": [

"clientTransferProhibited https://icann.org/epp#clientTransferProhibited",
"clientTransferProhibited   https://www.icann.org/epp#clientTransferProh
ibited".

],
"emails": "ABUSE@ENOM.COM",
"dnssec": "unsigned",
"name": "REDACTED FOR PRIVACY",
"org": "REDACTED FOR PRIVACY",
"address": "REDACTED FOR PRIVACY",
"city": "REDACTED FOR PRIVACY",
"state": "Michigan",
"zipcode": "REDACTED FOR PRIVACY",
"country": "US".

}

This above information collected from the preprocessing phase has been stored for feature extraction and generating patterns phase.

In the feature extraction and generating patterns phase, the features described in Table 1 have been extracted from the given URL, and the algorithms during development have followed from one previous study [24]. There are some built-in packages of python are used during the implementation including ipaddress [25], urllib.request [26], bs4 [27], socket [28], requests [29], googlesearch [30], whois [31], datetime [32], dateutil.parser [33]. The features are denoted as Feature 1 (F1): Having IP Address, Feature 2 (F2): URL Length, Feature 3 (F3): Shortening Service, Feature 4 (F4): Having '@' Symbol, Feature 5 (F5): Double Slash Redirecting, Feature 6 (F6): Prefix and Suffix, Feature 7 (F7): Having Sub Domain, Feature 8 (F8): SSLfinal State, Feature 9 (F9): Domain Registration Length, Feature 10 (F10): Favicon, Feature 11 (F11): Port, Feature 12 (F12): HTTPS token, Feature 13 (F13): Request URL, Feature 14 (F14): URL of Anchor, Feature 15 (F15): Links in tags, Feature 16 (F16): SFH, Feature 17 (F17): Submitting to email, Feature 18 (F18): Abnormal URL, Feature 19 (F19): Redirect, Feature 20 (F20): On mouseover, Feature 21 (F21): Right Click, Feature 22 (F22): Pop Up Window, Feature 23 (F23): Iframe, Feature 24 (F24): Age of Domain, Feature 25 (F25): Domain Name System (DNS) Record, Feature 26 (F26): Web Traffic, Feature 27 (F27): Page Rank, Feature 28 (F28): Google Index, Feature 29 (F29): Links pointing to another page and Feature 30 (F30): Statistical Report.

After performing the features extraction process a vector matrix of the provided URL has generated which is the pattern of the provided URL and ready to send it to the model so that model can detect whether it is phishing or not. The provide sample URL provides the pattern as like as follows:

Fig. 3 Output screenshot of the implemented framework

[1, −1, 1, 1, 1, −1, 0, 1, 1, −1, 1, −1, 1, −1, 1, 0, 1, −1, −1, −1, −1, −1, 1, 1,
1, −1, 1, 1, −1, 1]

The generated pattern represents the features as following sequences:

[F1, F2, F3, F4, F5, F6, F7, F8, F9, F10, F11, F12, F13, F14, F15, F16, F17, F18,
F19, F20, F21, F22, F23, F24, F25, F26, F27, F28, F29, F30]

Finally, the model which was saved by training from labeled data in the Model
Generation Phase has obtained the required format of the provided URL to detect.

3.3 Detection and Test Phase:

In this phase, the proposed model detects the URL whether it is phishing or not
from the generated pattern. The sample output was a "Phishing" screenshot of the
implemented framework in Fig. 3.

4 Experimental Results Analysis and Discussion

4.1 Environment Setup

The experimental environment has a machine whose configuration is Intel(R)
Core(TM) i5-6500 CPU @ 3.20 GHz processor, 64-bit PC, and 16 GB RAM. The
operating system is Ubuntu 18.04.1 LTS (Bionic Beaver). Python has been used to
implement the framework and also some packages which are available in python.

4.2 Dataset Used

For training or generating the model, a publicly available dataset has been used which is being collected from the UCI repository. Dataset has 4898 and 6157 are phishing and legitimate URLs respectively total 11,055 different types of URLs. A total of 30 features are considered to train the model [24, 34, 35]. The features or attributes used in the dataset are tabulated in Table 2.

4.3 Experiments, Results and Discussion

A detailed experiment during the implementation of the system has been performed. First of all, deep learning was in concern without any special reason, and from deep learning, it's been found that the system may be over fitted. As the used dataset is simple and the decision also has simplicity, deep learning with 200 epochs wasn't found as a well-fitted approach. Then, using backpropagation, the neural network performs well with the dataset for better training in terms of time and complexity.

It's been clearly and carefully looked into that there are multiple parameters that need to be in consideration during the development of any neural network architecture based system. Thus, there are 20 cases that have been tested and evaluated for determining the optimized parameters for the developed system tabulated in Table 3.

Here, the number of hidden layers and the number of epochs have been denoted as HL and EP respectively. HL2 means the number of hidden layers is 2 and EP100 means the number of epochs is 100. By following the determination rules of the hidden layer from Table 1, it has been identified that the dataset has better fitness to HL2. During the experiment, it has also been tested with HL200 in case of deep learning how the system works and found that the system overfits during the training time. Thus, it has been avoided during the final model generation. In one sentence, the model of the system has been tuned by optimized parameters before determining the final model generation.

Accuracy, mean squared error (MSE), and the error rate has been considered to evaluate the optimized parameters. However, backpropagation which represents the learning mode has been used to develop the neural network-based architecture. There are two hidden layers used. Each of the layers processes their own input and computes an output by using a formula:

$O = F(Wt * In)$, where O represents output, F represents a function, Wt and In represent weights and input respectively.

F is called the activation function which is a nonlinear function. It has been conveniently chosen between the sigmoid, hyperbolic tangent, or rectified linear unit for this function.

The formula of the network can be written by using the above formula is:

Table 2 Different attributes/features in the dataset

Address bar based	Abnormal based	HTML and JavaScript-based	Domain-based
An IP address is used as a substitute of a domain name	A high portion of "**anchors**" in a legitimate webpage	Phishers use "Status Bar Customization" to display a fake URL in the status bar	WHOIS database carries the feature of "Age of domain"
To hide the faltering part long URL is implemented	Request URL	To prohibit a user to view and save the webpage source code Phishers impose "Disabling Right Click"	DNS record
"TinyURL", a shortening web service, provides short aliases for redirection of long URLs	Use <Meta> , <Script> and <Link> tags in links	"Using pop-up window" to get user information	Website traffic
Ignoring the previous section of a URL using "@" symbol	"Server Form Handler" (SFH) carry suspicious empty string	IFrame redirection	Page rank
To redirect a user automatically to another page, used "//"	Provide data to email		Phishing web pages are not available in "Google index"
Adding prefix or suffix separated by "-" to the domain name	WHOIS database carries the feature of "Abnormal URL"		Legitimacy level can be assumed by "Number of Links Pointing to Page"
Sub Domain and Multi-Sub Domains			Statistical-reports based feature
Hyper text transfer protocol with secure socket layer (HTTPS)			

(continued)

Table 2 (continued)

Address bar based	Abnormal based	HTML and JavaScript-based	Domain-based
Leverage website favicon to detect phishing websites			
Domain registration length			
Utilizing non-standard port			
Presence of "HTTPs" token in the domain part of the URL			

Table 3 Evaluation parameters for assessment of the classifiers

Serial	Label	Epochs	Learning rate	Accuracy	MSE	Error rate
1	HL2_EP100	100	0.1	0.58	1.68	0.42
2	HL2_EP100	100	0.01	0.92	0.31	0.08
3	**HL2_EP100**	**100**	**0.001**	**0.93**	**0.30**	**0.07**
4	HL2_EP100	100	0.0001	0.89	0.46	0.11
5	HL2_EP200	200	0.1	0.58	1.68	0.42
6	**HL2_EP200**	**200**	**0.01**	**0.93**	**0.30**	**0.07**
7	HL2_EP200	200	0.001	0.92	0.30	0.08
8	HL2_EP200	200	0.0001	0.91	0.37	0.09
9	HL2_EP300	300	0.1	0.58	1.66	0.42
10	**HL2_EP300**	**300**	**0.01**	**0.93**	**0.28**	**0.07**
11	HL2_EP300	300	0.001	0.92	0.31	0.08
12	HL2_EP300	300	0.0001	0.91	0.34	0.09
13	HL2_EP400	400	0.1	0.58	1.68	0.42
14	**HL2_EP400**	**400**	**0.01**	**0.93**	**0.27**	**0.07**
15	HL2_EP400	400	0.001	0.92	0.31	0.08
16	HL2_EP400	400	0.0001	0.82	0.33	0.18
17	HL2_EP500	500	0.1	0.58	1.66	0.42
18	**HL2_EP500**	**500**	**0.01**	**0.93**	**0.29**	**0.07**
19	HL2_EP500	500	0.001	0.92	0.31	0.08
20	HL2_EP500	500	0.0001	0.92	0.33	0.08

$$Y = F\_3(Wt\_3 * F\_2(Wt\_2 * F\_1(Wt\_1 * X + B\_1) + B\_2) + B\_3)$$

Training this implemented neural network simply means optimizing the weights (Wt1, Wt2, Wt3) and the biases (B1, B2, B3) such that Y is as close to the expected output as possible. Where, Wt1, B1 and F1 refer to the first hidden layer, Wt2, B2, and F2 refer to the second and Wt3, B3 and F3 refer to the third hidden layer. For the activation function which is used to introduce non-linearities in the mix, sigmoid has been used which is one of the most popular activation functions. Sigmoid function squeezes the provided input within the 0 to 1 interval [36].

From the experiment and the testbed setup, it has been clearly visible in Table 3 that among all the assessment or evaluation parameters, the learning rate has a significant contribution to the performance of neural network-based systems. It's been found that with two hidden layers along with 400 epochs and 0.01 learning rate (HL2_EP400) provides the maximum 0.93 of accuracy, 0.27 of MSE, and 0.7 of error rate which is much applicable rather than all other.

In addition, deep learning was also being tested but got no improvement with the dataset. Moreover, randomly selected features of the training set also lead to an imbalance result every time. To solve the random selection problems seed has been used in coding. So that every time, the system will select the same features for training. After that, k-fold cross-validation has been performed to minimize the biases of the system during the model generation process.

Finally, the reliable and optimized parameters are found from more than 20 case studies based experiments such as two hidden layers, 400 epochs, and 0.01 learning rate. Using the optimized parameters then the final model has been stored and saved for further use. The saved model has the knowledge base of phishing URLs features and attributes and is also able to predict or detect the new URLs provided to the system which are not labeled. The list of URLs from openphish has been collected and tested to the proposed system. It's been obtained from the tested result that in case of maximum output, the system can detect or label the unknown URLs correctly compared to openphish, PhishTank, and some other databases that exist on the web but there are some exceptions as well.

5 Conclusion

It can be recapitulated that the implemented frameworks have significant value as the number of attackers' attention is increased in phishing. In order to develop anti-phishing approaches or systems using neural networks, it has been witnessed that parameter adjustment has a major role to play. Especially, the learning rate of the training model has a clear vision to enhance the overall performance of the neural network-based systems. In addition, a neural network-based system has been implemented and briefly evaluated in case of phishing URLs detection. The impact of configurable hyperparameters has been evaluated and witnessed during the development of the neural network architecture inspired anti-phishing approaches. The number of hidden layers should be in small amounts, usually 2–3 in case of a simple dataset. It has also been claimed that the number of data in a dataset has an impact on the learning base. From the experiments, it has been observed that the system sometimes provides misclassification of data during prediction or detection which can be overcome by training the model using more data as well as by enhancing the dimension of the dataset.

References

1. Abutair HY, Belghith A (2017) Using case-based reasoning for phishing detection. Procedia Comput Sci 109:281–288
2. Tan CL, Chiew KL (2014) Phishing website detection using URL-assisted brand name weighting system. In: Intelligent Signal Processing and Communication Systems (ISPACS), 2014 International Symposium. IEEE, pp 054–059
3. Shirazi H, Bezawada B, Ray I (2018) KnOw Thy Domaln Name: unbiased phishing detection using domain name based features. In: Proceedings of the 23rd ACM on Symposium on Access Control Models and Technologies. ACM, pp 69–75
4. Hutchinson S, Zhang Z, Liu Q (2018) Detecting phishing websites with random forest. In: International Conference on Machine Learning and Intelligent Communications. Springer, Cham, pp 470–479
5. Dong Z, Kapadia A, Blythe J, Camp LJ (2015) Beyond the lock icon: real-time detection of phishing websites using public-key certificates. In: Electronic Crime Research (eCrime), 2015 APWG Symposium. IEEE, pp 1–12
6. Benavides E, Fuertes W, Sanchez S, Sanchez M (2020) Classification of phishing attack solutions by employing deep learning techniques: a systematic literature review. In: Developments and advances in defense and security. Springer, Singapore, pp 51–64
7. Tran KN, Alazab M, Broadhurst R (2014) Towards a feature-rich model for predicting spam emails containing malicious attachments and URLs
8. Harikrishnan NB, Soman V, Annappa B, Alazab M (2019) Deep learning architecture for big data analytics in detecting malicious URL. In: Khalid at al. (ed) Big data recommender systems: recent trends and advances. Institution of Engineering and Technology (IET)
9. Huang Y, Yang Q, Qin J, Wen W (2019) Phishing URL detection via CNN and attention-based hierarchical RNN. In: 2019 18th IEEE International Conference On Trust, Security And Privacy In Computing And Communications/13th IEEE International Conference On Big Data Science And Engineering (TrustCom/BigDataSE). IEEE, pp 112–119
10. Cui Q, Jourdan GV, Bochmann GV, Couturier R, Onut IV (2017) Tracking phishing attacks over time. In: Proceedings of the 26th International Conference on World Wide Web, pp 667–676
11. Bahnsen AC, Bohorquez EC, Villegas S, Vargas J, González FA (2017) Classifying phishing URLs using recurrent neural networks. In: the 2017 APWG Symposium on Electronic Crime Research (eCrime). IEEE, pp 1–8
12. Le H, Pham Q, Sahoo D, Hoi SC (2018) URLnet: learning a URL representation with deep learning for malicious URL detection. arXiv preprint arXiv:1802.03162
13. Saxe J, Berlin K (2017) eXpose: a character-level convolutional neural network with embeddings for detecting malicious URLs, file paths, and registry keys. arXiv preprint arXiv:1702.08568
14. Rahman SSMM, Rafiq FB, Toma TR, Hossain SS, Biplob KBB (2020) Performance assessment of multiple machine learning classifiers for detecting the phishing URLs. In: Raju K, Senkerik R, Lanka S, Rajagopal V (eds) Data engineering and communication technology. Advances in intelligent systems and computing, vol 1079. Springer, Singapore. https://doi.org/10.1007/978-981-15-1097-7_25

15. Zhang Y, Hong JI, Cranor LF (2007) Cantina: a content-based approach to detecting phishing web sites. In: Proceedings of the 16th International Conference on World Wide Web, pp 639–648

16. Xiang G, Hong J, Rose CP, Cranor L (2011) Cantina+ a feature-rich machine learning framework for detecting phishing web sites. ACM Trans Inf Syst Secur (TISSEC) 14(2):1–28

17. Huang H, Qian L, Wang Y (2012) An SVM-based technique to detect phishing URLs. Inf Technol J 11(7):921–925

18. Li Y, Xiao R, Feng J, Zhao L (2013) A semi-supervised learning approach for detection of phishing webpages. Optik 124(23):6027–6033

19. Islam R, Abawajy J (2013) A multi-tier phishing detection and filtering approach. J Netw Comput Appl 36(1):324–335

20. Chen X, Bose I, Leung ACM, Guo C (2011) Assessing the severity of phishing attacks: A hybrid data mining approach. Decis Support Syst 50(4):662–672

21. Nishanth KJ, Ravi V, Ankaiah N, Bose I (2012) Soft computing based imputation and hybrid data and text mining: The case of predicting the severity of phishing alerts. Expert Syst Appl 39(12):10583–10589

22. LeCun Y, Bengio Y, Hinton G (2015) Deep learning. Nature 521(7553):436–444

23. Heaton J (2015) Artificial intelligence for humans. Neural Networks and Deep Learning, 1.0. Chesterfield, vol 3. Heaton Research Inc., USA

24. Mohammad RM, Thabtah F, McCluskey L (2014a) Intelligent rule-based phishing websites classification. IET Inf Secur 8(3):153–160

25. "An introduction to the ipaddress module", https://docs.python.org/3/howto/ipaddress.html https://docs.python.org/3/howto/ipaddress.html. Last accessed 19 March 2020

26. "urllib — URL handling modules", https://docs.python.org/3/library/urllib.html last accessed 19 March 2020

27. "beautifulsoup4 4.8.2", https://pypi.org/project/beautifulsoup4/. Last accessed 19 March 2020

28. "socket—Low-level networking interface", https://docs.python.org/3/library/socket.html. Last accessed 19 March 2020

29. "requests 2.23.0", https://pypi.org/project/requests/. Last accessed 19 March 2020

30. "Welcome to googlesearch's documentation!", https://python-googlesearch.readthedocs.io/en/latest/. Last accessed 19 March 2020

31. "whois 0.9.6", https://pypi.org/project/whois/. Last accessed 19 March 2020

32. "DateTime—Basic date and time types", https://docs.python.org/3/library/datetime.html. Last accessed 19 March 2020

33. "dateutil—powerful extensions to DateTime", https://dateutil.readthedocs.io/en/stable/. Last accessed 19 March 2020

34. Mohammad RM, Thabtah F, McCluskey L (2014b) Predicting phishing websites based on self-structuring neural network. Neural Comput Applic 25:443–458. https://doi.org/10.1007/s00521-013-1490-z

35. "Index of/ml/machine-learning-databases/00327", https://archive.ics.uci.edu/ml/machine-learning-databases/00327/'. Last accessed 16 March 2020

36. Introduction to Deep Learning—Sentiment Analysis, https://nlpforhackers.io/deep-learning-introduction/. Accessed 22 March 2020

Network Intrusion Detection for TCP/IP Packets with Machine Learning Techniques

Hossain Shahriar and Sravya Nimmagadda

Abstract To address the evolving strategies and techniques employed by hackers, intrusion detection systems (IDS) is required to be applied across the network to detect and prevent against attacks. Appropriately, each TCP/IP network layers has specific type of network attacks that means each network layer needs a specific type of IDS. Now-a -days Machine Learning becomes most powerful tool to deal with network security challenges given that the network level data generated is huge in volume and decision related to attacks need to be decided with high speed and accuracy. Classification is one of the techniques to deal with new and unknown attacks with network intrusion using machine learning. In this chapter, we detect the normal and anomaly attacks of the TCP/IP packets from publicly available training dataset using Gaussian Naive Bayes, logistic regression, Decision Tree and artificial neural network on intrusion detection systems. Using CoLab environment, we provide some experimental results showing that Decision tree performed better than Gaussian Naïve Bayes, Logistic regression and Neural Network with a publicly available dataset.

Keywords Intrusion detection · TCP/IP packets · Naive bayes · Logistic regression · Neural network

1 Introduction

In past few years, communication technology has developed tremendously. Networking is using widely in the industry, business and in our day to day life. Therefore, reliable network is an essential element for IT administrators. On the other hand, the rapid development of IT created serval challenges in order to build a network

H. Shahriar (✉) · S. Nimmagadda
Department of Information Technology, Kennesaw State University, Kennesa, Georgia
e-mail: hshahria@kennesaw.edu

S. Nimmagadda
e-mail: snimmag4@students.kennesaw.edu

© The Editor(s) (if applicable) and The Author(s), under exclusive license to Springer Nature Switzerland AG 2021
Y. Maleh et al. (eds.), *Machine Intelligence and Big Data Analytics for Cybersecurity Applications*, Studies in Computational Intelligence 919, https://doi.org/10.1007/978-3-030-57024-8_10

231

reliable. There are many types of attacks that can threaten the availability, integrity, confidentiality and non-repudiation of computer network. The Denial of service attack (DOS) is most common harmful attack [1]. Dos attack meant to shut down a machine or network and accomplish by flooding the target with traffic or sending it information that trigger a cash. Victims of DoS attack often target web servers of high-profile organizations such as banking, commerce and media companies, or government and trade organizations [2]. Attacks do not typically result in the theft or loss of significant information or other assets; they can cost victim a great deal of time and money to handle.

DOS attacks tend to momentarily deny several end-user services. This typically absorbs network resources in general and overloads the system with unnecessary requests. Thus, DOS serves as a large umbrella for all types of attacks aimed at accessing computer and network resources [3]. In 2000 Yahoo was the victim of a DOS attack and on the same data DOS also reported its first public attack. In 2018 February GitHub was affected by DOS attack and it was recording breaking 1.3 Tbsp of traffic that flooded its servers with 126.9 million packets of data each second. It was recorded as biggest DDoS attack and systems down for about 20 min [4]. Now Web service and social networking sites are target of DOS attacks. From another perspective, remote to local (R2L) attacks are another term for all type of attacks that are intended to have right permissions as the availability of some network resources is special to local users, such as file servers. There are several types of attacks with R2L, these kinds of attacks are aimed at making unauthorized access to network resources [1, 5].

To detect all kinds of attacks intrusion detection system (IDS) became important part of the network security. It will monitor the network traffic for suspicious activity and issues alert when such activity is occurred. There are two types of Intrusion detection systems 1. Signature based 2. Anomaly based. Signature based IDS detection of attacks by looking for specific patterns, such as byte sequence 1's and 0's in the network traffic and known malicious intrusion sequence used by malware. It is easy to detect the attacks whose patters are already exist in system, but it is difficult to find new attacks whose pattern is unknown [4]. In anomaly-based detection, the IDS for detecting both network and computer intrusion and misuse by monitoring system activity it as either normal or anomalous. The classification is based on heuristics or rules, rather than patterns or signatures and attempts to detect any type of misuse that falls out of normal system operation [6]. The two phases of most anomaly detection systems consist of the training phase and testing phase. Anomalies are detected in several ways, most often with artificial intelligence type techniques. Systems using artificial neural networks have been used to great effect. Another method is to define what normal usage of the system comprises using a strict mathematical model and flag any deviation from this as an attack. This is known as strict anomaly detection [6]. The example for anomaly based is in crucial stage of behavior determination is regarding the ability of detection system engine toward multiple protocols at each level. The IDS engine must be able to understand the process of protocols and its goal [5]. Even though the protocol analysis is very expensive in terms of computation, the benefits like increasing ruleset assist in lesser levels of false-positive alarms. Defining

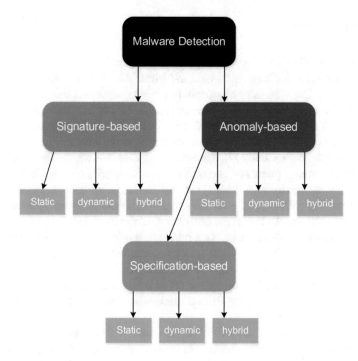

Fig. 1 Classification of IDS

the rule sets is one of the key drawbacks of anomaly-based detection. The efficiency of the system depends on the effective implementation and testing of rule sets on all the protocols. In addition, a variety of protocols that are used by different vendors impact the rule defining the process [5]. The classification of malware detection techniques is shown in Fig. 1.

Regarding the literature, Attacks detection considered as classification problem because the target had to conform whether the packet is normal or attack packet. Therefore, the model must accept as intrusion detection system (IDS) can be implemented based on significant machine learning algorithm. In this chapter, we are using Gaussian Naive Bayes, logistic regression, Decision tree and artificial neural network as machine learning algorithm [1]. Main aim is to build the intrusion detection, a predictive model capable of distinguishing between bad connection (Attacks) and good Connection (normal). By using the classification method of precision, recall, f1-score, support and model Accuracy we can find each connection is labelled as either normal or an attack with exactly one specific attack type.

The rest of the book chapter is organized as follows. Section II provides some related works. Section III introduces the KDD CUP datasets we used in experiment. Section IV discusses the classification techniques. Section V provide some the research results. Finally, Section V concludes the chapter.

2 Related Works

Few other researches have made comparison between different algorithm for classification problems. Szilveszter kovacs and Mean Alzubi [1], compared with J48, Random forest, Decision table algorithm in the network intrusion detection system by evaluate the efficiency and performance using accuracy and precision, recall. The result decision table classifier achieved the lowest value of false negative while the random forest classifier has achieved the highest average accuracy rate.

According to another study [7], the authors imported the KDD dataset and implemented the preprocess phases e.g. normalization of the attributes range to $[-1, 1]$ and converting symbolic attributes. Neural network feed forward was implemented in two experiments. The authors have concluded that neural network is not suitable enough for R2L and U2R attacks but on the other hand, it was recorded acceptable accuracy rate for DOS and PROBE attacks. As it relates to implement neural network against KDD intrusions, the effort of [1] the authors succeeded to implement the following four algorithms: Fuzzy ARTMAP, Radial-based Function, Back propagation (BP) and Perceptron-back propagation-hybrid (PBH). The four algorithms evaluated and tested for intrusions detection the BP and PBH algorithms recorded highest accuracy rate.

From another perspective, some of the researchers focus on attributes selection algorithms in order to reduce the cost of computation time. In [8] the authors are focused on selecting the most significant attributes to design IDS that have a high accuracy rate with low computation time. 10% of KDD was used for training and testing. They implemented detection system based on extended classifier system and neural network to reduce false positive alarm as much as possible. On the other hand, [9] the information gain algorithm was implemented as one of effective attributes selection. They implemented multivariate method as linear machine method to detect the denial of service intrusions.

Safaa Zaman and Fakhri Karry [10], compared with two algorithms (Support Vector Machine and Neural Network) to select best features set for each type of IDS by performance using Accuracy, Training Time and Testing time. The result indicates that each IDS type has different features set that can not only improve the overall performance of the IDS, but it also can improve its scalability.

In addition, the genetic algorithm was implemented to enhance detection of different types of intrusions. Meanwhile in [10] a methodology to detect different types of intrusions within the KDD is proposed. The proposed methodology aims to derive the maximum detection rate for intrusion types, at the same time achieved the minimum false positive rate. The GA algorithm used to generate several effective rules to detect intrusions. They succeeded to record 97% as accuracy rate based on this methodology. In some cases, if the single isolated machine learning algorithm used to handle all types of intrusions it would be derived by an unaccepted detection rate. In [11], the author used Naive Bayes algorithm to detect all intrusions types of KDD. He illustrated that the detection rate was not acceptable based on single machine learning algorithm.

There are some researchers focusing on specific type of attack such as [12] the authors proposed a system to collect new distributed denial of service dataset which includes the following types of attacks (http flood, smurf, siddos and udp flood) after the new DDOS dataset proposed. They implemented various machine learning algorithms to detect DOS intrusions, MLP algorithm recorded highest accuracy rate of 98.36%.

Other researches focusing on the machine learning algorithms in intrusion detection systems by comparing different supervised algorithms for the anomaly-based detection technique. The algorithms have been applied on the KDD99 dataset, which is the benchmark dataset used for anomaly-based detection technique. The result shows that not a single algorithm has a high detection rate for each class of KDD99 dataset. The performance measures used in this comparison are true positive rate, false positive rate, and precision.

3 Datasets

Our network intrusion detection datasets are from KDD Cup [13]. The datasets provide which consists of wide variety of intrusion simulated in a military network. It creates an environment to acquire raw TCP/IP dump data for a network by simulating a typical US Air Force LAN. In our experiment we are using both train and test datasets. These datasets contain 24 attacks types, which falls into four main classes, Denial of Service (DOS), Probe, User to Root (U2R), and Remote to Local (R2L). Both testing and training datasets huge data of network traffic connections and each connection represent with 41 quantitative and qualitative features are label from normal and attack data. This dataset has shown the normal and anomalous of class variable. The DOS attacks present 79% of KDD dataset while normal packets present 19% and other attacks types recorded 2% of existing. Based on the KDD datasets appears as an unbalanced dataset and the same time it includes the large number (41) of packets attributes. The screenshots for datasets are shown in Fig. 2.

These attributes categorized as a basic information which is collected using any connection implemented based on TCP/IP [1, 14]. Table 1 illustrates the fundamental attributes information for any connection implemented based on TCP/IP connection environment. The main contribution of this dataset is the introduction of 32 expert suggested attributes which help to understand the behavior of different types of attacks [1], In other word, the most significant attributes to detect DOS, R2L, U2R and PROBE are included.

In our experiment we are taking only training dataset. It contains 22,500 sample of which 13,449 are normal samples and 11,743 are anomaly samples. We are dividing the dataset into training and test sets respectively. The training set of the dataset accounts for 80% of each total sample, and the test set accounts for 20% of each total sample. And then we also test 75% (training) / 20% (testing) and 70% (training) and 30% (testing).

```
In [4]:   1  print(train.head(4))
          2
          3  print("Training data has {} rows & {} columns".format(train.shape[0],train.shape[1]))

             duration  protocol_type   service flag  src_bytes  dst_bytes  land  \
          0         0            tcp  ftp_data   SF        491          0     0
          1         0            udp     other   SF        146          0     0
          2         0            tcp   private   S0          0          0     0
          3         0            tcp      http   SF        232       8153     0

             wrong_fragment  urgent  hot  num_failed_logins  logged_in  num_compromised  \
          0               0       0    0                  0          0                0
          1               0       0    0                  0          0                0
          2               0       0    0                  0          0                0
          3               0       0    0                  0          1                0

             root_shell  su_attempted  num_root  num_file_creations  num_shells  \
          0           0             0         0                   0           0
          1           0             0         0                  '0           0
          2           0             0         0                   0           0
          3           0             0         0                   0           0

             num_access_files  num_outbound_cmds  is_host_login  is_guest_login  count  \
          0                 0                  0              0               0      2
          1                 0                  0              0               0     13
          2                 0                  0              0               0    123
          3                 0                  0              0               0      5

             srv_count  serror_rate  srv_serror_rate  rerror_rate  srv_rerror_rate  \
          0          2          0.0              0.0          0.0              0.0
          1          1          0.0              0.0          0.0              0.0
          2          6          1.0              1.0          0.0              0.0
          3          5          0.2              0.2          0.0              0.0

             same_srv_rate  diff_srv_rate  srv_diff_host_rate  dst_host_count  \
          0           1.00           0.00                 0.0             150
          1           0.08           0.15                 0.0             255
          2           0.05           0.07                 0.0             255
          3           1.00           0.00                 0.0              30
```

Fig. 2 KDD datasets

Table 1 The basic attributes of TCP/IP connections

Attributes	Types
Total durations of connections in seconds	Continuous
Total number of bytes from sender to receiver	Continuous
Total number of bytes from receiver to sender	Continuous
Total number of wrong fragments	Continuous
Total number of urgent packets	Continuous
Protocol type	Discrete
Type of service	Discrete
The status of the connection (normal or error)	Discrete
Label (1) if the connection established from to the same host. Otherwise label (0)	Discrete

4 Methodology

4.1 Gaussian Naive Bayes

Naïve Bayes is referring to the group of probabilities classifiers. It implements Bayes theorem for classification problem. First its classifiers to determine the total number of classes (Output) and calculate the conditional probability each dataset classes

(Input) [1]. Extension of the naïve Bayes is called Gaussian Naïve Bayes. Working with estimating the distribution of the data is easy because we can only need to estimate the mean and the standard deviation from the training data.

$$P(x_i|y) = \frac{1}{\sqrt{2\Pi\sigma_y^2}}\exp\left(\frac{-(x_i-\mu_y)^2}{2\sigma_y^2}\right)$$

where $P(x_i|y)$ assumption the continuous values associated with each class are distributed. Training data contains a continuous attribute x, First segment the data by the class and then compute mean and variance of x in each class. Let μ_y be the mean of the values in x associated with class y and let σ_y^2 be the Bessel corrected variance of the values in x associated with class y.

4.2 Logistic Regression

Logistics regression is a linear model for binary classification problem. A linear combination of product of independent variable $(x_1, x_2, x_3, \ldots x_n)$ and its corresponding weight $(w_1, w_2, w_3, \ldots w_n)$ and put these into sigmoid equations which is used to restrict the output to an interval between 0 and 1. The sigmoid formula is

$$S(z) = \frac{1}{1+e^{-z}}$$

The log function is monotone, maximum value of the probability function is same as maximum the value of the log probability function. The formula of the log probability function is

$$l(w) = \log(l(w)) = \sum_{i}^{n} \log(f(x_i|w))$$

Our aim is to maximize the log probability function and find an optimal weight w. we can use gradient descent algorithm to minimize this function by putting a minus sign in front of the log probability function. The cost function of logistic regression is

$$J(w) = -\sum_{i=0}^{n} y^{(i)}\log(S(h^{(i)})) + (1-y^{(i)})\log(1-S(h^{(i)}))$$

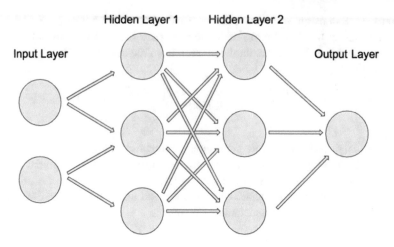

Fig. 3 Artificial neural network

4.3   Artificial Neural Network

Artificial neural network is a computational model based on the structure and function of biological neural networks. It has layers and each layer are made up of nodes. A node is a place for calculation, loosely modeled on a neuron in the human brain that is activated when given enough stimulation. When neural net is being trained, all of its weights and threshold are initially set to random values. Training data is fed to bottom layer- the input layer- and it passes through the succeeding layers, getting multiplied and added together in complex ways, until it finally arrives, radically transformed, at the output layer [15] (Fig. 3).

In our experiment we have 12 dense layers and each input node is connected to each output node. The classification problem is binary.

SoftMax Activation Function: It determines the output layer of the neural network for categorical target variables, the outputs can be interpreted as posterior probabilities. This is useful in classification as it gives a certainty measures on classification.

$$y_i = \frac{e^{x_i}}{\sum_{j=1}^{e} e^{x_j}}$$

4.4   Decision Tree

The decision tree classifier is one of the possible approaches to multistage decision making. A decision tree is composed of a root node, a set of interior nodes, and

```
Out[18]: DecisionTreeClassifier(class_weight=None, criterion='entropy', max_depth=None,
                max_features=None, max_leaf_nodes=None,
                min_impurity_decrease=0.0, min_impurity_split=None,
                min_samples_leaf=1, min_samples_split=2,
                min_weight_fraction_leaf=0.0, presort=False,
                random_state=0, splitter='best')
```

Fig. 4 Decision tree classifier

terminal nodes, called "leaves" [16]. The root node and interior nodes, referred to collectively as non-terminal nodes, are linked into decision stages. The terminal nodes represent final classification. The classification process is implemented by a set of rules that determine the path to be followed, starting from the root node and ending at one terminal node, which represents the label for the object being classified [16]. At each non-terminal node, a decision must be taken about the path to the next node. the main idea of this classifier is to build a lookup table, it helps to identify the predicted class of output. The Decision Tree classifier is shown in Fig. 4.

Gini measurement is the probability of a random sample being classified incorrectly if we randomly pick a label according to the distribution in a branch. To calculate Gini, we must have CC total classes and $p(i)p(i)$ is the probability of picking a datapoint with class ii [17], then the Gini Impurity is calculated as

$$G = \sum_{i=1}^{c} P(i) * (1 - P(i))$$

Entropy is degree of randomness of elements or in other words it is measure of impurity. Mathematically, it can be calculated with the help of probability of the items as. Where $p(x)$ is probability of item x.

$$H = -\sum p(x) \log p(x)$$

5　Evaluation

We are evaluating the datasets with the three algorithms. For network intrusion detections, there are serval evaluations metric can be used in a classification algorithm. In our experiment, the confusion matrix was generated for each machine learning classifiers. They are Gaussian Naïve Bayes, Logistic regression, Decision Tree and Neural network. Given our reduced dataset, we will start by scaling the data then splitting it into a test and train set. We calculated precision, recall, f1 score and support for both anomaly and normal attacks. To calculate the accuracy of model and confusion matrix., we need 4 measurements factors i.e., true positive (TP), true negative (TN, false positive (FP) and false negative (FN). Furthermore, the following performance metrics are below:

True Positive (TP): This value represents the correct classification attack packet as attacks.

True Negative (TN): This value represents the correct classification normal packets as normal.

False Negative (FN): this value illustrates that an incorrectly classification process occurs. Where the attack packet classified as normal packet, a large value of FN presents a serious problem for confidentiality and availability of network resources because the attackers succeed to pass through intrusion detection system.

False Positive (FP): this value represents incorrect classification decision where the normal packet classified as attack, the increasing of FP value increases the computation time, but on the other hand, it is considered as less than harmful of FN value increasing.

Precision: is one of the primary performance indicators. It presents the total number of records that are correctly classified as attack divided by a total number of records classified as attack. The precision can be calculated according to the following equation:

$$p = \frac{TP}{(TP + FP)}$$

Accuracy (ACC): Accuracy is one metric evaluating classification model, It is the fraction of predictions our model got right. Formally, accuracy has the following definition:

Accuracy = Number of correct predictions/total number of predictions.

Random forest is one of the classification trees algorithms, the main goal of this algorithm is to enhance trees classifiers based on the concept of the forest. In this chapter, random forest is used to fit in the training sets for feature selections, extract important features and plot as bar shown in Fig. 5.

In addition, the number of both the correctly and the incorrectly classified instances are recorded with respect to the time taken for proposed training model. During the testing phase, the following parameters were applied for the machine learning classifiers. Neural network classification is trainable params id 10,479 and applied autoencoder to the model of sequential to fit the training set to anomaly and normal packets with epochs = 500, batch_size = 256 and threshold value = 0.048. The anomaly detection −500 epochs are shown Fig. 6.

Table 2 shows the TP rates and precision values of the selected classifiers in the experiment of training and test sets for 70/30, 75/25/80/20. It can be concluded that Decision Tree classifier achieved highest TP rate of 1.00 for anomaly and 0.99 for normal for all 70%/30%, 75%/25% and 80%/20% of training and test sets. In other words, Gaussian Naïve Bayes reached the lowest value of 0.85 for anomaly attacks

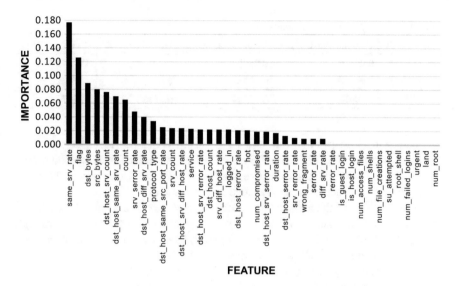

Fig. 5 Feature set selection

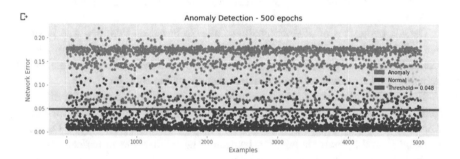

Fig. 6 Anomaly detection—500 epochs

Table 2 True positive and precision ratios

Classifiers	Recall	Precision
Gaussian naive bayes	0.85	0.94
Logistic regression	0.93	0.96
Decision tree	1.00	0.99
Neural network	0.87	0.94

classification process. From another perspective, the decision table classifier reached the highest precision value of 0.99 for anomaly and 0.94 for normal. Therefore, there is a large number of anomaly packets classified as attack packets.

In general, True positive and precision is important to perform the parameters for a Network Intrusion Detection, but in the same way, the most serious performance

Table 3 Accuracy comparison among classifiers

Dataset split	Gaussian naive bayes	Logistic regression	Neural network	Decision TREE
(80%/20%)	0.90176	0.95058	0.90176	0.99503
(75%/25%)	0.90266	0.95315	0.91171	0.99491
(70%/30%)	0.90672	0.95554	0.89732	0.99470

parameters id False positive rate and F1 Score. It concludes that Decision Tree Achieved highest value for F Score is 0.99 and Gaussian Naïve Bayes reached lowest value as 0.89. from another perspective, the Gaussian Naïve Bayes achieved highest FP rate of 0.15 and decision tress reached lowest value as 0.002.

Regarding Table 3 the logistic regression recorded the highest value 0.9555 based on the accuracy value and neural network presented as lowest value 0.8973. Through testing classification of 22,500 samples from the KDD Cup. The average accuracy rate is calculated by the following formula.

Average Accuracy Rate: $(TP + TN)/(TP + TN + FP + FN)$.

The accuracy values for Gaussian Naive Bayes, Logistic regression and Neural Network for training dataset for 80% and testing dataset for 20% and test 75% (training) / 20% (testing) and 70% (Training) and 30% (Testing) values are shown Table 3.

Dataset comprises of 50% anomalous data. So based on this fact, true positive rate and false positive rate were not calculated. Therefore, ROC curves were also not plotted because the generated performance values were approximately same for all train and test sets. We need more efficient data to get the right predictions.

The confusion matrix for the dataset is shown in Tables 4, 5 and 6 for different split of training and testing datasets of 80%/20%, 75%/25% and 70%/30% for 4 classification algorithm model of Logistic Regression, Gaussian Naïve Bayes, Decision Tree and Neural Network in machine learning. The result of the numerical examples can be concluded in the following points:

- The Decision Tree achieved highest accuracy rate 0.9950 and lowest value for False positive rate.
- The Neural Network classifier reached the lowest average accuracy rate for training and test sets of 70%/30% is 0.89.
- Regarding the average accuracy rate there is no big between Gaussian naïve Bayes and Neural network.
- Regarding the accuracy rate the all four models are all most same for training and test sets of 80%/20%, 75%/25% and 70%/30%.
- All machine learning classifiers present acceptable precision and recall rates for detecting anomaly and normal packets.
- Decision Tree classifier recorded the highest vale for detecting correctly the anomaly and normal packets
- Gaussian Naïve Bayes reached the lowest values for average accuracy rate, precision and recall to detecting anomaly and normal packets.

Table 4 Confusion matrix of network intrusion detection dataset (75% training, 35% testing)

Logistic regression	Preicted: No	Predicted: Yes	N = 6298
Actual: No	2750	179	
Actual: Yes	116	3253	

Gaussian naive bayes	Predicted:No	Predicted: Yes	N = 6289
Actual: No	2480	449	
Actual: Yes	164	3205	

Neural network	Predicted:No	Predicted:Yes	N = 6298
Actual:No	3111	260	
Actual:Yes	296	2631	

Decision tree	Predicted:No	Predicted: Yes	N = 6289
Actual: No	2915	14	
Actual: Yes	18	3351	

Table 5 Confusion matrix of network intrusion detection dataset (70% training, 30% testing)

Logistic regression	Predicted:No	Predicted: Yes	N = 7558
Actual: No	3298	200	
Actual: Yes	136	3924	

Gaussian naive bayes	Predicted:No	Predicted: Yes	N = 7558
Actual: No	2981	517	
Actual: Yes	188	3872	

Neural network	Predicted:No	Predicted:Yes	N = 7558
Actual:No	3494	573	
Actual:Yes	203	3288	

Decision tree	Predicted:No	Predicted: Yes	N = 7558
Actual: No	3483	15	
Actual: Yes	25	4035	

Table 6 Confusion matrix of network intrusion detection dataset (80% training, 20% testing)

Logistic regression			Gaussian naive bayes			Neural network			Decision tree		
Predicted:No	Predicted: Yes	N = 5039	Predicted:No	Predicted: Yes	N = 5039	Predicted:No	Predicted:Yes	N = 5039	Predicted:No	Predicted: Yes	N = 5039
2180	153	Actual: No	1974	359	Actual: No	2318	358	Actual:No	2180	153	Actual: No
96	2610	Actual: Yes	136	2570	Actual: Yes	137	2226	Actual:Yes	96	2610	Actual: Yes

- It can conclude the decision tree classifier can present an acceptable accuracy rate with lowest FN rate, which also increases the confidentiality and the availability of the network resources.

6 Conclusion

In this chapter, we used the KDD datasets to classify the network intrusion detection system by using Gaussian Naïve Bayes, Logistic Regression, Decision Tree and Neural network has evaluate the efficiency and performance of the machine learning model. The rates of different types of attacks In the KDD datasets are approximately 61% of normal attacks and 39% of anomaly attack. In the experiment 22,544 instance of records have been extracted as training data to build the training models for the selected machine learning classifiers.

The experimental result show that Decision Tree performed better than Gaussian Naïve Bayes, Logistic Regression and Neural Network in the efficiently all the types of attacks. The decision tree achieved the highest values for the Accuracy rate a and achieved the lowest FN rate is 0.002. On other hand Gaussian Naïve Bayes reached the lowest values for average accuracy rate, precision and recall to detecting anomaly and normal packets. Our future works include using other publicly available network datasets and comparing between supervised and unsupervised classifiers.

References

1. Almseidin M, Alzubi M, Kovacs S Evaluation of machine learning algorithm for intrusion detection system. Department of Information Technology,University of Miskolc, H-3515
2. Paloalto, Cyberpedia, What is denial of service attack? https://www.paloaltonetworks.com/cyberpedia/what-is-a-denial-of-service-attack-dos
3. Oke G, Loukas G Distributed defense against denial of service attacks: a practical view, Dept. of Electrical and Electronic Engineering Imperial College London SW72BT
4. Felter B, Vxchange, May 31 2019, 5 most famous recent DDos Attacks.
5. Veeremreddy J, Prasad KM, Anomaly-based Intrusion Detection System, June 2019.
6. Wikipedia, Anomaly- based Intrusion detection System
7. Basarsian MS, Bakir H (2019) Fuzzy logic and correlation- based hybrid classification on hepatitis disease data set. In: 2019 International Conference on Artificial Intelligence.
8. Alsharafat W (2013) Applying artificial neural network and extended classifier system for network intrusion detection. Int Arab J Inf Technol (IAJIT), 10(3)
9. Bhargava, Sharma G, Bhargava R, Mathuria M (2013) Decision tree analysis on j48 algorithm for data mining. Proce Int J Adv Res Comput Sci Softw Eng 3(6)
10. LaRoche P, Zincir-Heywood N (2009) Evolving TCP/IP packets: a case study pf port scans. In: 2009 IEEE Symposium on Computational Intelligence for Security and Defense Application.
11. Alkasassbeh M, Al-Naymat G, Hassanat AB, Almseidin M (2016) Detecting distributed denial of service attacks using data mining techniques. Int J Adv Comput Sci Appl 1(7):436–445
12. Bay SD (1999) The uci kdd archive [http://kdd. ics. uci. edu]. irvine, ca: University of california. Department of Information and Computer Science, vol 404, pp 405
13. Zaman S, Karray F TCP/IP Model and Intrusion Detection Systems, IEEE

14. Gervais H, Munif A, Ahmad T (2016) Using quality thersold distance to detect intrusion in TCP/IP network. In: 2016 IEEE International Conference on Communication, Network and Satellite.
15. Fleizach, Fukushima S (1998) A naive bayes classifier on 1998 kdd cup
16. Bahrololum M, Salahi E (2009) Machine learning techniques for feature reduction in intrusion detection systems. In: Fourth International Conference on Computer Science and Convergence Information Technology
17. Sahu S, Mehtra BM (2015) Network intrusion detection system using J48 decision tree. In: 2015 IEEE International Conference On Advance In Computing, Communication and Informatics.

Developing a Blockchain-Based and Distributed Database-Oriented Multi-malware Detection Engine

Sumit Gupta, Parag Thakur, Kamalesh Biswas, Satyajeet Kumar, and Aman Pratap Singh

Abstract In today's modern world, if there is one word that may strike fear within the heart of any mortal, especially the one who accesses the web or exchanges diskettes or other storage peripherals, then it has to be "malware". Malwares are software that are built to cause destruction and vandalize computers, servers, clients or an entire computer network. To deal with such challenges, myriad antivirus software are available in the market, but most of them are based on centralized systems. As an enhancement to the currently available solutions, the proposed work in this chapter aims to safeguard network devices against multiple malwares by designing and developing a decentralized and distributed database-oriented intrusion detection framework powered by three malware detection frameworks viz. signature-based, behavior-based and multi-antivirus-based engines. This detection system relies on Blockchain Technology and aims to classify the transferred executable files as either malign or benign in nature. A network is considered to comprise several general-purpose computers or nodes; either of low-end resource type or of high-end resource type. Whenever, a conveyable executables file reaches any node, it is broadcasted in the network. At this point, all the active nodes start scanning that very file individually. If the file is assessed as malign or malicious, then its file hash is added to the blockchain as a transaction along with its probability of being malicious. The node

S. Gupta (✉) · P. Thakur · K. Biswas · S. Kumar · A. P. Singh
Department of Computer Science & Engineering, University Institute of Technology, The University of Burdwan, Golapbag (North), Burdwan, West Bengal 713104, India
e-mail: sumitsayshi@gmail.com

P. Thakur
e-mail: idealparag.9471@gmail.com

K. Biswas
e-mail: bkamalesh99@gmail.com

S. Kumar
e-mail: satyajeetkumar499@gmail.com

A. P. Singh
e-mail: amanpratapsingh2000012@gmail.com

Y. Maleh et al. (eds.), *Machine Intelligence and Big Data Analytics for Cybersecurity Applications*, Studies in Computational Intelligence 919,
https://doi.org/10.1007/978-3-030-57024-8_11

or client that broadcasted the file can then go through the whole chain to get the final result; which is the weighted average of the probabilities for that very file hash. Thus, the proposed multi-malware detection engine uses only low-end resources and tends to achieve better and accurate results in terms of detecting and quarantining malware, without the requirement of specialized and expensive high-end resources.

Keywords Malware · Anti-malware detection engine · Signature-based detection · Multi-AV-based detection · Behavior-based detection · Blockchain technology

1 Introduction

The present digital era is known to be mostly dependent on inter-networking and the all-embracing usage of computing devices which make our lives easier and lavish. Due to the ubiquity of such devices and the predominant utilization of networking resources, the malign exertion and security concerns are on the escalation as well. Malign actions are performed with an ill-intention in an attempt to manipulate data, steal information and for impersonation. There are innumerable software deliberately designed to cause harm to a server, client, or a network. For instance, a software bug is a deficiency in the software program that is implanted voluntarily or due to some mistake, and phishing is a practice of sending an illegitimate email falsely claiming to be from a legitimate site in order to steal user's secret assets and information. Various types of malwares exist in literature. Some of the renowned ones include Trojan horses, computer viruses, ransomwares, worms, spyware, rootkits, adware, etc. It has become increasingly crucial to protect the digital infrastructure of the society against such malicious activities. To prevent these malicious activities, there are several anti-malwares that use either static or dynamic approaches to deal with the challenges posed by malwares.

Generally, the antivirus software uses some sort of hash matching algorithm or methodology from their known collection of organized hashes for providing security and protection. The major flaw of it is that if new threats or malign executables appear on the inter-connected network, then the antivirus fails to assure complete protection and sometimes tend to be most vulnerable to zero-day vulnerabilities. To resolve these types of issues, several research works have been conducted on zero-day vulnerabilities detection using different methodologies. Basically, a zero-day vulnerability or malware is a type of loophole that is not known to a patcher or a developer. However, these programs do not seem to be much efficient in real world conditions. The very fact is that almost all antivirus relies on the regular virus definition updates that depend on the cloud-based third party database; which is centralized and thus prone to cyberattacks i.e., mass attacks performed on those centralized stations that comprise the virus definitions. If somehow an attacker possibly gets access to the centralized data station, then he/she will have the basic access (or authority) to change or destroy the virus definitions, thus resulting in a compromising situation.

At present, mostly all the malware detection engines are based upon signature-based detection approach and it is only in the recent times that few antivirus have started adopting the techniques of anomaly-based or behavioral-based detection. The open-end challenges that can be taken up include the following:

(i)   A framework that does not depend on cloud-based centralized data storage for virus definitions.
(ii) A framework that has the potential to scan and detect all forms of malicious files under the malware classes like ransomware, computer viruses, Trojan horses, worms, rootkits, adware, spyware etc.
(iii) A framework that can scan an executables file by multi-malware detection engine, where each engine is implemented using different algorithms.
(iv) A framework that gives no or comparatively low false alarms.

Lately, Blockchain Technology has become the cynosure of the world due to the usage of cryptocurrencies like Bitcoin, but its popularity is not limited to only virtual currency. The notion of blockchain harbingers new trajectories for existing technological applications that when implemented can prove to be very much efficient and immune against many security concerns and challenges. The vast scope of blockchain framework has been the motivating factor in the development of a Blockchain-based malware detection framework having distributed signatures database characterized with decentralization, immutability, scalability and that too without any single-point failure. This framework has the potential to scan and detect all types of malicious files or malwares.

The rest of the chapter is organized as follows: Sect. 2 discusses about malware, few of the popular malwares, different components of malware, various approaches for detecting malware and taxonomy of malware detection techniques. Section 3 introduces the concept of blockchain and explains the working of a blockchain along with the different types of blockchain architectures known in literature. Section 4 explains some of previous research works done in the domain of malware detection along with a tabular comparative study. Section 5 explains the working of the proposed multi-malware detection engine through the system workflow and an algorithm. The implementation part is discussed in Sect. 6 along with the screenshots showing the scan results. Section 7 concludes the paper and Sect. 8 highlights on various avenues for future work.

2 Malware

The word malware was first coined by computer scientist and security researcher Yisrael Radai in 1990. Malware abides to be a concern with today's cutting-edge technologies. It anguishes all, from a newbie to a professional; everyone tends to be laid low with malware. But irrespective of why or how malware evolves, it is always bad if it lands up on one's personal computer or network. Few of the well-known malwares [1] have been discussed briefly here.

(i)   *Virus*: These are different from computer trojans and worms in some sense as they are dependent on other executables to be attached on them to infect. They are self-replicating malicious program.

(ii)   *Worm*: A worm is similar to computer viruses. It has the capability of replicating or duplicating itself over a network, thereby overloading the space, increasing the usage of bandwidth and causing the same damage as any virus.

(iii)   *Trojan Horse*: Trojan horse, or simply a Trojan, is a piece of software that can act as backdoor i.e., fr making a reverse TCP connection shell back to attacker, thus compromising the victim's machine. It will execute once the user installs or executes a program. It mostly acts as backdoors, spywares, or keyloggers.

(iv)   *Rootkits*: It is basically a program capable of hiding its existence by intercepting and malfunctioning the system API calls or may reside into other vulnerable running services. They are divided into three types viz. user mode (Ring 3), kernel mode (Ring 0) and hypervisor mode (Ring 1) rootkits.

(v)   *Ransomware*: A ransomware basically encrypts the data in the computer, thus making it inaccessible to the user and demands a ransom that needs to paid through bitcoins or other cryptocurrencies for regaining possession of or access to the infected files.

(vi)   *Keyloggers*: Keyloggers can be based on both software and hardware types. It basically provides the functionality of sharing the keyboard strokes remotely i.e., to record keys typed by a user. It leads in the leakage of sensitive information like security credentials, SSN, credit card details etc.

(vii)   *Adware*: Adware are like ads doping up on sites or apps. Hackers can install adware on user's machine to get these irritating advertisements and might make earnings too.

(viii) *Spyware*: Spyware is an unwanted package that infiltrates the user's machine, steals the net usage information and other sensitive data. Spyware gathers personal data like Mastercard or bank account details and passes it to ads, data centers or external users.

2.1   Components of Malware

Malware authors and attackers create malware using the components which will help them achieve their goals. They use malware to steal information, delete data, change system settings, provide access or merely multiply and occupy the space. Malware is capable of propagating and functioning secretly. The essential components of most malware programs are as follows:

(i)   *Crypter*: It refers to a software program which will conceal the existence of malware. Attackers use this software to elude antivirus detection. It protects malware from undergoing reverse engineering or analysis, thus making it difficult to detect using the installed safety mechanism. The method of remaking an existing component, subassembly, or product, without the help of drawings, documentation, or computer model is understood as reverse engineering [2].

(ii)   *Downloader*: It is a kind of Trojan that downloads other malwares or malicious codes and files from the web on to the PC or device. Usually, attackers install downloader after they first gain access to a system.

(iii)   *Dropper*: Attackers must install the malware program or code on the system to create it run, and this program can do the installation task covertly. The dropper can contain unidentifiable malware code undetected by the scanners and can download additional files that are needed for executing the malware on a target system.

(iv)   *Exploit*: It is a part of the malware that contains code or sequence of commands which will make the most of a bug or vulnerability in an exceedingly digital system or device. It is the code that the attackers use for breaching the system's security through software vulnerabilities. They can be either local or remote in nature.

(v)   *Injector*: This program injects the exploits or malicious code available within the malware into other vulnerable running processes and changes the way of execution to cover or prevent its removal.

(vi)   *Obfuscator*: It is a program that hides the malicious code of malware via various techniques, thus making it hard for security mechanisms to detect or remove it.

(vii)   *Packer*: This software compresses the malware files to convert the code and data of malware into an unreadable format. The packers use compression techniques for packing the malware.

(viii) *Payload*: It is a part of the malware that performs desired activity and compromises safety when activated. The payload could also be used to delete files, modify files, affect system's performance, open ports, change settings etc.

(ix)   *Malicious Code*: It is a code snippet that defines the essential functionality of the malware and comprises commands leading to security breaches.

2.2 Malware Detection Approaches

A malicious activity can be detected if the user's computer behaves abnormally such as abnormally slow speed of computer, sudden increase in storage, freezing and crashing of computer, unwanted popups and ads, sudden increase in bandwidth consumption, opening of non-configured ports or remote connection to unknown IP. There are many malware detection approaches known. Some of the popular ones are discussed below.

(i)   *Scanning*: A scanner is a very crucial component of anti-malware software for detecting malwares. In absence of a scanner component, there is a huge risk that the computer is going to be attacked by hackers. Scanner basically scans the file to know whether it is a malicious file or not by different detection techniques. These techniques have been thoroughly discussed in the next section.

(ii) *Integrity Checking*: In this approach, reading and recording of the integrated data is initiated to determine a baseline for those files and system areas. The main loophole of integrity checking is that it cannot determine the cause of file corruption i.e., whether it is caused by a bug or anything else. There are many improved integrity checkers available that can analyze and identify the kinds of changes that viruses make.

(iii) *Interception*: Interceptors are primarily used for diverting logic bombs and Trojans. It controls request to the OS for the process that causes a threat to the program or network access. If it manages to find such an invitation, the interceptor pops up a prompt to ask for user's action for permitting it or not.

(iv) *Code Emulation*: Anti-malware executes a virtual machine to mimic CPU and memory actions. Here malware is executed on the virtual machine or a sandbox system rather than the processor. Code emulation deals efficiently with the encrypted and polymorphic virus. After running the emulator for a protracted time, the decrypted malware body eventually presents itself to a scanner for detection.

(v) *Heuristic Analysis*: This approach helps in detecting new or unknown malwares. It may be static or dynamic in nature. In static analysis, the anti-malware analyses the file format and code structure to find out whether the code is infected or not, whereas in dynamic analysis, it performs a code emulation of the suspicious code to check for infections. It is susceptible to too many false positives result.

2.3 Malware Detection Techniques

Techniques used for malware detection can be broadly categorized into three classes, namely Signature-based, Behavior-based and Specification-based detections. Each of these classes follows the three analytical framework viz. Static Analysis, Dynamic Analysis and Hybrid Analysis [3–6].

2.3.1 Signature-Based Detection

In this technique, a file hash is matched with the list of available malicious file hashes. If matching is found, the corresponding file is labeled as malign or a malicious file, otherwise it is labeled as a benign file. The general workflow of signature-based malware detection and its detailed examination is explained in [5]. As already discussed, most of the antivirus tools are based on the signature-based detection approach. A database of known file hashes or signatures is updated by the antivirus software authority so that it can detect the presence of existing malwares without any mistake. The prominent pros of this technique are that it can detect known sign of malware without any error and fewer amounts of computer resources are required for detecting the malware. Its drawback is that it is unable to detect new and unknown signs of malware or any zero-day malware because such signatures are not present in the database.

2.3.2 Behavior-Based Detection

It is also known as heuristic or anomaly-based detection. It is primarily used to investigate the behavior of both existing and new malwares. Behavioral parameter includes various factors like the source or destination internet protocol (IP) address of malware, kinds of attachments, and other statistical parameters. It always occurs in two phases: training phase and detection phase. During the training phase, the behavior of the system is observed within the absence of malware attack and machine learning technique is employed to form a profile of such normal behavior. In the detection phase, the baseline is compared against this behavior and differences are flagged as potential attacks [6]. The advantage of this system is that it can detect new as well as unknown signs of malware and it focuses on detecting zero-day attacks. The disadvantage of this system is that it must update the information describing the system behavior and the statistics in a normal profile but it tends to be large. It need more resources like CPU time, memory and disk space and also the level of false positive is high.

2.3.3 Specification-Based Detection

It is an offshoot of behavior-based detection that tries to beat the standard high warning rate related to the latter. It depends on the program specifications that determine the expected behavior of complex and acute security programs. It involves program executions and detects deviation of their behavior from the specification, instead of detecting the circumstance of that very attack patterns. This system is analogous to anomaly detection but the difference is that rather than looking forward to machine learning techniques, it supports manually developed specifications that capture legitimate system behavior [6].

2.3.4 Static Analysis Detection

It is also known as code analysis that involves going through the executable binary code without actually executing it to have a better understanding of the malware and its purpose. It also gathers the information about malware functionality and collects technical pointers or simple signatures it generates. Such pointers include file name, MD5 checksums or hashes, file type and file size. It is the procedure of investigating an executable file without running or installing it. Some of the static malware analysis techniques are file fingerprinting, local and online malware scanning, performing strings search, identifying packing or obfuscation methods, finding the portable executables (PE) information, identifying the file dependencies and malware disassembly.

2.3.5   Dynamic Analysis Detection

It is also known as behavioral analysis that involves the execution of the malware code to know how it interacts with the host system and its impact on it after infecting the system. Dynamic analysis involves execution of malware to examine its conduct, operations and identifies technical signatures that confirms the malware intent. It is the process of studying the behavior of the malware by running in a monitored environment such as virtual machines and sandboxes to determine the spreading of malware. It involves the function of system baselining and host integrity monitors. The main advantage of dynamic analysis is that it accurately analyses the zero-day malware.

2.3.6   Hybrid Analysis Detection

This approach deals with the combination of two techniques viz. static and dynamic analysis [6]. Initially, it checks for any malware hash or signature and if matching is found, then in the next stage it observes the nature of the code. Hence this technique combines the benefits of both the static and dynamic analysis frameworks.

Figure 1 depicts the different types of malware detection techniques.

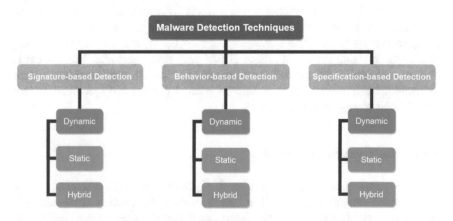

Fig. 1 Classification of malware detection techniques

3 Blockchain Technology

Blockchain and a set of connected protocols, have recently taken the world of Finance and Technology by storm through its ground-breaking application in the form of the Bitcoin (a cryptocurrency) and a whole lot of innovative applications. Although Bitcoin has been the most talked-about application of the Blockchain Technology, but new applications like Smart Contracts have tried to take advantage of a lot of abstract nature of the platform as well. It is presumed to be both tempting and significant for making certain increased privacy and security concerns for various applications in several alternative domains and also within the Internet of Things (IoT) eco-system. Blockchain has been enforced in many non-monetary systems such as distributed storage systems, proof-of-location, healthcare, decentralized selection etc. Recent analysis of articles, applications and projects were surveyed to assess the implementation of Blockchain for increased security, to spot associated challenges and to propose solutions for Blockchain enabled increased security systems [7].

3.1 How Does a Blockchain Work?

When a block stores new data it is appended to the blockchain. Blockchain, as its name emphasis, consists of multiple blocks set up along to form a chain. For a block to be appended to the blockchain network, the following four things must take place:

(i)   Occurrence of a new transaction.
(ii) Verification of the occurred transaction.
(iii) Storing the verified transaction in the block.
(iv) Generating hash for the stored block.

Figure 2 demonstrates the working of a blockchain using a sequence diagram.

3.2 Types of Blockchain Architecture

All blockchain architecture fall into one of the following three classes:

(i)   *Public Blockchain Architecture*: A public blockchain architecture means that the data and system access is available to anyone who is eager at participating e.g. Litecoin, Bitcoin, and Ethereum are public blockchain systems.
(ii) *Private Blockchain Architecture*: A private blockchain architecture is controlled only by users of any particular association or by genuine users who have an invitation to participate.
(iii) *Consortium Blockchain Architecture*: A consortium blockchain architecture comprises a group of associations and organizations. In a consortium, functions are set up and controlled by the primary users assigned for this task.

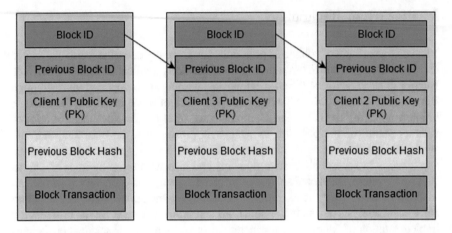

Fig. 2 Blockchain sequence diagram

4 Previous Related Works

This section presents some works by different researchers pertaining to the domain of malware detection.

Raje et al. in paper [8] have developed a decentralized firewall system powered by a new malware detection engine that was built using Blockchain Technology. A Deep Belief Network (DBN) was employed. During the approach, the files were modeled as grayscale images and the DBN was used to classify those images into two classes. An intensive dataset of 10,000 files was used for training the DBN. Validation testing was done using 4000 files previously unexposed to the network. The ultimate results of whether to allow a file or block it was obtained by arriving at a symbol of labor-based consensus within the blockchain network.

Talukder et al. [9] have proposed an Anti-malware database management system using customized Blockchain. The work aimed at enhancing security of a system by initiating distributed malware prevention codes. Instead of employing a new consensus algorithm, the authors have utilized their existing user verification algorithm for every user willing to communicate with the distributed ledger for prevention. The reason behind using their algorithm was the light weight and no requirement of utmost computational power; that is usually seen in traditional consensus algorithms. The first usage of the user verification using consensus algorithm is when a user's system finds out a replacement malware that does not exist within the database and also the user's system uploads the hash and other information to form an update within the database.

In the work by Cha et al. [10], a unique anti-malware system called SplitScreen has been developed that executes an additional screening step, before initiating the signature matching phase prevalent in existing approaches. The screening step filters out most of the non-infected files and also identifies those malware signatures that do not seem to be crucial. This approach significantly improves the end-to-end performance because safe files can be quickly identified, thus requiring no further processing, and malware files can subsequently be scanned using only the signatures that are necessary and evident. It naturally ends up in a network-based anti-malware solution during which clients only receive signatures they seek, not every malware signature ever created as in the existing approaches. The SplitScreen has been implemented as an extension to ClamAV [11]; one of the most popular open source anti-malware software. For the present number of signatures, the proposed implementation is $2\times$ faster and requires $2\times$ less memory than the first ClamAV. The gaps increase with increase in the number of signatures.

Fuji et al. [12] have developed a blockchain-based malware detection method for sharing the signatures of suspected files between users and allowing them to rapidly answer increasing malware threat. For the purpose of evaluation, real-world simulation was performed for predicting the detection accuracy. Compared to other heuristic or behavior-based methods, the proposed system was found to improve the false negative rate and also the false positive rate.

Several researches are ongoing for detecting zero-day malwares or anomaly-based malwares. For instance, Sun et al. in paper [13] have proposed a probabilistic approach and implemented ZePro, a prototype for identifying the path of a zero-day attack. A zero-day exploit is difficult to detect in the initial stage as it is flaw that reveals the susceptibilities of hardware and software, thereby creating havoc on the system without creating much fuss [14]. In the process, a dependency graph named the object instance graph was built by analyzing system calls so that a zero-day attack can be detected beforehand.

As per Gandotra et al. [15], a zero-day malware can be best detected by adopting an integrated approach whereby both the static and dynamic analysis features of mal-ware are incorporated along the process of Machine Learning. The proposed model was tested on a real-world dataset of malicious files and the results show that the integrated approach yielded better accuracy.

Lin et al. in their work [16] have emphasized the importance of a virtual time control mechanics-based method over performing dynamic analysis. The proposed method used a modified Xen hypervisor, during which a virtual clock source was generated consistent with the predefined speed ratio and accelerated the sandbox system. This approach neither modifies package kernels nor intercepts system function calls, thereby making it compatible with other OSs.

Kim et al. [17] have designed a transferred Deep-Convolutional Generative Adversarial Network (tDCGAN) that generates fake malware and then learns how to differentiate it from the actual malwares. tDCGAN achieved an average classification accuracy of 95.74% which was better than other state-of-art models.

The work by McConaghy et al. [18] focused on the efficacy of BigchainDB within the decentralized eco-system. The various advantages of such a framework include the 1 million writes per second throughput, petabytes of knowledge storage, sub-second latency, distributed database, decentralized control, immutability, and creation and movement of digital assets. Moreover, it derives characteristics of recent distributed databases such as linear scaling in throughput and capacity with the quantity of nodes, a full-featured NoSQL, efficient querying, and applying permission.

Avasarala et al. [19] have discussed the strategy for automated machine learning based zero-day malware detection. The proposed system performs training on a dataset comprising both malign and benign file samples, partitions the dataset into a plurality of categories, and then trains category-specific classifiers for classifying the files into two separate classes of malign and benign files.

A prototype implementation of a complete consensus algorithm on the primary layer of Blockchain has been done by Aniello et al. [20]. The performance of the proposed method was enhanced in terms of availability and scalability by incorporating a Byzantine Fault Tolerant consensus along with a Distributed Hash Table.

Dorri et al. [21] have developed a tiered Lightweight Scalable Blockchain (LSB) to meet the requirements of an IoT system in an optimal fashion. The authors have explored and explained the utility of LSB by taking a case study of a smart home setting as a sample IoT application. It is known that low resource devices at homes enjoy a centralized manager responsible for establishing shared keys for communication and for processing all incoming and outgoing requests. LSB achieves the benefits of decentralization by constructing an overlay network where high resource devices can jointly manage a public Blockchain along with ensuring end-to-end privacy and security. The overlay is organized as distinct clusters to cut back overheads and therefore the cluster heads are responsible for managing the general public Blockchain. The proposed LSB framework has been designed to incorporate various optimizations such as the lightweight consensus algorithm, distributed trust and throughput management. Based on the qualitative analysis, it can be said that LSB is resilient to several security attacks and based on simulation results it is proved that LSB decreases packet overhead and delay and increases Blockchain scalability, thus performing better compared to relevant baselines.

The existing research works associated with Blockchain primarily focuses on a transparent ledger-based decentralized system that eliminates the requirement of a third party and adapts to the world of Web 3.0 or the third generation web. Web 3.0 is that generation of internet services for websites and applications which can focus on employing a machine-based understanding of data to provide a data-driven and semantic web. The objective of Web 3.0 is to form more intelligent, robust, connected and open websites [22]. Table 1 provides a comparative study of a number of the previous related works within the domain of malware detection.

Table 1 Comparative study of some previous related works on malware detection

Author (Paper ID)	Publication month, year	Paper title	Keywords	Objective	Subject(s)	Results
Fuji et al. [12]	Aug., 2019	Blockchain-Based Malware Detection Method Using Shared Signatures of Suspected Malware Files	Blockchain (Ethereum), heuristic or behaviour-based malware detection system	Sharing the signatures of suspected files between users, allowing them to rapidly respond to increasing malware threats	Malware detection	Improved the false negative rate and the false positive rate
Talukder et al. [9]	May, 2019	An approach for an Distributed Anti-Malware System Based on Blockchain Technology	Blockchain, Anti-malware Database Management, Blockchain based Distributed System	Manage anti-malware signature database efficiently using distributed ledger	Malware detection	Better data management without involving any third party
Kim et al. [17]	April, 2018	Zero-day malware detection using transferred generative adversarial networks based on deep autoencoders	Malicious software, Zero-day attack, Generative adversarial network, Autoencoder, Transfer learning, Robustness to noise	Develop a transferred deep-convolutional generative adversarial network, which generates fake malware and learns to distinguish it from real malware	Malware detection	95.74% average classification accuracy

(continued)

Table 1 (continued)

Author (Paper ID)	Publication month, year	Paper title	Keywords	Objective	Subject(s)	Results
Sun et al. [13]	March, 2018	Using Bayesian Networks for Probabilistic Identification of Zero-day Attack Paths	Zero-day Attack Path, Bayesian Networks	Develop a probabilistic approach and implement a prototype system for zero-day attack path identification	Malware detection	Probability threshold of recognizing high-probability nodes is 80%
Dorri et al. [21]	Dec., 2017	LSB: A Lightweight Scalable Blockchain for IoT Security and Privacy	Internet of Things, Blockchain, Security, Privacy, Smart home	Develop a comprehensive tiered framework based on Blockchain technology for preserving security and privacy for IoT that is lightweight	IoT security	6 leading zeros takes 2.3 s. Increasing the length of zeros to 7, increases the processing time to 29.22 min
Aniello et al. [20]	Dec., 2017	A Prototype Evaluation of a Tamper-resistant High Performance Blockchain-based Transaction Log for a Distributed Database	Blockchain, Cloud Federation, BFT, DHT	Implementation and an experimental evaluation of a prototype of layered blockchain-based architecture, which employs a total consensus algorithm on the first layer blockchain	Distributed database	Throughput—500 op/s Response—500 op/s

(continued)

Table 1 (continued)

Author (Paper ID)	Publication month, year	Paper title	Keywords	Objective	Subject(s)	Results
Lin et al. [16]	Nov., 2017	Efficient dynamic malware analysis using virtual time control mechanics	Dynamic analysis, Virtual time control, Information entropy, Anti-analysis, Hypervisor	Virtual time control mechanics based method to detect of stealthy malware attacks in suspicious files efficiently	Malware detection	Logged record size increased by up to 42% compared with conventional sandboxes
Raje et al. [8]	Nov., 2017	Decentralised Firewall for malware detection	Malware, Blockchain consensus, Portable Executable, Deep belief network, Restricted Boltzmann machine	Design and development of a decentralized firewall system powered by a novel malware detection engine based on DBN	Malware detection	DBN^2—Accuracy is 89.28% and TPR is 0.9826 DBN^3—Accuracy is 88.14% and TPR is 0.9789
Gandotra et al. [15]	July, 2017	Zero-day malware detection	Malware detection, Static malware analysis, Dynamic malware analysis, Feature selection, Machine learning	Static and dynamic malware analysis is being used along with machine learning algorithms for malware detection and classification	Malware detection	Static and dynamic features considered together provide high accuracy for distinguishing malware binaries from clean ones
Avasarala et al. [19]	May, 2017	System and method for automated machine-learning, zero-day malware detection	Machine Learning, Zero-day-malware detection	Implement a zero-day malware detection based on ML to distinguish a malign/benign file	Malware detection	TP rate increased from 80 to 90% FP rate reduced from 18 to 7%

(continued)

Table 1 (continued)

Author (Paper ID)	Publication month, year	Paper title	Keywords	Objective	Subject(s)	Results
McConaghy et al. [18]	June, 2016	BigchainDB: A Scalable Blockchain Database	Blockchain, distributed database, NoSQL, IPFS, DNS	Design a database based on Blockchain technology	Malware signature sharing, distributed database	1 million writes per second throughput, storing petabytes of data, and sub-second latency
Cha et al. [10]	April, 2011	SplitScreen: Enabling Efficient, Distributed Malware Detection	Distributed System, feed-forward bloom filter	Develop a two-phase scanning that enables fast and memory-efficient malware detection that can be decomposed into a client/server process that reduces the amount of storage on, and communication	Malware detection	2 times faster and less memory than the original ClaimAV software

5 Proposed Methodology

The proposed multi-malware detection engine is intended to analyze the behavior of the files or executables that has been downloaded from the web or received from the storage peripherals by external means. Whenever any kind of distrustful activity is found, it will cautiously handle that explicit file and alert all the clients or users and automatically update the anti-malware distributed database using the Blockchain Technology by running as a background process. The core functionality lies in consuming resources of all the machines in an exceedingly local network for each incoming file to be tested.

Any arriving file to a client or node within the inter-connected network is first checked from the prevailing hash present within the distributed database and if the hash is not present, then it is broadcasted to all the nodes or any specific node (or machine) within the local network. This broadcast would not be disbursed within the blockchain. At every client or node, this received file is scanned and executed using the anti-malware detection engine deployed at that node. Each machine analyses the broadcasted file and then determines the weighted percentage of the file being malign.

It is worth noting that each one the nodes possess distinctive anti-malware detection engines. This can be hashed by the node's own key and added to the blockchain as a transaction. The node that broadcasted the file can then go through the whole chain to get the final result, which is the weighted average of the probabilities for that very file hash.

The resultant value in percentage of file being malicious is a direct measure of trust for that particular node in the inter-connected network. Since, Blockchain Technology is being used here, therefore any unauthorized changes to the blockchain node is not going to happen because the hash of that block is different and synchronized. Hence the results are immune to tampering or alteration.

Algorithm 1 demonstrates the total working of the proposed multi-malware detection engine.

Algorithm 1: Proposed Multi-Malware Detection Engine

```
start
LOAD: load files for scanning
select a file to scan
compute its MD5 file hash
if file hash is present in distributed database then
  file flagged as malign;
  move file to quarantine;
  alert every node in the network;
else
  file is broadcasted to all the available nodes in the network;
    if file is found to be malign in at least one node then
      file is flagged as malign;
      move file to quarantine;
      alert every node in the network ;
      append the obtained result from each malware detection engine to
      the blockchain network;
      update the distributed database to a new state;
    else
      file is flagged as safe;
    end if
  end if
repeat LOAD
end
```

Algorithm 2 and Algorithm 3 show the steps to be executed during the Signature-based and multi-AV-based malware detection processes respectively.

Algorithm 2: Signature-based Detection Engine

```
start
f: File
signature_based_detection(f)
  signature=md5_hash(f);
  /*here n is the entry number in the distributed database.*/
  for i=0 in range(n) do
    if(signature==hash(existing)) then
      /*File moved to quarantine*/
      base64_encode(f);
      notify all the active nodes;
    else
      file is safe;
    end if
  end for
end function
end
```

Algorithm 3: Multi-AV-based Detection Engine

```
start
f: File
apiKey: Your API key
MultiAV_detection(f)
  result=VirusTotalAPI_request(api, f);
  if file is found malign then
    /*Average of all antivirus results*/
    percent=weighted_probability(result);
    signature=md5_hash(f);
    file is flagged as malign;
    /*File moved to quarantine*/
    base64_encode(f);
    alert every node in the network;
    append the obtained result from each malware detection engine to
    the blockchain network;
    update the distributed database to a new state;
  else
    file is labelled as safe;
  end if
end function
end
```

Figure 3 shows the workflow of the proposed multi-malware detection engine.

First of all, before getting started a user will have to install the anti-malware detection engine. So, for this an installer will select what detection engine is suitable for the machine according to various parameters such as its processor speed, RAM, ROM etc. Installer will generate the system report and based on that report, the user will be assigned a malware detection engine from the list of available detection engines. Table 2 shows the list of three malware detection engines used in the proposed framework and Table 3 shows the list of clients to whom the engines have been assigned during implementation of the proposed system.

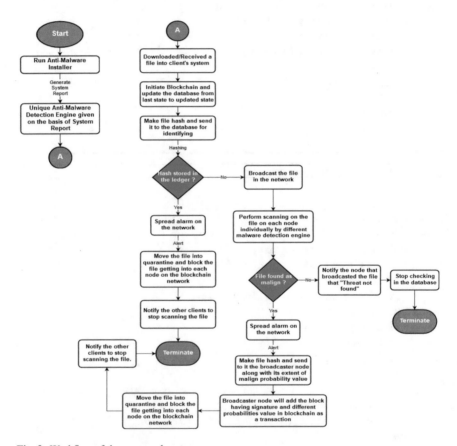

Fig. 3 Workflow of the proposed system

Table 2 List of anti-malware detection engines	Detection engine	Engine ID
	Signature-based detection engine	A
	Behavior-based detection engine	B
	Multi-AV-based detection engine	C

Client's ID	Assigned engine ID
40d123e5f316bef78bfdf5a008837577	B
35d91262b3c3ec8841b54169588c97f	A
ff6626c69507a6f511cc398998905670	C

Table 3 Assignment of detection engine to clients

Suppose a user receives a new file on his or her system from the Internet or by any physical means. It is quite possible that a large number of malwares can infiltrate into a user's system from the Internet via email or through malicious websites. Whenever a new file gets into the user's machine or node, the proposed system will make a hash of that file by using the MD5 (Message-Digest 5) hash function generation algorithm that produces a 128-bit hash value. In addition to this, a micro-process also runs through which the blockchain network in activated, the anti-malware engine is launched automatically and the distributed database is updated from last state to new state with more number of signatures as compared to the last state of the database. Basically, this is done automatically after a scheduled time. After obtaining the MD5 hash of the incoming file, its entry is checked with the content of the distributed database. If the hash exists, then the node will generate a threat alert about the malicious activity and move the file into quarantine and immediately block the incoming file from further propagation into the blockchain network. So, if the identified malware tries to compromise or infect any of the nodes or clients in the inter-connected network, it will be instantly blocked because its hash is already present in the database.

Now, if the file hash is not present in the distributed ledger or distributed database, then there is a need to broadcast the file to all the nodes in the local network on which this system is deployed and it is carried out in the blockchain network. After broadcasting the incoming file in the network, at every active node (which are available for performing the scan operation), this file is scanned through the anti-malware present at that node. In this proposed system, the anti-malware detection engines that have been used are of three types viz. signature-based detection based on static analysis detection, multi-AV-based detection by using different antivirus software and behavior-based detection based on dynamic analysis technique.

After scanning the broadcasted file on each node, the file is stated as either malign or benign. If the file is a malign, then its hash is supplied to the node which broadcasted the file along with its probability of being malicious. This is done on every node where the file is broadcasted. Now after receiving all the signatures from each available node, it adds that block in blockchain as a transaction. The structural design of the transaction block is shown in Fig. 4.

It is noted that virus signature is added to the database after encrypting through certain private key which is shared among the nodes in the blockchain network by digital signature key exchange technique. After that, the node or client that broadcasted the file can then go through the whole chain to get the final result, which is the weighted average of the probabilities for that very file hash. The trust of the node is

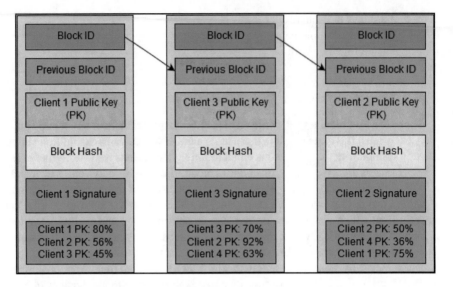

Fig. 4 Structure of the transaction block in a blockchain

determined directly by the weights or value of the probability. The weighted average of the probability value is calculated by the given formula (see Eq. 1) which is as follows:

$$P = \sum_{i=0}^{Clients\ PK} P(i) \tag{1}$$

Here, P is the total average probability by summing up the individual probabilities of clients in corresponding blocks in the blockchain and *Clients PK* is the public key of each client in the corresponding blocks where public key is the unique identification of the client/node in the blockchain network.

6 Implementation and Results

For the purpose of implementation, malware samples and signatures have been collected from the repository by VirusShare.com that contains a list of plain text files with one hash per line. The files numbered 0–148 are 4.3 MB in size with 131,072 hashes each and the files numbered 149 and above are 2.1 MB in size with 65,536 hashes each [23]. Currently, the anti-malware engine designed in this chapter has been successfully implemented to perform malware detection using signature-based detection and multi-AV-based detection only.

Figure 5 shows the scan result of the proposed signature-based malware detection engine that has been implemented using the Tkinter library of the python programming language together with os, glob, threading, sys, urllib.request, time, hashlib and base64 libraries.

The multi-AV-based malware detection engine was developed using the VirusTotal REST API v3 which is powered by over 70 Antivirus scanners available online like Quick Heal, Avast, Bit Defender, Avira, AVG, McAfee, Norton, etc. [24, 25]. The scan result obtained was in json format. Figure 6 shows the scan result and Fig. 7 shows the detailed scan result of the proposed multi-AV-based malware detection engine that has been implemented using the Tkinter library of the python programming language together with json, requests, os, glob, threading, sys, urllib.request, time, argparse, hashlib and base64 libraries. Figure 8 shows the report of the multi-AV-based malware detection engine.

The blockchain App was developed using Node.js, Socket.io and Express. This blockchain is used to hold the result of scanned files. Figure 9 shows the blockchain network where 4 clients are connected with each other by a blockchain. Whenever any user adds a new block, it will be received by other nodes, and then they will update their hash tables accordingly.

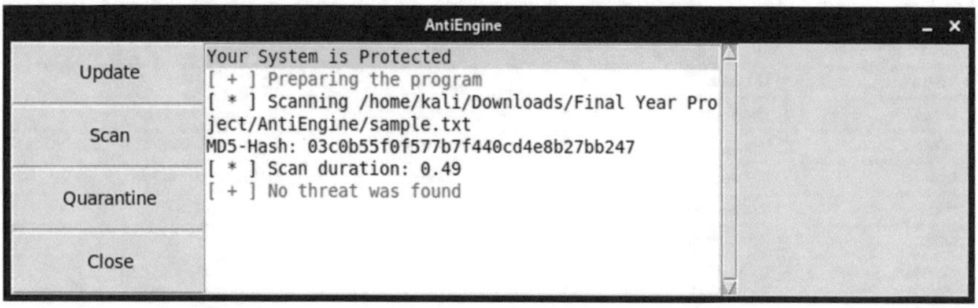

Fig. 5 Screenshot showing the scan result of the signature-based malware detection engine

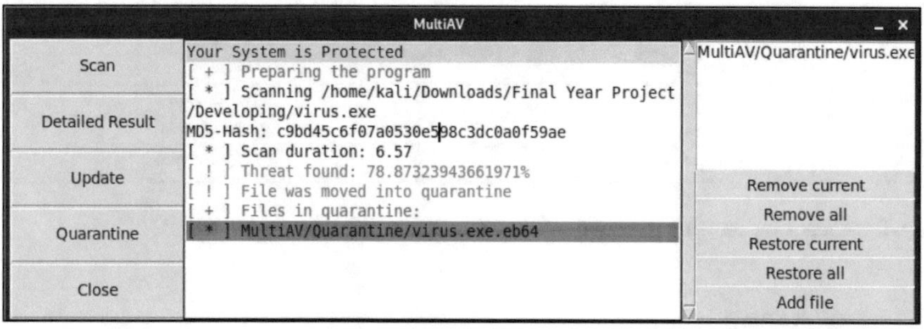

Fig. 6 Screenshot showing the scan result of the multi-AV-based malware detection engine

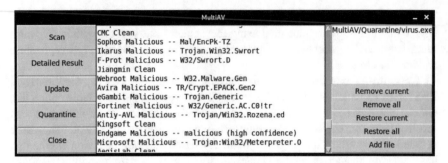

Fig. 7 Screenshot showing the detailed scan result of the multi-AV-based malware detection engine

Acronis	ⓘ Suspicious	Ad-Aware	ⓘ Trojan.CryptZ.Gen
AhnLab-V3	ⓘ Trojan/Win32.Shell.R1283	ALYac	ⓘ Trojan.CryptZ.Gen
Antiy-AVL	ⓘ Trojan/Win32.Rozena.ed	SecureAge APEX	ⓘ Malicious
Arcabit	ⓘ Trojan.CryptZ.Gen	Avast	ⓘ Win32:SwPatch [Wrm]
AVG	ⓘ Win32:SwPatch [Wrm]	Avira (no cloud)	ⓘ TR/Crypt.EPACK.Gen2
BitDefender	ⓘ Trojan.CryptZ.Gen	BitDefenderTheta	ⓘ Gen:NN.ZexaF.34108.eq1@aanlexci
Bkav	ⓘ W32.FamVT.RorenNHc.Trojan	CAT-QuickHeal	ⓘ Trojan.Swrort.A
ClamAV	ⓘ Win.Trojan.MSShellcode-7	Comodo	ⓘ TrojWare.Win32.Rozena.A@4jwdqr
CrowdStrike Falcon	ⓘ Win/malicious_confidence_100% (D)	Cybereason	ⓘ Malicious.cdce4e
Cylance	ⓘ Unsafe	Cyren	ⓘ W32/Swrort.D
DrWeb	ⓘ Trojan.Swrort.1	eGambit	ⓘ Trojan.Generic
Emsisoft	ⓘ Trojan.CryptZ.Gen (B)	Endgame	ⓘ Malicious (high Confidence)
eScan	ⓘ Trojan.CryptZ.Gen	ESET-NOD32	ⓘ A Variant Of Win32/Rozena.ED
F-Prot	ⓘ W32/Swrort.D	F-Secure	ⓘ Trojan.TR/Crypt.EPACK.Gen2
FireEye	ⓘ Generic.mg.190714dcdce4e77!	Fortinet	ⓘ W32/Generic.AC.C0!tr
GData	ⓘ Trojan.CryptZ.Gen	Ikarus	ⓘ Trojan.Win32.Swrort
K7AntiVirus	ⓘ Trojan (004c49f81)	K7GW	ⓘ Trojan (004c49f81)
Kaspersky	ⓘ HEUR:Trojan.Win32.Generic	Malwarebytes	ⓘ Trojan.Rozena
MAX	ⓘ Malware (ai Score=81)	MaxSecure	ⓘ Trojan.Malware.121218.susgen
McAfee	ⓘ Packed-FDA!190714DCDCE4	McAfee-GW-Edition	ⓘ BehavesLike.Win32.Swrort.lh
Microsoft	ⓘ Trojan.Win32/Meterpreter.O	NANO-Antivirus	ⓘ Trojan.Win32.Shellcode.ewfvwj
Qihoo-360	ⓘ HEUR/QVM20.1.0474.Malware.Gen	Rising	ⓘ Malware.Heuristic!ET#100% (RDMK:cm...
Sangfor Engine Zero	ⓘ Malware	SentinelOne (Static ML)	ⓘ DFI - Malicious PE
Sophos AV	ⓘ Mal/EncPk-TZ	Sophos ML	ⓘ Heuristic
SUPERAntiSpyware	ⓘ Trojan.Backdoor-Poisonivy	Symantec	ⓘ Packed.Generic.347
Trapmine	ⓘ Malicious.high.ml.score	TrendMicro	ⓘ BKDR_SWRORT.SM
TrendMicro-HouseCall	ⓘ BKDR_SWRORT.SM	VIPRE	ⓘ Trojan.Win32.Swrort.B (v)

Fig. 8 Screenshot showing the multi-AV-based malware detection engine report

7 Conclusion

The work presented in this chapter focuses on developing a multi-malware detection
engine that is based on the Blockchain Technology. It provides a non-centralized
and distributed database-oriented networking environment without any involvement

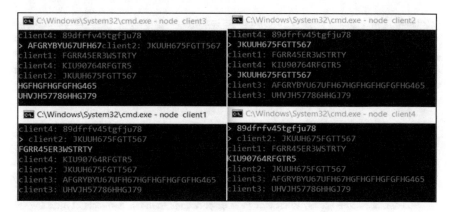

Fig. 9 Screenshot showing different clients of the blockchain network

of third parties. Presently, software companies developing antivirus products collect malicious hashes of the file from the system and update its centralized database after verification. Later, after a certain period of time, consumers get notifications to update their virus definitions to a new state. This process is time consuming and causes loss of bandwidth. Moreover, the antivirus software available today consumes more disk space which is ever-increasing with updates and latest releases.

Thus, the proposed multi-malware detection engine strives to enhance security of user-end devices, provides protection from malware, does away with the notion of third-party by using Blockchain framework, maintains a distributed database thereby avoiding any single point failure and ensures faster scanning of malicious files with less memory usage as compared to normal conventional antivirus programs. It also enhances the capability of malware detection for low-end resources, provides options to download or use the proposed framework and based on the resources available in the PC, suggests the most appropriate malware detection engine to the user that can perform high end scanning too thus giving comparatively high positive results. Last but not the least, whenever any node or client faces a malign activity, all the other nodes on the inter-connected network will be alarmed immediately, thereby making the network immune to that attack or breach.

8   Future Work

Since security-providing software are generally used on personal computers, servers and mainframes, they fail to provide security to IoT devices because of resource scarcity and storage space limitations, research can be carried out in future to provide security not only for the personal or commercial computers and mainframes server but also for the smart phones and Internet-enabled devices. As the proposed architecture focuses on small organizational unit connected in local area network, it can be extended from smaller organizational units to large corporations to impact the global market. Furthermore, Machine Learning approaches can be explored to achieve better performance. The work of implementing the behavior-based malware detection is already in the pipeline. The proposed work can be embellished by detecting the source of malware i.e., from where the attack actually originated.

References

 1. Malwarebytes (2019) What is Malware? (online). Available at https://www.malwarebytes.com/malware. Accessed 30 Dec 2019
 2. Npd-solutions.com (2019) What is reverse engineering? (online). Available at https://www.npd-solutions.com/reverse-engineering.html. Accessed 30 Dec 2019
 3. Louk M, Lim H, Lee H (2014) A classification of malware detection techniques (online). hindawi.com. Available at https://www.hindawi.com/journals/tswj/2014/983901. Accessed 5 Aug 2014
 4. Van Hung P (2011) An approach to fast malware classification with machine learning technique. Keio University, 5322 Endo Fujisawa Kanagawa 252-0882 JAPAN
 5. Tian R (2011) An integrated malware detection and classification system. Changchun University of Science and Technology, thesis
 6. Robiah Y, Rahayu SS, Zaki MM, Shahrin S, Faizal M, Marliza R (2009) A new generic taxonomy on hybrid malware detection technique. Int J Comput Sci Inf Secur (IJCSIS) 5(1)
 7. Lastovetska A (2019) Blockchain architecture basics: components, structure, benefits & creation (online). Available at https://mlsdev.com/blog/156-how-to-build-your-own-blockchain-architecture. Accessed 30 Dec 2019
 8. Raje S, Vaderia S, Panigrahi R, Weilson N (2017) Decentralised firewall for malware detection. arXiv preprint arXiv:1711.01353v1
 9. Talukder S, Roy S, Mahmud TA (2019) An approach for an distributed anti-malware system based on blockchain technology. In: 2019 11th international conference on communication systems & networks (COMSNETS). IEEE
10. Cha S, Moraru I, Jang J, Truelove J, Brumley D, Andersen D (2011) SplitScreen: enabling efficient, distributed malware detection. J Commun Netw 13(2):187–200
11. Kojm T (2019) ClamavNet (online). Available at http://www.clamav.net. Accessed 30 Dec 2019
12. Fuji R, Usuzaki S, Aburada K, Yamaba H, Katayama T, Park M, Shiratori N, Okazaki N (2019) Blockchain-based malware detection method using shared signatures of suspected malware files. Available at https://doi.org/10.1007/978-3-030-29029-0_28
13. Sun X, Dai J, Liu P, Singhal A, Yen J (2018) Using Bayesian networks for probabilistic identification of zero-day attack paths. IEEE Trans Inf Forensics Secur 13(10):2506–2521
14. FireEye (2019) What is a zero-day exploit?|FireEye (online). Available at https://www.fireeye.com/current-threats/what-is-a-zero-day-exploit.html. Accessed 30 Dec 2019
15. Gandotra E, Bansal D, Sofat S (2016) Zero-day malware detection. In: 2016 sixth international symposium on embedded computing and system design (ISED). IEEE, pp 171–175
16. Lin C, Pao H, Liao J (2018) Efficient dynamic malware analysis using virtual time control mechanics. Comput Secur 73:359–373
17. Kim J, Bu S, Cho S (2018) Zero-day malware detection using transferred generative adversarial networks based on deep autoencoders. Inf Sci 460–461:83–102
18. McConaghy T, Marques R, Muller A, De Jonghe D, Mc-Conaghy T, McMullen G, Henderson R, Bellemare S, Granzotto A (2016) Bigchaindb: a scalable blockchain database. White paper, BigChainDB

19. Avasarala BR, Day JC, Steiner D, Bose BD (2016) System and method for automated machine-learning, zero-day malware detection. US Patent 9,292,688

20. Aniello L, Baldoni R, Gaetani E, Lombardi F, Margheri A, Sassone V (2017) A prototype evaluation of a tamper-resistant high performance blockchain-based transaction log for a distributed database. In: 2017 13th European dependable computing conference (EDCC). IEEE, pp 151–154

21. Dorri A, Kanhere SS, Jurdak R, Gauravaram P (2017) LSB: a lightweight scalable blockchain for IOT security and privacy. arXiv preprint arXiv:1712.02969

22. WhatIs.com (2019) What is Web 3.0? A definition by WhatIs.com (online). Available at https://whatis.techtarget.com/definition/Web-30. Accessed 30 Dec 2019

23. VirusShare.com. Available at https://virusshare.com/hashes.4n6. Accessed 30 Dec 2019

24. VirusTotal. API. Available at https://support.virustotal.com/hc/en-us/articles/115002100 149-API. Accessed 31 Jan 2020

25. Crunchbase. VirusTotal. Available at https://www.crunchbase.com/organization/virustotal. Accessed 29 Feb 2020

Ameliorated Face and Iris Recognition Using Deep Convolutional Networks

Balaji Muthazhagan⊙ and Suriya Sundaramoorthy

Abstract Biometric systems which are both secure and reliable is imperative for the verification and identification of individual subjects. Such systems also need to respond with superior accuracy for proof of identity and concurrently ensure ease of access. In this chapter we propose approaches using deep convolutional networks which give extremely accurate results with substantially smaller processing time for face and iris recognition. Two approaches based on transfer learning using VGG-16 and VGG-19 is considered: using the pre-trained models as feature extractors and fine tuning their existing architectures. The accuracy across a multitude of datasets is evaluated for these ameliorated versions of face and iris recognition using both the techniques.

Keywords Biometric systems · Face recognition · Iris recognition · Deep convolutional neural networks · SVM classification · Transfer learning

1 Introduction

The field of biometrics aid in the identification of an individual through a collated set of behavioral attributes such as voice, keystrokes, signature etc. and physical attributes such as the face, fingerprint etc. [1]. Jain et al. [2] published seven characteristics which identifies an attribute to qualify for biometric recognition:

- Universality—The attribute chosen should be present among a large set of the population who will engage with the system. For example, let us say that the attribute under consideration is a scar, we cannot expect most members of the

B. Muthazhagan (✉) · S. Sundaramoorthy
PSG College of Technology, Coimbatore, Tamil Nadu 641004, India
e-mail: balajimuthazhagan@gmail.com

S. Sundaramoorthy
e-mail: suriyas84@gmail.com

Y. Maleh et al. (eds.), *Machine Intelligence and Big Data Analytics for Cybersecurity Applications*, Studies in Computational Intelligence 919, https://doi.org/10.1007/978-3-030-57024-8_12

277

population to house a scar. Thus, it does not qualify as a good attribute for biometric recognition.

- Uniqueness—The attribute chosen must be unique to a given individual. For example, if we consider age as an attribute, it cannot uniquely identify an individual and thus does not qualify for biometric recognition. On the other hand, an attribute such as a fingerprint is unique to everyone.
- Permanence—The attribute chosen must not change drastically over time. It should be able to resist changes over time.
- Measurability or collectability—The attribute chosen should be easily accessible and uncomplicated to acquire. For example, let us take the case of a footprint as an attribute. When the user must interact with the system, the user must remove the shoes and other layers of clothing to interact. This process is highly inefficient.
- Performance—The system which consumes the attribute chosen must ensure accuracy, speed, and a low error rate.
- Acceptability—The system which consumes the attribute must be easily acceptable to the users of the system. It should not be abstruse, slow, or less comfortable to use.
- Circumvention—The system should not easily fall prey to false biometric identifiers. Examples of this include face spoofing, gummy finger etc. (Fig. 1).

Biometric recognition dates to the 1970s where agencies used such systems for pinpointing criminals based on their fingerprints and today the inclusion of such systems can be found in objects of everyday usage like mobile phones. In the

Fig. 1 Classification of biometric systems

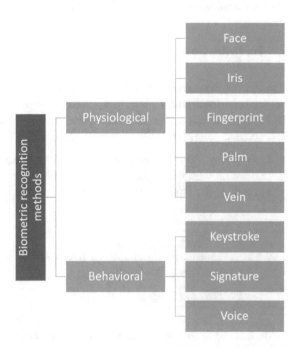

emerging world, this is one of the most popular means of personal authentication since token based and alphanumeric methods can easily be emulated and forgotten [3]. Given that 81% of the data breaches deal with passwords, there is a need to establish proof of identity. Biometric recognition involves the identification of millions, if not billions of individuals and the factors which distinguish them may be very tenuous. Real word data which will be fed as the input will contain a considerable amount of distortion and noise. Also, the biometric which is being identified might undergo change with the passage of time [4]. Multiple samples need to be collected and documented which clearly indicate the variations in features. The input data belongs to a class of personal information therefore the security should be close to impregnable and ensure accuracy at the same time. Thus, there is a need to address these issues with an efficient and reliable methodology which can solve these issues. In this chapter we focus on two biometric attributes: face and iris. These attributes were chosen because of their suitability with respect to the seven characteristics.

Face based biometric recognition is one of the most frequently used amongst all biometric recognition systems [5]. This is because the face is one of the most accessible parts of the human body and biometric systems incorporating face detection require very less participation from the subject. Algorithms involving the face must overcome the main challenge of variance in appearance. The face which is given as an input is subjected to a lot of changes due to change in pose, lighting, exposure, expression or even age. Occlusions in front of the face as well (such as spectacles) and dynamic facial characteristics which change throughout time such as hair length, hair color and color of the skin also increase the complexity. Iris based biometric recognition is one of the most trusted authentication methods. This is because it contains a lot of features which do not change prominently over time which makes it extremely hard to impersonate. It is a well-protected piece of muscle with distinctive patterns like rings, furrows, crypts, and freckles. It also possesses a distinguishable color which is invariant over time (Fig. 2).

Machine learning methods is currently integrated with a multitude of applications and has continued to increase its relevance in consumer-oriented products. Earlier machine learning algorithms were restricted in their ability to process input data in

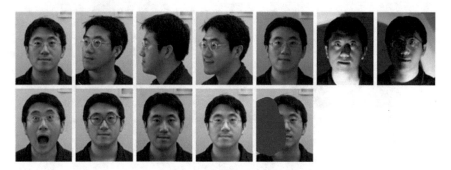

Fig. 2 Variance in face appearance

its original form since it required considerable transformation of the data to under-standable vectors but with the advent of deep learning techniques where features are automatically learnt from the data, classification accuracies have continued to increase [6]. This ongoing research on developing quality deep learning algorithms has resulted in better judgment and prediction. Since biometric identification essen-tially boils down to a classification problem where the input is the biometric to be identified, with increased accuracy and lower real-time processing time, deep learning algorithms seem to be a good fit [7] in solving this problem. In this chapter, we propose a recognition system for face and iris-based biometrics using two vari-ants of transfer learning from existing deep learning architectures and compare their accuracies across a multitude of datasets.

2 Related Works

2.1 Face Based Biometric Recognition

Facial recognition algorithms are largely classified into three types [8]:

- Local approaches which treat only some facial features
- Holistic approaches which treat the entire face without extracting any facial features
- Hybrid approaches which encompass both local and holistic approaches (Fig. 3).

Local approaches contain local appearance-based and key points-based approaches. Local appearance-based technique considers the face as a geometric representation and highlights the prominent details of a face such as eyes, nose, lips, ears, and hair as patches. LBP (Local Binary Pattern) is one popular technique which does this. The face is split into spatial arrays and then a square matrix is slid across

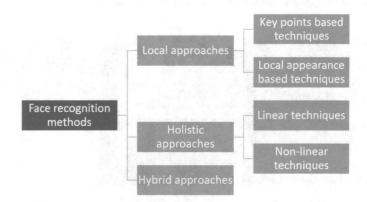

Fig. 3 Classification of face recognition systems

the splits. Based on the neighborhood pixels and a threshold value of the center pixel of the square, a histogram is obtained. A swift face recognition system, proposed by Khoi et al. [9], used LBP in their core architecture. The system had an accuracy of 90.95% on the Labelled Faces in the Wild (LFW) dataset. Xi et al. [10] proposed an LBP based network which had a similar topology to that of a CNN and achieved an accuracy of 94.04% on the LFW dataset. Kambi and Guo [11] proposed a system which was a combination of LBT and K-NN. This system was successful in solving the main challenges involved in the variance in appearances such as illumination, occlusions etc. and achieved an accuracy of 85.71% on the LFW dataset.

The next prominent technique in this category is HOG (Histogram of oriented approaches) which is primarily used for extracting the edges and shapes. Correlation filters have also given good results with respect to accuracy and discrimination in this area. A robust face recognition system was developed by Karaaba et al. [12] which used a multitude of histograms, but this system had an exceptionally low accuracy of 23.49% on the LFW dataset. Arigbabu et al. [13] proposed a system which used the pyramid histogram of gradient descriptor and achieved an accuracy of 88.50% on the LFW dataset.

Key points-based techniques can be broken down into two steps: key-point identification and followed by extraction of features. They basically attribute some information to the identified geometric features such as distance between nose and the lips, size of the nose, distance between eyes etc. and match faces. This can be achieved by defining feature specific descriptors. Scale invariant feature transform (SIFT) is one such method which identifies this. The idea behind this technique was to first transform the image into a representable form containing the areas of interest. Lenc and Král [14] proposed a system which used SIFT and achieved an accuracy of 98.04% on the LFW dataset. Speeded-up robust features (SURF) evolved from SIFT but utilizes wavelets to ameliorate the performance. The architecture developed by Du et al. [15] use SURF and achieved an accuracy of 95.60% on the LFW dataset. A robust system was built by Vinay et al. [16] using both SIFT and SURF, however this also suffered from a low accuracy of 78.86% on the LFW dataset (Fig. 4).

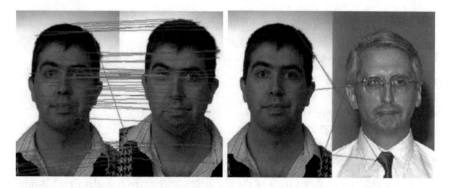

Fig. 4 Face matching through key points-based techniques

Holistic approaches can broadly be classified into linear approaches and non-linear approaches. Eigenfaces is an extremely popular and successful method in the linear approach domain. Eigenfaces use Principal component analysis (PCA) to transform images into principal components. This helps in largely reducing the dimensions of the data which is given as input. Thereafter eigenvectors are calculated, and the images are nothing but linear combinations of principal components. Seo et al. [17] proposed a system which uses PCA and achieved an accuracy of 85.10% on the LFW dataset. Annalakshmi et al. [18] used LDA coupled with independent component analysis to achieve an accuracy of 88% on the LFW dataset. These can also be described as dimensionality reduction techniques. Thereafter the emergence of Gabor filters established that features can be extracted by their frequency and scale. Hussain et al. [19] came up with a system which used Gabor filters and achieved an accuracy of 75.3% on the LFW dataset. Non-linear techniques involve Kernel principal component analysis (KPCA) and Kernel linear discriminant analysis which are improved versions of PCA and LDA. Lu et al. [20] proposed a system which uses KPCA and achieved an accuracy of 48% on the UMIST face dataset.

Hybrid approached combine the advantages of local as well as holistic approaches. A hybrid system was proposed by Fathima et al. [21] which used linear discriminant analysis and Gabor wavelets. The system achieved an accuracy of 88% on the AT&T face dataset. Ding and Tao [22] proposed a system which used multimodal deep face representation (MM-DFR). This uses convolutional neural networks (CNNs) to extract the features and an auto-encoder is used to reduce the dimensionality of the features generated. This achieved an accuracy of 99% on the LFW dataset. Sun et al. [23] proposed a system which used Long short-term memory (LSTM) in CNNs to correctly classify human activity recognition. This was tested on the OPPORTUNITY dataset and achieved an accuracy of 90.60%. The dataset contains 46,495 training sequence samples and 9894 testing sequence samples. Taigman et al. [24] proposed a DeepFace architecture which achieved an accuracy of 97.35% on the LFW dataset (Fig. 5).

2.2   Iris Based Biometric Recognition

The generalized steps involved in an iris recognition system are as follows:

- Iris image acquisition
- Segmentation of the iris image
- Normalization of the image
- Extraction of features from the image
- Finding correlation between features identified
- Mapping to subjects.

John Daughman et al. was one of the earliest publishers of a modern algorithm in which he describes a method based on 2D Gabor wavelet transform [25]. The system takes the iris image from a connected camera as the input and distinguishes two

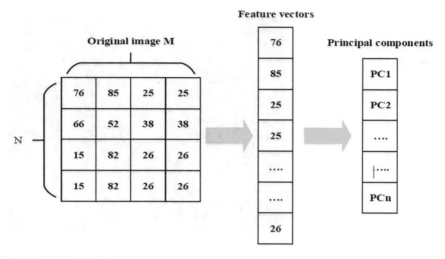

Fig. 5 Example of PCA

periphery boundaries: outer sclera and the inner pupillary boundary. This is achieved with the help of an Integro-differential operator. To account for the offset of the pupils from the centre of the iris, there is a projection of the pseudo polar coordinates to real coordinates. This is done by analysing the annular rings. This method considers images which do not have any occlusions such as eyelashes, eyelids etc. A 2D Gabor wavelet transform is performed, and the features are extracted. The output is 8 rows of 256-bit code. Since the format of this code remains the same, it is easier to make comparisons and draw results. Normalized hamming distance is used to map this code to a subject. Doughman's method achieved an equal error rate (EER) of 0.08%.

In 1997, Wildes [26] used LED point source along with the standard camera to obtain the image of the iris from among 40 subjects. Using Hough transform, the boundaries of the inner and outer iris is calculated. Based on this and a derivation from Laplacian of Gaussian, a signature template is developed. Thereafter correlation measures are used to check for similarities. Wilde's system had an EER of 1.76%. Boles and Boashash [27] developed a system which used wavelet transform zero-crossing for the representation of features at different resolution levels. Single dimensional signals are obtained, and the zero-crossing representation is derived. The image which has the edge detected is used to estimate the diameter and the center of the iris. Once the center has been identified, virtual circles are developed from the center, and is normalized to maintain uniformity across data points. These representations are stored as iris signatures. Thereafter dissimilarity measures across images of the same person and images of different people are calculated. This system achieved an accuracy of EER of 8.13%. Ko et al. [28] developed a system which was based on Daughman's approach and used cumulative sum-based change points. The image after being normalized to 64×300 is processed for feature extraction using the cumulative sums method on cells of dimensions 3×10. This process is done

Fig. 6 Identification of
features and generation of 8
rows of 256-bit code using
Daughman's method

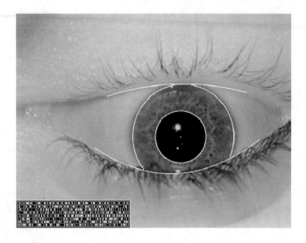

vertically and horizontally and the minimum and maximum of these values is noted. Thereafter thresholding is applied on the summation values. This system achieved an accuracy of 98.21% on the CASIA iris dataset (Fig. 6).

Huang et al. [29] developed a system which was based off on Independent component analysis. The first step in the process is image acquisition which is performed with different noise and illumination levels. Integro differential operator and curve fitting is used for iris localization. N concentric circles are projected with M samples. These projections are transformed to a matrix of size N × M. The components which are independent are estimated from the coefficients of the features. Thereafter patterns are recognized using the Euclidean distance classifier. The system achieved an accuracy of 93.8% on images which had varying illuminations and 62.5% on images which had noise interference.

Most of the approaches which involve machine learning encompass a two-step approach: features are identified from the images and then a classifier is used to recognize them. Kumar and Passi [30] tailored an approach based on the amalgamation of Haar wavelet, Log-Garbor, DCT and FFT. This resulted in a particularly good accuracy. In [31], Farouk elastic graph matching coupled with Daughman's [25] approach. The idea proposed by Minaee et al. [32] used multi-layer scattering based convolutional neural networks. This iris images were broken down using wavelets of different sizes, scales, and orientations. Thereafter the features were used for classification. These algorithms even though they achieve good outputs with sustained accuracy involve an arduous step of pre-processing. This pre-processing step includes the likes of iris segmentation and laying out the same on a rectangular area. They are also generally personalized to the identification of a few hand-crafted features which is not optimal considering the variety in datasets which are prepared under different conditions and resolutions. This issue of pre-processing can be eliminated with the use of deep learning networks. Also, the features that are learned from a previously trained deep learning network can be transferred to another task as shown by Minaee et al. [32].

3 Proposed System

Transfer learning is a type of learning in machine learning where a model developed for a task can be reused for other tasks. This results in a multitude of advantages like the reduction in time to train these models and decrease in the scale of compute resources [33] (Fig. 7).

There are two common approaches to apply transfer learning in machine learning tasks:

- Approach 1: Use the pre-trained networks as feature extractors and then supplement it with the help of a classifier.
- Approach 2: Replacing the fully connected layers of an existing pre-trained network with new layers and fine tuning the weights.

3.1 VGG-16 and VGG-19 Architectures

VGG [34] is a convolutional neural network architecture developed by the Visual Geometry Group at Oxford. The model achieved a top-5 test accuracy of 92.7% in ImageNet.

The architecture of VGG-16 and VGG-19 is described as follows:

- The first and the second layer are convolutional layers containing 64 filters of dimensions 3×3 having a stride of 1. The input dimensions which was fed into the model was 224×224.

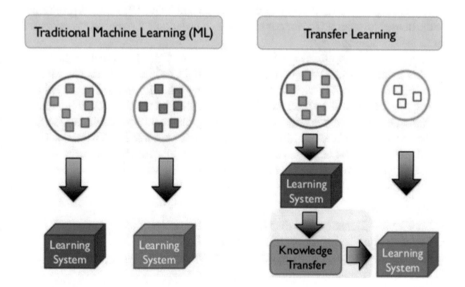

Fig. 7 Difference between traditional machine learning and transfer learning

- This is followed by a pooling layer where the dimensions reduce from $224 \times 224 \times 64$ to $112 \times 112 \times 64$.
- This is followed by 2 convolutional layers containing 128 filters of dimensions 3×3 having a stride of 1 which makes the new dimension to be $112 \times 112 \times 128$.
- This is followed by a pooling layer where the dimensions reduce from $112 \times 112 \times 128$ to $56 \times 56 \times 128$.
- This is followed by two convolutional layers containing 256 filters of dimensions 3×3 having a stride of 1 which makes the new dimension to be $56 \times 56 \times 256$.
- This is followed by a pooling layer where the dimensions reduce from $56 \times 56 \times 256$ to $28 \times 28 \times 256$.
- For VGG-16, this is followed by 3 convolutional layers containing 512 filters of dimensions 3×3 having a stride of 1 which makes the new dimension to be $28 \times 28 \times 512$. For VGG-19, there are 4 convolutional layers having the same set of filters as in VGG-16.
- This is followed by a pooling layer where the dimensions reduce from $28 \times 28 \times 512$ to $14 \times 14 \times 512$.
- For VGG-16, this is followed by 3 convolutional layers containing 512 filters of dimensions 3×3 having a stride of 1 which retains the dimension of $14 \times 14 \times 512$. For VGG-19, there are 4 convolutional layers having the same set of filters as in VGG-16.
- This is followed by a pooling layer where the dimensions reduce from $14 \times 14 \times 512$ to $7 \times 7 \times 512$.
- This is followed by 2 fully connected layers with 4096 units and 1 fully connected layer with 1000 units.
- The final layer is a softmax output with 1000 classes (ImageNet dataset) (Fig. 8).

In this chapter, we shall consider the VGG-16 and VGG-19 networks pre-trained on the ImageNet dataset [35] for our tasks. We consider both the approaches proposed for transfer learning.

In approach 1, we directly consider VGG-16 and VGG-19 as feature classifiers and this is fed into an SVM classifier. First the fully connected layers are removed from VGG-16 and VGG-19. The layers preceding this will produce an output of $7 \times 7 \times 512$ which will be used as quantified measure of the features in the image. Each image in the given dataset is passed through the architecture without the fully connected layers, and the resulting feature vector will be stored in hdf5 format. These stored values are later fed into a classifier and the results are observed (Fig. 9).

In approach 2, we replace the fully connected layers with layers of our own. The layers that we are adding to these models are as follows:

- Flatten layer
- Core layer—Dense layer with 512 units and activation function as ReLu
- Regularization layer—Dropout layer with 0.5 rate
- Core layer—Dense layer with output class number of units (Fig. 10).

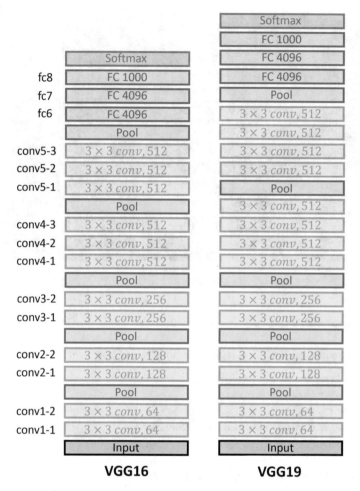

Fig. 8 VGG-16 and VGG-19 architecture

3.2 Face Based Biometric Recognition

The datasets considered for face recognition are listed in Table 1. 75% of the data was used for training and 25% was used for evaluation (Figs. 11 and 12).

3.2.1 Approach 1

The fully connected layers in VGG-16 and VGG-19 are removed and the features generated are fed into an SVM classifier (Fig. 13).

The accuracy for approach 1 across various datasets is illustrated in Table 2.

Fig. 9 Sample images from ImageNet dataset

3.2.2 Approach 2

The fully connected layers are replaced with layers of our own:

- Flatten layer
- Dense layer (512)
- Dropout layer (0.5)
- Dense layer (output class) (Fig. 14; Table 3).

3.3 Iris Based Biometric Recognition

The datasets considered for iris recognition are listed in Table 4. 75% of the data was used for training and 25% was used for evaluation (Fig. 15).

3.3.1 Approach 1

The fully connected layers in VGG-16 and VGG-19 are removed and the features generated are fed into an SVM classifier (Fig. 16; Table 5).

The accuracy across various datasets are illustrated in Table 5.

3.3.2 Approach 2

The fully connected layers are replaced with layers of our own:

```
Model: "vgg16"

Layer (type)                 Output Shape              Param #
=================================================================
input_1 (InputLayer)         [(None, 224, 224, 3)]     0
_____
block1_conv1 (Conv2D)        (None, 224, 224, 64)      1792
_____
block1_conv2 (Conv2D)        (None, 224, 224, 64)      36928
_____
block1_pool (MaxPooling2D)   (None, 112, 112, 64)      0
_____
block2_conv1 (Conv2D)        (None, 112, 112, 128)     73856
_____
block2_conv2 (Conv2D)        (None, 112, 112, 128)     147584
_____
block2_pool (MaxPooling2D)   (None, 56, 56, 128)       0
_____
block3_conv1 (Conv2D)        (None, 56, 56, 256)       295168
_____
block3_conv2 (Conv2D)        (None, 56, 56, 256)       590080
_____
block3_conv3 (Conv2D)        (None, 56, 56, 256)       590080
_____
block3_pool (MaxPooling2D)   (None, 28, 28, 256)       0
_____
block4_conv1 (Conv2D)        (None, 28, 28, 512)       1180160
_____
block4_conv2 (Conv2D)        (None, 28, 28, 512)       2359808
_____
block4_conv3 (Conv2D)        (None, 28, 28, 512)       2359808
_____
block4_pool (MaxPooling2D)   (None, 14, 14, 512)       0
_____
block5_conv1 (Conv2D)        (None, 14, 14, 512)       2359808
_____
block5_conv2 (Conv2D)        (None, 14, 14, 512)       2359808
_____
block5_conv3 (Conv2D)        (None, 14, 14, 512)       2359808
_____
block5_pool (MaxPooling2D)   (None, 7, 7, 512)         0
=================================================================
Total params: 14,714,688
Trainable params: 14,714,688
Non-trainable params: 0
```

Fig. 10 Base model of VGG-16 used in approach 1

Table 1 Face datasets

Dataset name	No of images	Unique faces	Image size
Georgia Tech face	700	50	131×206
Labeled faces in the wild	700	50	250×250
YouTube face	700	50	320×240
AR face database	4000	126	768×576
Face recognition data, University of Essex	7900	395	180×200

- Flatten layer
- Dense layer (512)
- Dropout layer (0.5)
- Dense layer (output class) (Fig. 17).

The accuracy across various datasets is illustrated in Table 6. To see which parts of the image were largely considered for the model to achieve this we use a sliding window approach. The image is split and is first transformed into a square. It is then demarcated with equal number of rows and equal number of columns. Then we select a square of interest, black it out and see whether this results a change in the observation. If it results in a change, then it means that the square selected is responsible for one of the features used in classification, else it can be ignored (Fig. 18).

4 Conclusion and Future Work

Face and iris based biometric recognition have garnered a good deal of attention in terms of incorporation into daily objects and scientific research. This chapter highlighted two different approaches by which transfer learning can be applied on deep learning architectures of VGG-16 and VGG-19 and how they significantly bettered the accuracy of the systems over conventional methods. We considered an end-to-end approach to eliminate human crafted errors which can be introduced. The proposed system can be extended to other biometric systems as well, especially systems which have a lower number of class labels. Future work revolves around iterating the fine-tuning process across a multitude of layers to further increase the accuracy of the models. To eradicate dataset bias, the focus should also steer toward creating a dataset consisting of varied age ranges, genders, and cultures. Occlusions should be part of the dataset because they ideally represent real world scenarios. Also, deeper convolutional networks such as the ResNet [36] should be explored for transfer learning.

Model: "vgg19"

Layer (type)	Output Shape	Param #
input_2 (InputLayer)	[(None, 224, 224, 3)]	0
block1_conv1 (Conv2D)	(None, 224, 224, 64)	1792
block1_conv2 (Conv2D)	(None, 224, 224, 64)	36928
block1_pool (MaxPooling2D)	(None, 112, 112, 64)	0
block2_conv1 (Conv2D)	(None, 112, 112, 128)	73856
block2_conv2 (Conv2D)	(None, 112, 112, 128)	147584
block2_pool (MaxPooling2D)	(None, 56, 56, 128)	0
block3_conv1 (Conv2D)	(None, 56, 56, 256)	295168
block3_conv2 (Conv2D)	(None, 56, 56, 256)	590080
block3_conv3 (Conv2D)	(None, 56, 56, 256)	590080
block3_conv4 (Conv2D)	(None, 56, 56, 256)	590080
block3_pool (MaxPooling2D)	(None, 28, 28, 256)	0
block4_conv1 (Conv2D)	(None, 28, 28, 512)	1180160
block4_conv2 (Conv2D)	(None, 28, 28, 512)	2359808
block4_conv3 (Conv2D)	(None, 28, 28, 512)	2359808
block4_conv4 (Conv2D)	(None, 28, 28, 512)	2359808
block4_pool (MaxPooling2D)	(None, 14, 14, 512)	0
block5_conv1 (Conv2D)	(None, 14, 14, 512)	2359808
block5_conv2 (Conv2D)	(None, 14, 14, 512)	2359808
block5_conv3 (Conv2D)	(None, 14, 14, 512)	2359808
block5_conv4 (Conv2D)	(None, 14, 14, 512)	2359808
block5_pool (MaxPooling2D)	(None, 7, 7, 512)	0

Total params: 20,024,384
Trainable params: 20,024,384
Non-trainable params: 0

Fig. 11 Base model of VGG-19 used in approach 1

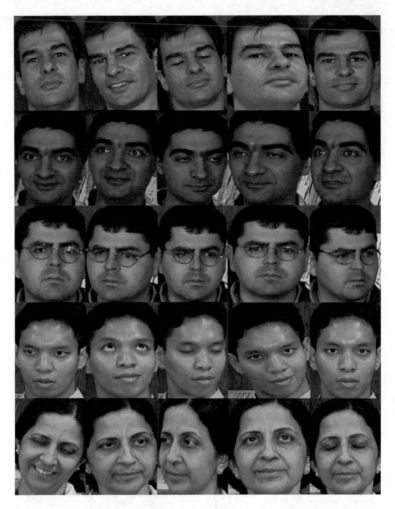

Fig. 12 Sample images from the Georgia Tech dataset

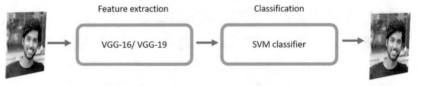

Fig. 13 VGG-16/VGG-19 architecture using SVM classifier for face recognition

Table 2 Accuracy of approach 1 across face datasets

Dataset name	Prediction accuracy VGG-16 (%)	Prediction accuracy VGG-19 (%)
Georgia Tech face	96.47	97.13
Labeled faces in the wild	97.70	97.81
YouTube face	96.89	96.43
AR face database	96.57	97.83
Face recognition data, University of Essex	96.31	96.38

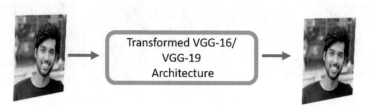

Face with class label

Fig. 14 Transformed VGG-16/VGG-19 architecture for face recognition

Table 3 Accuracy of approach 2 across face datasets

Dataset name	Prediction accuracy VGG-16 (%)	Prediction accuracy VGG-19 (%)
Georgia Tech face	97.22	98.03
Labeled faces in the wild	97.79	98.12
YouTube face	97.04	97.31
AR face database	97.47	97.89
Face recognition data, University of Essex	97.88	97.91

Table 4 Iris datasets

Dataset name	No of images	Unique iris	Image size
IIT Delhi iris database	1120	224	320 × 240
CASIA-iris-interval	2639	249	320 × 280

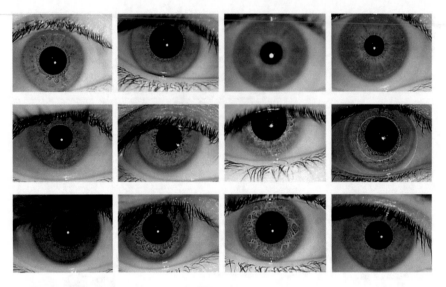

Fig. 15 Sample images obtained from the IIT Delhi Iris database

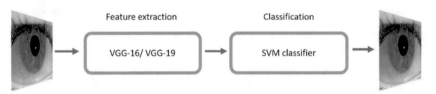

Fig. 16 VGG-16/VGG-19 architecture using SVM classifier for iris recognition

Table 5 Accuracy of approach 1 across iris datasets

Dataset name	Prediction accuracy VGG-16 (%)	Prediction accuracy VGG-19 (%)
IIT Delhi iris database	93.29	93.44
CASIA-iris-interval	94.57	94.59

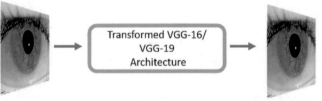

Fig. 17 Transformed VGG-16/VGG-19 architecture for iris recognition

Table 6 Accuracy of approach 2 across iris datasets

Dataset name	Prediction accuracy VGG-16 (%)	Prediction accuracy VGG-19 (%)
IIT Delhi iris database	94.11	94.67
CASIA-iris-interval	95.06	95.39

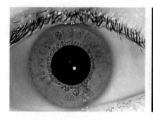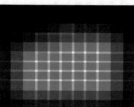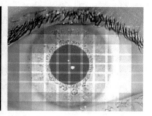

Fig. 18 Regions of interest showcasing important features

References

1. Phillips P, Martin A, Wilson C, Przybocki M (2000) An introduction evaluating biometric systems. Computer 33:56–63. https://doi.org/10.1109/2.820040
2. Jain AK, Bolle R, Pankanti S (1999) Biometrics: personal identification in networked society. Kluwer, Boston
3. Imran J, Raman B (2019) Deep motion templates and extreme learning machine for sign language recognition. The Vis Comput. https://doi.org/10.1007/s00371-019-01725-3
4. Ali M, Monaco J, Tappert C, Qiu M (2016) Keystroke biometric systems for user authentication. J Signal Process Syst 86:175–190. https://doi.org/10.1007/s11265-016-1114-9
5. Jain AK, Ross AA, Nandakumar K (2011) Introduction to biometrics. https://doi.org/10.1007/978-0-387-77326-1
6. Litjens G, Kooi T, Bejnordi B et al (2017) A survey on deep learning in medical image analysis. Med Image Anal 42:60–88. https://doi.org/10.1016/j.media.2017.07.005
7. Liu L, Ouyang W, Wang X et al (2019) Deep learning for generic object detection: a survey. Int J Comput Vision 128:261–318. https://doi.org/10.1007/s11263-019-01247-4
8. Chihaoui M, Elkefi A, Bellil W, Amar CB (2016) A survey of 2D face recognition techniques. Computers 5:21. https://doi.org/10.3390/computers5040021
9. Khoi P, Huu L, Hoai V (2016) Face retrieval based on local binary pattern and its variants: a comprehensive study. Int J Adv Comp Sci Appl. https://doi.org/10.14569/ijacsa.2016.070632
10. Xi M, Chen L, Polajnar D, Tong W (2016) Local binary pattern network: a deep learning approach for face recognition. In: 2016 IEEE International Conference on Image Processing (ICIP). https://doi.org/10.1109/icip.2016.7532955
11. Kambi Beli IL, Guo C (2017) Enhancing face identification using local binary patterns and k-nearest neighbors. J Imaging 3(3):37
12. Karaaba M, Surinta O, Schomaker L, Wiering MA (2015) Robust face recognition by computing distances from multiple histograms of oriented gradients. In: 2015 IEEE symposium series on computational intelligence. https://doi.org/10.1109/ssci.2015.39
13. Arigbabu OA, Ahmad SMS, Adnan WAW, Yussof S, Mahmood S (2017) Soft biometrics: gender recognition from unconstrained face images using local feature descriptor. *arXiv preprint* arXiv:1702.02537
14. Lenc L, Král P (2015) Automatic face recognition system based on the SIFT features. Comput Electr Eng 46:256–272. https://doi.org/10.1016/j.compeleceng.2015.01.014
15. Du G, Su F, Cai A (2009) Face recognition using SURF features. MIPPR 2009: pattern recognition and computer vision. https://doi.org/10.1117/12.832636

16. Vinay A, Hebbar D, Shekhar VS et al (2015) Two novel detector-descriptor based approaches for face recognition using SIFT and SURF. Procedia Comput Sci 70:185–197. https://doi.org/10.1016/j.procs.2015.10.070
17. Seo HJ, Milanfar P (2011) Face verification using the LARK representation. IEEE Trans Inf Forensics Secur 6:1275–1286. https://doi.org/10.1109/tifs.2011.2159205
18. Annalakshmi M, Roomi SMM, Naveedh AS (2018) A hybrid technique for gender classification with SLBP and HOG features. Cluster Comput 22:11–20. https://doi.org/10.1007/s10586-017-1585-x
19. Hussain SU, Napoléon T, Jurie F (2012) Face recognition using local quantized patterns. Proceedings of the British machine vision conference 2012. https://doi.org/10.5244/c.26.99
20. Lu J, Plataniotis K, Venetsanopoulos A (2003) Face recognition using kernel direct discriminant analysis algorithms. IEEE Trans Neural Networks 14:117–126. https://doi.org/10.1109/tnn.2002.806629
21. Fathima AA, Ajitha S, Vaidehi V et al (2015) Hybrid approach for face recognition combining gabor wavelet and linear discriminant analysis. In: 2015 IEEE international conference on Computer Graphics, Vision and Information Security (CGVIS). https://doi.org/10.1109/cgvis.2015.7449925
22. Ding C, Tao D (2015) Robust face recognition via multimodal deep face representation. IEEE Trans Multimedia 17:2049–2058. https://doi.org/10.1109/tmm.2015.2477042
23. Sun J, Fu Y, Li S et al (2018) Sequential human activity recognition based on deep convolutional network and extreme learning machine using wearable sensors. J Sens 2018:1–10. https://doi.org/10.1155/2018/8580959
24. Taigman Y, Yang M, Ranzato M, Wolf L (2014) DeepFace: closing the gap to human-level performance in face verification. In: 2014 IEEE conference on computer vision and pattern recognition. https://doi.org/10.1109/cvpr.2014.220
25. Daugman J (2007) New methods in iris recognition. IEEE Transactions on systems, man and cybernetics. Part B (Cybernetics) 37:1167–1175. https://doi.org/10.1109/tsmcb.2007.903540
26. Wildes R (1997) Iris recognition: an emerging biometric technology. Proc IEEE 85:1348–1363. https://doi.org/10.1109/5.628669
27. Boles W, Boashash B (1998) A human identification technique using images of the iris and wavelet transform. IEEE Trans Signal Process 46:1185–1188. https://doi.org/10.1109/78.668573
28. Ko J-G, Gil Y-H, Yoo J-H (2006) Iris recognition using cumulative SUM based change analysis. In: 2006 International symposium on intelligent signal processing and communications. https://doi.org/10.1109/ispacs.2006.364885
29. Huang Y-P, Luo S-W, Chen E-Y (2002) An efficient iris recognition system. Proceedings international conference on machine learning and cybernetics. https://doi.org/10.1109/icmlc.2002.1176794
30. Kumar A, Passi A (2010) Comparison and combination of iris matchers for reliable personal authentication. Pattern Recogn 43:1016–1026. https://doi.org/10.1016/j.patcog.2009.08.016
31. Farouk R (2011) Iris recognition based on elastic graph matching and Gabor wavelets. Comput Vis Image Underst 115:1239–1244. https://doi.org/10.1016/j.cviu.2011.04.002
32. Minaee S, Abdolrashidiy A, Wang Y (2016) An experimental study of deep convolutional features for iris recognition. In: 2016 IEEE Signal Processing In Medicine And Biology Symposium (SPMB). https://doi.org/10.1109/spmb.2016.7846859
33. Pan J (2017). Review of metric learning with transfer learning. https://doi.org/10.1063/1.4992857
34. Simonyan K, Zisserman A (2014) Very deep convolutional networks for large-scale image recognition. CoRR, abs/1409.1556.
35. Deng J, Dong W, Socher R et al (2009) ImageNet: a large-scale hierarchical image database. In: 2009 IEEE conference on computer vision and pattern recognition. https://doi.org/10.1109/cvpr.2009.5206848
36. He K, Zhang X, Ren S, Sun J (2016) Deep residual learning for image recognition. In: 2016 IEEE conference on Computer Vision and Pattern Recognition (CVPR). https://doi.org/10.1109/cvpr.2016.90

Presentation Attack Detection Framework

Hossain Shahriar and Laeticia Etienne

Abstract Biometric-based authentication systems are becoming the preferred choice to replace password-based authentication systems. Among several variations of biometrics (e.g., face, eye, fingerprint), iris-based authentication is commonly used in every day applications. In iris-based authentication systems, iris images from legitimate users are captured and certain features are extracted to be used for matching during the authentication process. Literature works suggest that iris-based authentication systems can be subject to presentation attacks where an attacker obtains printed copy of the victim's eye image and displays it in front of an authentication system to gain unauthorized access. Such attacks can be performed by displaying static eye images on mobile devices or iPad (known as screen attacks). As iris features are not changed, once an iris feature is compromised, it is hard to avoid this type of attack. Existing approaches relying on static features of the iris are not suitable to prevent presentation attacks. Feature from live Iris (or liveness detection) is a promising approach. Further, additional layer of security from iris feature can enable hardening the security of authentication system that existing works do not address. To address these limitations, this chapter introduces iris signature generation based on the area between the pupil and the cornea. Our approach relies on capturing iris images using near infrared light. We train two classifiers to capture the area between the pupil and the cornea. The image of iris is then stored in the database. This approach generates a QR code from the iris. The code acts as a password (additional layer of security) and a user is required to provide it during authentication. The approach has been tested using samples obtained from publicly available iris database. The initial results show that the proposed approach has lower false positive and false negative rates.

Keywords Presentation attack · Haar cascade classifier · Local binary pattern · Liveness detection

H. Shahriar (✉) · L. Etienne
Department of Information Technology, Kennesaw State University, Marietta, Georgia
e-mail: hshahria@kennesaw.edu

L. Etienne
e-mail: letienn3@students.kennesaw.edu

Y. Maleh et al. (eds.), *Machine Intelligence and Big Data Analytics for Cybersecurity Applications*, Studies in Computational Intelligence 919,
https://doi.org/10.1007/978-3-030-57024-8_13

297

1 Introduction

Security is based on three principal elements commonly known under CIA trad: Confidentiality, Integrity, and Availability. Authentication is a security control that is used to protect the system with regard to the CIA properties. Authentication is an essential step for accessing resources and/or services. Authentication is an essential step for giving access to resources to authorized individuals and prevent leakage of confidential information while maintaining the integrity of a system. There are many forms of biometrics that are currently being used for authentication such as fingerprint matching, facial recognition, shape of ear, iris pattern recognition and gait movement [1]. Among all, iris pattern recognition is a widely used biometric-based authentication approach [2, 3]. In an iris-based authentication system, iris images are captured from users, and features are extracted to be matched at a later stage for authentication. Iris is unique for everyone. It has distinct textures and patterns that can be used for authentication. Iris-based authentication can overcome the limitations of traditional password based authentication systems that are vulnerable to brute force and dictionary-based attacks. Several iris-based commercial tools are available, including Iridis [4] and Eyelock [5]. The research literature shows a rise in the application of iris-based authentication systems in areas such as immigration and border control [6], healthcare, public safety, point of sales and ATM [1], and finance and banking [7].

Recently, iris spoofing attacks have emerged as a significant threat against traditional iris-based authentication systems. For example, an attacker may obtain a printed copy of the iris of a victim or using a reconstructed iris image sample and display the image in front of an authentication system to gain unauthorized access (known as presentation attack) [8, 9]. Such attack can be performed by displaying static eye images on mobile devices or iPad (known as screen attack) [10]. This attack would lead to the risk of the wrong person gaining access or being misidentified; therefore, render security vulnerability. There are approaches to prevent presentation attacks [8, 11–13]. However, most of them rely on static features of the iris. Feature from live Iris (or liveness detection) is a promising approach [14–16], where iris images are taken with high quality camera and features are extracted. Further, additional layer of security from iris feature can enable hardening the security of authentication system that existing works do not address.

This chapter proposes iris code generation between the area of the pupil and the cornea. Figure 1 shows the red and yellow circles, which represent the area of cornea and iris. Our approach analyzes live images taken in a camera in infra-red light.

Haar-Cascade [17] and LBP classifiers [18] are used to capture the area between the pupil and the cornea. The captured area is stored in database repository for future matching purpose. The approach generates QR code from the iris image. The code is then used as a password. During authentication, iris images are matched, and the user is required to provide the QR code to be authenticated. The combination of the QR code and the iris images make hacking harder. I A prototype has been implemented using OpenCV library. The approach has been tested using samples of iris images

Fig. 1 Iris area between
cornea and pupil

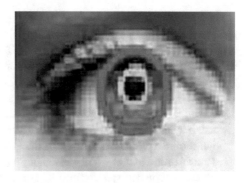

obtained from publicly available iris dataset [9]. The initial results show that the proposed approach has lower false positive and false negative rates. Furthermore, Haar Cascade classifier works better than LBP classifier [19, 20].

This chapter is organized as follows. Section 2 discusses related work that detect attacks against iris-based authentication systems. Section 3 provides an overview of Haar-Cascade and LBP classifiers. Section 4 discusses the proposed framework in detail. Section 5 highlights the implementation details and evaluation of results. Finally, Sect. 6 concludes the paper and discusses future work.

2 Background and Related Works

2.1 Attacks on Iris-Based System

It has been found that Media based Forgery and Spoofing are the most common kind of attacks in biometric based authentication system. Similarly, we find replay attack against iris is common [21]. Those kinds of attack method can be detected as liveness detection. Liveness detection allows system to validate the authentication process of valid user by real biometric identifiers. Below we define several attack types that this chapter is intended to mitigate.

a. **Media based forgery**: Media based forgery is one of the common intrusion methods to deceive any biometric based authentication or processing system. Intruder can present printed images or frames of images of authenticated user and slip out of liveness detection to get authenticated user's access in the system. For finger print authentication system, attackers can use authenticated user's printed finger print in polymer plastic to authenticated access in the system.

b. **Spoofing**: Spoofing is a method of biometric liveness attack against identification system where a dummy artificial object of a user is presented by an intruder to the system to imitate the identification feature which the process is designed to check so that it can allow authentication to attacker. It is like using the cloned biometric part of any authenticated user and apply a biometric part to get access in the system. Spoofing is mostly used by most attackers in biometric authentication attack. In context of our topic we can do face spoofing attack by using printed iris image or any cosmetic contact lens. These kinds of attacks can be crucial and alarming points for system authentication and cause a serious damage to system.

c. **Fake Iris**: Iris recognition system uses data stored in the system that are merely bits of code in binary form. Reverse engineering is possible to obtain the actual image of the iris. Genetic algorithm can be used to make different attempts using synthetic iris to be recognizable to iris detection. It takes about 100 to 200 iterations to produce a similar iris image that is stored in iris recognition system.

d. **Presentation attacks**: The presentation of biometric spoof is called presentation attack. Biometric spoof could be some image, video instead of a live person; or fake silicon or gelatin fingerprints or fake synthetic iris instead of real eye. Recognition system should be equipped with liveliness detection systems. It detects whether the presentation is alive or a spoof.

2.2 Related Work

In this section we describe related work and the approached used to detect attacks on iris-based authentication systems.

We searched in IEEE and ACM digital libraries with keywords "iris liveness detection" during year 2000 and 2019, which resulted in 67 papers. We further narrow down the list of papers that are intended for presentation attack detection and removed survey papers from the list. This led to the list of papers shown in Table 1. The list may not be exhaustive but represents the common cited works from the literature.

Pacut et al. [8] detect liveness of iris by analyzing the frequency spectrum as it reveals signatures within an image. Ratha et al. [11] split images of biometric fingerprints known as shares. These shares are stored in different databases. During authentication, one of the shares acts as an ID while another share is retrieved from the central database to be matched with a known image. Andreas et al. [12] rely on PRNU which the difference between the response of a sensor and the uniform response from light is falling on camera sensor. This approach captures the noise level information (irrelevant data) from iris images. Given that a new iris image is required to authenticate, the PRNU fingerprints from stored images are compared with the given one.

Puhan et al. [16] detect iris spoofing attacks using texture dissimilarity. As the illumination level is increased to an open eye, the pupil size decreases. Printed iris does not demonstrate such change of the pupil. High value of normalized Hamming distance between a captured image and known image results in warning of spoofed image. Adam et al. [13] detect live iris based on amplitude spectrum analysis. In this approach, a set of live iris images are analyzed to obtain the amplitude levels while performing Fourier transformation. A fake iris image has dissimilar amplitude levels compared to the real iris image.

Table 1 Summary of related work

Work	Approach	Feature type	Performance (FP, FN)
Pacut et al. [8]	Analysis of frequency of Iris im-ages	Static	2.8%, 0%
Ratha et al. [11]	Splitting of data	Static	N/A, N/A
Andreas et al. [12]	Camera photo response non-uniformity (PRNU) fingerprint	Dynamic	[0.21–23.26%], [0.21–23.26%]
Puhan et al. [16]	Liveness detection based on texture dissimilarity of Iris for contact lens	Static	N/A, N/A
Adam et al. [13]	Liveness detection based on amplitude spectrum analysis	Static	N/A, 5%
Karunya et al. [22]	Image quality assessment	Static	N/A, N/A
Thavalengal [14]	Liveness detection based on multi spectral information	Static	N/A, N/A
Huang et al. [23]	Pupil constriction	Dynamic	0.3–1.4%, N/A
Kanematsu et al. [15]	Liveness detection based on variation of brightness	Dynamic	N/A, N/A
Mhatre et al. [24]	Feature extraction and encryption using bio-chaotic algorithm (BCA)	Static	N/A, N/A
Le-Tien et al. [26]	Modified convolutional neural network (CNN) for feature extraction combined with softmax classifier	Static	4%, N/A
Şahin et al. [27]	Convolutional neural network based deep learning for iris-sclera segmentation	Static	N/A, N/A
Our work	Iris code and QR code generation	Static and dynamic	5.3%, 4.2%

Karunya et al. [22] assess captured iris image quality to detect spoofing attacks. color, luminance level, quantity of information, sharpness, general artifacts, structural distortions, and natural appearance are qualities that can be used to differentiate between real images from fake images. Thavalengal [14] detects liveness of iris based on multi spectral information. This method exploits the acquisition workflow for iris biometrics on smartphones using a hybrid visible (RGB)/near infrared (NIR) sensor. These devices are able to capture both RGB and NIR images of the eye and iris region in synchronization. This multi-spectral information is mapped to a discrete feature space. The NIR image detects flashes in a printed paper and no image in case of a video shown for authentication. If a 3D live model is shown, an image shows 'red-eye' effect which could be used to detect iris liveness.

Huang et al. [23] rely on pupil constriction to detect iris liveness detection. The ratio of iris and pupil diameters is used as one of the considerations during authentication. Liveness prediction is evaluated based Support Vector Machine (SVM) classifier. A database of fake irises, printed images, and plastic eye balls is built for training and testing of SVM classifier. As the intensity of light increases, the pupil size decreases. The SVM can differentiate the real iris from a fake one.

Kanematsu et al. [15] detect liveness based on variation of brightness. This approach relies on the variation of iris patterns induced by a pupillary reflex for various brightness levels of light. Like anti-virus programs that include database of viruses, this approach relies on database of fake irises to detect fake authentication attempts.

Mhatre et al. [24] extract features and encrypt with Bio-Chaotic Algorithm. The input image is divided into parts to apply the Bio-Chaotic algorithm. An image is segmented and randomly one block of image is selected to hide a secret message using a unique key. The entire image is encrypted. The graph of both original and encrypted iris image is generated so that one can see the difference after the encryption process. Only authorized user knows about the random block selected and the key so an attacker fails to fraud. The decryption process is the reverse of encryption process.

Gowda et al. [25] propose a CNN architecture modeling a robust and reliable biometric verification system using traits face (ORL dataset) and iris (CASIA dataset). The datasets are divided into small batches, then processed into the network. In the experiment, they resize the image to $60 \times 60 \times 1$ from the original size and use two convolution layers. The output of first convolution layer is the input for the next. After using suitable filters and the convolution process done, the rectified linear unit (ReLU) and Max pooling operations are carried out in each layer. The CNN framework architectures proposed performs feature extraction in just two convolution layers using a complex image.

Xu et al. [19] propose a deep learning approach to iris recognition using an iterative altered Fully Convolutional Network (FCN) for iris segmentation and a modified resnet-18 model for iris matching. The segmentation architecture is built upon FCNs that have been modified to accurately generate pixel-wise iris segmentation prediction. There are 44 convolutional layers and 8 pooling layers in this architecture. Two datasets (UBIRIS.v2 and CASIA-Iris-Interval) in this experiment where they show that generating a more accurate iris segmentation is possible by combining networks such as FCN and resnet-18. The results show that the architecture proposed outperforms prior methods on several datasets.

Le-Tien et al. [26] propose an iris-based biometric identification system using a modified CNN used for feature extraction combined with Softmax classifier. The system is based on the CNN model Resnet50 where the CASIA Iris Interval dataset is used as an input. The iris recognition consists of 2 separate processes: feature extraction and recognition. to obtain the normalized image with dimensions 100×100

and 150×150 pixels as the input image of CNN, the system starts by image preprocessing. During the image preprocessing, the system uses a threshold algorithm to estimate location of pupil regions and Hough transform after performing equalize histogram algorithm to calculate pupil center, pupil's radius and iris boundary's radius, iris boundary's center. After image preprocessing, CNN and a Softmax classifier are combined to feature extraction and classification.

Şahin et al. [27] applied traditional and convolutional neural network based deep learning methods for iris-sclera segmentation. They compare performance on two distinct eye image datasets (UBIRIS and self-collected data). Their results show that deep learning based segmentation methods outperformed conventional methods in terms of dice score on both datasets. Our appraoch is difference in the sense we design an iris-based authentication system instead.

Table 1 shows a summary of related works and their characteristics, approaches, feature type, and performance measures (false positive and false negative rate). As illustrated, most works rely on static features of image, whereas we rely on dynamic response to light in the pupil area to generate iris code and subsequently the QR code.

3 Classifier for Iris Detection System

In this section we discuss the two classifier that we use to detect iris patterns from images. These classifiers are Haar-Cascade and Local Binary Pattern. We choose these two classifiers as they are readily available with OpenCV development environment to access. Other classifiers can be used for evaluation as future work plan.

3.1 Haar-Cascade Classifier

Haar-cascade classifier is popular for iris detection as it can be trained to achieve higher accuracy. We rely on the classifier built in OpenCV platform to train 1000 positive samples images having eyes and 1000 negative sample images that are not related to eyes. More specifically, we configured the parameters of the classifier to achieve the highest level of accuracy to identify the iris region. The classifier is divided by three key contributors.

Integral Image: It allows fast computation and optimization to recognize objects of interests. For example, in Fig. 2, the sum within D can be calculated using Eq. (1).

$$W(D) = L(4) + L(1) - L(2) - L(3) \qquad (1)$$

Fig. 2 Representation of
haar like feature

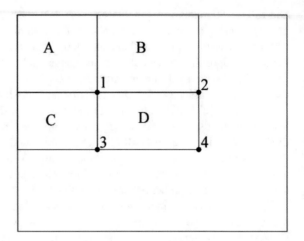

In Eq. (1), $W(D)$ represents the weight of the image and $L(i)$ is the value of color
level at the ith point. The sum of pixel values over rectangular regions are calculated
rapidly using integral images.

Learning Features: A minimum number of visual features are selected from a large
set of pixels. Three common features are recognized: edge feature, line feature, and
center-surround feature.

Cascade: It allows excluding background regions that are discarded based on integral image and learning features. The detection process generates a decision tree
by boosted process (known as *cascade*). Figure 3 shows that each image is being
processed by positive and negative images and having the similarity result by
choosing True or False. The learning algorithm keeps matching to next available
positive image until a match is found with a given image.

A positive result introduces the evaluation of second classifier which is adjusted to
achieve high detection rates. A negative result leads to immediate rejection of images.
Currently, the process uses Discrete Ada boost and a decision tree as basic classifier.

Fig. 3 Representation of
cascade decision tree

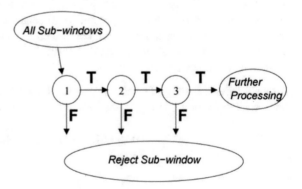

The classifier builds a decision tree for the image environment. Cascade stages are built by training classifiers using Discrete Ada Boost [17]. Then it is adjusted for the threshold to minimize false negative rates. In general, a lower threshold yields to higher detection rates from positive examples and higher false position rates from negative examples. After the cascade classifier training is fully accomplished, it can be applied as a given reference to detect objects from new images.

3.2 LBP Classifier

Local Binary Patterns (LBP) [28] are visual descriptors for texture classification. It combines Histogram of Oriented Gradients (HOG) descriptor used for detection and recognition of objects. Figure 4 explains three neighborhoods to define texture and calculate local binary pattern as per given steps. Steps for LBP cascade classifier feature calculation is given below:

Divide the image under consideration into cells (small units). The more the cells, the more possibilities of detection.

Compare the pixel value of the center with each of the 8 neighboring pixels in a cell. If the center pixel value is greater than the neighbor's value, consider "0". Otherwise, "1". This gives an 8-digit binary number.

Determine the histogram of the frequency of each "number" over the cell. This histogram can be seen as a 256-dimensional feature vector.

Concatenate histograms of all cells. This gives a feature vector for the entire window.

Like Haar-Cascade classifier, we trained LBP classifiers with a set of negative and positive image samples. The feature vectors used were from OpenCV platform.

Fig. 4 Pixel calculated by LBP classifier

4 IRIS Signature Generator Framework

At the heart of our proposed approach, we generate iris code using the classifiers discussed in Sect. 3. The iris code is generated by enrolling real world users and the code is saved in a repository. The code is generated again from a new image during authentication for matching. We first discuss the authentication process followed by code generation process in Sects. 4.1 and 4.2, respectively.

4.1 Authentication Process

Figure 5 shows the authentication process. In the proposed approach, there are two databases for each user; one for iris code and another for assigned user code. First, a camera is used to take images of the iris detection and recognition. Features are extracted from captured iris images and the user provides QR code (as a password). If there is a match between the iris of the user and the database of iris code, and user code matches the provided QR code, then the user is granted access.

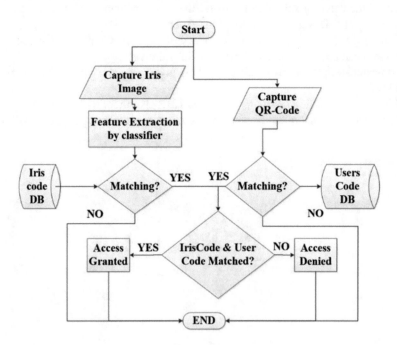

Fig. 5 Flowchart of iris code and QR code-based authentication

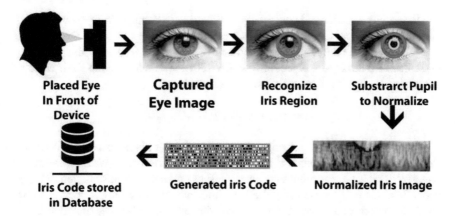

Fig. 6 Iris code generation process for authentication

4.2 Iris Code and QR Code Generation

Here we discuss how we generate iris code (used as user ID) and the QR code (used as password) from given iris images. Figure 6 shows iris code generation process from live eye. Iris is the situated colored ring of muscle around the eye pupil which controls the diameter and the size of the pupil and the amount of light that could reach the retina. Using an iris scanner (a camera for scanning iris), a person's eye is scanned. The data of the iris is unique to each person.

The camera takes a picture in infrared light. Most cameras (e.g., laptop camera) now support infrared lights have longer wavelengths than normal red lights and are not visible to the human eye. The infrared light helps to reveal unique features for dark colored eyes which cannot be detected by normal light.

We implemented a prototype [28] using OpenCV [29] platform that detects iris region with pupil (using classifiers). Next, we identify the pupil area in the center of iris region and normalize the iris area image in black and white mode. We then subtract the iris area from the pupil area (which reflects the area based on pupillary response for current illumination level). An iris code is generated using the pupillary response area, which is a 512-digit number. The iris code is stored in the database for a new user during enrollment. It is checked for matching during the authentication process. For matching, we rely on Hamming distance between the two images. Hamming distance computes the number of dissimilar bits among two codes assuming the code length for both images is the same. For example, if image A = 1001, and image B = 1100, the H(A, B) = 2 (as the second and fourth bits of A and B are dissimilar).

One limitation of storing only iris code and relying on it for authentication is that the approach is vulnerable to presentation attack. If an attacker can obtain the printout of the iris image under correct illumination level, then the attacker would obtain access to the system. To prevent this, we generate a QR code to act as a password. Unlike traditional text-based password, the QR code is an image representation, it can be read by a reader and converted to a bit string to compare with known strings. We now discuss our proposed approach of generating the QR code. From the iris image, we separate the Red, Green, and Blue color planes. The color information is presented as matrix (Mat object in OpenCV [30]). We then generate Hash value by combining hashes for each of the planes as follows:

$$H = H(R) \; XOR \; H(G) \; XOR \; H(B)$$

Here, H(R) is the hash generated from the Red color plane matrix, and XOR the Boolean operator. The length of the hash is 128 bits (16 bytes). We apply Message Digest (MD5) hash algorithm to generate hashes out of matrix information. We then generate a micro QR code using the hash information. A micro QR code can have 25 alphanumeric characters (for error correction level M [31]). The provided length is sufficient to our goal.

5 Implementation and Evaluation

We implemented a prototype using OpenCV platform [28] to detect iris recognition and spoofing attack detection using the proposed framework. We collected a dataset of iris images from [9] to evaluate our approach. This dataset is commonly used by other literature works. It contains 2854 images of authentic eyes and 4705 images of the paper printouts collected from 400 sets of distinct eyes. The photographed paper printouts have been applied to successfully forge iris recognition system. For our evaluation, we randomly selected 300 samples from authentic eyes to train the classifiers, and then applied it to 200 samples of printed iris images.

Figure 7 shows a sample of images from the dataset where (a) real eye image, (b) printed image of the iris of same eye.

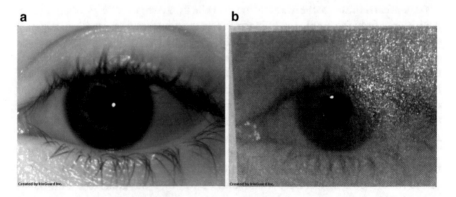

Fig. 7 **a** Real eye image **b** printed eye image from dataset

Figure 8 shows a set of results where (a) sample eye image, (b) iris recognition output of Haar-Cascade classifier (the yellow circle), and LBP classifier (red circle), (c) result of iris center and its radius, (d) converting to iris code by normalization of the iris image. Figure 9 shows a sample of QR code.

Table 2 shows a summary of the evaluation. Among 300 samples used for training, the reported false positive rate for Haar-cascade and LBP classifiers is 4.5% and 5.7%,

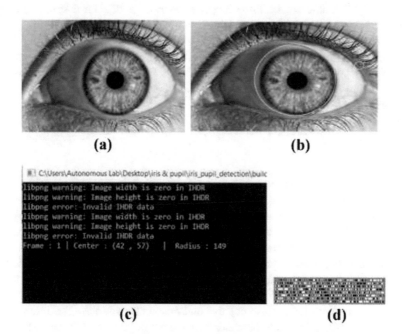

(a)   **(b)**

C:\Users\Autonomous Lab\Desktop\iris & pupil\iris_pupil_detection\builc

```
libpng warning: Image width is zero in IHDR
libpng warning: Image height is zero in IHDR
libpng error: Invalid IHDR data
libpng warning: Image width is zero in IHDR
libpng warning: Image height is zero in IHDR
libpng error: Invalid IHDR data
Frame : 1 | Center : (42 , 57)  |  Radius : 149
```

(c)   **(d)**

Fig. 8 Screenshots of classifier output (top row) and iris code (bottom row)

Fig. 9 Screenshots of micro QR code

Table 2 Summary of evaluation

Classifier	# of authentic samples	FP (%)	# of paper samples	FN (%)
Haar-cascade	300	4.5	200	3.6
LBP	300	5.7	200	4.6
Avg.	300	5.2	200	4.3

respectively. The last row of Table 2 shows the average of Haar-cascade and LBP classifier FP rate (5.2%). The paper printed samples were replayed to test the system for attacks. The FN rate for Haar-cascade and LBP classifiers is 3.6% and 4.6%, respectively. The micro QR code could prevent this false acceptance of images as defense in depth. The underlying cause of FP and FN is due to classifier parameter tuning which can be improved further by considering large number of samples and other machine learning approaches.

6 Conclusion

Iris spoofing attacks have emerged as a significant threat against traditional iris-based authentication systems. In this chapter, an iris-based authentication framework has been developed which extracts iris patterns from live image followed by QR code. The information can be used to detect presenation attacks. The iris pattern recognition applied two common machine learning approaches namely Haar Cascade and Local Binary Pattern. A prototype tool using OpenCV library has been developed. The approach has been evaluated with a publicly available dataset and the initial results look promising with lower false positive and negative rates. The initial results look promising with lower false positive and false negative rates. The future work plan includes evaluating with more samples and employing other machine learning techniques.

References

1. Thakkar D (2019) An overview of biometric iris recognition technology and its application areas. https://www.bayometric.com/biometric-iris-recognition-application/
2. Boatwright M, Luo X (2007) What do we know about biometrics authentication? In: Proceedings of the 4th annual conference on information security curriculum development, Sept 2007
3. Sheela S, Vijaya P (2010) Iris recognition methods-survey. Int J Comput Appl 3(5):19–25
4. Iridis. http://www.irisid.com/productssolutions/technology-2/irisrecognitiontechnology
5. Eyelock. https://www.eyelock.com/
6. Daugman J, Iris recognition at airports and border-crossings. Accessed http://www.cl.cam.ac.uk/~jgd1000/Iris_Recognition_at_Airports_and_Border-Crossings.pdf
7. Roberts J (2016) Eye-scanning rolls out at banks across U.S., June 2016. Accessed from http://fortune.com/2016/06/29/eye-scanning-banks/
8. Pacut A, Czajka A (2006) Aliveness detection for iris biometrics. In: Proceedings 40th annual 2006 international carnahan conference on security technology, Oct 2006, pp 122–129
9. Czaikja A (2015) Pupil dynamics for iris liveness detection. IEEE Trans Inf Forensics Secur 10(4):726–735

10. Raghavendra R, Raja KB, Busch C (2015) Presentation attack detection for face recognition using light field camera. IEEE Trans Image Process (TIP) 24(3):1060, 1075
11. Ratha NK, Connell J, Bolle R (2001) Enhancing security and privacy in biometrics-based authentication systems. IBM Syst J 40(3):614–634
12. Uhl A, Holler Y (2012) Iris sensor authentication using camera PRNU fingerprints. In: Proceedings of 5th IARP international conference on biometric (ICB)
13. Czajka A (2013) Database of iris printouts and its application: development of liveness detection method for iris recognition. In: 18th International conference on methods and models in automation and robotics (MMAR), pp 28–33
14. Thavalengal S, Nedelcu T, Bigioi P, Corcoran P (2016) Iris liveness detection for next generation smartphones. IEEE Trans Consumer 62(2):95–102
15. Kanematsu M, Takano H, Nakamura K (2007) Highly reliable liveness detection method for iris recognition. In: Proceedings of 46th annual conference of the society of instrument and control engineers of Japan (SICE), pp 361–364
16. Puhan N, Sudha N, Hegde S (2011) A new iris liveness detection method against contact lens spoofing. In: Proceedings of 15th IEEE international symposium on consumer electronics (ISCE), pp 71–74
17. Zhao Y, Gu J, Liu C, Han S, Gao Y, Hu Q (2010) License plate location based on haarlike cascade classifiers and edges. In: 2010 Second WRI global congress on intelligent systems. https://doi.org/10.1109/gcis.2010.55
18. Li C, Zhou W (2015) Iris recognition based on a novel variation of local binary pattern. Visual Comput 31(10):1419–1429
19. Shahriar H, Haddad H, Islam M (2017) An iris-based authentication framework to prevent presentation attacks. In: 2017 IEEE 41st annual computer software and applications conference (COMPSAC), pp 504–509
20. Etienne L, Shahriar H (2020) Presentation Attack Mitigation. In: Proceedings of IEEE computer software and applications conference (COMPSAC), July 2020, 2 pp (to appear)
21. Menotti D, Chiachia G, Pinto A, Schwartz WR, Pedrini H, Falcao AX, Rocha A (2015) Deep representations for iris, face, and fingerprint spoofing detection. IEEE Trans Inf Forensics Secur 10(4):864–879
22. Karunya R, Kumaresan S (2015) A study of liveness detection in fingerprint and iris recognition systems using image quality assessment. In: Proceedings of international conference on advanced computing and communication systems, pp 1–5
23. Huang X, Ti C, Hou Q, Tokuta A, Yang R (2013) An experimental study of pupil constriction for liveness detection. In: Proceedings of IEEE workshop on applications of computer vision (WACV), pp 252–258
24. Mhatre R, Bhardwaj D (2015) Classifying iris image based on feature extraction and encryption using bio-chaotic algorithm (BCA). In: IEEE International conference on computational intelligence and communication networks (CICN), pp 1068–1073
25. Types of biometrics (2020) https://www.biometricsinstitute.org/what-is-biometrics/typesof-biometrics/
26. Le-Tien T, Phan-Xuan H, Nguyen-Duy P, Le-Ba L (2018) Iris-based biometric recognition using modified convolutional neural network. In: 2018 International conference on advanced technologies for communications (ATC), Ho Chi Minh City, pp 184–188
27. Şahin G, Susuz O (2019) Encoder-decoder convolutional neural network based iris-sclera segmentation. In: 2019 27th Signal processing and communications applications conference (SIU), Sivas, Turkey, pp 1–4
28. Adrian Rosebrock, Local binary patterns with python and OpenCV. https://www.pyimagesearch.com/2015/12/07/local-binary-patterns-with-python-opencv/
29. OpenCv. Accessed from http://opencv.org/opencv-3-2.html
30. OpenCV basic structure. Accessed from http://docs.opencv.org/2.4/modules/core/doc/basic_structures.html
31. Mini QR code. Accessed from http://www.qrcode.com/en/codes/microqr.html

Classifying Common Vulnerabilities and Exposures Database Using Text Mining and Graph Theoretical Analysis

Ferda Özdemir Sönmez

Abstract Although common vulnerabilities and exposures data (CVE) is commonly known and used to keep vulnerability descriptions. It lacks enough classifiers that increase its usability. This results in focusing on some well-known vulnerabilities and leaving others during the security tests. Better classification of this dataset would result in finding solutions to a larger set of vulnerabilities/exposures. In this research, vulnerability and exposure data (CVE) is examined in detail using both manual and computerized content analysis techniques. Later, graph theoretical techniques are used to scrutinize the CVE data. The computerized content analysis made it possible to find out 94 concepts associated with the CVE records. The author was able to relate these concepts to 11 logical groups. Using the network of the relationships of these 94 concepts further in the graph theoretical analysis made it possible to discover groups of contents, thus, the CVE items which have similarities. Moreover, lacking some concepts pointed out the problems related to CVE such as delays in the review CVE process or not being preferred by some user groups.

Keywords Content analysis · Text mining · Graph theoretical analysis · Leximancer · Pajek · CVE · Common vulnerabilities and exposures

1 Introduction

Common Vulnerabilities and Exposures (CVE) dictionary [1], which is also called as dataset or database in some sources, is a huge set of vulnerabilities and exposures data which is considered as the naming standard for vulnerabilities and exposures in numerous security-related studies, books, articles and by the vendors of security-related products including Microsoft, Oracle, Apple, IBM, and many others. Despite

F. Ö. Sönmez (✉)
Informatics Institute Middle East Technical University, Ankara, Turkey
e-mail: ferdaozdemir@gmail.com

© The Editor(s) (if applicable) and The Author(s), under exclusive license to Springer   313
Nature Switzerland AG 2021
Y. Maleh et al. (eds.), *Machine Intelligence and Big Data Analytics for Cybersecurity Applications*, Studies in Computational Intelligence 919,
https://doi.org/10.1007/978-3-030-57024-8_14

its widespread use, the information provided does not have sufficient classification qualities. This lack of proper classification results in immature or inadequate use of this database. The CVE number and other fields do not provide any kind of classification for the data.

The author is of the opinion that, even the most advanced security-related tools can be improved by better digesting the CVE database knowledge. The coverages of security tests can also be enlarged through the use of the knowledge of relationships of existing vulnerabilities and exposures. If the user can conduct this effort for more vulnerabilities simultaneously, or if the vendors can create tools that deal with more issues rather than a single or a few then this would increase the overall efficiency of the security tasks.

When the CVE data is examined with bare eyes, it can be discovered that there are vulnerabilities which are resulted due to very close reasons or related to the same origins. An example is two vulnerabilities resulted because of a wrong setup in a configuration file. While the security analyst checks the software/system for one of these vulnerabilities (a probably more common one) either manually or using automated tools she/he may neglect the other which may be discovered during the later phases such as in the production phase. Another example set may include two vulnerabilities which may be caused due to similar activities. For example, think of a set of authentication problems for a vendor product. The tester or even the developer may neglect some vulnerabilities, even if they were entered the CVE dataset.

The motivation of this study includes by better classifying the CVE data, decreasing rework for comparative vulnerabilities that may be inspected together utilizing the same data sources or same technologies which otherwise may cause more effort that have to be spent in planning, data collection, and data preparation tasks and more endeavor on technology setup, education, and dissemination of knowledge.

Another motivation target is to reduce the redesign of similar tools or having multiple tools achieving related tasks that have the potential to cover more situations by better examination of CVE. Since the set of vulnerabilities are increasing, there is a need for continuous design and implementation. When the vendors can not benefit from the CVE data for their specific tools or similar tools from other vendors, this may cause late responses to newly detected vulnerabilities and exposures. Moreover, the redesign of similar tools would result in improper usage of money, material, and time resources. When the number of tools increases unnecessarily by the redesign of similar tools, this would cause more maintenance costs for the vendors and more educational costs for the users.

Besides these financial complications, when security analysis and monitoring tools exhibit less information, the users had to use multiple tools for the analysis of a single security data file. They also had to apply more effort in the analysis to remember, merge, and compare information coming from multiple tools. They may need more sophisticated approaches and even implementing their own code to better handle some situations.

Designing and developing a security analysis tool or conducting security tests requires thorough preparation, including deciding on target vulnerabilities and exposures, collecting and preparing security-related data, and establishing the environment that will be used for the study including the tools and technologies. Not examining the CVE database, not forming a consolidated and up to date vulnerability information, and not injecting this information into the work along with using contemporary technologies results in numerous inefficiencies. Since in its current form, the CVE data does not provide enough classifiers, there had been previous attempts to classify this dataset.

There are two main problems with classifying the CVE dataset. The classification should rely on the textual descriptions which are not prepared based on any standard or format. Some of the classification efforts use the Common Vulnerability and Enumeration (CWE) [2] system in conjunction with CVE. This results with better accuracy, however, not all CVE records are associated with the CWE system. The second problem is the taxonomies provided so far, in general focuses on categorizations of vulnerabilities, or security targets (confidentiality, integrity, availability). This categorization may be beneficial for some security tests, but they will not help when optimizing the efforts when working with security data.

This study involved both using textual content analysis of CVE data and using graph theoretical analysis techniques for the concepts discovered during the content analysis to scrutinize the relationship of these concepts. It will not be wrong to say that, in general, existing security analysis studies focus on the most well-known vulnerabilities. Examining CVE data may improve the vulnerability or exposure coverage level of these designs by finding vulnerabilities that may be detected using similar technologies or data sources. At least, it may enable finding gaps in terms of vulnerabilities and may result in novel designs.

The initial incentive for the examination of the CVE dataset emerged when attempting to discover a gap through the vulnerabilities to provide a security analysis prototype. The existing form of the dataset did not provide a hint regarding the relations of these vulnerabilities. This concluded with limited if no understanding of necessary implementations, and the current status.

There are very few studies that used the content analysis over the CVE dataset. In the author's knowledge, although there are a few studies which deal with related data such as sender data to find out relations of CVE contributors and Common Vulnerability Scoring System (CVSS) data, there is no study that has taken graph theoretical analysis techniques over the CVE dataset concepts yet. The contribution of this study over existing studies is having the premier focus on the examination of CVE data rather than solving a security problem and using novel techniques over this dataset which has not been applied before leading a better examination of the dataset. This contribution depends on taking novel categorization criteria. Using the outputs from this examination or repeating a similar examination may result in improved coverages of security issues and may provide detailed domain specific information that may be valuable in developing new designs for the field.

The objectives of these study include to examine, understand, and group the CVE data using automated tools and graphical analysis techniques so that they may be classified in a manner which best suits to the categorization of technologies and associated data sources. The number of vulnerabilities neglected in security tests can be reduced this way. In the long term, this would affect newly created security testing and monitoring tools in a way that increases efficiency.

The scope of this study is limited to providing a summary of the dataset using the automated concept analysis tool. Following this, having this concept map network information, conducting the applicable graph theoretical data analysis and classification techniques which are limited with available information in the network data to find out the groups, subgroups, global and local relationships in the data.

This paper is organized as follows. Section 2 describes the common vulnerabilities and exposures concept. It contains the summary of literature focusing on the content analysis of CVE data and the use of graph theoretical analysis techniques for CVE and security domain at large. Section 3 has the data and methodology description. Section 4 is the results section. Finally, Sect. 5 presents the discussions and conclusions.

2 State of Art

2.1 Common Vulnerabilities and Exposures

In this section, a background for the CVE database is provided. In the following two sections, a recall for two major techniques used in this study, content analysis with text mining and graph theoretical analysis, exist. This recall includes relevant studies from the literature either directly using CVE data or other security-related data when the number of studies using CVE is very low for a technique.

CVE is simply a dictionary of names of commonly known cybersecurity vulnerabilities and exposures [1]. It enables the vendors and users of tools, such as networking tools, and database tools use the same language. Before CVE, each vendor was giving a different name to the same vulnerability causing numerous communication and understandability problems [3]. The use of the same common dictionary of vulnerabilities also empowers the comparison of products that claim to be doing similar tasks. The description of the vulnerabilities includes information related to the environment and conditions in which the vulnerabilities are mostly identified or expected, such as the operating system, the application name, data source types, and related user/system actions. Sample tuples including name, status, description columns of CVE items are given in Table 1.

Developing a taxonomy of any form for categorization was not aimed during the creation of the CVE database. It is believed to be beyond the scope of the efforts. The developer organization also decided it would bring more complexity to the database which will cause maintenance issues. Having this simple approach allowed the database to continuously grow since from the start. The aim was to provide an index for each vulnerability/exposure and enough information to distinguish it from

Table 1 Samples of CVE Items

Name	Status	Description
CVE-1999-0315	Entry	"Buffer overflow in Solaris fdformat command gives root access to local users."
CVE-1999-0419	Candidate	"When the Microsoft SMTP service attempts to send a message to a server and receives a 4xx error code, it quickly and repeatedly attempts to redeliver the message, causing a denial of service."
CVE-1999-0204	Entry	"Sendmail 8.6.9 allows remote attackers to execute root commands, using ident."
CVE-1999-0240	Candidate	"Some filters or firewalls allow fragmented SYN packets with IP reserved bits in violation of their implemented policy."

other similar records. The intention was to possess all the vulnerabilities/exposures in itself. Other than the naming and indexing, status, and description information CVE contains a maintenance extension (CMEX) mainly designed to be used internally. CMEX contains administrative data containing a version number, category (which does not correspond to a vulnerability taxonomy but includes items such as software, configuration etc.), reference which contain URLs to enable more descriptive information for some vulnerabilities, and keywords. CMEX does not provide categorization and purely designed for internal usages.

The maintenance and validation of the CVE database are conducted by the CVE editorial board members who meet regularly. The proposals, discussions, and votings are done through an electronic mail list. The whole process starts with the assignment phase when a number is assigned to a potential problem. This record still is not validated by the board. The second phase is the proposal phase. The candidate item is proposed to the board at this phase. Voting takes place as a part of the proposal phase. Some members of the editorial board vote, the others stay as observers. After an amount of discussion or after getting sufficient votes, the moderator starts the interim decision phase. The next phase is the final decision phase which is followed by the publication phase. During the publication phase the record will be announced as a new entry if accepted or will be recorded in the candidate database if rejected during the decision phase. For further information related to CVE attributes and decision mechanisms please refer to Baker et al. [4].

In its current form, the CVE dataset assigns a unique identifier for each item which consists of a numerical value ordered by the acceptance date. Encapsulating new categorization criteria in some particular way would eventually increase the usability of the dictionary. There have been some earlier studies using the CVE dataset for various purposes including classification. CVE data is used as the main data or as a control data for these earlier studies. The CVE data has become a single data source or combined with information from other vulnerability databases.

2.2　Content Analysis Through Text Mining

The aim of computerized content analysis is to find out the themes and the relationships among them through text mining. Text mining is a field of artificial intelligence that converts unstructured big data into normalized, structured data suitable for further analysis. The resulting structured data can also be used for machine learning algorithms to fulfill various targets. Typically, text mining depends on activities including text preprocessing, text transformation, parsing, stop-word removal, tokenization, information extraction, and filtering [5].

Automatic content analysis through text mining provides a convenient alternative to manual analysis to gather domain knowledge and to create domain ontologies [6]. Repeating content analysis through time allows the examination of the change in the concept networks and track the modifications of the important terms. There are various ways of doing text mining. Information retrieval focuses on facilitating information access rather than analyzing information. Natural language processing combines artificial intelligence and linguistics techniques with the aim of understanding human natural language. Information extraction from text is conducted to extract facts from structured or unstructured text documents. Finally, text summarization provides a descriptive summary of large textual files to provide an overview of the data [5].

Earlier studies have various objectives, including, classification, prediction, data summary, and use various techniques. Guo and Wang [7] created an ontology definition using the Protégé [8] ontology tool for CVE data for better security content management rather than classifying the vulnerabilities. The creators of CVE also proposed a categorization system for CVE data called, Common Weakness Enumeration, (CWE) (CWE). In the author's knowledge, this categorization system is not directly associated with all the CVE items yet. Chen et al. [9] proposed a framework for the categorization of the CVE dataset [9]. In Chen et al.'s framework, the descriptions of CVE are taken as a bag of words, and based on the frequency of each word, numerical values are assigned to each word. The pairs of the word- numerical value forms a vulnerability vector. Later, these vectors are used for the categorization of the dataset items using supervised learning methods, including Support Vector Machines (SVM's) [10]. Wen et al. [11] took a similar approach and used SVMs for automatic classification of vulnerabilities data. Wen et al. used a classification framework on (National Vulnerability Database) NVD and (Open Source Vulnerability Database) OSVDB vulnerability databases. This framework can also be utilized to classify the CVE dataset. In Wen et al.'s study, the accuracy of the categorization is checked by comparison with the CWE categorizations. Na et al. [12] used Naive Bayes classification methodology to classify uncategorized vulnerability documents. Bozorgi et al. [13] used SVM to classify the combined data coming from both CVE and Open Source Vulnerability Database, (OSVDB).

Another classification of CVE dataset study has been conducted by DeLooze [14] using Self-Organizing Map's (SOM's) [15]. DeLooze used the textual description of the CVE items to point out vulnerabilities and exposures having similar features.

Wang et al. [16] data mining on the CVE data to mine security requirements for agile projects. In their approach the CVE data is used as a repository. Wang et al. demonstrated how the outputs of data mining can be integrated to other agile operations. Subroto et al. [17] used CVE data as a part of a threat prediction system that is created from social media data. Subroto et al. created a descriptive summary of the CVE data using text clouding, histogram, and dendrogram to find out the most frequent occurrences. They compared the outputs of the predictive model created using Twitter data with the CVE outputs to validate the predictive model.

Mostafa and Wang [18] mined CVE dataset to find out keywords and weights. Later, Mostafa and Wang used these keywords and weights as a part of a semi-supervised learning framework that identifies bugs automatically from bug repositories of RedHat and Mozilla. CVE data has been used for text mining along with other data sources as a part of a proactive cyber security design [19]. Chen et al. suggests the use of concept maps and inputting the resulting information to a risk recommendation system. Due to several factors, the proposed study is distinct from Chen et al.'s study. The first factor is using security data sources as root concepts. Since the proposed study aims to classify the concepts to find groups of vulnerabilities/exposures that should be handled together, the choices of alternative security data sources have been input to the text mining as root concepts. The second factor is the use of a different methodology. In the proposed study the provided concept maps are not used as is, instead, several mathematical, and graph theoretical analyses are conducted using the outputs of the content analysis which resulted in various approaches for grouping vulnerability data.

There are various categorization criteria used in the earlier classification efforts. In general, this criterion embraces the categorization of vulnerabilities. They do not have a specific aim to categorize the vulnerabilities based on technologies or data sources.

2.3 Graph Theoretical Analysis

Graph theoretical analysis has a history going to the Harvard researchers who seek for cliques in the 1930s using interpersonal relations data. A while later, Manchester anthropologists investigated the structure of community relations in tribal and village societies. These efforts have been a basis for contemporary graph theoretical analysis [20]. Graph structures allow calculations of various metrics such as symmetry/asymmetry and reciprocity and various analysis types such as analysis of cliques and analysis of influencers.

A network is a special kind of graph, which has vertices, directional and directionless lines between the vertices and additional information related to either vertices or links. A vertex is the smallest unit in a network and the line is the tie connecting these vertices. While a directed line is called an arc an undirected one is named as an edge. The values related to the lines may indicate for example the order or the strength of the relationship. Additional values that are not directly related to the lines are called attributes.

Ruohonen et al. [21] used graph theoretical techniques when they examine the contributors to the CVE data and time delays during the CVE process. They used the CVE coordination information sent to the MITRE organization as a part of CVE proposals. Although the use of CVE related data is limited, graph theoretical analysis techniques are applied to numerous security-related studies in the literature. In general, graph theoretical techniques are as well useful in classifying and clustering security-related data. These techniques also become convenient when examining network activities and thrust relationships in the security domain.

Deo and Gupta [22] applied these techniques to the world wide web. In their model, a node represented a web page and an edge is used to represent a hyperlink. The study aimed to improve web searching and crawling, and ranking algorithms. Özdemir [23] examined the effects of networks in the systemic risk within the banking system of Turkey. Zegzhda et al. [24] used graph theory to model cloud security. Sarkar et al. [25] used information from Dark Web Hacker forums to predict enterprise cyber incidents through social network analysis. Wang and Nagappan [26] (Preprint) used social network analysis to characterize and understand software developer networks for security development.

Increasing the usability of the CVE dictionary is aimed at both the content analysis and graph theoretical analysis focused earlier work mentioned so far. The proposed study also uses the same inputs as with the majority of the earlier text mining studies, the textual description of the CVE items.

3 Methodology

3.1 Data Set

As of the start of this study, the CVE dataset gathered from the CVE web site [27] included 95574 vulnerability and exposure records. For each of the items, seven attributes are stored, which are: name, status, description, references, phase, votes, and comments. The "Name" attribute consists of values in the form "CVE" + "−" + Year + "−o" + Number. The "Status" column may be either "Entry" or "Candidate". Candidate items are not reviewed and accepted by CVE editorial boards yet or temporary. "Description" column includes the information which characterizes the vulnerability or exposure. "Reference" column points out either short names of the related products or URL's which include additional information related to the CVE item, such as product web site. "Phase" may include terms, such as "Interim", "Proposed" and "Modified". "Vote" includes information related to the responses of the CVE editorial team. "Phase", "Vote" and "Comments" fields are blank for the entry records. They hold information related to the acceptance or rejection causes for the candidate ones.

During the computerized content analysis and application of graph theoretical techniques, only the "Entry" items were used (Candidates are eliminated), which resulted in 3053 items. However, prior to computerized content analysis, during the exploration of the popular or highlighting security analysis related terms in the database, both candidate and entry data were used to expand the amount of targeted vulnerability data.

Eliminating the "Candidate" vulnerabilities and using only the "Entry" vulnerabilities was a decision made by the author after an initial examination of the whole dataset based on three reasons. Some of the candidate vulnerabilities do not have complete descriptions such as the ones starting as "Unknown vulnerability". All of them are either not even proposed to the editorial board and marked with the sentence "Not proposed yet" or in the middle of the process having markers such as "DISPUTED". Some of the candidate vulnerabilities which are actually rejected but not cleaned from the database also have "REJECTED" markers in the description. But this does not mean that other "Candidate" items are not already rejected, cause leaving a marker in description text is optional. There are also some vulnerabilities which are marked as "RESERVED", again this group does not have descriptions but these CVE number groups are probably reserved by some vendors. In total, the number of vulnerabilities which suit to the described groups in this paragraph is about 26,300 based on Excel filtering.

Other candidate vulnerabilities have descriptions without markers, but again these are also subject to change and rejection or already rejected by the board. Some of the records are in the "Candidate" situation for more than even 10 years. The number of Candidate records which have CVE dates earlier than 2016 is 83,435. For the listed reasons, the paper focuses only on the "Entry" dataset which is reviewed and accepted to be part of the CVE dataset by the reviewers of the CVE editorial board. These are the actual vulnerabilities used by both vendor companies and in the relevant security documents.

3.2   Content Analysis of CVE Database

Computerized content analysis techniques make it possible to examine large sets of unstructured data. The most important advantage of using this technique is due to its ability to provide a summary view of data with low subjectivity. The size of CVE makes it impractical to analyze the content manually. For this purpose, first, a semi-computerized content analysis has been made to investigate the frequency of occurrence of important security data sources related terms in the CVE data knowing that data is the genesis of all kind security analyses using keywords.

During this analysis, the output from an earlier study [28] has been input. Although this earlier study focused on security visualization requirements, it involved a survey in which the most popular security data sources used in the security analysis methods in the enterprises was questioned. The most commonly possessed infrastructure elements and most commonly used enterprise applications were also inquired in this survey. Briefly, the participants of the survey were 30 security experts either from the private sector or academia. The participants had hands-on experience in the field as a part of an enterprise security team and/or holding reputable security certificates. The survey was conducted online. Although the survey included other questions, the results of the three questions (security analyses data, sources, enterprise applications, infrastructure elements) have been input for the content analysis.

The keyword inputs coming from the survey have been used during the semi-computerized content analysis to find out associated registered vulnerabilities and exposures. This effort yielded partially understanding the CVE contents and their relationships. Later, a computerized content analysis has been made to find out frequent concepts that may be related to security analysis/monitoring studies by either pointing data sources, attack types or technologies and the relations between them.

During the semi-computerized content analysis, the CVE dataset has been filtered using the Excel filtering mechanism to query concepts that came out during the requirement analysis study. At this step, the subgroup of CVE items that correspond to a specific keyword is taken independently and among that group, a frequency analysis of words has been made to point out the terms which take place more than once or which commonly take place in that specific group.

During the computerized content analysis work, Leximancer [29] tool has been used to ascertain frequent concepts and relations among them. This tool finds out relational patterns from the supplied text. It employs two stages, semantic and relational having statistical algorithms and employing non-linear dynamics and machine learning [30]. Once a concept is identified by using supervised or unsupervised ontology discovery, a thesaurus of the words is formed by the tool to find out relationships. The textual input is tagged with multiple concepts for classification purposes. The output concepts and their co-occurrence frequencies form a semantical network. The tool has a web-based user-interface and allows the examination of concept maps to discover indirect relationships. Although the author used this user interface to examine data visually multiple times, generated graphics are too complicated, thus not included in this paper. These complex exhibits of data also lead to the decision to accomplishing graph theoretical analysis using a specific tool that better handles complex network relationships.

Running the Leximancer computerized content analysis tool multiple times through the web interface resulted in three subsequent decisions including

- selection of concept seeds,
- filtering of data based on word types (noun like words/verb like words) and
- consolidating similar concepts to form compound concepts.

Detailed graphical analysis of the generated network is done in the next phase using the Pajek tool [31]. Leximancer provided a set of the selected terms and the frequency and prominence relationships between them in word matrix forms. This frequency matrix holding the most prominent concepts is used for further analysis. The tool also provides a set of CVE records that are associated with each term.

Leximancer can execute in unsupervised or supervised modes. Initially, unsupervised execution of the tool using only the CVE data is conducted, which resulted in associations of data that may not be useful when the aim is to find groups of security data sources, technologies, attack types, and vulnerabilities. During the content analysis, Leximancer allows inputting a set of seed terms that should be included in the resulting terms, in the supervised mode. The tool combines this initial set with the auto-discovered terms. During the autodiscovery phase the terms which are "noun like" and/or "verb like" can be selected. Leximancer allows determining the percentage of "noun-like", and "verb-like" concepts in the resulting concept set, such as %60 non-like concepts and %40 verb like concepts. In this study, since the main aim is to find the relationships of technologies, data sources, attack types, after some trial, 100% noun-like concepts are included and verb like concepts are excluded in the resulting semantic concept network. The reason for excluding verb like concepts is caused due to the fact that the verb like clauses were not reciprocating to technologies, analyses types, data sources or the names of malicious activities. After filtering, the operation resulted in discovering the mostly occurred concepts, and the relationships among them.

Finally, concepts that point out similar items are grouped in compound concepts to eliminate redundancies. Compound concepts formed by joining uppercase and lowercase forms of concepts such as Ftp and FTP, concepts and their abbreviated forms such as Simple Mail Transform Protocol and SMTP, and the concepts which point out the same set of technologies such as different versions of Windows operating system. The process model of the analyses is shown in Fig. 1.

3.3   Applying Graph Theoretical Analysis Techniques on CVE Concepts

The numerical results gathered from the computerized content analysis indicating the relationships of concepts have been used in graph theoretical techniques to further clarify the relations of vulnerabilities and exposures within each other. The results, which are presented as a frequency matrix by the Leximancer tool, consist of concepts as the nodes and edges which correspond to the frequency of occurrence of each term together in a common vulnerability and exposure description. The concepts which are connected to each other with higher line values are more related to each other.

It is common to use graph theoretical techniques to investigate the spread of a contagious idea and/or a new product. It is also used to evaluate research courses and traditions, and changing paradigms. In this study, some of these techniques are used to scrutinize the relationships of concepts discovered through the use of content analysis techniques.

During the graph theoretical analysis, the following steps are taken. First, the density of the network is calculated and the whole network is visualized using the Pajek tool. Since the number of vertices is very high in the provided network, the graphics generated this way using Pajek had a similar level of complexity to the Leximancer outputs. Later, the degree of each vertex, the number of lines incident

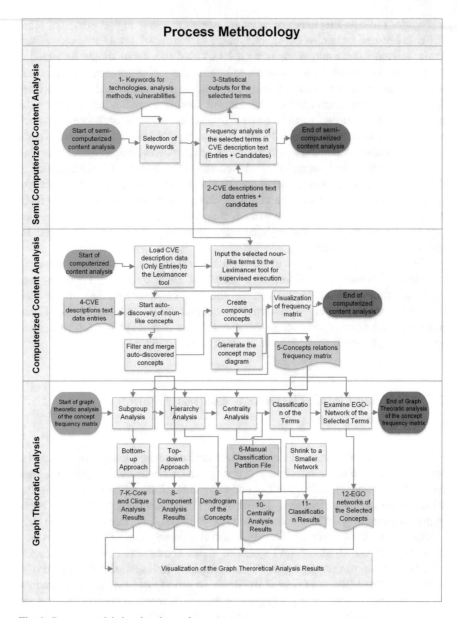

Fig. 1 Process model showing the analyses steps

with the vertice, is calculated. Subgroup analyses and centrality analysis followed this initial investigation.

Several approaches are taken to find out the subgroups of concepts. The bottom-up, node-centric, approaches are mainly based on the degree, the number of connections to the vertice. These approaches define the characteristics of the smallest substructures A clique is a connected set of vertices in which all the vertices are adjacent to all the other vertices and a k-core is a set of vertices in which all vertices has at least k neighbors. There are also various types of relaxed cliques. An N-clique is a type of clique where N represents the allowed path distance among the members of a subgroup, such that a friend of a friend is also accepted as a part of a clique for a 2-Clique sub-network. A p-clique, on the other hand is related to the average number of connections among a group where each vertice is connected to another vertice with a probability of p $(0 < p < 1)$.

The reason for doing subgroup analysis is to search for groups of concepts that share common properties and which are more homogeneous within each other. As a bottom-up approach, the author checked the network to find out k-cores, cliques, and relaxed cliques as well. As a top-down approach, the components i.e. maximal connected networks that have more than two vertices are searched. Top-down approaches are network-centric and mainly rely on node similarity, blockmodeling, and modularity maximization. They involve finding the paths (walk, semi-walk, cycle) between the vertices and searching for nodes that have higher structural equivalence among each other. Structural equivalence occurs between nodes having similar structural properties such as having similar ties between themselves and between their neighbor concepts. One way of measuring the dissimilarity of vertices is the number of neighbors that they don't share. Using this dissimilarity metric, a dendrogram of the vertices is formed to provide a hierarchical clustering of the CVE concepts.

Later, a classification of vertices are made resulting in having a partition matrix with the following classifications: (1) protocols, (2) operating systems, (3) end-user of middleware applications, (4) browsers, (5) protection systems and related terms, (6) host machines and related terms, (7) network traffic and related terms, (8) network components, (9) format, (10) attacks/exposures, and (11) vulnerability.

Although the centrality analysis and subgroups (both clustering and classification) of the data are conducted, sometimes a few concepts which are not very central may have interesting or unexpected relations. For this reason, the EGO networks of the selected terms are formed to expose these relationships.

4 Results

4.1 Semi Structured Content Analysis Results Through Keywords

Using keywords on the CVE data, and making frequency analysis allowed to make a smooth introduction to the dataset contents. Since data transfer and data sharing are important sources of many vulnerabilities, first, technologies related to sharing data are examined finding noticeable data types, technologies, and components as Sharepoint, Microsoft, Windows, HTML, library, URL, SQL, Linux, MAC, and Vmware. Elaborating more may yield interesting results. Within the author's knowledge, there is no security analysis method or tool study that focuses on the Sharepoint tool sharing mechanism or any specific analysis related to the data flow among multiple virtual machines, such as Vmware. The most popular words related to the dangers of sharing resources were: denial, (conceivably pointing out denial of action of sharing), XSS, Trojan, and Cross-site. Interestingly, none of the descriptions which encapsulate "share" involve the term "malware" in the CVE database.

When we look at the security analysis methods, we see that selection of data source dominates these analyses types, thus, checking for those data sources or subgroups of them such as data related to some protocols in CVE dataset would provide the level of coverage of associated vulnerabilities and exposures for them. Figure 2 shows selected security-related data sources and/or subgroups of them and the results of the corresponding content analysis made using the CVE dataset. This figure demonstrates that the number of vulnerabilities for some security data

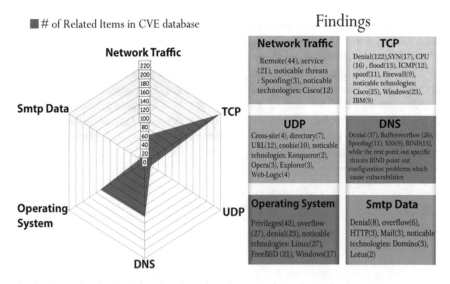

Fig. 2 Content analysis results related to selected commonly used security data sources

sources is very low, supporting that the database is mostly used for network-related vulnerabilities. Among the network-related vulnerabilities, TCP protocol dominates. The corresponding keywords found based on frequency analysis mostly point out some vulnerabilities which are more common for that specific data source such as SMTP and denial pairs, or some more vulnerable technology related to a specific data source, or may not be meaningful for that data source at all.

The survey results list the most popular enterprise applications as "Static Web Pages", "Dynamic Web Application", "Enterprise Resource Planning (ERP)", "Supply Chain Management (SCM)", "Customer Relationship Management (CRM)", and "Other" systems. Figure 3 shows the amount of using these applications in the organizations and the corresponding content analysis results made using the CVE data. Although some enterprise applications such as ERP and SCM systems are widely used, no corresponding recorded vulnerabilities are found in the database. When the keywords are examined, in the database, a low level of existence for two specific vendor products SugarCRM and Microsoft Business Solutions is identified.

Each IT system component can be a target for a security attack or may have specific vulnerabilities that make them potential subjects for security analysis tasks. Use of "File Sharing Server", "Web Server", "Mail Server (Internal)", "Mail Server (External)", "Application Server", "Database Server", "Cloud Storage",

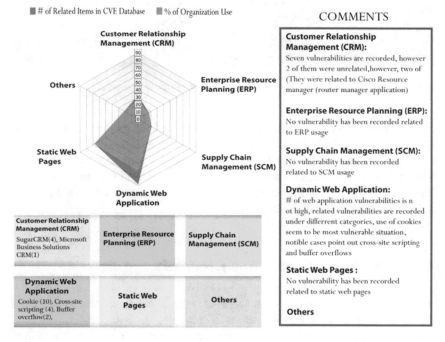

Fig. 3 Content analysis results for selected enterprise software systems

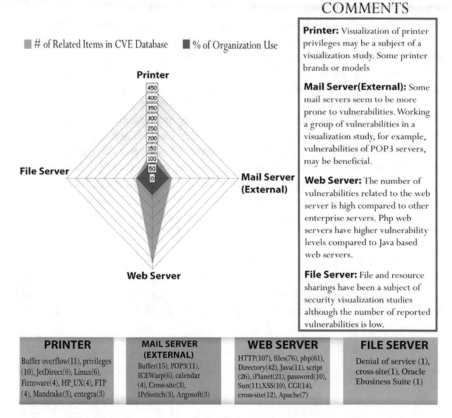

Fig. 4 Content analysis results for selected enterprise infrastructure elements

"Other Cloud Services", "External Router", "Internal Switch or Router", "Wireless Network", Printer", "E-Fax", and "Other" systems have been questioned during the survey. The most popular systems and corresponding content analysis results are listed in Fig. 4. This picture shows the vulnerabilities related to some server types which commonly exist in the enterprises, such as File Server or Mail Server are not included in the vulnerability database.

4.2 Computerized Content Analysis Results

During the content analysis, Leximancer allows the input of a set of seed terms that should be included in the resulting terms. It combines this initial set with the auto-discovered terms. The tool also allows determining the percentage of "noun-like" and "verb-like" concepts in the resulting concept set. In this study, since the main aim is to find the relationships of security analysis technologies, and data sources, after some trial 100% noun-like concepts were included during the analysis.

While semi-computerized content analysis allowed to determine relationships to some technologies and keywords, the fully computerized content analysis made by the Leximancer tool allowed to have upper-level concept relationships by providing a set of concepts. The tool also provides the pairs of concepts and a numeric value, frequency, which indicate the number of times of appearance in the same vulnerability and/or exposure description for each pair, Fig. 5. The list of concepts provided by Leximancer is shown in Fig. 6 in a grouped manner.

Knowing these upper-level vulnerability concept relationships may help to make better decisions while designing a new security analysis task or product as described in Sect. 1. Leximancer tool revealed 94 concepts which were all noun-like words. They correspond to either data sources or technologies. Later, to use in the graph theoretical analysis technique these concepts are classified into the following classes:

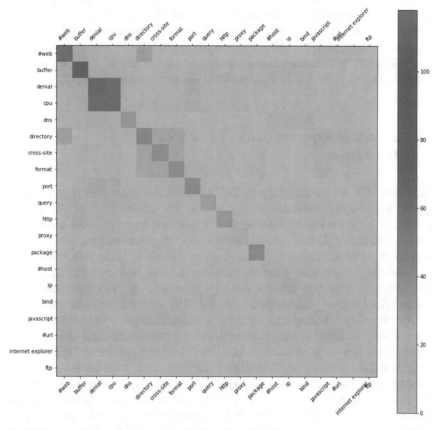

Fig. 5 Concepts that are most frequently used together with other concepts in the same vulnerability description

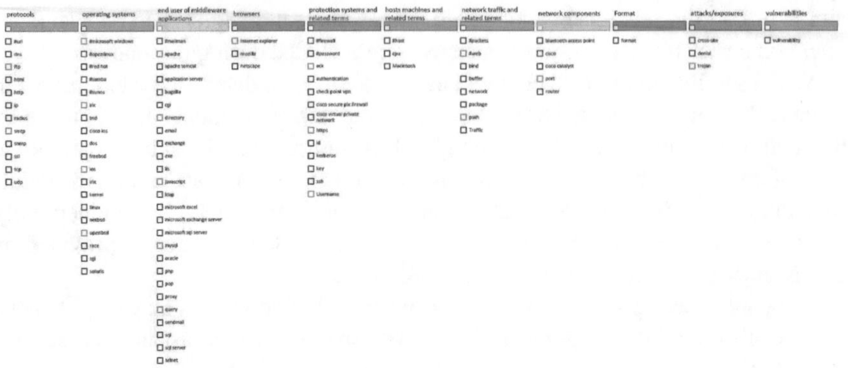

Fig. 6 Concepts that are revealed through computerized content analysis

(1) protocols, (2) operating systems, (3) end-user of middleware applications, (4) browsers, (5) protection systems and related terms, (6) hosts machines and related terms, (7) network traffic and related terms, (8) network components, (9) format, (10) attacks/exposures, and (11) vulnerability, as shown in Fig. 6. The tool also provided data in matrix form showing the frequency and prominence relationships of these concepts. In Fig. 5, the top 20 concepts from this concept matrix are presented.

4.3 Results of Applying Graph Theoretical Analysis Techniques

As described in the methodology section, several network analysis techniques are applied to the concept network. Before starting the detailed analysis, the network is visualized and the structure is examined. The basic properties of the network are summarized in Fig. 7. This network is not a very dense network, the density of the network is calculated around 0.44 which means about %44 of the potential connections exist in the provided network. Based on the weighted centrality calculation (line values are taken into consideration), the top twenty vertices are the same as the computerized content analysis results, illustrated in a sorted matrix shown in Fig. 6.

The discovery for subgroups in directed and undirected networks differs. The concept network is an undirected network (having edges rather than arcs). For this kind of network weak components are searched first (as suggested by Pajek network analysis tool developers), which resulted in having a single large component encapsulating all the vertices. Later, k-core analysis is made results of which is shown in Fig. 8. Looking at these outputs, there are 21 vertices that make 34 core, meaning 21 vertices have 34 neighbors. From this result we understand that a high number of concepts are related to a numerous of other concepts. In order to find out the concepts which are not in touch to that many other concepts but related to some fewer ones, a visualization is generated. During k-core analysis results visualization, the vertices

```
Number of vertices (n): 94
-------------------------------------------------------------------

                                               Arcs          Edges
-------------------------------------------------------------------

Number of lines with value=1                     0           1198
Number of lines with value#1                     0            786
-------------------------------------------------------------------

Total number of lines                            0           1984
-------------------------------------------------------------------

Number of loops                                  0             94
Number of multiple lines                         0            945
-------------------------------------------------------------------


Density [loops allowed] = 0.43843368
Average Degree = 42.21276596
```

Fig. 7　Summary of concept network structure

```
Input core partition of N2 (94, core=34)
===================================================================

Dimension: 94
The lowest value:   6
The highest value: 34


Frequency distribution of cluster values:

   Cluster      Freq      Freq%   CumFreq   CumFreq%  Representative
   ----------------------------------------------------------------
         6         2     2.1277         2    2.1277  pop
         8         3     3.1915         5    5.3191  exe
        10         4     4.2553         9    9.5745  apache tomcat
        12         2     2.1277        11   11.7021  #samba
        14         3     3.1915        14   14.8936  #openlinux
        16         4     4.2553        18   19.1489  microsoft excel
        18         6     6.3830        24   25.5319  cisco virtual private network
        20         4     4.2553        28   29.7872  #red hat
        22         5     5.3191        33   35.1064  vulnerability
        24         2     2.1277        35   37.2340  ssl
        26         2     2.1277        37   39.3617  path
        28        10    10.6383        47   50.0000  cgi
        30         7     7.4468        54   57.4468  solaris
        32        19    20.2128        73   77.6596  package
        34        21    22.3404        94  100.0000  #web
   ----------------------------------------------------------------
        Sum       94   100.0000
```

Fig. 8　K-core subgroup analysis results

which have higher connectivity (between k-core 23 and k-core 34) are removed to find out subtle sub-groups as shown in Fig. 9.

　　Another subgroup analysis technique is based on similarities. In this analysis, a dissimilarities matrix of concepts based on the line values and connectivity of the concepts is generated applying graph theoretical analysis techniques to the concept

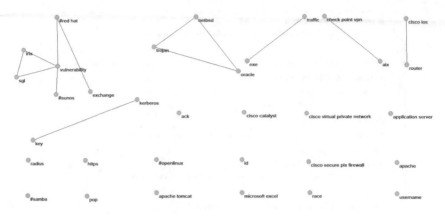

Fig. 9 K-core results including vertices between having 1–22 cores

relationships data. Later, a dendrogram, a tree structure utilized to demonstrate the arrangements of clusters of items is created using the dissimilarities information. In order to demonstrate, some sample subgroups of the dendrogram are marked using letters in the alphabet, as shown in Fig. 10a.

In this graph, group "A" corresponds to concepts related to Cisco networking, group "B" corresponds to mainly protection systems, group "C" corresponds to web application development, group "D" corresponds to Linux type operating systems, group "E" corresponds to browsers, group "F" corresponds to network traffic protocols, and group "G" corresponds to another set of operating systems which may be merged with group D. While the concepts that are mostly related (most central) with other concepts might be observed in Fig. 6, dendrogram view provides an alternative perspective and way to find out subgroups of the concepts.

As a continuation of these efforts ego networks of the selected concepts are generated. An ego network corresponds to a sub-group of a network where the selected vertex and its adjacent neighbors and their mutual links are included. In this way, it is possible to observe local relationships of concepts that are not centralized most in the whole network. A sample ego network, created for the "application server" concept is shown in Fig. 10b.

Finally, the concepts are classified using the partition matrix which groups the concept vertices in 11 groups. Following this, this classified network is shrunk to present top-level relationships of the concept groups such as application, browser, and network traffic. Figure 11 presents the resulting network. In this view, the line weights are proportional to the line values which indicate simultaneous existence in a vulnerability record. This picture shows that CVE consists of records mostly related to relations of network system/traffic to the end-user of middleware applications and protocols. Similarly, exposures related to the host machines and applications are relatively higher than in other groups.

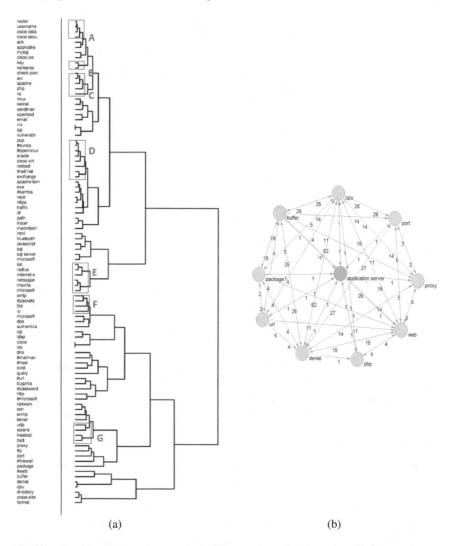

(a)   (b)

Fig. 10 **a** Dendrogram hierarchy results of CVE concepts **b** EGO network of "Application Server

5 Discussion

The size of CVE data eliminates the possibility of examining it manually. The semi-computerized analysis using keywords, computerized analysis, and graph theoretical analysis provided an in-depth knowledge for the large common vulnerabilities and exposures dataset. The techniques that have been used have made it easier to access details gradually, which can not be gathered through manual ways.

The conclusions of this study can be grouped into two parts: the resolutions related to the analysis methods and the resolutions related to the dataset. The closeness

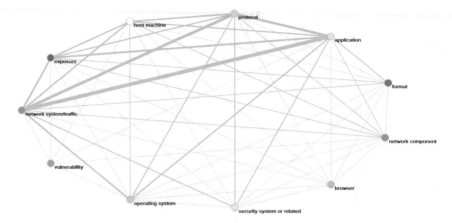

Fig. 11 Shrunk network based on classification partition

information for each concept was captured during the computerized content analysis phase. Creation of subgroups using graph theoretical analysis provided comprehensive knowledge on these concepts. Several subgroup analysis methods are utilized. This ended up some methods having more logical results compared to others for this data.

Manual analysis for the selected keywords associated to the enterprise systems and software and security datasets provided a summary of the data and distribution of vulnerabilities for these selected groups, Figs. 2, 3 and 4. Computerized content analysis allowed finding out top-level concepts for the large dataset and grouping the corresponding vulnerability records automatically, applying graph theoretical techniques resulted in being able to analyze the CVE data comprehensively. Each analysis type is powerful in its unique way and provided different perspectives of the data. K-cores is one of the techniques that created a clustering of data. Although, in general, only removing the k-cores with lower values may be meaningful, in this case removing the upper portion resulted in discovering more subtle relationships, Fig. 9. Dendrogram provided a hierarchical clustering view of the concepts, which is an upper-level perspective that allows examining lower-level hierarchies as well, Fig. 10a. Dendrogram analysis may be repeated by removing the concepts having a similarity level lower than a threshold value, which will have results with higher accuracy. On the other hand, ego networks presented a localized view for specific concepts. These localized close concept relationships point out vulnerabilities that can be worked on together, Fig. 10b. Primarily, the vulnerabilities related to nodes that have stronger connections with each other may be grouped to optimize the time and effort given to handle them. Although validating this is out of the scope of this study, these close relationships most probably point out the same data or same platforms that may be handled together during manual or automatic security tests. Consequently, collecting data and having a test setup may be relatively easy when the vendors or analysts make this optimization. Lastly, reducing the initial set to 94 concepts made it possible to manually classification of the CVE data to 11 groups. Visualization of these upper-level classification results, Fig. 11, also provided a totally new perspective which was not possible prior to this study. This figure shows the top level classifications. It presents an overall summary of all the CVE entries. In this picture, it is clear that CVE is full of entries related to vulnerabilities which are related to both network system/traffic and applications. The group of vulnerabilities which are related to both network system/traffic and protocols comes next.

At the start of this study, it was admitted that the CVE data lacked enough classifiers. Conducting these analyses resulted also in knowing the content and its problems better. There are numerous indications showing the content is outdated in various discourses. One indicator is not consisting of new technologies. In other words, it looks like the CVE database is not considered as a platform for reporting and storing weaknesses related to the newest technologies. For example, although there are many browsers, the fact that no output concept related to the most widely used Google Chrome is a sample which points out the problem. This doesn't mean that there are no records, but not having a concept shows at least the number of existing records is below a threshold value. One of the reasons for being outdated is the delays caused by the vulnerability evaluation process. Because when the candidate records are checked, one may encounter to some newer technologies which are staying in candidate status for a long time.

Another problem related to the content is the existence of the records related to numerous technologies that are not currently actively used in other words that aredeprecated. Among these, some of Linux operating system versions can be listed. Perhaps archiving this type of outdated vulnerabilities data in another data store and clearing the list will make it more popular among new users and increase its overall usability.

Considering the part of the concepts discovered through the keywords (manual content analysis), one can see that the number of corresponding vulnerabilities and exposures is very low for some keywords. Disclosing these low values may result in thinking that focusing on a single concept or a few concepts may not increase the efficiency of novel security solutions. However, there are many security designs, both academic and commercial, which focus on a single type of vulnerability and lack very similar other ones. Thus, even when the low number of vulnerability records per concept is taken into account during the creation of novel designs, this may lead to an increase in the vulnerability coverages for them. These low numbers also indicate lacking vulnerabilities for some important systems in the database.

When we further examined the concepts and their relationships with threats and technologies, it is logical to say that some of these associations are less meaningful. However, still, a few of the associations discovered during the content analysis resulted in the novel security design ideas which have not been discovered in the literature or encountered in product design. For example, analysis of printer privileges of the users, analysis of Share point application structure, and visualization of traffic between multiple virtual servers residing on the same machine are three of them.

As mentioned in the previous paragraphs, the examination of the content analysis showed that CVE data lacks vulnerabilities related to some of the enterprise security data sources that correspond to commonly used enterprise software or infrastructure elements. For example, although ERP systems and SCM systems are commonly used in the enterprises, the number of vulnerabilities related to these are very low or even none in the existing CVE data.

6   Conclusions

This study pointed out several future studies. Some brands or technologies are more prone to vulnerabilities compared to their competitors. For example, PHP based web servers have a higher number of vulnerabilities compared to Java-based web servers. Giving more priority to these technologies when designing novel security designs may be more profitable. Other examples of technologies that are more prone to vulnerabilities compared to similar ones are a few operating systems. Vulnerability lists for both development languages and operating systems are available in numerous other sources, such as security-related forums, and vendor websites. However, reaching similar results during this examination was encouraging to repeatedly conduct similar analyses on the data.

This examination showed that there are many vulnerabilities that arise associated with the wrong configuration of the systems. Visualization is a method to classify the malware files which take both binary and code versions of the files as input. Visualization of the configuration files and settings which are more prone to errors may be a future study topic to detect the errors in the configuration files. This may be an example of visualization of static data which may be beneficial for the enterprises.

As another future study subjects, the concepts that are discovered in the computerized content analysis of the CVE data can be used in a backward content analysis study. This time, other similar resources may be examined using the concepts captured from this study. In this analysis, what percentage of the concepts found using the CVE description text are covered in security products can be examined.

Knowing the CVE concepts may also help in finding recent directions of the hacker communities. If recent popular vulnerabilities known among these communities can be found by searching the concepts, again in deep web forums of such communities, this information can be disclosed through security visualization focused new studies covering those specific vulnerability groups.

References

1. Corporation TM (2017) Common vulnerabilities and exposures. Common vulnerabilities and exposures: http://cve.mitre.org

2. CWE (2017) Common weakness enumeration. 06 28 2017 tarihinde. https://nvd.nist.gov/cwe.cfm
3. Martin RA (2002) Managing vulnerabilities in networked systems. Computer 34(11):32–38. https://doi.org/10.1109/2.963441
4. Baker DW, Christey SM, Hill WH, Mann DE (1999) The development of a common enumeration of vulnerabilities and exposures. In: Second international workshop on recent advances in intrusion detection, Lafayette, IN, USA
5. Allahyari M, Safaei S, Pouriyeh S, Trippe ED, Kochut K, Assefi M, Gutierrez JB (2017) A brief survey of text mining: classification, clustering and extraction techniques. In: Conference on knowledge discovery and data mining, Halifax, Canada
6. Collard J, Bhat TN, Subrahmanian E, Sriram RD, Elliot JT, Kattner UR, Campbell C, Monarch I (2018) Generating domain ontologies using root- and rule-based terms. J Washington Acad Sci 31–78
7. Guo M, Wang J (2009) An Ontology-based approach to model common vulnerabilities and exposures in information security. In: ASEE Southeast section conference
8. Musen MA (2015) The protégé project: a look back and a look forward. AI Matters 1(4):4–12. https://protege.stanford.edu/
9. Chen Z, Zhang Y, Chen Z (2010) A Categorization framework for common computer vulnerabilities and exposures. Comput J 53(5)
10. Cortes C, Vapnik V (1995) Support-vector networks. Mach Learn 20(3):273–297
11. Wen T, Zhang Y, Wu Q, Yang G (2015) ASVC: an automatic security vulnerability categorization framework based on novel features of vulnerability data. J Communs 10(2):107–116
12. Na S, Kim T, Kim H (2016) A study on the classification of common vulnerabilities and exposures using naïve bayes. In: International conference on broadband and wireless computing, communication and application
13. Bozorgi M, Saul LK, Savage S, Voelker GM (2010) Beyond heuristics: learning to classify vulnerabilities and predict exploits. In: Proceedings of the 16th ACM SIGKDD international conference on knowledge discovery and data mining, Washington, DC, USA
14. DeLooze L (2004) Classification of computer attacks using a self-organizing map. In: Proceedings from the fifth annual IEEE SMC information assurance workshop, West Point, IEEE, New York, s 365–369. https://doi.org/10.1109/iaw.2004.1437840
15. Kohonen T (1998) The self-organizing map. Neurocomputing 21(1–3):1–6. https://doi.org/10.1016/S0925-2312(98)00030-7
16. Wang W, Gupta A, Niu N (2018) Mining security requirements from common vulnerabilities and exposures for agile projects. In: 1st International workshop on quality requirements in agile projects, Banff, Canada, IEEE, s 6–9
17. Subroto A, Apriyana A (2019) Cyber risk prediction through social media big data analytics and statistical machine learning. J Big Data 50–69
18. Mostafa S, Wang X (2020) Automatic identification of security bug reports via semi-supervised learning and CVE mining
19. Chen H-M, Kazman R, Monarch I, Wang P (2016) Predicting and fixing vulnerabilities before they occur: a big data approach. In: IEEE/ACM 2nd international workshop on big data software engineering, Austin, IEEE, TX, USA, s 72–75
20. Nooy W, Mrvar A, Batagelj V (2011) Exploratory social network analysis with pajek. Cambridge University Press, Cambridge
21. Ruohonen J, Rauti S, Hyrynsalmi S, Leppänen V (2017) Mining social networks of open source CVE coordination. In: Proceedings of the 27th international workshop on software measurement and 12th international conference on software process and product measurement, Gothenburg, Sweden: ACM, s 176–188
22. Deo N, Gupta P (2003) Graph-theoretic analysis of the world wide web: new directions and challenges. Mat Contemp 49–69
23. Özdemir Ö (2015) Influence of networks on systemic risk within banking system of Turkey. METU, Ankara, Turkey

24. Zegzhda PD, Zegzhda DP, Nikolskiy AV (2012) Using graph theory for cloud system security modeling. In: International conference on mathematical methods, models, and architectures for computer network security, St. Petersburg, Springer, Russia, s 309–318
25. Sarkar S, Almukaynizi M, Shakarian J, Shakarian P (2019) Predicting enterprise cyber incidents using social network analysis on dark web hacker forums. Cyber Defense Rev 87–102
26. Wang S, Nagappan N (2019) Characterizing and understanding software developer networks in security development. York University, York, UK
27. CVE (2016) Download CVE list. Common vulnerabilities and exposures: https://cve.mitre.org/
28. Özdemir Sönmez F, Güler B (2019) Qualitative and quantitative results of enterprise security visualization requirements analysis through surveying. In: 10th International conference on information visualization theory and applications, Praque, IVAPP 2019, s 175–182
29. Leximancer (2019) Leximancer. Brisbane, Australia. https://info.leximancer.com/
30. Ward V, West R, Smith S, McDermott S, Keen J, Pawson R, House A (2014) The role of informal networks in creating knowledge among health-care managers: a prospective case study. Heath Serv Delivery Res 2(12)
31. Pajek (2018) Analysis and visualization of very large networks. Pajek/PajekXXL/Pajek3XL: http://mrvar.fdv.uni-lj.si/pajek/

Machine Intelligence and Big Data
Analytics for Cybersecurity Applications

A Novel Deep Learning Model to Secure Internet of Things in Healthcare

Usman Ahmad, Hong Song, Awais Bilal, Shahid Mahmood, Mamoun Alazab, Alireza Jolfaei, Asad Ullah, and Uzair Saeed

Abstract Smart and efficient application of DL algorithms in IoT devices can improve operational efficiency in healthcare, including tracking, monitoring, controlling, and optimization. In this paper, an artificial neural network (ANN), a structure of deep learning model, is proposed to efficiently work with small datasets. The contribution of this paper is two-fold. First, we proposed a novel approach to build ANN architecture. Our proposed ANN structure comprises on subnets (the group of neurons) instead of layers, controlled by a central mechanism. Second, we outline a prediction algorithm for classification and regression. To evaluate our model experimentally, we consider an IoT device used in healthcare i.e., an insulin pump as a proof-of-concept. A comprehensive evaluation of experiments of proposed solution

U. Ahmad (✉) · H. Song · A. Bilal · U. Saeed
School of Computer Science and Technology, Beijing Institute of Technology, Beijing 100081, China
e-mail: usmanahmad@bit.edu.cn

H. Song
e-mail: songhong@bit.edu.cn

A. Bilal
e-mail: awaisbilal@bit.edu.cn

U. Saeed
e-mail: uzairsaeed@bit.edu.cn

S. Mahmood
School of Computing, Electronics and Mathematics, Coventry University, Coventry, UK
e-mail: mahmo136@coventry.ac.uk

M. Alazab
Charles Darwin University, Darwin, Australia
e-mail: mamoun.alazab@cdu.edu.au

A. Jolfaei
Macquarie University, Sydney, Australia
e-mail: alireza.jolfaei@mq.edu.au

A. Ullah
School of Information and Electronics, Beijing Institute of Technology, Beijing 100081, China
e-mail: engr.asad@bit.edu.cn

© The Editor(s) (if applicable) and The Author(s), under exclusive license to Springer Nature Switzerland AG 2021
Y. Maleh et al. (eds.), *Machine Intelligence and Big Data Analytics for Cybersecurity Applications*, Studies in Computational Intelligence 919,
https://doi.org/10.1007/978-3-030-57024-8_15

and other classical deep learning models are shown on three small scale publicly available benchmark datasets. Our proposed model leverages the accuracy of textual data, and our research results validate and confirm the effectiveness of our ANN model.

Keywords Artificial neural network (ANN) · Deep learning · Internet of Things (IoT) · Healthcare · Security · Small datasets

1 Introduction

The Internet of Things (IoT) revolution is reshaping the service environment by integrating the cyber and physical worlds, ranging from tiny portable devices to large industrial systems. IoT brings a new wave of intelligent devices connected to the internet for the aim of exchanging information. The rapid development of IoT industry is facilitating various domains. Typical applications of IoT technologies include healthcare, intelligent transportation, smart home/cities, agriculture, and finance, etc. By 2025, Huawei's Global Industry Vision (GIV 2019) predicts that 100 billion connected devices with the billions of massive connections will be used worldwide [1].

With the rapid advancement in the cyber attacking tools, the technical barrier for deploying attacks become lower. Moreover, the IoT industry brings new security issues due to the changing service environment. Security and privacy of IoT devices became one of the most paramount research problems. Extensive surveys of security threats and current solutions in different layers of the IoT system are published in [2, 3]. Khan et al. [4] outlined nineteen different types of security attacks on IoT which were categorized in three broader classes: low-level, intermediate-level, and high-level security issues.

Deep learning algorithms are inspired by the structure and information processing of the biological system called Artificial Neural Networks (ANNs). Some of the major deep learning architectures are convolutional neural networks (CNN), recurrent neural networks (RNN), deep neural networks (DNN), deep belief networks (DBN), and hybrid neural networks [5]. Deep learning has given a great deal of attention over the last several years in the domain of IoT security where it shows the potential to rapidly adjust to new and unknown threats and provides a significant solution against zero-day attacks [6, 7]. Generally, deep learning models aim to enhance the performance in detecting a security attack with the help of learning from training dataset. For example, the task of deep learning in intrusion detection system is to classify system behavior, whether benign or malicious. The learning can be supervised, semi-supervised, and unsupervised.

The small dataset contains specific attributes used to determine current states or conditions. For example, smart devices attached to drones or deployed on wind turbines, valves or pipes collect small datasets in real time environments such as temperature, pressure, wetness, vibration, location, or even an object is moving or not. In spite of the faster growth of big data, small data studies continue to perform a vital role in various research domains due to their utility in solving the targeted problems [8, 9]. In many IoT use cases small dataset is more important than the big dataset. For example, the insulin pump system is a small device to automatically inject insulin into the body of diabetic patient. The insulin pump system continuously monitors the glucose level of the diabetic patient to manage sugar level by injecting the insulin when it is required. Security attacks are deployed to disrupt the functionality of insulin pump system by injecting the lethal dose and endanger the lives of patients. We need effective security mechanisms ensure the correct dosing process of insulin pump system. Deep learning is an effective solution by predicting the thresh hold value of insulin to be injected based on the log of insulin pump system [10].

Deep learning has shown the potential to rapidly adjust to new and unknown threats, over the traditional methods [6, 7]. However, training the deep learning model from small datasets is surprisingly scarce and does not work well [11, 12]. Deep learning models need to guarantee high performance on the small dataset. In this paper, we proposed a data-intensive approach to build an artificial neural network (ANN) to efficiently work with small datasets. The contribution of this paper is as follows:

(1) We proposed a novel approach to build a supervised ANN model. Our proposed ANN structure comprises on subnets (group of neurons) instead of layers, controlled by a central mechanism. We put forward a strong hypothesis based on which we construct the architecture of our ANN model, holding the dataset values (illustrated in Sect. 3).
(2) We proposed a prediction algorithm for classification and regression.There are several activation functions used by the traditional ANN algorithms. We did not use any activation function; instead, we proposed a novel prediction algorithm. We evaluated our model on textual data using three small scale publicly available benchmark datasets and provide a comparative analysis with Multilayer Perceptron's (MLPs) and Long Short-Term Memory (LSTM) recurrent neural network models.
(3) We outline the experimental setup to evaluate our model using Arduino (open source platform). We consider the insulin pump device from healthcare domain as a proof-of-concept.

2 Related Work

Extensive surveys present the security threats and state-of-the-art solutions in IoT [2, 3]. Khan et al. [4] outlined eighteen different types of security solutions in IoT. The insulin pump system is a wearable device to automatically inject insulin into the body. Security attacks are deployed to disrupt the functionality of insulin pump system. In [10], the authors proposed a solution to secure the insulin pump system based on recurrent neural network (LSTM) using the log of insulin pump system. In [13], the author proposed a framework based on deep learning approach for intrusion detection in the IoT, called DFEL and present the significant experimental results. In [14], the author investigated the security attacks to the IEEE 802.11 network and proposed a solution based on a deep learning approach for anomaly detection. Another security mechanism based on deep learning approach is proposed to the detection of botnet activity in the IoT devices and networks [15]. In [16], the author proposed a solution based on recurrent neural network to detect attacks in IoT devices connected to the home environment.

We discussed an example from the automotive industry in the paragraph to reflect the importance of small data in IoT security. IoT is advancing the old-fashioned ways of locking/unlocking and starting cars. Passive keyless entry and start (PKES) systems allow drivers to unlock and start the cars by just possessing the key fob in their pockets. Despite the convenience of PKES, it is vulnerable to security attacks. In [17, 18], the authors exploited the PKES system security mechanism and demonstrated the practical relay attacks. In [19], Ahmad et al. proposed a solution to secure the PKES system, based on machine learning approaches using last three months log of the PKES system.

A MEC-oriented solution in 5G networks to anomaly detection is proposed, which is based on the deep learning approach [20]. The author proposed and deep learning method to detect the security attacks in IoT [21]. They extracted a set of features and dynamically watermark them into the signal. Das et al. [22], proposed a solution based on a deep learning approach to authenticate the IoT and tested on the low poser devices. Ahmed et al. [23], proposed a present a deep learning architecture to address the issue of person re-identification.

Training the deep learning model from small datasets is surprisingly scarce and does not work well. Researcher published the literature to improve the performance of deep learning model on small datasets. In [11], the authors how that the performance can be improved on small datasets by integrating prior knowledge in the form of class hierarchies. In [12], the author demonstrated the experimental results showing that the cascading fine-tuning approach achieves better results on small dataset. A deep learning based solution is proposed to classify the skin cancer on a relatively small image dataset [24].

3 Materials and Methods

The biological neural network is one of the most complex systems on the planet, and the study of human memory is still in its infancy. A list of questions remains unanswered about how the data is determined and moved from neuron to neuron. The researchers of neuroscience also rely on the hypotheses and assumptions to understand the shape and working of biological neural network [25, 26]. Hypotheses and assumptions encourage the critical approach and can be a starting point of the revolutionary research [27]. This section presents the architecture of our proposed ANN model based on a strong hypothesis and the prediction algorithm.

3.1 ANN Architecture

The biological neural network is actively engaged in the functions of memorization and learning. Human memory is capable of storing and processing the massive data with details from the image [28, 29]. In [30], the author presented the strong foundation that if ANN truly inspired by the biological network, then it must learn by memorizing the training data for prediction. *So we put forward and evaluates a strong hypothesis that the ANN model must have the memory and hold the dataset in it, as the biological neural network has capability of storing data.* In traditional ANN models, a neuron is a mathematical function called the activation function that produces the output based on the given input or set of input. But, the neurons are the memory cells in our ANN model that hold the dataset values.

3.1.1 Mesh of Subnets

Our ANN structure is the grouping of neurons into subnets instead of layers in a manner that we refer as the *mesh of subnets*. Usually, the textual dataset is structured in the tables; but, our model organizes the dataset in the subnets wherein each attribute value of dataset is kept in a separate subnet. Neurons in the ANN model are spread in the subnets, and the collection of neurons in a particular subnet holds the data of the one single attribute of the dataset.

3.1.2 Connections and Weights

New subnets, neurons, and the connections between them are created when data is inserted to ANN model during the training. The neurons are interconnected. The connections between neurons are established based on the flow of incoming training data, and each connection has an initial weight value 1. The connections between neurons become stronger (i.e., updating weight), depend on the occurrence of duplicate input data values during the training process. If the data (neuron) already exists in the subnet, then only weight value is updated by 1 and data is not repeated to avoid data duplication in a subnet. As a result, no two neurons in a subnet can hold the same data value. The weight value expresses how solid connection two neurons have with each other. So, weight is updated on each occurrence of the same input data making the connection stronger on each iteration.

Figure 1 shows that how we structure the data into subnets. The values of attribute 1 are stored in subnet 1. Value 10 is repeated 3 times in attribute column, but subnet 1 has one single neuron holding value 10. Similarly, the values of attribute 2 are stored in subnet 2. Value 29 is repeated 2 times in attribute column and subnet 2 have one neuron holding value 29, and so on. Connections are established between the neurons of subnet 1 and subnet 2, based on the frequency of input data. The first and fourth records have the same data, so our ANN model updates the weight value by 1 and

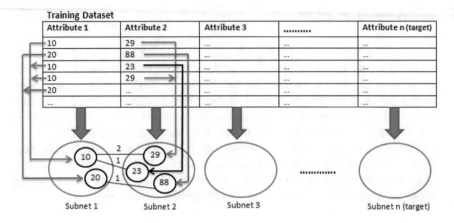

Fig. 1 Structuring the training dataset to subnets

does not repeat the data. Accordingly, the connection between neurons containing value 10 of subnet 1 and neuron holding value 29 of subnet is 2 have weight value 2, as shown in Fig. 1.

3.1.3   Central Mechanism

We have a central point of connection of all neurons, like a nucleus of our ANN model. Each neuron in the ANN model has a connection with the central point through the subnet. This ensures that each and every neuron is in connection and has direct access to all the neurons in the ANN via the central point. The central point also contains the neurons (along with connections and weights) and subnets. This central mechanism plays two major roles as below:

- Interconnect all neurons of the ANN model, so provide direct access to all the neurons through subnet.
- It has the capability to add biasness by changing the weight values of connection between the central mechanism's neurons and ANN model's neurons.

3.1.4   Memory Requirement

Our model avoids the data repetition in the subnet. If the data already exists in the subnet's neuron, then our model does not repeat the data in a subnet but updates the weight values. Let the training dataset have n number of attributes a,then the total number of subnets S are calculated as below:

$$S = \sum_{i=1}^{n} a_i \qquad (1)$$

Total number of neurons N in a subnet are calculated as below:

(1) Iterate through all values in the attribute once: $O(n)$
(2) For each value seen in the attribute, check to see if it's in the Subnet $O(1)$, amortized

(a) If not, create a neurons with the value and weight value is as below:

$$Initial\ weight\ value = 1 \qquad (2)$$

(b) If so, update the weight value as below:

$$weight = weight + 1 \qquad (3)$$

- **Space**: $O(nU)$, where n is the number of attributes and U is the number of distinct values in an attribute.

3.2 Prediction Algorithm

Input $I_1, I_2, I_3, \ldots, I_{n-1}$ are input data values for each record
   Output I_n: the class attribute

Theorem 1 *Let we have n number of attributes, then $I_1, I_2, I_3, \ldots, I_n$ are given values for each record. We have subnets $S_1, S_2, S_3, \ldots, S_n$ for input data $I_1, I_2, I_3, \ldots, I_n$, respectively, where S_n is the target subnet.*

(1) Forward the input data $I_1, I_2, I_3, \ldots, I_{n-1}$ to the $S_1, S_2, S_3, \ldots, S_{n-1}$ subnets, respectively.
(2) if value I_1 exists in subnet S_1 then Select the neuron containing value I_1 from subnet S_1 else Find the closest value.
(3) List all connected neurons with selected neurons in step 2 meeting the following three conditions:

   (a) Neurons $\in S_2, S_3, S_4 \ldots S_{n-1}$
   (b) Neurons must have the maximum weight value
   (c) Neuron must be connected to the same neurons in the subnet S_n as our selected neuron in step 2 is connected.

(4) if value I_2 exists in the listed neurons in step 3 and value $I_2 \in S_2$ then Select the neuron containing value I_2 else Find the closest value.
(5) List all neurons of target subnet S_n along with weight values in L_1, which are connected to the selected neuron in step 4.

(6) *Find the neuron N_2 having the maximum weight value in the list L_1 as below:* **if** *classification* **then** *Select the neuron having the maximum weight value in the list L_1.* **else if** *regression* **then** *Calculate the weighted average of the list L_1.*

$$Weighted\ Average = \frac{\sum_{i=1}^{n} w_i . a_i}{\sum_{i=1}^{n} w_i} \tag{4}$$

(7) *Repeat step 4 to 6 for input values $I_3, I_4, I_5, \ldots, I_{n-1}$. So, we get neurons $N_3, N_4, N_5, \ldots, N_{n-1}$ against the data value I_1.*
(8) *Perform step 6 on selected neurons $N_2, N_3, N_4, \ldots, N_{n-1}$, so we get single value V_1 against the input value I_1.*
(9) *Repeat step 1 to 8 for all reaming input values $I_2, I_3, I_4, \ldots, I_{n-1}$, we get $V_2, V_3, V_4, \ldots, V_{n-1}$ against the input values $I_2, I_3, I_4, \ldots, I_{n-1}$, respectively. So, the total number of values calculated for each test record Rec are as below:*

$$Rec = \sum_{i=2}^{n-1} I_i \cdot \sum_{j=1}^{n-1} V_j \tag{5}$$

(10) *Perform step on $V_2, V_3, V_4, \ldots, V_{n-1}$ and get one value against each record Rec.*
(11) *Repeat step 1–10 for each test record.*
(12) *Calculate the accuracy % and the prediction error rate (RMSE) of the test dataset.*

$$RMSE = \sqrt{\frac{\sum_{i=1}^{n} (P_i - O_i)^2}{n}} \tag{6}$$

4 Results and Discussion

Healthcare is one of the distinctive domains of IoT technologies, and the security threat in healthcare can result in the loss of life. To evaluate our model experimentally, we consider the insulin pump system to automatically inject insulin into the body of diabetic patients. Security attacks are deployed to disrupt the functionality of insulin pump system by injecting the lethal dose and endanger the lives of patients. In [10] , the authors proposed a machine learning based solution to secure the insulin pump system using last three months log of insulin pump system. We evaluate our model on publicly available diabetes datasets for concrete comparisons. We used two public datasets of diabetes patients, similar to the log of insulin pump system: first, Pima Indian diabetes dataset [31] for classification and second, diabetes dataset (data-01) [32] for regression. We also considered another small scale benchmark dataset to validate our model i.e., Iris dataset [33].

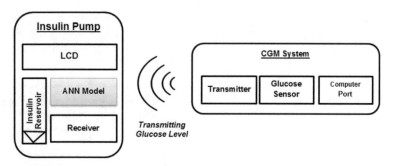

Fig. 2 Block diagram of our proposed solution for the insulin pump system

4.1 Testing Environment

This section presents the experimental setup to evaluate our model using Arduino (an open source platform). Arduino supports both, programmable microcontroller and a software programming language. The insulin pump system consist of two separate physical devices; first, Continuous Glucose Monitoring (CGM) system, second the insulin pump itself. CGM measures the glucose level from blood and send it to insulin pump. Insulin pump receives and analyses this glucose level and injects insulin against accordingly. Arduino UNO boards are used to implement the insulin pump system. One board performs as an insulin pump and second for the CGM system. We have attached the RF modules (433 MHz AM transmitter and a receiver) with both devices for communication between the insulin pump and CGM. We have used RC Switch Arduino library to transmit and receive data over an RF medium. Figure 2 illustrates the block diagram of insulin pump system and CGM with our proposed solution.

CGM system checks and transmits the glucose level to the insulin pump. Insulin Pump receives and calculates the insulin amount on behalf of glucose level. Man in the middle attack can be deployed to disrupt the functionality of insulin pump system by injecting the lethal dose and endanger the lives of patients. Our ANN model predicts the threshold value of insulin. Insulin pump compares if the insulin amount is greater than the predicted threshold insulin, and then generates an alarming situation of an attack. If the insulin amount is less than the predicted threshold insulin, then proceed further and inject the insulin. The work flow of insulin pump with our proposed solution is presented in Fig. 3.

4.2 Results

There are several built-in libraries and packages available to implement the deep learning models, for instance, tensorflow, Keras, Caffe, and Theano, etc. The structure

Fig. 3 Work flow of our
proposed solution for the
insulin pump system

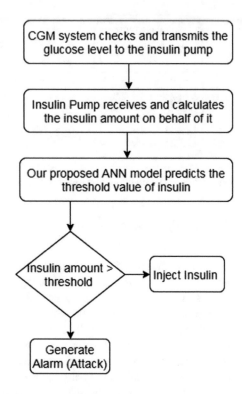

of our model is totally different from the traditional deep learning models. We are not using the layered model and activations functions, so no built-in deep learning library is used in this research. We write the code from scratch in Python programming language. Our proposed solution is evaluated using three datasets. Iris flower dataset and Pima Indian diabetes dataset used for classification and diabetes dataset from UCI machine learning repository used for regression. We have used 20% of the dataset for testing and the remaining 80% for training. We outline the comparison of proposed solution and two classical deep learning models that work better on textual data. We create a fully-connected sequential MLPs and Recurrent Neural Network (RNN) LSTM structures with three, four, and five layers using Keras libraries to compare the results of our proposed model. We have evaluated our model on Pima Indian diabetes dataset as follow:

(1) *True Positive (TP)*. Correctly forecasted the patient has diabetes
(2) *True Negative (TN)*. Correctly forecasted the patient don't have diabetes
(3) *False Positive (FP)*. Incorrectly forecasted the patient has diabetes
(4) *False Negative (FN)*. Incorrectly forecasted the patient don't have diabetes
(5) *Accuracy*. Percentage of the correct forecast, that is,

$$Accuracy = \frac{TP + TN}{TP + TN + FP + FN}. \tag{7}$$

Table 1 Classification: comparison of accuracy rates on Pima Indian diabetes and Iris datasets

Proposed model		Pima Indian diabetes dataset	Iris dataset
		64 %	97 %
MLPs	3-Layers	69 %	35 %
	4-Layers	78 %	33 %
	5-Layers	66 %	29 %
RNN (LSTM)	3-Layers	65 %	92 %
	4-Layers	65 %	94 %
	5-Layers	65 %	94 %

Table 2 Regression: comparison of error rates (RMSE) on diabetes datasets

Proposed model		Diabetes dataset
		81
MLPs	3-Layers	108
	4-Layers	108
	5-Layers	108
RNN (LSTM)	3-Layers	85
	4-Layers	87
	5-Layers	84

The experimental results for classification are summarized in Table 1. Our network achieved 97% accuracy on the Iris dataset, where RNN (LSTM) with 3 layers has 92% accuracy. The accuracy rate is slightly improved if we increase the layers. RNN (LSTM) with 4 and 5 layers structure has maximum 94% accuracy rate. MLP structure did not perform well on Iris dataset and achieved 35% accuracy rate with 3 layers structure. The accuracy graph of MLP is decreased to 33 and 29% with 4 and 5 layers, respectively. On the other hand, the 4-layer MLP structure performed well and achieved 78% accuracy on Pima Indian diabetes dataset where our model achieved comparatively low accuracy that is 64% accuracy. The accuracy of RNN (LSTM) on Pima Indian diabetes dataset is 65%. The experimental results for regression are summarized in Table 2. Our model achieved better error rate, i.e., 81 as compared to RNN(LSTM) and MLP on diabetes dataset, where RNN (LSTM) has high error rate of 84. Finally the MLPs have highest error rate of 108.

5 Conclusion

We proposed a deep learning model that efficiently works on small datasets. The contribution of this paper is three-fold. First, we proposed a novel approach to build ANN architecture. Our ANN model is a combination of subnets under control of a central mechanism, where a subnet is a collection of neurons. A neuron is a memory cell that holds the dataset values. Second, we outline a comprehensive prediction algorithm for classification and regression. We evaluated our model on three small scale publicly available benchmark datasets. We also performed a comparative analysis with classical deep learning models. As future work, we plan to implement our model on large textual and image datasets to prove that our ANN model can also efficiently work on larger datasets.

References

1. GIV Huawei (2019) Touching an intelligent world, Huawei Technologies. [Online]. Available: https://www.huawei.com/minisite/giv/Files/whitepaper_en_2019.pdf
2. Yang Y, Wu L, Yin G, Li L, Zhao H (2017) A survey on security and privacy issues in internet-of-things. IEEE Internet Things J 4(5):1250–1258
3. Chen L, Thombre S, Järvinen K, Lohan ES, Alén-Savikko A, Leppäkoski H, Bhuiyan MZH, Bu-Pasha S, Ferrara GN, Honkala S et al (2017) Robustness, security and privacy in location-based services for future iot: a survey. IEEE Access 5:8956–8977
4. Khan MA, Salah K (2018) Iot security: review, blockchain solutions, and open challenges. Future Gener Comput Syst 82:395–411
5. Liu W, Wang Z, Liu X, Zeng N, Liu Y, Alsaadi FE (2017) A survey of deep neural network architectures and their applications. Neurocomputing 234:11–26
6. Mohammadi M, Al-Fuqaha A, Sorour S, Guizani M (2018) Deep learning for iot big data and streaming analytics: a survey. IEEE Commun Surv Tutor 20(4):2923–2960
7. Kwon D, Kim H, Kim J, Suh SC, Kim I, Kim KJ (2019) A survey of deep learning-based network anomaly detection. Cluster Comput 22(1):949–961. [Online]. Available: https://doi.org/10.1007/s10586-017-1117-8
8. Kitchin R, Lauriault TP (2015) Small data in the era of big data. GeoJournal 80(4):463–475
9. Kitchin R (2014) The data revolution: Big data, open data, data infrastructures and their consequences. Sage
10. Ahmad U, Song H, Bilal A, Saleem S, Ullah A (2018) Securing insulin pump system using deep learning and gesture recognition. In: 17th IEEE international conference on trust, security and privacy in computing and communications/12th IEEE international conference on big data science and engineering (TrustCom/BigDataSE). IEEE, pp 1716–1719
11. Barz B, Denzler J (2019) Deep learning on small datasets without pre-training using cosine loss. CoRR, vol. abs/1901.09054. [Online]. Available: http://arxiv.org/abs/1901.09054
12. Ng H-W, Nguyen VD, Vonikakis V, Winkler S (2015) Deep learning for emotion recognition on small datasets using transfer learning. In: Proceedings of the 2015 ACM on international conference on multimodal interaction. ACM, pp 443–449
13. Zhou Y, Han M, Liu L, He JS, Wang Y (2018) Deep learning approach for cyberattack detection. In: IEEE INFOCOM 2018-IEEE conference on computer communications workshops (INFOCOM WKSHPS). IEEE, pp 262–267
14. Thing VL (2017) Ieee 802.11 network anomaly detection and attack classification: a deep learning approach. In: IEEE wireless communications and networking conference (WCNC). IEEE, pp 1–6
15. McDermott CD, Majdani F, Petrovski AV (2018) Botnet detection in the internet of things using deep learning approaches. In: 2018 International joint conference on neural networks (IJCNN). IEEE, pp 1–8
16. Brun O, Yin Y, Gelenbe E (2018) Deep learning with dense random neural network for detecting attacks against iot-connected home environments. Procedia Comput Sci 134:458–463

17. Francillon A, Danev B, Capkun S (2011) Relay attacks on passive keyless entry and start systems in modern cars. In :Proceedings of the network and distributed system security symposium (NDSS). Eidgenössische Technische Hochschule Zürich, Department of Computer Science
18. Choi W, Seo M, Lee DH (2018) Sound-proximity: 2-factor authentication against relay attack on passive keyless entry and start system. J Adv Transp
19. Ahmad U, Song H, Bilal A, Alazab M, Jolfaei A (2018) Secure passive keyless entry and start system using machine learning. In: Wang G, Chen J, Yang LT (eds) Security, privacy, and anonymity in computation, communication, and storage. Lecture notes in computer science. Springer International Publishing, Cham, pp 304–313
20. Maimó LF, Celdrán AH, Pérez MG, Clemente FJG, Pérez GM (2019) Dynamic management of a deep learning-based anomaly detection system for 5g networks. J Ambient Intell Humanized Comput 10(8):3083–3097
21. Ferdowsi A, Saad W (2018) Deep learning-based dynamic watermarking for secure signal authentication in the internet of things. In: 2018 IEEE international conference on communications (ICC). IEEE, pp 1–6
22. Das R, Gadre A, Zhang S, Kumar S, Moura JM (2018) A deep learning approach to iot authentication. In: 2018 IEEE international conference on communications (ICC). IEEE, pp 1–6
23. Ahmed E, Jones M, Marks TK (2015) An improved deep learning architecture for person re-identification. In: Proceedings of the IEEE conference on computer vision and pattern recognition, pp 3908–3916
24. Fujisawa Y, Otomo Y, Ogata Y, Nakamura Y, Fujita R, Ishitsuka Y, Watanabe R, Okiyama N, Ohara K, Fujimoto M (2019) Deep-learning-based, computer-aided classifier developed with a small dataset of clinical images surpasses board-certified dermatologists in skin tumour diagnosis. Br J Dermatol 180(2):373–381
25. Edelman S (2016) The minority report: some common assumptions to reconsider in the modelling of the brain and behaviour. J Exp Theor Artif Intell 28:751–776
26. Thompson RH, Swanson LW (2010) Hypothesis-driven structural connectivity analysis supports network over hierarchical model of brain architecture. Proc Natl Acad Sci USA 107(34):15235–15239
27. Nkwake AM (2013) Why are assumptions important? Springer, New York, NY, pp 93–111. [Online]. Available: https://doi.org/10.1007/978-1-4614-4797-9_7
28. Amin H, Malik AS (2013) Human memory retention and recall processes. Neurosciences 18(4):330–344
29. Brady TF, Konkle T, Alvarez GA, Oliva A (2008) Visual long-term memory has a massive storage capacity for object details. Proc Natl Acad Sci 105(38):14325–14329
30. Ahmad U, Song H, Bilal A, Mahmood S, Ullah A, Saeed U (2019) Rethinking the artificial neural networks: a mesh of subnets with a central mechanism for storing and predicting the data. CoRR, vol abs/1901.01462, 2019. [Online]. Available: http://arxiv.org/abs/1901.01462
31. Pima Indian diabetes database. www.ics.uci.edu/~mlearn/MLRepository.html
32. Diabetes data set. https://archive.ics.uci.edu/ml/datasets/diabetes
33. Fisher RA (1936) The use of multiple measurements in taxonomic problems. Ann Eugenics 7(2):179–188

Secure Data Sharing Framework Based on Supervised Machine Learning Detection System for Future SDN-Based Networks

Anass Sebbar, Karim Zkik, Youssef Baddi, Mohammed Boulmalf, and Mohamed Dafir Ech-Cherif El Kettani

Abstract Securing Data-sharing mechanism between Software Defined Networks (SDN) nodes represent one of the biggest challenges in SDN context. In fact, attackers may steal or perturb flows in SDN by performing several types of attacks such as address resolution protocol poisoning, main in the middle and rogue nodes attacks. These attacks are very harm full to SDN networks as they can be performed easily and passively at all SDN layers. Furthermore, data-sharing permit to an attacker to gather all sensitive flows and data from SDN architecture. In this chapter, we will propose a framework for secure data sharing that detect and stop intrusions in SDN context while ensuring authentication and privacy. To do so, we propose a defense mechanism that detect and reduce the risk of attacks based on advanced machine learning techniques. The learning and data pre-processing steps was performed by using a constructed data set dedicated to SDN context. The simulation results show that our framework can effectively and efficiently address sniffing attacks that can be detected and stopped quickly. Finally, we observe high accuracy with a low false-positive for attack detection.

Keywords SDN · Data-sharing · Sniffing · MitM · Random forest · SDN dataset · Arp-poisoning · Anomaly detection

1 Introduction

Software Defined Networks (SDN) is a new flexible, automated and dynamic network architecture that abstractly manages network services and provides more networking functionality. To do so, SDN separates the control layer from the data layer which

A. Sebbar · K. Zkik · M. Boulmalf
Université Internationale de Rabat, TICLab, Rabat, Morocco

A. Sebbar (✉) · M. D. Ech-Cherif El Kettani
ENSIAS-Mohammed V Rabat University, Rabat, Morocco

Y. Baddi
STIC, ESTSB-Chouaib Doukkali University, El Jadida, Morocco

Y. Maleh et al. (eds.), *Machine Intelligence and Big Data Analytics for Cybersecurity Applications*, Studies in Computational Intelligence 919,
https://doi.org/10.1007/978-3-030-57024-8_16

facilitate management and promote flexibility and automation [1–3]. However, SDN suffer from many security issues related to its centralized architecture and the separation of the control and data planes. Thus, there are several security and privacy concerns and issues regarding data sharing between SDN nodes and SDN storage center especially at the level of the data plane. In fact, SDN architectures are vulnerable to various security threats such as mitm, DDoS and arp poisoning attacks that aims to steal encryption keys and sensitive data and to disrupt flows and poison communications [4, 5].

In the literature, at the best of our knowledge there are just some few researches that propose secure framework for data sharing in SDN context. Xiaoning et al recommend a mechanism to protect the sharing of flow entries in the SDN, in order to minimize the total number of flow entries while guaranteeing the survival of traffic against a communication failure [6]. In addition, Klaedtke et al. provide a mechanism for protecting network flows, presenting an access control system to express various policies on who can access the OpenFlow switch tables and on how the author's present scheme accounts for various user requirements, security, including data sharing [7]. These research propose some useful framework to secure flows in SDN architectures. However, these research present classic solutions that have many limitations related to the automation of rules and performance. In addition, they don't propose any solution to deal with zero-days attacks or to reduce false positive and false negative rates.

To fill these gap, we propose in this chapter a secure and efficient data-sharing mechanism based on supervised machine learning techniques in order to detect and stop anomalies that affect communications and flows in SDN. To do so, we provides a secure data sharing framework between SDN nodes and controller while respecting network security requirements.

The contribution of our chapter presented below:

- Generating various manipulation flows (of-switches, Odl controller) to create a large scale SDN dataset,
- Performing various new attacks on SDN network to demonstrate the limitation of classical solution such as firewalls and intrusion detection and prevention systems,
- Designing a secure data-sharing framework based on machine learning model for early detection of attacks attempts such us arp poisoning and man-in-the-middle attacks.

This chapter is organized as follows. Section 2 discusses literature review, describing the security issues in SDN architecture, and used machine learning detection techniques for SDN architectures. Section 3 proposed secure data-sharing framework based on machine learning model. Section 4 illustrates experimental environment and simulation results. Finally, we conclude by conclusion in Sect. 5.

2 Literature Review

2.1 Security Issues in SDN Architecture

Network infrastructures over the world use traditional networks, as they are the established standard with years and years of security and threat mitigation iterations. Thus, it would be extremely challenging to convince service providers and businesses to drop their reliable networking infrastructure and replace it with an SDN technology. In order to encourage companies to deploy and use SDN services, its necessary to perform tests and study on large scale SDN scenarios with unpredictable traffic patterns and security threats [8, 9].

Furthermore, security is consider by many network experts the biggest challenging factor that affect and slow down the adoption of SDN. In fact, SDN inheriting most of the traditional networks vulnerabilities and its brings a new set of vulnerabilities because the layers separation and the centralized control plane. Figure 1 presents the potential attack vectors, the main existing security threats on SDN networks and mitigation approaches extracted from literature.

Being the central point where the routing logic is handled, it has access to information about the whole network, Fig. 1 illustrates and allows the Controller to have a global view of the Network Topology. Network Apps running on top have access to a simplified view of the topology through the use of the abstraction layer, and can make informed routing decisions without prior knowledge of the topology. The SDN

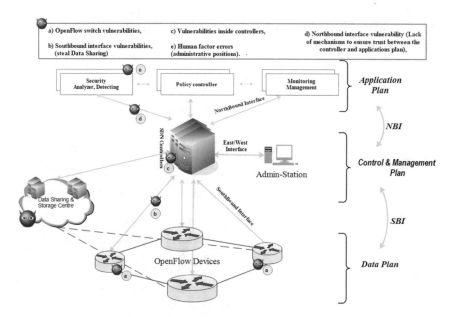

Fig. 1 Global view of software defined networking attacks

controller offers a number of abstraction layers, or Interfaces, which depends on the type of communication on that level. There is a total of four Interfaces known as the Northbound, Southbound, Eastbound and Westbound interfaces [10].

SDN is similar in a lot of ways to traditional networking, and as such, the angles of attack from traditional networking apply to SDN as well. SDN also have additional security improvement and issues compared to traditional networks, due to the nature of the controller and it being the central piece of the topology. Attack vectors on SDN networks fit under one of the following categories shown in Table 1. These examples of security issues are each affecting one or more layers of SDN at a time, but the majority are focused on the Control plane and the Data plane, as well as their underlying SBI interface.

The attacks on SDN networks are categorized by type and depends on the SDN layer/Interface they affect. There are seven main categories used to classify Security Issues and/or attack over SDN networks.

Mitigation for the identified attack vectors on SDN are being tested/documented continuously, but two of the seven attack vectors on SDN have yet to get a proposed solution to this date. Those vectors are the Data Leakage and the Data Modification vectors, while the Unauthorized Access and Configuration Issues are the vectors that were the most active in terms of proposed mitigation solutions [11, 12].

The OpenFlow switch specification describes the use of TLS for mutual authentication between controllers and OF switches [13], but it is not enabled by default. This makes it possible for arp poisoning and Man in the middle attacks that can halt the operations in the network and cause a damage by stilling information specifically data sharing. It is of paramount importance that the communication between the control and data plane is using the proper authentication mechanisms to avoid any added security issues, because the controller is the central piece of the SDN topology. These attacks are a types of attack sample to execute and difficult to detect, when an

Table 1 Attack vectors on SDN networks

Category	Examples of issues
Unauthorized access	Unauthorized controller access
	Unauthorized/unauthenticated application
Data leakage	Flow rule discovery
	Credentials management leak
Data modification	Unauthorized flow rules modification (MitM attack)
Malicious/compromised applications	Fraudulent rule insertion
Denial of service	Controller or switch communication flooding
	Flow table flooding
Configuration issues	Lack of TLS adoption
	Lack of secure provisioning
System-level SDN security	Lack of visibility of network state

attacker secretly relays passively to targets and if necessary changes the connection between parties, who believes that they are directly connected and come danger when attacker use this power actively by infecting malicious packets to targets [14]. Therefore, it is a method of compromising a communication channel in southbound SDN interface SBI which an attacker, having connected to the channel between counter parties' controller and infrastructure layers, intervenes in the transmission protocol, deleting or distorting information [15]. So, Data modification attack part of an SDN controller capabilities is the function to program network devices in order to control network traffic. If an attacker manages to seize control of the controller, they effectively gain control over the whole network, as they can add or modify rules in the tables of the underlying OF switches, shaping the traffic in a way advantageous to them. Indeed, we propose a mitigation for stopping data modification and duplicate packets by using machine learning for early detection of these kind of attack, we propose a framework that combine between firewall and IDS/IPS with a machine learning model to give a quick anomalies alert [16].

2.2 Machine Learning Anomalies Detection for SDN Architecture

Intrusion is an activity that violates the security policy of an information system, hence intrusion detection is based on the assumption that the intruder's behavior will be significantly different from normal behavior, which will ensure the detection of a large number of unauthorized actions. Intrusion detection systems are generally used in conjunction with other security systems, such as access control and authentication, as additional protection for information systems. There are many reasons why intrusion detection is an important part of the overall security system. First, many existing systems and applications have been designed and built without regard to security requirements. Second, computer systems and applications may have flaws or errors in their configuration, which may be used by attackers to attack systems or applications. Thus, the preventive method may not be as effective as expected. Intrusion detection systems can be divided into two classes: signature detection systems and anomaly detection systems. Signature detection systems identify patterns of data traffic or applications that are considered malicious, while anomaly detection systems compare the activity to normal behavior.

The steps presented in the framework Anomaly Detection model will vary depending on the method used. During detection, the system created in the simulation step is compared to the selected parameterized data block. Threshold criteria will be selected to determine abnormal behavior [17, 18]. Machine learning can automatically create the required model based on certain training data. The application of this approach requires the necessary preparation of the data, but this task is less complicated compared to the calculation of the abnormal model [3]. With increasing complexity and the number of different attacks [19], machine learning methods that allow you to create and maintain Anomaly Detection Systems with less human intervention are the only practical approach for creating the next generation of intrusion detection systems. Applying machine learning methods for intrusion detection will automatically create a model based on a training data set that contains data instances described using a set of attributes (functionalities) [20].

The attributes can be of different types. Different algorithms for anomaly detection have been considered, and Table 2 presents the advantages and disadvantages of each of them. Anomaly detection includes both controlled and uncontrolled methods. A comparative analysis has shown that controlled training methods are significantly superior to uncontrolled methods if the test data do not contain unknown attacks. Among the controlled methods, the best performance is obtained with non-linear methods such as SVM, multilayer perception, and rule-based methods. Uncontrolled methods such as SVM and RF, model show better performance than other methods, although they differ in the detection efficiency of all classes of attacks [17, 18].

3 Proposed Framework Based on Machine Learning Techniques to Secure Data Sharing in SDN

In this section, we present a defense framework based on a machine learning technique for securing data-sharing in SDN context. As stated earlier, we focus on data-sharing for analysis nodes purposes. We explain each module of the framework and explain how the framework satisfies detection objectives. A flowchart view of the framework is presented in Fig. 2. We will base ourselves on a set of rules presented as a predefined security policy, and effective pre-processing techniques that lead to attack discovery. By taking into account Man in the middle attacks, arp poisoning and attacks that are based on SDN nodes vulnerabilities.

The Framework is divided into three steps. First the data collection phase and pre-processing based on pre-defined rules, through an evaluation process for attack classification and attribute optimization, after selecting the best features (this metric is evaluated by attack types) for attack detection. The second phase, is the training of our SDN dataset which is characterized by the testing of new enchantments of the packet filtering over time, and temporarily the classification of the anomalies whether it is an attack or not. The third phase is the test using the new enchantments data to validate the efficiency of our model.

The steps of proposed detection framework for SDN networks based on random forest algorithm are as follows:

Step 1—Data Collection Phase: is generated from various manipulation flows normal and abnormal in our SDN environment, after collecting data from SDN nodes using tshark, we pass to pre-processing is divided on two steps cleaning and data treatment. The data cleanup operations is to delete unnecessary information in order to deleting duplicate or erroneous entries. The data treatment operations on the columns

Table 2 Benefits and disadvantage of machine learning techniques

Methods	Benefits	Limitation
K-nearest neighbors (KNN)	• Easy to implement when there are multiple predictors • It is used to create models that process with non-standard data types such as the test	• Great memory needs • Depends on the choice of similarity function that is used to compare instances • The absence of a fundamental method of choice except by cross-validation or similar method • Expensive computer science
Neural network	• A neural network neuron can perform spots that a linear program cannot perform • When an item doesn't handle the task, the method can continue to work due to parallel data processing • A neural network neuron doesn't need to be reprogrammed	• The neural network needs training • It can be implemented in any application • High processing time for large neural networks
Decision tree	• Easy to implement • Requires a bit of data preparation • Ability to process types of digital and other data • Uses a white box model • The ability to test the model using statistical tests • Works with big data in no time	• The problem of learning the optimal decision tree, as you know is NP-complete in several aspects of optimality and even for simple problems • There are stains that cannot be displayed by the decision tree as it does describe it was completely
Random forest	• For a single decision tree to give accurate results • Uses a "committee" of randomly created decision trees with different sets of attributes • Fast-learning mechanism for discovering relationships within a dataset	• When creating a decision tree, the non-optimal and very complex trees that don't process the data well
Support vector machine	• Finding the optimal separation from the hyper plane • Gere a large dimension of data • Generally, works very well	• Needs positive and negative examples • You have to choose a good function of the core • It requires a lot of memory and processor time • There are problems of digital stability when solving the QP constraint

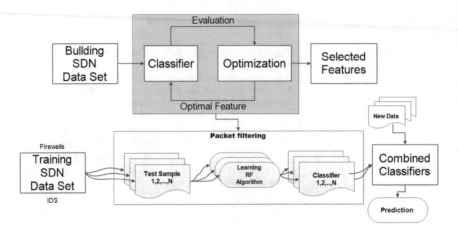

Fig. 2 Anomaly detection steps model framework

in order to prepared dataset, and creating new features based on rules. To do so, data collection is based on the following rules and functionalities:

– **Rule 1**—If the anomaly firewall and IDS detects traffic as normal, the traffic will be normal (true positive and false negative).
– **Rule 2**—If anomaly detection detects an attack and the abuse detection does not detect any attack, it is not an attack, it is rather an incorrect classification.
– **Rule 3**—If anomaly detection detects an attack and the abuse detection detects an attack, it is an attack and determines the attack class.

Step 2—Training Phase: use categorical data for machine classification, by convert a set of features and associated labels to 0 or 1 means normal or abnormal illustrated on Algorithm 1, in order to the random forest model is prepared to answer questions related to feature vectors. Consequently, the choice of our model is based on a controlled learning algorithm used for classification, feature selection and regression. For this, we choose the random forest (RF) model because the construction of a forest is good enough to solve the problem of anomaly detection for large-scale data. Whatever the quality assessment of the algorithm be very accurate and not at the expense of the learning and testing separation. So, the answers give on the learning sample and the importance of the attributes that be evaluated. In each branch of random forest, the tree breaks down the selection into several roots A1, A2, ... for each sample A and T which is divided into several parts is randomly selected with repetitions n observations from n source. The classification tree formula is presented as follows:

$$
\begin{cases}
I_G(Tree) & = \sum_{j=1} P(A_j).I_G(A_j) \\
I_G = 1 - \sum_{i=0} T_i^2 : \forall i \in N
\end{cases}
\tag{1}
$$

$$\begin{cases} I_E(Tree) & = \sum_{j=1} P(A_j).I_E(A_j) \\ \quad I_E = -\sum T_i ln T_i : \forall i \in N^* \end{cases} \tag{2}$$

where $P(A_j)$ sample size A_j divided by the total number of observations, and I_E entropy measure sample.

Step 3—Test Validation Phase: using a new sample from our dataset to test sets for training an analytical model. The RF model is built on the basis of the training set and is tested on the validation set. The procedure is repeated N times, where N is the number of iterations.

Algorithm 1: Algorithm anomaly classification

---

1 **Begin**
2 A_i [] ← NULL   ▷ *Samples dataset*
3 **for** *each data in node$_i$* **do**
4   observe P(A$_i$) where data in node$_i$
5   train_test_split(X=A, y, test_size=0.3) ;
6   RandomForestClassifier()
7   **if** *all A$_i$ == False* **then**
8   Return Assign corresponding class = Normal or Abnormal ▷ Attack classification ;
9   **else**
10   Return $New\_Class to node_i$ ▷ *Suspicious node$_i$*
11 **End**

Finally, our defense framework steps based on random forest model: starting by proceeding to the collection of basic data on the generation of normal and abnormal traffic in SDN context, next selecting random samples from a given set of traffic. Then, this algorithm will build a decision tree for each sample after training. It will then receive the forecast result of each decision tree. At this stage, voting will take place for each projected result by testing set. Finally, select the forecast result with the most votes as the final forecast result.

4 Experimental Environment and Results

4.1 Environment

In this section, we will present our testbed, then discuss the data collection procedure in the SDN topology, in order to analyze and include our security measures based on the machine learning technique, to reduce the impact of potential attacks targeting data sharing.

Table 3 Experiment setup

	Operating system	CPU	RAM	Storage	Network adapter card
Opendaylight Controller	Ubuntu 18.04	2	4	4	1
Pfsense Firewall IDS/IPS	FreeBSD 64-bit	2	1	4	2
OVS/Mininet	Ubuntu	2	2	4	1
Attacker machine	Kali Linux	2	2	4	1
Tools	Ettercap, SSLStrip, dsniff, Tshark				

Special attention must be paid to securing the controller and SDN nodes, as these are the most important and sensitive points that can cause catastrophic damage. We will choose to install our testbed in a virtual machine-based environment to check all configuration options in case we need to modify anything to fit our SDN testbed.

We use a base installation of the Opendaylight controller, OVS Switches in Mininet emulator, and the Pfesense firewall with a snort as IDS/IPS. To generate the different attacks tested in our study using Ettercap, SSLStrip, dsniff tools on the kali-Linux virtual machine illustrated in Table 3.

Figure 3 presents our TestBed architecture and also how to collect data with pfsense firewall and IDS/IPS. Specifically, we will use the pfsense and snort firewall as a checkpoint between the controller and the SDN nodes.

We used pfsense firewall and snort IDS/IPS to filter capacities by blocking access to private IPs coming from outside the network as well as to the IPs of the BOGON network. We have also deployed a Proxy with an authentication system and a private CA, as well as an SSL interception system to block arp poisoning and MitM attack attempts.

The packet filtering features are included in the baseline and are accessible under Firewall at the top, then Rules. Figure 4 illustrates some of the filtering rules we've implemented for our SDN network. Finally, we tested the IPS capabilities by trying a Nmap port scan attack using the Sparta network stress test utility, and using Ettercap and SSLstrip we succeeded to perform a man-in-the-middle and arp poisoning attack to steal data sharing confidentiality and integrity, this presents false positive and true negative alerts in Snort. Therefore, two different processes are used: the generation of normal traffic in the first place, and the creation of several scenarios of attacks such us MitM, arp poisoning. Tshark and firewall, IDS/IPS helps us to establish our SDN dataset. SO, by analyzing our SDN dataset, we will train it using the Random forest model in order to classify normal to abnormal traffic, that test other samples collected to identify arp poisoning and MitM, then predict susceptible and vulnerable SDN nodes. Table 4 presents a detailed description of the features to be used.

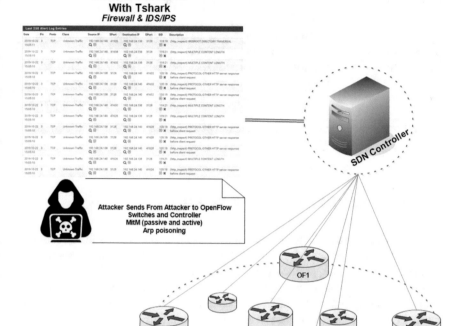

Fig. 3 Testbed architecture and data collection

	States	Protocol	Source		Port	Destination	Port	Gateway	Queue	Schedule	Description	Actions
✖	0 / 0 B	*	Reserved Not assigned by IANA		*	*	*	*	*		Block bogon networks	⚙
✔	1 / 2 KiB	IPv4 *	*		*	*	*	*	none			⬇ ✎ 🗐 ⊘ 🗑

Rules (Drag to Change Order)

	States	Protocol	Source	Port	Destination	Port	Gateway	Queue	Schedule	Description	Actions
✔	1 / 2.06 MiB	*	*	*	LAN Address	443 80	*	*			Anti-Lockout Rule	⚙
✔	1 / 8 KiB	IPv4 *	LAN net	*	*	*	*	none			Default allow LAN to any rule	⬇ ✎ 🗐 ⊘ 🗑
✔	0 / 0 B	IPv6 *	LAN net	*	*	*	*	none			Default allow LAN IPv6 to any rule	⬇ ✎ 🗐 ⊘ 🗑

Fig. 4 Filtering SDN rules in place on the pfSense firewall

Table 4 Attributes listing of the filtering rules and functionalities

Attribute based rules	Description
Private IP block	Blocked packets incoming to the network with private IP addresses
Bogon Networks	Blocked the packets with IPs not assigned by the IANA
Force HTTPS	Enabled a setting on the firewall that makes it only accept HTTPS requests
Proxy setup	Set a proxy server with accounts and an internal CA for validating select websites
SSL intercept	Set up a system within firewall to intercept SSL MitM attacks on the network

4.2 Implementation Framework Results

Packet filtering is "a technique used to control network access by monitoring outgoing and incoming packets and allowing them to pass or halt based on the source and destination IP addresses, protocols and/or ports" [21, 22]. This function is generally performed by a firewall, Intrusion Detection Systems and enables the possibility of blocking potentially harmful traffic from entering our network. It is crucial to have as many deployed security measures as possible in order to minimize the risk of an attack that could damage or alter the way our SDN network operates.

After a successful configuration of the SDN testbed with the Firewalls, an analysis is performed using the machine learning model to detect and predict an arp poisoning and man-in-the-middle attacks (Fig. 5).

Machine learning strategies use certain metrics to calculate the binary classification problem. The measures are listed from the matrix of confusion.

Figure 6 present the easiest way to measure the performance of a classification task when the output can be two or more types of classes. The MC is a table with two dimensions, namely. "Actual class" and "Predictable,". Therefore, this table filled with True Positive (TP), True Negative (TN), False Positive (FP), False Negative (FN), as shown below.

The explanations of the terms associated with the confusion matrix are as follows:

- **True Positives (TP)** is the case when the actual class and the predicted class of the data point are 1.
- **True Negatives (TN)** is the case when the actual class and the predicted class of the data point are 0.
- **False positives (FP)** is the case when the actual data point class is 0 and the predicted data point class is 1.
- **False Negatives (FN)** is the case when the actual data point class is 1 and the predicted data point class is 0.

The accuracy metric presented on classification report table (Table 5). Probably the simplest and most intuitive measure of classifier performance. We a count of the number of times we predicted the correct class from the total number of predictions

Last 250 Alert Log Entries									
Date	Pri	Proto	Class	Source IP	SPort	Destination IP	DPort	SID	Description
2019-10-22 15:05:11	3	TCP	Unknown Traffic	192.168.24.140 Q ⊞	41926	192.168.24.138 Q ⊞	3128	119:18 ⊞ ✕	(http_inspect) WEBROOT DIRECTORY TRAVERSAL
2019-10-22 15:05:10	3	TCP	Unknown Traffic	192.168.24.140 Q ⊞	41634	192.168.24.138 Q ⊞	3128	119:21 ⊞ ✕	(http_inspect) MULTIPLE CONTENT LENGTH
2019-10-22 15:05:10	3	TCP	Unknown Traffic	192.168.24.140 Q ⊞	41632	192.168.24.138 Q ⊞	3128	119:21 ⊞ ✕	(http_inspect) MULTIPLE CONTENT LENGTH
2019-10-22 15:05:10	3	TCP	Unknown Traffic	192.168.24.138 Q ⊞	3128	192.168.24.140 Q ⊞	41632	120:18 ⊞ ✕	(http_inspect) PROTOCOL-OTHER HTTP server response before client request
2019-10-22 15:05:10	3	TCP	Unknown Traffic	192.168.24.138 Q ⊞	3128	192.168.24.140 Q ⊞	41632	120:18 ⊞ ✕	(http_inspect) PROTOCOL-OTHER HTTP server response before client request
2019-10-22 15:05:10	3	TCP	Unknown Traffic	192.168.24.138 Q ⊞	3128	192.168.24.140 Q ⊞	41632	120:18 ⊞ ✕	(http_inspect) PROTOCOL-OTHER HTTP server response before client request
2019-10-22 15:05:10	3	TCP	Unknown Traffic	192.168.24.140 Q ⊞	41630	192.168.24.138 Q ⊞	3128	119:21 ⊞ ✕	(http_inspect) MULTIPLE CONTENT LENGTH
2019-10-22 15:05:10	3	TCP	Unknown Traffic	192.168.24.140 Q ⊞	41628	192.168.24.138 Q ⊞	3128	119:21 ⊞ ✕	(http_inspect) MULTIPLE CONTENT LENGTH
2019-10-22 15:05:10	3	TCP	Unknown Traffic	192.168.24.138 Q ⊞	3128	192.168.24.140 Q ⊞	41628	120:18 ⊞ ✕	(http_inspect) PROTOCOL-OTHER HTTP server response before client request
2019-10-22 15:05:10	3	TCP	Unknown Traffic	192.168.24.138 Q ⊞	3128	192.168.24.140 Q ⊞	41628	120:18 ⊞ ✕	(http_inspect) PROTOCOL-OTHER HTTP server response before client request
2019-10-22 15:05:10	3	TCP	Unknown Traffic	192.168.24.138 Q ⊞	3128	192.168.24.140 Q ⊞	41628	120:18 ⊞ ✕	(http_inspect) PROTOCOL-OTHER HTTP server response before client request
2019-10-22 15:05:10	3	TCP	Unknown Traffic	192.168.24.140 Q ⊞	41626	192.168.24.138 Q ⊞	3128	119:21 ⊞ ✕	(http_inspect) MULTIPLE CONTENT LENGTH
2019-10-22 15:05:10	3	TCP	Unknown Traffic	192.168.24.138 Q ⊞	3128	192.168.24.140 Q ⊞	41624	120:18 ⊞ ✕	(http_inspect) PROTOCOL-OTHER HTTP server response before client request

Fig. 5 Resulting alerts from the previous Nmap scan

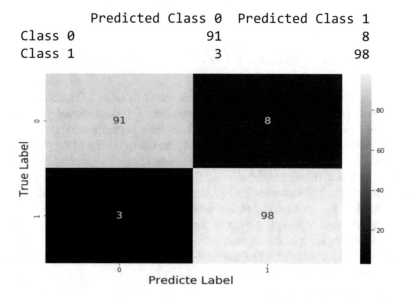

	Predicted Class 0	Predicted Class 1
Class 0	91	8
Class 1	3	98

Fig. 6 Confusion matrix

Table 5 Classification report

	Precision	Recall	F1-score
Class 0	0.97	0.92	0.94
Class 1	0.92	0.97	0.95
Accuracy			0.94
Macro avg	0.95	0.94	0.94

(TNP), and Recall present the percentage of true positive cases that have been classified, per to percentage of positive classifications that are truly positive (Precision) to detect MitM frauds, and the F1 score is the harmonic mean of accuracy and recall, these metrics are calculated using the following formulas:

$$Accuracy = \frac{TP + TN}{TNP} \tag{3}$$

$$Recall = \frac{TP}{TP + FN} \tag{4}$$

$$Precision = \frac{TP}{TP + FP} \tag{5}$$

$$F1\text{-}score = \frac{2 * Precision * Recall}{Precision + Recall} \tag{6}$$

the main results of the random forest model are shown in Table 5. It reports that those have a good percentage of positive classification on each class normal presented by class 0 or abnormal presented by the class 1, so the examination RF mode base on 80% of training and 20% of testing. Therefore, the interpretation of the performance results shows a good and efficient classification attending 94% accuracy, that present a quick prediction (alert) and detection of the attacks. In the same way, the precision, recall (Fig. 7), and F1-score analysis present's respectively 95%, 94%, and 94%.

To validate and confirm our classification results, we use the cross-validation 5 fold method, these technique permit to the re-sampling procedure used to evaluate our model on the limit of the dataset, which leads to a less biased or less optimistic assessment of the quality of the RF model than other methods such as training/testing. The cross-validation work as follows:

1. Shuffle the dataset randomly
2. Divide the dataset into five groups
3. For each unique sample:

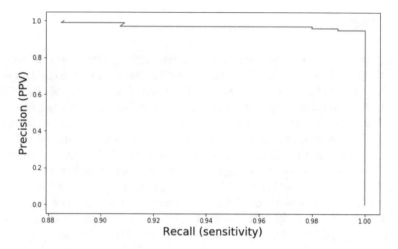

Fig. 7 Precision-recall curve

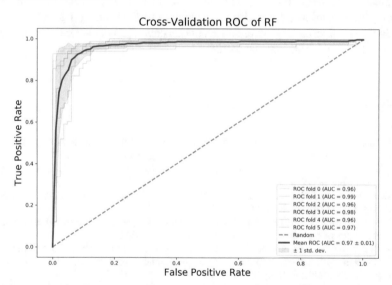

Fig. 8 ROC cross validation curve

- Take a group as a dataset test
- Take the rest of the groups as sample training data
- Prepare a model on training samples and evaluate it on a test sample
- Keep the model score and drop the model.

Figure 8 illustrates the cross validation ROC curve, the results show that the early detection of the attacks on SDN structure is proved successfully by using random forest machine learning technique that conduct by using this framework we contribute to secure effectively the SBI interface and the control data layers.

5 Conclusion

In this paper, we present a secure and efficient data-sharing mechanism based on the supervised machine learning technique. Under the premise of ensuring data security, the proposed framework aims to classify abnormal nodes functionality in SDN context. Including the initialization phase by building and managing data collection using label data. To do so, security stress-test by analyzing several arp poisoning and MitM attack scenarios to enrich our port data set using, ettercap, SSLstrip, and Sparta software. In this mitigation, we detect and stop SDN nodes anomalies to improve secure data-sharing. Consequently, the performance advantages of the proposed framework in terms of time and cost show a very good efficiency classification with an accuracy of 94% and can effectively improve the efficiency of data sharing.

Future work and perspectives would imply the implementation of the studied topology on physical switches and separate peripherals for the hosting of controllers in order to have a better idea of the performance of SDN topologies under realistic workloads. Another interesting perspective would be to implement a hybrid architecture, the transition from traditional networks to SDN being a gigantic task, it would also familiarize us with the different translation protocols for communication between SDN and traditional networks.

References

1. Sebbar A et al (2019) New context-based node acceptance CBNA framework for MitM detection in SDN architecture. Procedia Comput Sci 160:825–830
2. Alsmadi I (2016) The integration of access control levels based on SDN. Int J High Perform Comput Netw 9(4):281–290
3. Benzekki K, El Fergougui A, Elalaoui AE (2016) Software-defined networking (SDN): a survey. Secur Commun Netw 9(18):5803–5833
4. Ali AF, Bhaya WS (2019) Software Defined Network (SDN) security against address resolution protocol poisoning attack. J Comput Theor Nanosci 16(3):956–963
5. Zkik K et al (2019) An efficient modular security plane AM-SecP for hybrid distributed SDN. In: 2019 international conference on wireless and mobile computing, networking and communications (WiMob). IEEE, pp 354–359
6. Zhang X et al (2017) Flow entry sharing in protection design for software defined networks. In: GLOBECOM 2017—2017 IEEE global communications conference. IEEE, pp 1–7
7. Klaedtke F et al (2015) Towards an access control scheme for accessing flows in SDN. In: Proceedings of the 2015 1st IEEE conference on network softwarization (NetSoft). IEEE, pp 1–6
8. Dacier MC et al (2017) Network attack detection and defense—security challenges and opportunities of software-defined networking. Dagstuhl Rep 6:1–28
9. Kreutz D et al (2014) Software-defined networking: a comprehensive survey. arXiv preprint arXiv:1406.0440
10. Open Networking Foundation (2017) SDN definition. https://www.opennetworking.org/sdn-definition/. Accessed June 2017
11. Hong S et al (2015) Poisoning network visibility in software-defined networks: new attacks and countermeasures. NDSS 15:8–11

12. Lu Z et al (2017) The best defense strategy against session hijacking using security game in SDN. In: 2017 IEEE 19th international conference on high performance computing and communications; IEEE 15th international conference on smart city; IEEE 3rd international conference on data science and systems (HPCC/SmartCity/DSS). IEEE, pp 419–426
13. Sebbar A et al (2019) Using advanced detection and prevention technique to mitigate threats in SDN architecture. In: 15th international wireless communications & mobile computing conference (IWCMC). IEEE, pp 90–95
14. Brooks M, Yang B (2015) A Man-in-the-Middle attack against OpenDayLight SDN controller. In: Proceedings of the 4th annual ACM conference on research in information technology. ACM, pp 45–49
15. Sebbar A et al (2018) Detection MITM attack in multi-SDN controller. In: IEEE 5th international congress on information science and technology (CiSt). IEEE, pp 583–587
16. Sebbar A et al (2020) MitM detection and defense mechanism CBNA-RF based on machine learning for large-scale SDN context. J Ambient Intell Hum Comput
17. Ahmed T, Oreshkin B, Coates M (2007) Machine learning approaches to network anomaly detection. In: Proceedings of the 2nd USENIX workshop on tackling computer systems problems with machine learning techniques. USENIX Association, pp 1–6
18. Sultana N et al (2019) Survey on SDN based network intrusion detection system using machine learning approaches. Peer-to-Peer Netw Appl 12(2):493–501
19. Yan Q, Gong Q, Deng FA (2016) Detection of DDoS attacks against wireless SDN controllers based on the fuzzy synthetic evaluation decision-making model. Ad Hoc Sens Wirel Netw 33(1–4):275–299
20. Belhadi A et al (2020) The integrated effect of Big Data Analytics, Lean Six Sigma and Green Manufacturing on the environmental performance of manufacturing companies: the case of North Africa. J Clean Prod 252:119903
21. Tu H et al (2014) A scalable flow rule translation implementation for software defined security. In: Network operations and management symposium (APNOMS), 2014 16th Asia-Pacific. IEEE, pp 1–5
22. Anonyme (2017) What is packet filtering? Definition from techopedia. https://www.techopedia.com/definition/4038/packet-filtering. Accessed June 2019

MSDN-GKM: Software Defined Networks Based Solution for Multicast Transmission with Group Key Management

Youssef Baddi, Sebbar Anass, Karim Zkik, Yassine Maleh,
Boulmalf Mohammed, and Ech-Cherif El Kettani Mohamed Dafir

Abstract Multicast communication is an important requirement to support many types of applications, such as, IPTV, videoconferencing, group games. Recently this multicast applications type emerges fast, in one side more application provider proposed many applications, in other side Internet research community has proposed many different multicast routing protocols to support efficient multicast application. Therefore, the necessity of secure mechanism to provide the confidentiality and privacy of communications are more and more insistent. In current standardized IP multicast architecture, any host can join a multicast group, as source or receiver, without authentication, because no host identification information is maintained by routers, this situation leads clearly to many security risks issues. For security enhancement in multicast communication, in this paper an SDN based multicast solution with Group Key Management (GKM) approach was introduced. Our proposal solution, MSDN-GKM, includes many SDN modules to support multicast functions, group key generation, Group key exchange, storage, use, and keys replacement if any multicast group membership occurs. To prove the efficiency of our proposal solution a prototype is implemented in our SDN platform. The test-bed result proves that our proposal solution is greater to the traditional IP multicast proposed in the literature, which is reflected in two aspects: firstly, multicast metrics performance in terms of end-to-end delay, tree construction delay and delay variation. Secondly, the multicast group key management performance in terms of storage overhead and time processing.

Y. Baddi (✉)
ESTSB-Chouaib Doukkali, STIC, El Jadida, Morocco
e-mail: baddi.y@ucd.ac.ma

S. Anass · K. Zkik · B. Mohammed
Université Internationale de Rabat, TICLAB, Sala Al Jadida, Morocco

S. Anass · E.-C. El Kettani Mohamed Dafir
ENSIAS, Mohammed V University, ESIN, Rabat, Morocco

Y. Maleh
LaSTI, University Sultan Moulay Slimane, Beni Mellal, Morocco

© The Editor(s) (if applicable) and The Author(s), under exclusive license to Springer 373
Nature Switzerland AG 2021
Y. Maleh et al. (eds.), *Machine Intelligence and Big Data Analytics for Cybersecurity Applications*, Studies in Computational Intelligence 919,
https://doi.org/10.1007/978-3-030-57024-8_17

1   Introduction

The rapid propagation and development of the Internet, 80% of which is high-speed multimedia applications such as video conferencing, audio and multimedia. However, the traditional unicast and broadcast mode of communication is not optimal for these applications. This poses a huge and rapid consumption of network bandwidth. In particular, multicasting allows a sender of one or more multicast sources to send a message to a plurality of receivers in the network.

Deering [18], proposes in his research thesis addressing IP multicast transmission, reveals the attribute of data duplication to the core network instead of the border nodes. The source of the IP multicast transmission and the receivers and the lower network between the two multicast media must follow a process: (1) the host takes the sending and receiving of IP multicast by establishing a TCP/IP connection. (2) The IGMP protocol (v1, v2) takes group management to join, leave and query. (3) IP address allocation and mapping the IP multicast address (layer 3) to the MAC address (layer 2). At this point, the main purpose of the multicast communication mode is to provide data to a set of selected receivers in an efficient way: the application in the sending node must send a single copy of each packet and address it to the group of computers involved; the central network takes care of duplicating the messages to the receivers in the group. As a result, IP multicasting is considered a green technology known as bandwidth conservation technology, IP multicasting by nature reduces the total number of flooded packets in a network and, by the way, bandwidth consumption.

Software Defined Networking (SDN) is a centralized technology that allows network nodes to be monitored and managed. This new technology configuration under development offers some interesting technical, scalability, adaptability and agility features, making it easier to support traditional networks, Internet services and applications. SDN is represented by the separation of infrastructure and control planes, by a few explicit network nodes (controllers, OpenFlow switches) and by a few protocols (OpenFlow, ForCES, OVSDB, BGP) for information flow and card information. The SDN design consists of two parts: the SDN switches and the SDN controller. The SDN switches are only responsible for the data transmission and the SDN controller is responsible for the deployment of the flow management rules. This centralized architecture will then allow us to have an overview of the entire network and the calculation of the multicast trees will be done once and centrally at the control plane level.

In this paper, we propose an SDN multicast-based solution with group key management scheme to secure multicast communication in an SDN context. In fact, the proposed multicast SDN Controller is responsible for routing, multicast tree computing, handling joining and leaving events, user authentication, and many multicast group key management functions. The proposed SDN multicast controller includes five modules: Group Key Management, the Group Management Module, the Multicast Signalization Message Dispatcher Module, The Multicast Tree Management Module, the Multicast Tree Computing Module and the Multicast Member Management Module.

This chapter is organized as follows. Section 2 describes the background and start of art, including multicast IP technology, Multicast routing protocols, an overview of multicast Trees, Group Key Management, SDN technology, in this section we introduce existing solutions and implementation of group key management. In Sect. 3 we present and detail our proposed solution. Section 4 reports and discussed the network topology and parameters used in the simulation, also the results of the simulation study. Finally, Sect. 5 provides concluding remarks.

2 Related Works and Research Scopes

2.1 Multicast IP

The IP multicast paradigm and protocols are standardized by the Internet Engineering Task Force (IETF) under the Network Working Group. First proposed by Deering [19] in 1991 as a thesis project, the multicast IP paradigm is designed as a technique which support Group based Application communication. IP multicast is emerging to be the most used vehicle of delivery for multimedia and group-based applications on the Internet, with the guarantee of reaching the millions of users on the Internet.

The main architectural component on this paradigm is the multicast routing protocol that delivers multicast data packets (data stream) to the group members exclusively, following the basic IP multicast model proposed in [18]. Multicast IP, in the third layer of TCP/IP stack, is a routing approach to ensure one-to-many and many-to-many communication. The Multicast IP routing protocol duplicates IP multicast packets at routers level and forward them to the intended receivers. Multicasting expects to deliver data to a set of selected receivers in an effective manner: application sender acting as a multicast server needs to send only one single copy of each packet and address it to an identified multicast group of involved computers, acting as receivers; the network deals with message's duplication to the receivers of the group. Thusly, IP Multicast abstains from handling overheads related with replication at the source and spares the system data transmission.

The fundamental task of a multicast routing protocol in a multicast overflow is building a logical optimal multicast tree under the network topology, which all multicast sessions and packets will pass to reach all multicast receivers and execute the multicast packet replication operation, this problem of building a multicast tree is known by the minimum Steiner tree problem MST [33]; this issue is a proved NP-hard problem in many works in the literature [15, 28, 43, 56], since it tries to find a low-cost tree spanning all multicast groups at once, including all sessions, receivers and sources, by minimizing the multicast tree cost, the transmission delay, and the delay variation between group receivers, which needs using a heuristic algorithm. Multicast routing protocols are divided in two categories (SBT and ST).

Source-based Tree SBT is an intersection of the all shortest paths between the multicast source and each multicast receivers of the multicast group [10]. The main motivations behind the use of the Source-based Tree SBT are the simplicity of building in a distributed manner using only the unicast routing information [28, 55], which help in the optimization of transmission delay between the source and each multicast receiver. However, the Source-based Tree SBT needs more additional costs to maintains the tree, otherwise the stats to be stored in each intermediates notes is very high, with a complexity equal to $O(S*G)$, where S and G are sources set size and the number of groups set in the topology respectively [24, 28]. Source Based Tree SBT is used by several standardized multicast routing protocols by IETF, such as DVMRP [52], MOSPF [35], PIM-DM [21].

Shared Tree, or Core-Based Tree depend on the used protocol, can be constructed using a shared RP tree: It requires the selection of a central router called "Rendezvous point" RP in the PIM-SM [22] protocol and "Core" in Core-Based Tree [9] protocol. Shared tree is more appropriate when there are many multicast sources in specific multicast group. Under this approach, the global tree is composed by two separates parts by one selected node: sources tree and receivers' tree. One node in the network is selected as the center, and all sources of all multicast groups forward messages to the selected center node [20]. As the SBT tree, a shortest path multicast tree is constructed rooted at the selected center node, between all sources and this center node. In addition, a shared multicast tree routed also at the selected center node is build spanning all multicast receivers. With this architecture, only routers on the logical shared tree need to maintain information related to group members. SBT proves good performance in terms of the amount of state information to be stored in the routers and the whole cost of the multicast routing tree [21, 35, 52]. Joining and leaving a group member operation are achieves explicitly in a hop-to-hop way along the shortest path from the LAN router directly connected to the receiver node to the selected core router resulting in less control over-head, efficient management of multicast path created in changing group memberships, scalability and performance [9, 22]. Source Based Tree SBT is used by several standardized multicast routing protocols by IETF, such as Protocol Independent Multicasting-Sparse mode PIM-SM [22] and Core-Based Tree (CBT) [9].

2.2 Group Key Management

Many researches work to secure multicast communications are conducted in last years, many group key management protocols and architectures have been proposed in the literature to address the security issue in multicast group communication [25]. Many survey papers studies are published, to cite this group key management protocols and architectures [2, 11, 32, 40, 48]. Almost of these surveys cites and classifies group key management protocols and architectures in traditional IP network (wired and wireless). Traditional group key management protocols are generally classified into centralized, decentralized and distributed protocols or architectures.

In a centralized system such as in [1, 17, 31, 38, 54] a single designated entity, called group manager, is employed for controlling the whole group and it does not have to rely on any auxiliary entity to perform key distribution. Centralized key distribution uses a dedicated key server, responsible for computation and distribution of the TEK to all multicast members, which resulting in simpler protocols. However, centralized methods with only one managing entity fail entirely once the server is failed; the central server is a single point of failure and the attack target. In the remainder of this section, we summarize important researches in centralized group key management.

We start by One of the first algorithms, the Naïve group Key Management Scheme with a centralized group controller GC, which it shares a secret key with all multicast group members. The Naïve Group Key Management scheme works as follows.

First the scheme proposes using two keys: one key for each group member and one group key. Each time a new member joins the multicast group, the group controller generates a new group key, encrypts the new key with the old key, and then sent using the multicast tree an update to each existing group member. This new group key is then sending, by the group controller, the new key to the new joining member via the secure channel between the joining member and the group controller. In the leaving case, whenever a member leaves the multicast group, the group controller generates a new group key and sends with unicast mode the key to each of the remaining members one by one.

To perform this scheme, each group member stores two keys (a shared key with the controller and the group key), and the controller stores $n + 1$ keys (one key for each client, and the group key).

The head advantages of this scheme are in its simple process without any complexity, easy to implement and does not require any specialized, underlying infrastructure, for example, in PIM-SM [22], the Rendezvous point can be a good group controller agent. However, this scheme scales poorly in terms of both group size and group dynamics. All operations are executed by one agent, the group controller is naturally a high single point of failure. It represents also a performance bottleneck in situations where the group controller also performs the task of rekeying the keys on membership change. The entire group would be affected if the security of the group controller is compromised [11].

Pairwise Keys or N Root/Leaf pairwise Keys approach is a brute force method, first proposed by Wallner [54] in RFC 2627. The pairwise Keys approach works as follows: The approach uses a new entity as a group initiator, named Group Controller (GC), which attributes and distributes a separate secret key to each group member. Wallner [53] called this key as a Key Encryption Key (KEK) as it's used to crypt the group key used to encrypt multicast data, this group key also named as Traffic Encryption Key (TEK). The KEK secret key is used to establish a unicast secure channel between the GC and each member.

As presented in Fig. 1, we can distinct two membership event: member group join (a) and leave (b). When a member leaves the group, the GC generates a new TEK and sends it to each residual member via the secure channel. When a member leaves the group, the Group Controller GC creates a new TEK and sends it the encrypted with KEKs to remaining members.

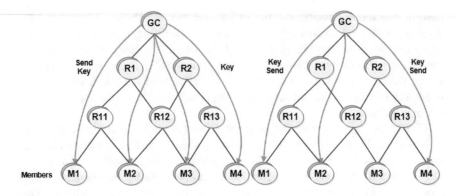

Fig. 1 pairwise keys

Harney and Muckenhirn proposed the Group Key Management Protocol (GKMP) standardized in RFC 2093 [27] and then in RFC 2094 [26]. GKMP protocol uses a entity called key server (KS), responsible to generate a Group Key Packet (GKP) that contains two keys: a Group TEK (GTEK) and a Group KEK (GKEK).

The GKEK is used to secure the distribution of the new GKP, and each time required the data traffic is encrypted by the GTEK. The protocol starts, when a first group member join the multicast session, this first new member helps the KS to create a Group Key Packet (GKP) that contains a group traffic encryption key (GTEK) and a group key encryption key (GKEK).

Each time a new member joins the session, the key server KS generates a new GKP, which contains a new GTEK to assure back-ward secrecy, and sends it securely to the new member encrypted with the KEK established with this new member, and sends it to the other members encrypted with the old GTEK. The key server refreshes the GKP periodically and uses the GKEK for its forwarding to the group members. When a member leaves the group, the key server SK generates a new GKP and sends it to each remaining member encrypted with the KEK that it shares with each member.

As you can conclude, GKMP requires O(n) re-key messages for each leave from the group. Consequently, this solution does not scale to large groups with highly dynamic members.

Presented in many surveys as one of the most widely used centralized group key management schemes, Logical Key Hierarchy (LKH) protocol proposed at same time by Wong et al. [54] and Wallner et al. [1, 17, 31, 38]. In the LKH scheme, a hierarchy of keys is used to reduce the required number of TEK update messages induced by re-keying after membership changes to the order of $log(n)$.

Wallner et al. [1, 17, 31, 38, 54] introduced two type of group key: Traffic Encryption Key (TEK) and Key Encryption Key (KEK). The first one is a symmetric key, TEK is used to encrypt and decrypt the multicast data and Key Encryption Key (KEK) to encrypt the group key TEK. In this approach, Wallner et al. [15] also introduce a server called key distribution center (KDC) responsible to maintain a tree of keys, the leaves of the tree correspond to group members and each leaf holds a KEK associated with that one member. Each group member receives and maintains a copy of the KEK associated with its leaf and the KEKs corresponding to each node in the path from its parent leaf to the root of the logical tree.

To improve the scalability of the centralized approach and to minimize the problem of concentrating the work in a single node and area, many protocols are proposed to divide key management process of a large group among subgroup managers. In a Decentralized scheme such as in [8, 16, 37, 39, 51], the large group is divided into several small subgroups, different controllers are used to manage each subgroup. This approach success to minimize the problem of concentrating the work on a single node and then reduces rekeying overheads while providing scalability. The key distribution function is shared on a set of sub-controllers or sub-agents who are responsible for managing the keys within the affected subgroup. In the remainder of this section, we summarize important researches in decentralized group key management (Fig. 2).

In a Distributed scheme such as in [10, 12–14, 30, 42, 50], the security mechanisms are distributed across multiple entities previously authenticated. There is no explicit group key manager GKM, and the members themselves cooperate to establish a group key and do the key generation. Such schemes improve the reliability of the overall system and reduces the bottlenecks in the network in comparison to the centralized approach. However, they create new faces of synchronization and conflict resolution.

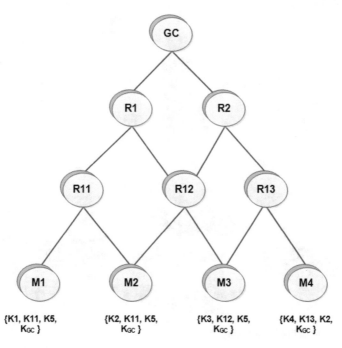

Fig. 2 Logical key tree protocol

2.3   Multicast and Software-Defined Networking SDN Integration

It is difficult to monitor traffic links and make a global management on routing to accommodate new group members and multicast groups.

Software-defined networking (SDN) is a centralized network control and management technology that offers programmability and flexibility. For this, the SDN architecture separates the control plane from the data plane and unifies control into a single external control software called "Controller" [10], which can manage the entire network using several programmable services and APIs, many solution-based on SDN technology are proposed in the literature to solve many difficult issues in traditional networks [29, 44–47, 57].

As shown in Fig. 3, the SDN architecture has divided the network into three layers: application layer, control layer and infrastructure layer. As a result, Technical SDN provides new power for multicast tree conditioning, which helps monitor traffic and perform overall management and adjustment on routing to accommodate new group members and multicast groups.

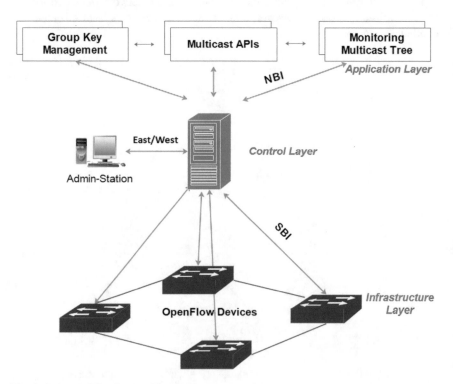

Fig. 3 Software defined networking architecture

SDN can transport many simultaneous multicast sessions. Considering each multicast session in isolation can cause congestion on some links and reduce network usage. For this, the optimized sharing of resources of SDN nodes and the links between several coexisting multicast trees, so the network tries to support all multicast groups by optimizing the use of resources.

3 Proposal Solution

In this section, the detailed architecture of the proposed secure group key management models based in Software-defined Network are described. Our proposal model is based in PIM-SM protocol, we use a set of PIM signaling messages [23], also many controller modules acts as a multicast PIM router [23].

Any proposed multicast SDN Controller with a key management scheme enabled must satisfy the forward and backward secrecy requirement:

- Forward secrecy: The multicast key used to crypt the multicast data and signaling messages must be changed to ensure that a departing member cannot decrypt data transmissions after he/she has left the multicast group.
- Backward secrecy: The multicast key must be changed to ensure that a new member cannot decrypt data transmitted before he/she joined the multicast group.

The proposed scheme also guaranties that if a specific group member node can't use its access information, the key, to deduce another group member's access information.

3.1 General Architecture

In Fig. 4 we present our proposal architecture as a general overview of the SDN multicast solution with a group key management schema.

One or multiple multicast sources sends multicast packet to directly connected OpenFlow switch, generally because of using PIM-SM protocol, the multicast data is sent in a PIM-register packet, encapsulate in a Packet-In message by the OpenFlow Switch, if a related multicast is already existing, the multicast packets are forwarded to existing receivers, which are waiting to receive these multicast packets. If the multicast tree is not existing because of the empty set of receivers the multicast SDN controller notify the OpenFlow switch directly connected to every multicast source to stop sending multicast packets.

On other side, to receive this multicast packets of a specific multicast group identified by a multicast address, the receivers send Internet group management proto-col (IGMP) report message in Ipv4 or MLD in IPv6 to the directly connected OpenFlow switch. Then, the OpenFlow switch which received IGMP or MLD message sends

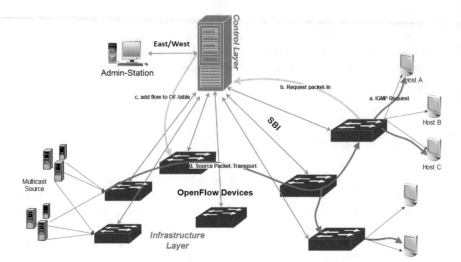

Fig. 4 Multicast scenarios based on SDN context

Packet-In message to the SDN controller so that controller knows the multicast group join and the location of receiver.

Our multicast SDN controller use the basic Topology and discovery modules to identify the status of the network configuration, and to keep track of links between the SDN controllers to acquire topology information through link layer discovery protocol (LLDP), this information is used by the multicast SDN controller to build the graph for calculating an optimal multicast tree, selection of optimal core OpenFlow Switch and provide an in stored record of links currently in the networks.

3.2 Multicast Tree Computing Mathematic Modeling

The main goal of our design is to propose an algorithm which produces multicast trees with low cost, multicast delay, multicast delay variation, and number of multicast key to save.

Along this entire article the network is modeled as a simple, bidirectional, and connected graph $G = (N, E)$, where N is the set of nodes and E is the set of edges (or links). The nodes represent the OpenFlow Switch and the edges represent the network communication links connecting the OpenFlow Switches.

Let $|N|$ be the number of networks OpenFlow Switches and $|E|$ the number of network links. An edge $e \in E$ connecting two adjacent OpenFlow Switch $u \in N$ and $v \in N$ will be denoted by $e(u, v)$, the fact that the graph is directional, implies the existence of a link $e(v, u)$ between v and u. Each edge is associated with two positive real value: a cost function $C(e) = C(e(u, v))$ represents link utilization (may be either monetary cost or any measure of resource utilization), and a delay function $D(e) = D(e(u, v))$ represents the delay that the packet experiences through passing that link including switching, queuing, transmission and propagation delays. We associate for each path $P(v_0, v_n) = (e(v_0, v_1), e(v_1, v_2), ..., e(v_{n-1}, v_n))$ in the network two metrics:

$$C(P(v_0, v_n)) = \sum_{0}^{n-1} C(P(v_i, v_{i+1})) \tag{1}$$

$$D(P(v_0, v_n)) = \sum_{0}^{n-1} C(P(v_i, v_{i+1})) \tag{2}$$

Our SDN multicast proposition is based in a Shared multicast Tree ST model, more specifically Shared Tree with Rendezvous Point node. In this model, a multicast tree $T_M(S, RP, D)$ is a sub-graph of G spanning the set of sources node $S \subset N$ and the set of destination nodes $D \subset N$ with a selected Rendezvous Point RP. Let $| S |$ be the number of multicast destination nodes and $| D |$ is the number of multicast destination nodes.

Practically, all sources node needs to transmit the multicast information to selected Rendezvous Point RP via unicast routing, then it's will be forwarded to all receptors in the shared tree, to model the existence of these two parts separated by Rendezvous RP, we use both cost function and delay following:

$$C(T_M(S, RP, D)) = \sum_{s \in S} C(s, RP)) + \sum_{d \in D} C(P(RP, d)) \tag{3}$$

and

$$D(T_M(S, RP, D)) = \sum_{s \in S} D(s, RP)) + \sum_{d \in D} D(P(RP, d)) \tag{4}$$

Because of the concurrency nature of multicast application, the delay variation is an importance metric to be optimized, the Delay Variation (7) function is defined as the difference between the Maximum (5) and minimum (6) transmission delays along the multicast tree from the sources to all destination nodes and is calculated as follows:

$$MAX_D(T_M(S, RP, D)) = \max_{s \in S, d \in D} D(s, RP, d) \tag{5}$$

$$MIN_D(T_M(S, RP, D)) = \min_{s \in S, d \in D} D(s, RP, d) \tag{6}$$

$$DV(T_M(S, RP, D)) = MAX_D(T_M(S, RP, D)) - MIN_D(T_M(S, RP, D)) \tag{7}$$

In our solution to optimize the number of group keys generate by the SDN multicast controller for each OpenFlow Switch, we denote Min_GK as the minimum number of the Group Key to save in each OpenFlow Switch. The general problem is modeled as to fin the optimal multicast tree in the network with an optimal function Opt_F as follow:

$$Opt\_F(RP, T_M) = \begin{cases} \min C(T_M(S, RP, D)) \\ D(T_M(S, RP, D)) < \alpha \\ DV(T_M(S, RP, D)) < \beta \\ Min\_GK < \gamma \end{cases} \tag{8}$$

3.3 Controller SDN

The multicast controller is the main component in our work and considered in the network architecture the core of multicast network and responsible for any multicast procedures, including routing, multicast tree computing and building, joining and leaving events management, user authentication and multicast group management, and group key management.

In this section we present a design and implementation of the proposed controller as a multicast controller and the module related to the group key management functions.

As shown in Fig. 5, our controller design involves of six modules: multicast signalization message dispatcher, multicast tree computing, multicast member management, multicast tree Management, multicast group key generation, and multicast group key management.

The multicast member management module (1) is divided in two sub-modules: multicast sources management sub-module and multicast receiver management sub-

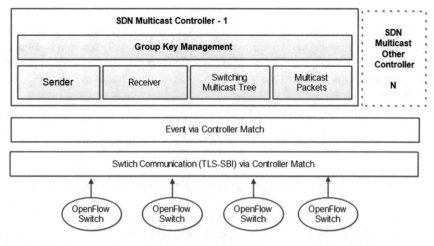

Fig. 5 SDN

module. Each one of these sub-modules create a database to store and update a list of multicast members according to the multicast group membership events (joint and leave).

The Group management module (2) in this case is responsible to restore and update information of multicast members (sources and receivers) for providing the full knowledge of group members to a network administrator. When the controller receives a PIM-register message from the Open-Flow switch, the multicast sources management sub-module stores a state related to this source and the related multicast group address (S, M). This store event triggers a notification to the multicast tree management to create a multicast tree to forward data to receivers, if exist, otherwise the sub-module, notify the source to stop forwarding multicast data.

In other side multicast receiver management sub-module stores a state related to receivers each time an OpenFlow switch receive an IGMP (for IPv4 stack or MLD if the receiver uses IPv6 stack) request from the receiver. As any multicast IP solutions, any nodes want to join the multicast group, send explicitly a join message to the first OpenFlow Switch asking to the join the multicast routing tree. In the sub-module level, when a new receiver is added to the receiver's database, this sub-module triggers a notification to the multicast tree management to create a path from the new receiver to the existing multicast tree to forward data to receivers.

The main component of any multicast routing protocol is the building of the multicast routing tree, in our solution this task is implemented in the **multicast tree computing module (3)**. This module handle any notification sent by the two sub-modules of multicast member management module (multicast sources management sub-module and multicast receiver management sub-module).

The Multicast Tree Management module (4) follow any change events, including notifications from the multicast member management module, multicast tree computing submodule and the Topology module.

The multicast signalization message dispatcher module (5) deal with two types of messages: IGMP/MLD and PIM messages. IGMP/MLD packets which can be identified by class D address in IPv4 stack with IGMP or the IPv6 address with the prefix ff00::/8, which is equivalent to the IPv4 multicast address 224.0.0.0/4.

3.4   The Multicast Signalization Message Dispatcher Module

The main function to be handled by the multicast signalization message dispatcher module is to translate and dispatch packets in Packet-In messages received from OpenFlow Switches to the appropriate SDN multicast controller modules. The multicast signalization message dispatcher module deals generally with two types of messages in Packet-In messages: IGMP/MLD and PIM messages.

3.5   The Multicast Member Management Module

Using the multicast member management avoid many problems existing in traditional solution, such as, we don't need to renew the multicast group keys when a group member node joins, leaves, and hands off a multicast group if the node is mobile. In this case the multicast controller and especially the multicast member management generate one multicast key for the specific multicast member affected by the multicast membership event.

When controller receives a joining group event, the event is forwarded first to the User authentication module, if the user authentication is enabled, this module is responsible to identity of user and updates the group membership access functions.

With our proposal solution, we identify two scenarios: the leave events can be initiated by the participant herself or be determined by the multicast controller.

Every time a group member explicitly sent an IGMP/MLD request to leave the multicast group, the corresponded OpenFlow Switch forward a message to the SDN multicast controller to deletes the secret keys related to this user from the multicast member stat database in the multicast member management module.

3.6   The Group Management Module

The Group management module in our solution is responsible to restore and update information of multicast members (sources and receivers) for providing the full knowledge of group members to a network administrator.

The Group management module work in collaboration with the Multicast Tree Management module and Multicast tree computing module, based in the notifications received from this modules and multicast group membership messaged received from multicast group members, the Group management module updates the group membership states database in the SDN multicast controller, each state include Openflow switches and ports where receivers and senders are connected and use to send multicast group membership messages (IGMP or MLD).

Group management module Events and changes in the group membership states database in the SDM multicast controller, triggers a notification to the Multicast tree computing module to update the multicast tree if necessary (the update can optimize the multicast tree).

3.7   Multicast Tree Computing Module

In SDN architecture, the main objective is to make the network intelligence centralized in software-based SDN controllers, which maintain a global view and topology of the network. In our solution, the multicast routes computing module uses PGRASP

[3–7] algorithm to calculate a minimum spanning tree (MST) centralized on core OpenFlow switch, this node is called Rendezvous Point router in PIM-SM protocol [23] or core router in CBT protocol []. Choosing an optimal multicast tree with all shortest path will minimize the number of group key will be saved in the states data base.

Algorithm 1 PGRASP Algorithm

Begin
S [] ← 0;   ▷ *Candidat_Element$_i$*
if $state_Candidate_Element_i$ == *False* **then**
   Build the Restricted Candidate List (RCL);
   Select();   ▷ Selection Element S from RCL
   for i ← *0; i < Max_Candidate_Element; i++* **do**
      S ← $S \cup m_i, m_j$;   ▷ Solution
      B_S ← Local_Search(S);   ▷ Best Solution
   end
end
Return B_S;
End

Contrary to all local search meta-heuristics based on deterministic local search methods, we will us the proposed algorithm by Feo and Resende (1999), the proposed algorithm is called Greedy Randomized Adaptive Procedure GRASP [41].

The implemented version in our solution, is the PGRASP algorithm, which is a parallel version of RASP algorithm. The GRASP heuristic has an inherent parallel nature [41], since iterations are independent from each other. GRASP iterations may be easily shared among processors forming so an effective parallel implementation of GRASP Algorithm; this implementation is called Parallel GRASP (PGRASP). Each PGRASP branch can be regarded as a search in some region of the feasible space not requiring any information from others iterations. The basic PGRASP algorithm is described in Fig. 6.

The main idea of every GRASP algorithm branch is to create after every iteration step a new optimal solution, independent of previous ones, where each iteration consists of two phases: first one called construction phase using a randomized greedy algorithm, the second phase is a local search phase using any local search algorithm, such as Hill Climbing, Simulated annealing, tabu search algorithm.

The construction phase is a non-deterministic phase allows to diversify the search and to produce an initial feasible solution that is used as the starting point for the local search phase. The construction phase is also responsible to create and update of a restricted candidate list (RCL) formed by the best starting solutions.

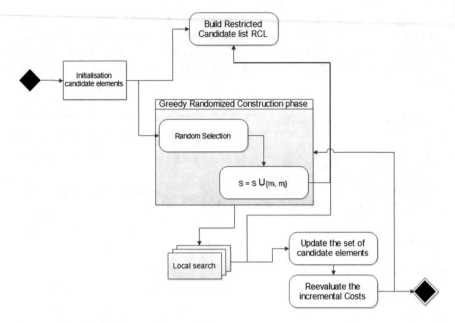

Fig. 6 2DV-PGRASP-CR algorithm execution

4 Implementation and Results

4.1 *Experimental Environment*

In this section, in the object to demonstrate the performance of our solution, we will present our test-bed and prototype experiment implemented in our SDN platform, describes all used tools and the technical specifications and topologies of the built architecture.

The studied scenario was designed in order to be large enough to provide realistic results and to be handled randomly and efficiently within used tools. The network topology is shown in Fig. 7, the topology contains three independent parties: the core network, the network extension and nodes in the multicast groups. The random networks topologies are generated using script generator and we adopt Waxman [] as the graph model. Our studies were performed on a set of 100 random networks. The values of $\alpha = 0.2$ and $\beta = 0.2$ were used to generate networks with an average degree between 3 and 4 in the mathematical model of Waxman. Table 1 show a set of used parameters in our studies.

We adopt Mininet 2.2.0 [34, 49] to simulate SDN topologies and select ODL [36] as the SDN controller extended by our multicast module. Table 2 illustrate a configuration setup, it will be deployed on a set of virtual machines VMs in order to establish our SDN environment.

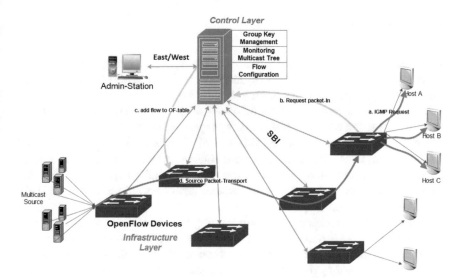

Fig. 7 Multicast scenarios based on SDN context

Table 1 Table of parameters

Parameter	Value
Network size	[800–1000]
Network core size	10%
Network extension size (include end devices)	20%
Multicast Group size	Random $\in [1 - 10]$
Multicast receivers size	20% of end devices
Multicast sources size	5% of end devices
Traffic type	UDP CBR
Waxman parameters	$\alpha = 0.2$ and $\beta = 0.2$
Average node degree	Between 3 and 4

Table 2 Setup configuration

Actor	OS	System	CPU/RAM
Multicast source	Ubuntu 16.04	VNC	2/2
Multicast receiver	Ubuntu 16.04	Mtest	2/2
SDN multicast controller	Ubuntu 16.04	OpenĐalight	4/2
Mininet	Ubuntu 16.04	OpenFlow 1.*	2/2
	Emulator V2.2.0		

4.2   Experimental Results

In this section, we compare some multicast and security parameters, such as computation cost, communication complexity and storage complexity of proposed scheme
with various existing group key management schemes.

We compare our proposed solution, MSDN-GKM, with a native implementation
of multicast session with PIM-SM protocol as specified by RFP7761 [23] and implemented in pimd daemon [] (native-PIM), GKMP [26], and PIM-SM with LKH [31]
(PIM-SM with LKH). Our scheme is more efficient in term of multicast IP metrics
and secure in term of Group Key Management as it optimize delay, delay variation,
multicast Tree cost, and preserves the forward and backward secrecy in multicast
group key management.

Multicast communications differentiate multiple type of delays, we can site: Join
delay, end to end delay transmission. The join delay is an important QOS parameter
for evaluate the performance of any multicast communication solution, in our solution
the join delay is sent directly to the SDN multicast controller, instead of being handled
by the routers in the network. We consider also the end-to-end delay transmission,
which is defined as the required time to transmit multicast packets from source node
to the furthest receiver node in the multicast group after group key management
processes is established. Figure 4 shows the multicast end-to-end delay for a network
with a size of 10–140 end device nodes. The multicast group size is between 10 and
80% of the overall nodes of the network. Simulation results show that multicast trees
build by our proposed algorithm have an average multicast delay better than native-
PIM [23] and GKMP [26], and PIM-SM with LKH [31] solutions and support more
multicast members (Fig. 8).

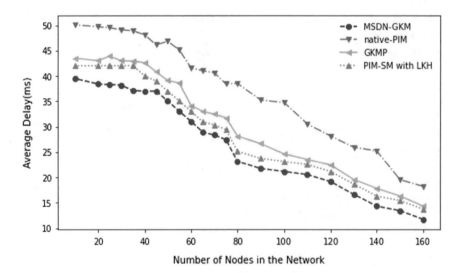

Fig. 8 Average delay transmission versus network size

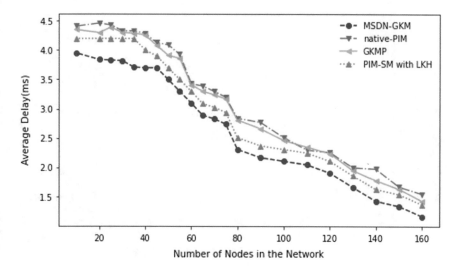

Fig. 9 Average delay variation transmission versus network size

Delay Variation is the difference between the first time of the reception of a multicast packet by a receiver of the multicast group and the last reception of the same multicast packet by another receiver. This metric present if the architecture supports reel time application and the group key management process chose an optimal multicast tree. In Fig. 9 the Delay Variation is plotted as a function of the number of nodes in the network topology, the network size contains 10–160 end device nodes. The multicast group size is between 10% and 80% of the overall nodes of the network. Simulation results show that multicast trees build by our proposed algorithm have an average multicast delay variation better than native-PIM [23] and GKMP [26], and PIM-SM with LKH [31] solutions.

Multicast tree cost is computed with function defined in formula (6) with $alpha \in [0-50], \beta \in [0-15]$, and we accept $\gamma \in [1-30]$. Figure 10 presents a comparison study of multicast tree Cost generated by each tested solution.

The comparison of storage overhead with the related schemes is shown in Fig. 11. The storage Overhead needed to manage all group keys of our proposed scheme at the SDN multicast Controller and OpenFlow Switches is much less than the native-PIM [23] and GKMP [26], and PIM-SM with LKH [31] solutions.

We have evaluated how much our proposed method optimize the processing time of group membership changes (join and leave), multicast tree management, and all failure recovery functions in the SDN multicast controller. We measured the processing time to add a new receiver, generate group key, add path to the multicast tree, and receive the first multicast data by this receiver from the multicast source. The comparison of storage overhead with the related schemes is shown in Fig. 12.

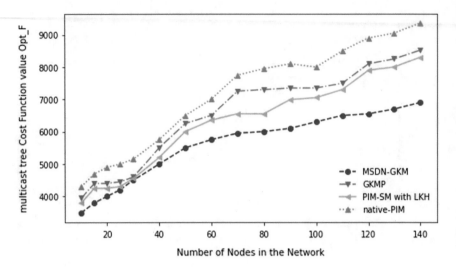

Fig. 10 Multicast tree cost function versus network size

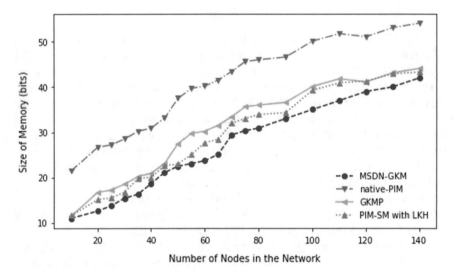

Fig. 11 Storage overhead versus network size

5 Conclusion and Future Work

Ensure a wide deployment and secure confidential multicast group communications
is important topic, the fact that recent network applications and protocols are based
in multicast IP communications. In this paper, we presented a survey in the multicast IP, group key management schemes, the multicast IP and SDN integration.
Current real implemented multicast sessions are based in Deering model [18, 19],

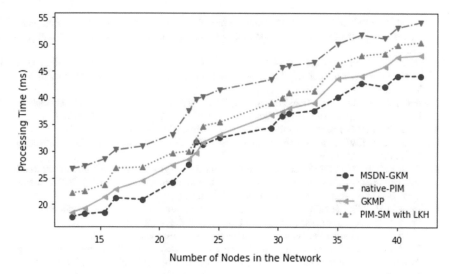

Fig. 12 Time processing versus network size

which members, sources and receivers, can join or leave the group dynamically. We reviewed several proposals, especially centralized solution which is more comparable to our proposed solution based in SDN technology. As a contribution of this paper, we propose a multicast scheme based on SDN and design a new multicast SDN controller which is responsible for many multicast functions, such as routing, multicast tree computing, handling join and leave events, group members management, and multicast group key management. Based in a set of test-beds we demonstrate that our new multicast scheme improves the multicast efficiency and performance, and multicast security requirements. Our future work is focused on proposing a new efficient group key management more adapted to high dynamic membership events, where the members are mobile.

References

1. Agee R, Wallner D, Harder E (1999) Key management for multicast: issues and architectures
2. Baddi Y, Ech-Cherif El Kettani MD (2013) Key management for secure multicast communication: a survey. In: 2013 national security days (JNS3), pp 1–6
3. Baddi Y, Ech-Cherif El Kettani MD (2013) Parallel grasp algorithm with delay and delay variation for core selection in shared tree based multicast routing protocols. In: Third international conference on innovative computing technology (INTECH 2013), pp 227–232
4. Baddi Y, Dafi M, El Kettani E-C (2013) Parallel greedy randomized adaptive search procedure with delay and delay variation for RP selection in PIM-SM multicast routing. In: Proceedings of the 2013 eighth international conference on broadband and wireless computing, communication and applications, BWCCA' 13, USA, 2013. IEEE Computer Society, pp 481–487

5. Baddi Y, El Kettani MDEC (2013) Parallel grasp algorithm with delay and delay variation for rendezvous point selection in PIM-SM multicast routing. J Theor Appl Inf Technol 57(2):235–243

6. Baddi Y, El Kettani MDE-C (2014) PIM-SM protocol with grasp-RP selection algorithm based architecture to transparent mobile sources in multicast mobile IPV6 diffusion. J Mob Multimed 9(3–4):253–272

7. Baddi Y, Ech-Cherif El Kettani MD (2014) QOS-based parallel grasp algorithm for RP selection in PIM-SM multicast routing and mobile IPV6. Int Rev Comput Softw (IRECOS) 9(7)

8. Ballardie A (1996) Scalable multicast key distribution. RFC 1949 (Experimental)

9. Ballardie A (1997) Core based trees (CBT version 2) multicast routing—protocol specification. RFC Editor, United States

10. Becker K, Wille U (1998) Communication complexity of group key distribution. In: Proceedings of the 5th ACM conference on computer and communications security, CCS '98, New York, NY, USA. Association for Computing Machinery, pp 1–6

11. Hossein H (2006) Handbook of information security, information warfare, social, legal, and international issues and security foundations, vol 2. Wiley, New York

12. Boyd C (1997) On key agreement and conference key agreement. In: ACISP

13. Brickell EF, Lee PJ, Yacobi Y (1988) Secure audio teleconference. In: Pomerance C (ed) Advances in cryptology–CRYPTO '87, Berlin, Heidelberg. Springer, Berlin, Heidelberg, pp 418–426

14. Burmester M, Desmedt Y (1994) A secure and efficient conference key distribution system (extended abstract). In: EUROCRYPT

15. Calvert KL, Zegura EW, Donahoo MJ (1995) Core selection methods for multicast routing

16. Chaddoud G, Chrisment I, and Schaff A (2001) Dynamic group communication security. In: Proceedings. Sixth IEEE symposium on computers and communications, pp 49–56

17. Wong CK, Gouda M, Lam SS (2000) Secure group communications using key graphs. IEEE/ACM Trans Netw 8(1):16–30

18. Deering SE (1988) Multicast routing in internetworks and extended LANs. Technical report, Stanford University, Stanford, CA, USA

19. Deering SE, Cheriton DR (1990) Multicast routing in datagram internetworks and extended LANs. ACM Trans Comput Syst 8:85–110

20. Estrin D, Handley M, Helmy A, Huang P, Thaler D (1999) A dynamic bootstrap mechanism for rendezvous-based multicast routing. In: INFOCOM '99. Eighteenth annual joint conference of the IEEE computer and communications societies. Proceedings. IEEE, vol 3, pp 1090–1098

21. Farinacci D, Li T, Hanks S, Meyer D, Traina P (2005) Protocol independent multicast—dense mode (PIM-DM): protocol specification (revised)

22. Fenner B, Handley M, Holbrook H, Kouvelas I, Parekh R, Zhang Z, Zheng L (2016) Protocol independent multicast—sparse mode (PIM-SM): protocol specification (revised). Technical report

23. Fenner B, Handley MJ, Holbrook H, Kouvelas I, Parekh R, Zhang ZJ, Zheng L (2016) Protocol independent multicast—sparse mode (PIM-SM): protocol specification (revised). RFC 7761

24. Grad D (1997) Diffusion et Routage: Outils de Modélisation et de Simulation. In: Actes du CNRIUT97, Congrès National de la Recherche en IUT, Toulouse, 10 pp

25. Hardjono T (2000) Router-assistance for receiver access control in PIM-SM. In: Proceedings ISCC 2000. Fifth IEEE symposium on computers and communications, pp 687–692

26. Harney H, Muckenhirn C (1997) Group key management protocol (GKMP) architecture. RFC 2094

27. Harney H, Muckenhirn C (1997) Group key management protocol (GKMP) specification. RFC 2093

28. Karaman A, Hassanein H (2006) Core-selection algorithms in multicast routing—comparative and complexity analysis. Comput Commun 29(8):998–1014

29. Karim ZKIK, Sebbar A, Baddi Y, Boulmalf M (2019) Secure multipath mutation SMPM in moving target defense based on SDN. Procedia Comput Sci 151:977–984

30. Kim Y, Perrig A, Tsudik G (2000) Simple and fault-tolerant key agreement for dynamic collaborative groups. In: Proceedings of the 7th ACM conference on computer and communications security, CCS '00, New York, NY, USA, 2000. Association for Computing Machinery, pp 235–244

31. Aswani Kumar Ch, Sri Lakshmi R, Preethi M. Implementing secure group communications using key graphs. Defence Sci J 57(2):279–286

32. Mapoka Trust T (2013) Group key management protocols for secure mobile multicast communication: a comprehensive survey. Int J Comput Appl 84:28–38

33. Mehlhorn K (1988) A faster approximation algorithm for the Steiner problem in graphs. Inf Process Lett 27(3):125–128

34. Mininet, 19 June 2020 [online]. Available at: http://mininet.org/

35. Moy J (1994) MOSPF: analysis and experience. Request for comments, United States

36. Opendaylight, 19 June 2020 [online]. Available at: https://www.opendaylight.org/

37. Oppliger R, Albanese A (1996) Distributed registration and key distribution (DiRK). In: Proceedings of the 12th international conference on information security IFIP SEC'96, Hall

38. Pande AS, Thool RC (2016) Survey on logical key hierarchy for secure group communication. In: 2016 international conference on automatic control and dynamic optimization techniques (ICACDOT), pp 1131–1136

39. Rafaeli S, Hutchison D (2002) Hydra: a decentralised group key management. In: Proceedings. Eleventh IEEE international workshops on enabling technologies: infrastructure for collaborative enterprises, pp 62–67

40. Rafaeli S, Hutchison D (2003) A survey of key management for secure group communication. ACM Comput Surv 35(3):309–329

41. Resende MGC, Ribeiro CC (2005) Parallel greedy randomized adaptive search procedures

42. Rodeh O, Birman K, Dolev D (2000) Optimized group rekey for group communication systems

43. Salama HF (1996) Multicast routing for real-time communication of high-speed networks. PhD thesis

44. Sebbar A, Boulmalf M, Ech-Cherif El Kettani MD, Baddi Y (2018) Detection MITM attack in multi-SDN controller. In: 2018 IEEE 5th international congress on information science and technology (CiSt). IEEE, pp 583–587

45. Sebbar A, Karim ZKIK, Baddi Y, Boulmalf M, Ech-Cherif El Kettani MD (2019) Using advanced detection and prevention technique to mitigate threats in SDN architecture. In: 2019 15th international wireless communications & mobile computing conference (IWCMC). IEEE, pp 90–95

46. Sebbar A, Karim ZKIK, Baddi Y, Boulmalf Y, Ech-Cherif El Kettani MD (2020) MITM detection and defense mechanism CBNA-RF based on machine learning for large-scale SDN context. J Ambient Intell Hum Comput 2020

47. Sebbar A, Zkik K, Boulmalf M, Ech-Cherif El Kettani MD (2019) New context-based node acceptance CBNA framework for MITM detection in SDN architecture. Procedia Comput Sci 160:825–830

48. Seetha R, Saravanan R (2015) A survey on group key management schemes. Cybern Inf Technol 15(3):3–25

49. Sood M, Sharma KK (2014) Mininet as a container based emulator for software defined networks

50. Steiner M, Tsudik G, Waidner M (1996) Diffie-Hellman key distribution extended to group communication. In: Proceedings of the 3rd ACM conference on computer and communications security, CCS '96, New York, NY, USA, 1996. Association for Computing Machinery, pp 31–37

51. Cain B, Hardjono T, Monga I (2000) Intra-domain group key management protocol. INTERNET-DRAFT

52. Waitzman D, Partridge C, Deering SE (1988) RFC 1075: distance vector multicast routing protocol

53. Wallner D, Harder E, Agee R (1999) Key management for multicast: issues and architectures. RFC 2627 (informational)

54. Wallner D, Harder E, Agee R (1999) Rfc2627: key management for multicast: issues and architectures
55. Wei L, Estrin D (1994) A comparison of multicast trees and algorithms. Technical report
56. Zappala D, Fabbri A (2001) An evaluation of shared multicast trees with multiple active cores. Springer, London, UK, pp 620–629
57. Zkik K, Sebbar A, Baddi Y, Belhadi A, Boulmalf M (2019) An efficient modular security plane AM-SecP for hybrid distributed SDN. In: 2019 international conference on wireless and mobile computing, networking and communications (WiMob). IEEE, pp 354–359

Machine Learning for CPS Security: Applications, Challenges and Recommendations

Chuadhry Mujeeb Ahmed, Muhammad Azmi Umer,
Beebi Siti Salimah Binte Liyakkathali, Muhammad Taha Jilani,
and Jianying Zhou

Abstract Machine Learning (ML) based approaches are becoming increasingly common for securing critical Cyber Physical Systems (CPS), such as electric power grid and water treatment plants. CPS is a combination of physical processes (e.g., water, electricity, etc.) and computing elements (e.g., computers, communication networks, etc.). ML techniques are a class of algorithms that learn mathematical relationships of a system from data. Applications of ML in securing CPS is commonly carried out on data from a real system. However, there are significant challenges in using ML algorithms as it is for security purposes. In this chapter, two case studies based on empirical applications of ML for the CPS security are presented. First is based on the idea of generating process invariants using ML and the second is based on system modeling to detect and isolate attacks. Further several challenges are pointed out and a few recommendations are provided.

1 Introduction

The enormous growth in Artificial Intelligence (AI) has impacted almost every sector of life. Particularly, ML which is a subset of AI has shown its efficacy in various domains, such as, in healthcare [1], self-driving cars [2], and cyber-security [3].

C. Mujeeb Ahmed (✉) · B. S. S. Binte Liyakkathali · J. Zhou
Singapore University of Technology and Design, Singapore, Singapore
e-mail: chuadhry@mymail.sutd.edu.sg

B. S. S. Binte Liyakkathali
e-mail: liyakkathali@sutd.edu.sg

J. Zhou
e-mail: jianying_zhou@sutd.edu.sg

M. A. Umer
DHA Suffa University, Karachi, Pakistan
e-mail: muhammadazmiumer@yahoo.com

M. A. Umer · M. T. Jilani
Karachi Institute of Economics and Technology (KIET), Karachi, Pakistan
e-mail: m.taha@pafkiet.edu.pk

© The Editor(s) (if applicable) and The Author(s), under exclusive license to Springer
Nature Switzerland AG 2021
Y. Maleh et al. (eds.), *Machine Intelligence and Big Data Analytics for Cybersecurity Applications*, Studies in Computational Intelligence 919,
https://doi.org/10.1007/978-3-030-57024-8_18

These systems are often distributed in nature, therefore, cloud computing seems to be a more viable choice for the development of such systems. However, the rapid development of such systems and their integration with cloud infrastructure introduces more vulnerabilities, for example, cyber attacks on Maroochy water services [4], Ukrainian power plants [5], and as well as Stuxnet [6] have shown the serious threats to critical infrastructures.

Recently researchers have started to apply ML for cyber-security [7]. One example is malware detection where current techniques mostly rely on creating malware signatures using domain experts [8]. Once these malware signatures get published, they become obsolete. Since malware developers quickly adapt their attacks, there is a need for an automatic malware signature generation mechanism. This is possible using ML as discussed in [8]. Likewise, ML-based solutions have been deployed in CPS that range from utility to the medical industry. Sensors are the integral component of these systems. These sensors usually generate noisy data and ML helps to make sense of the data [9]. ML techniques are also being used to detect anomalous data [10]. It has been demonstrated to be useful for anomaly detection ranging from the application layer to kernel events [11]. As different events happening from the application layer to the kernel layer get recorded in system logs and traces, these logs and traces are very helpful for anomaly detection in the system. But these traces and logs are huge in a real-time system. Therefore, ML techniques are quite helpful for online anomaly detection. The scope of ML applications is quite broad and in the interest of brevity, we focus on applying ML in CPS security in the rest of this chapter.

The primary role of a CPS is to control the underlying process in critical infrastructure (CI). Such control is effected through the use of computing and communication elements such as Programmable Logic Controllers (PLCs) and Supervisory Control and Data Acquisition systems (SCADA), and communications networks. The PLCs receive data from sensors, compute control actions, and send these over to the actuators for effecting control over the process. The SCADA workstations are used to exert high-level control over the PLCs, and the process, and provide a view into the current process state. Each of these computing elements is vulnerable to cyber-attacks as evident from several widely reported successful attempts such as those reported in [5, 12, 13]. Such attacks have demonstrated that while air-gapping a system might be a means for securing a CI, it does not guarantee to keep attackers from gaining access to the CPS.

An example of a CPS is shown in Fig. 1. It shows the high-level architecture of an electrical power system. This is composed of electricity generation (power plants), transmission (electric grid system) and end-users (smart home). As one can imagine this power system is composed of a multitude of devices and physical processes. Power generation and transmission depend on the demand from the utilities and the users. To meet the requirements of the energy demand the critical infrastructure is utilized to ensure a continuous supply of power. Each of the processes in the critical infrastructure is a complex engineering system and needs a sophisticated control to achieve its desired objectives. For example, at the generation stage, we have generators, Intelligent Electronic Devices (IEDs) also incorporating electric

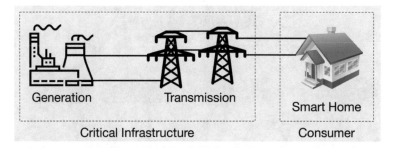

Fig. 1 A generic electrical power system as an example of CPS

relays, all these devices are autonomously controlled by the PLCs. This means that we have a lot of sensors monitoring the physical process, actuators/generators and the physical infrastructure that communicate the current physical states with each other and with the PLCs.

Successful attacks on CI have led to a surge in the development of defense mechanisms to prevent, contain, and react to cyber-attacks. One such defense mechanism is the anomaly detector that aims at raising an alert when the controlled process in a CI moves from its normal to an unexpected, i.e. *anomalous*, state. Approaches used in the design of such detectors fall into two broad categories: design-centric [14] and data-centric [9, 15, 16]. The focus of this chapter is on the data-centric approaches that rely on well-known methods for model creation such as those found in the system identification [17] and ML literature.

The use of ML to create anomaly detectors becomes attractive with the increasing availability of data and advanced computational resources. However, recent attempts [16, 18, 19] to create anomaly detectors and test them in a real water treatment plant, point to several challenges that must be overcome before such detectors can be deployed with confidence in a live plant.

In this chapter, we start by introducing the basics of ML so that an interested reader without sufficient background could understand the rest of the chapter. A detailed discussion is carried out on the challenges and practical aspects in the design of anomaly detectors using real plant data. To address the challenges brought up by earlier research efforts [20], two case studies are taken up to discuss how to solve those challenges. Despite best efforts, there are still some open challenges related to using ML in CPS security. We provide recommendations to be considered when designing future intrusion detection systems.

2 Machine Learning Preliminaries

Before going into further depth of security issues, it is necessary to have some basic understanding of ML and its types. ML can be broadly categorized into four categories i.e. Supervised Learning, Semi-supervised Learning, Unsupervised Learning,

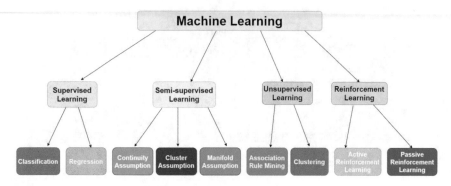

Fig. 2 Types of machine learning

and Reinforcement Learning as described in Fig. 2. The first three categories are highly functional in the literature, while the last one is mushrooming at a steady pace. We have defined each category in the following subsections.

2.1 Supervised and Semi-supervised Learning

In Supervised there is a feature vector $X_{i=1}, \ldots, X_n$, and a class variable 'Y'. Relationship of feature vector X and class variable Y is described below:

$$Y = f(X) \tag{1}$$

Every feature vector 'X' has a corresponding label in the class variable 'Y' as shown in Fig. 3. Based on the class variable 'Y' described in Eq. 1, supervised learning can be further classified into the regression and classification problem. If 'Y' is a real-valued attribute, then it would be considered as a regression problem. If 'Y' is a discrete-valued attribute, then it would be a classification problem.

In Semi-supervised learning, the model is trained using both labeled and unlabeled data. In the first phase, the model is trained using labeled data. In the second phase, labels are assigned to unlabeled data using the model trained in the earlier phase. In the third phase, both earlier labeled data and new assigned labeled data is used to train the model.

2.2 Unsupervised Learning

In unsupervised learning, there is a feature vector $X_{i=1}, \ldots, X_n$, but there is no class variable. Unsupervised learning can be broadly classified into clustering and association rule mining. In clustering, transactions of the dataset are grouped into different

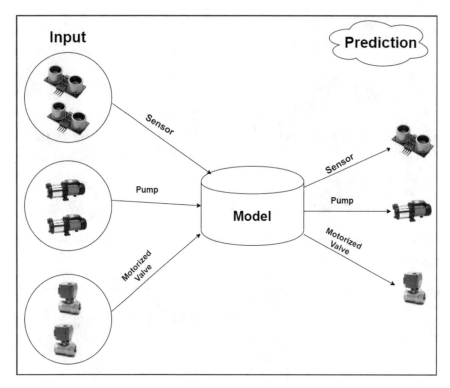

Fig. 3 Supervised learning

clusters based on some similarity measure, as shown in Fig. 4, while association rule mining is a rule-based ML approach. It is further described in Sect. 7.1.

2.3 Reinforcement Learning

This type of ML is quite different than traditional ML techniques. It works on the principle of action and reward. Here, there is an agent 'A' percepts the state of environment 'E'. It then acts on the environment 'E'. Based on its action, the agent receives a reward 'R' as described in Fig. 5. This reward helps the agent in evaluating its action. Reinforcement learning can be classified into active and passive reinforcement learning approaches. In passive reinforcement learning, the agents' policy is fixed. It performs actions and learns how good is that policy. While in active reinforcement learning, the agent decides which action needs to be taken in the current situation. Therefore, it is necessary to learn the complete model with possibilities related to the outcome of all actions.

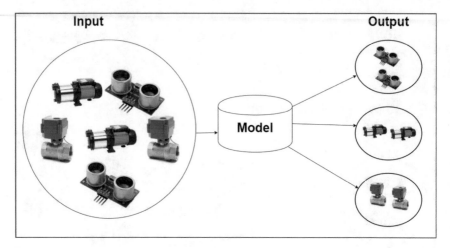

Fig. 4 Unsupervised learning

Fig. 5 Reinforcement
learning

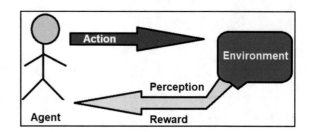

The basic discussion regarding ML above shall be enough to understand the topics discussed in the following. Moreover, it enables us to differentiate between the problems associated with each type of ML, particularly while applying it to securing the CPS.

3   ML Phases: Modeling, Training and Deployment

A physical process is controlled based on the sensor measurements. A physical process follows certain design requirements and patterns. For example, a water distribution system is driven by the demand from the users. To fulfill the estimated demand it must be able to hold a certain amount of water in the reservoir but the reservoir's capacity is limited therefore, a controller ensures that the water level is never below the minimum water requirement and not high than the capacity else flooding might happen. The autonomously operating systems under precise control are still susceptible to disruptions either driven by a fault or more recently due to an

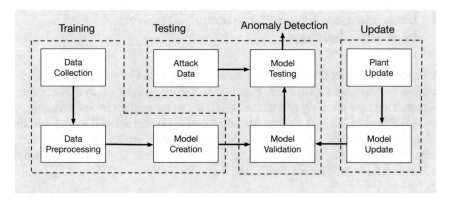

Fig. 6 Three stages in the development of an anomaly detector: model creation (training phase), model deployment (testing phase), and retraining (update phase). Real-time data from the host plant needs to be collected to create attacks for testing the anomaly detector

attack by a malicious entity. It is desired to design an anomaly detector that operates autonomously and notify the operators in case of an anomaly.

An anomaly detector can either be a black box model being fed by the raw data and learning from it or a white box model meaning extracting the context-aware features and learning from those. In this work, the focus is on ML models. There can be a range of attacks on critical infrastructure but as discussed in the threat model, the focus of this work is to look into the attacks on the physical process. Figure 6 shows the steps involved in developing a ML model. We start with the data collection process. The biggest challenge to perform research in data-based models is the lack of data availability especially data collection from a real industrial plant. In this study, we had access to real water treatment testbed (SWaT). The second challenge is the lack of attack samples, that is overcome by launching a range of attacks on the SWaT testbed. Once the model has been created and validated then the next phase is to test the performance for a range of attacks. Another important feature to incorporate is to update the model in case there is an update in the plant, for example, design parameters change or some devices are replaced over time.

The model becomes increasingly useless if it generates false positives that annoy the operators and waste their time in debugging the process that is otherwise operating normally. Thus, an anomaly detector must have an ultra-high detection rate as well as an ultra-low rate of false alarms. There are no widely accepted numbers associated with such rates though we believe that a detection rate of at least 99%, and a false alarm rate of less than one false alarm in 6-months, is needed for an anomaly detector deployed for real-time monitoring of a city-scale plant or power grid. Thus, as the plant components degrade, or the plant is upgraded, the anomaly detector must adapt itself to the new reality.

4   Design of Learning-Based Anomaly Detectors: Practical Challenges

A typical process to design an anomaly detector consists of model creation, validation, testing, and deployment. Figure 6 shows the activities during these three phases. There are a variety of ML methods that can be used for model creations on the real-time data [21–23]. System identification [17] is another well-known method often used by control engineers to build the state-space model of a system using state observations [24, 25]. Based on the different phases of learning, the following outlines challenges associated with each stage.

4.1   Model Creation

Challenges faced during the creation of an anomaly detector are described next.

Supervised versus unsupervised learning: Recently, there have been studies where supervised ML is used for attack detection [26–28]. The challenge of using supervised learning is the lack of labeled data. Firstly, labeled data for the attacks is hard to get and secondly, it would not detect unknown or unseen attacks. Supervised learning is not suitable for detecting zero-day vulnerabilities in CPS.

Recent studies have used unsupervised or semi-supervised ML algorithms to detect attacks in an CPS [9, 18, 29]. In particular, data was obtained from Secure Water Treatment (SWaT) testbed [30]. Generally, unsupervised learning models are designed based on the normal operation of the plant's behavior wherein any observation that deviates from the "normal" is termed as an anomaly. In [9] the authors compared models derived using both supervised and unsupervised learning. In this study it was observed that models created using unsupervised learning perform better in attack detection though, due to sensitivity to noise in the data, they lead to a higher rate of false alarms than those derived using supervised learning. Unsupervised learning can detect unknown attacks but can also increase the number of false alarms.

Model localization: A CPS in large systems is mostly a complex distributed control system. For example, a water treatment process consists of several stages and sub-processes. These separate physical processes might be connected logically and physically. An important consideration here is whether to create a ML model for the entire process or one each for different stages. Considering the distributed model versus a model for the whole system, or having a cluster of models, is an important design consideration that can influence detector performance. Global and local invariants were derived from SWaT in [15, 16]. The use of global invariants makes the model strong against the distributed attacks. Because these global invariants make it possible to detect multi-point attacks over different stages of the plant.

Scalability: Take the SWaT testbed [30] as an example. There is a multitude of sensors including level, flow, pressure, and chemical sensors for measuring the water quality and quantity. Studies have reported results from using models derived using supervised and semi-supervised learning [31] on the SWaT testbed. It has been observed that supervised learning lacks scalability due to the lack of availability of the labeled data. On the other hand, unsupervised algorithms can be trained for a large process plant without the need of having a labeled dataset. An interesting example of the scalability of one class classifiers is found in [31] for the case of sensor fingerprinting. The idea is that by using a one-class classifier for each sensor, a unique fingerprint is created to detect intrusion without the need to train the classifier based on the labeled data from all the sensors. The limitation for supervised learning, in that case, is in the event of an increase or decrease in the number of sensors; the models would need to be retrained but using a one-class classifier the models would be retrained only for the affected sensors.

4.2 Testing and Updating

Due to unforeseen deviations, a ML-based technique starts to behave differently from what it was tested and validated for. A few of such challenges are listed below.

Component Degradation: It is possible that in production alarms start appearing even for the system validated before the deployment. The behavior of the process might still be fine but for detectors, it appears to be in attack. It is challenging to incorporate drifts due to component degradation.

Noisy data: It has been demonstrated that an attacker can "hide" in the noise distribution of the data [9, 32]. In [33] the authors conclude that often ML algorithms miss the attacks in the noisy process data. For such a stealthy attacker it is important to consider the process noise distribution to train the detector [34]. The challenge arises because the noise is specific to the particular state of the process. For example, a water tank filling process in a tank would exhibit different noise profile as compared to the water tank emptying process.

Attack Localization: It has been reported that even though detectors using ML algorithms can detect anomalies, they fail to provide hint relevant to the location in the plant where the anomaly may have originated. One solution to this problem is to use a model for each sensor. However, doing so may miss anomalies created due to coordinated multi-point attacks.

Plant's Operational Specifications Update: The operational configuration of the plant might change over time, e.g., the amount of product it produces. For example, for a water storage tank there are set points such as high (H) and low (L) that represent normal operating levels. A change in these parameters makes the prior trained models useless, hence, it is a challenge to design an algorithm that could work well for the modified parameters.

Attack Detection Speed: The speed at which a process anomaly is detected is of prime concern due to the safety of the plant. The earlier the anomaly is detected, and reported, the sooner appropriate actions to mitigate the impact could be undertaken. The speed of anomaly detection is an important parameter to consider while designing ML-based intrusion detection for CPS.

5   Experimental Evaluation on SWAT Testbed

In the following, we will consider two case studies carried out on the SWaT testbed. We will first briefly outline the salient features of the testbed and then summarize the case studies along with challenges solved by undertaking those studies.

The Secure Water Treatment (SWaT) plant is a testbed at the Singapore University of Technology and Design [35]. SWaT as seen in Fig. 7, has been used extensively by researchers to test defense mechanisms for CI [36]. A brief introduction is provided in the following to aid in understanding the challenges described in this work.

SWaT is a scaled-down version of a modern water treatment process. It produces 5 gallons/min of water purified first using ultrafiltration followed by reverse osmosis. The CPS in SWaT is a distributed control system consisting of six stages. Each stage is labeled as P_n, where n denotes the nth stage. Each stage is equipped with a set of sensors and actuators. Sensors include those to measure water quantity, such as, level in a tank, flow, and pressure, and those to measure water quality parameters such as pH, oxidation reduction potential, and conductivity. Motorized valves and electric pumps serve as actuators.

Stage 1 processes raw water for treatment. Chemical dosing takes place in stage 2 to treat the water depending on the measurements from the water quality sensors. Ultrafiltration occurs in stage 3. In stage 4 any free chlorine is removed from water before it is passed to the reverse osmosis units in stage 5. Stage 6 holds the treated

Fig. 7 SWaT testbed is used for the reported case studies

water for distribution and cleaning the ultrafiltration unit through a backwash process. Data from the sensors and actuators is communicated to the PLCs through a level 0 network. PLCs communicate with each other over a level 1 network.

6 Threat Model

In this section, we introduce the types of attacks launched on our secure water treatment testbed (SWaT). Essentially, the attacker's model encompasses the attacker's intentions and capabilities. The attacker may choose its goals from a set of intentions [37], including performance degradation, disturbing a physical property of the system, or damaging a component. We define different threat models to evaluate the study reported in Sect. 7. It includes under-flowing and over-flowing of water tank, to burst the pipes, to intentionally waste the water by passing it to the drain, and to unnecessarily reduce the water in tank. A sample of such attacks is presented in Table 1.

It is assumed that the attacker has access to $y_{k,i} = C_i x_k + \eta_{k,i}$ (i.e., the opponent has access to ith sensor measurements). Also, the attacker knows the system dynamics and the control inputs and outputs.

Data Injection Attacks: For data injection attacks, it is considered that an attacker injects or modifies the real sensor measurement. The attacker's goal is to deceive the control system by sending incorrect sensor measurements. In this scenario, the level sensor measurements are increased while the actual tank level is invariant. This makes the controller think that the attacked values are true sensor readings, and hence, the water pump keeps working until the tank is empty and cause the pump to burn out. The attack vector can be defined as,

$$\bar{y}_k = y_k + \delta_k. \tag{2}$$

Table 1 A sample of attacks launched on SWaT

Attack #	Start time	End time	Attack point	Start state	Attack	Attacker intention
1	11:50 AM	11:56 AM	MV101	MV101=OPEN & P101=ON	MV101=CLOSE	To underflow the tank
2	12:17 PM	12:21 PM	MV101	MV101=CLOSE & P101=OFF	MV101=OPEN	To overflow the tank
3	4:36 PM	4:38 PM	P602	P602=OFF	P602=ON	To burst the pipe
4	4:44 PM	4:46 PM	MV303	MV303=CLOSE	MV303=OPEN	To intentionally drain the water
5	4:50 PM	4:51 PM	MV301, 302	MV301=OPEN & MV302=CLOSE	MV301=OPEN & MV302=OPEN	To burst the pipe
6	4:54 PM	4:57 PM	MV101	MV301=OPEN & MV101=OPEN	MV301=OPEN & MV101=CLOSE	To unnecessarily reduce the water in tank

where y_k is the sensor measurement, $\bar{y}_k$ is the sensor measurement with attacked value and δ_k is the bias injected by the attacker.

7 Case Study-1: Invariant Generation Using Data-Centric Approach

As discussed in the challenges, supervised learning needs attack data. Lack of attack data creates a bottleneck for anomaly detection problems. Unsupervised learning does not require attack data. Association rule mining was used to mine the invariants. An invariant is a normal condition of a physical plant that holds during its operation when it is in a given state. Invariants were mined using the benign data of SWaT. There are different types of invariants. Using the aforementioned approach, the following type of invariants was mined:

$$X \implies Y \tag{3}$$

7.1 Association Rule Mining

Association Rule Mining (ARM) [38] is an unsupervised ML technique. It was used for anomaly detection in the given case study as discussed in Algorithm 1. ARM is a rule-based ML technique. Traditionally it was used for market basket analysis. But, it also has applications in bioinformatics, intrusion detection, predicting customer behavior, etc. It has various algorithms including Apriori, FP-growth, Eclat, etc. FP-growth was used in this study using the Orange-Associate library in Python. It first generates frequent itemsets from the dataset. Association rules are mined using these frequent itemsets.

7.1.1 Frequent Itemsets

An itemset is either a single or a combination of multiple attributes in the dataset. To qualify as a frequent itemset, the itemset has to achieve the minimum support level requirement.

Support: Support for an itemset 'I' in dataset 'D' can be calculated using the presence of 'I' in the transactions or rows 'r' of 'D'.

$$S(I) = \frac{|r \in D; I \in r|}{|D|} \tag{4}$$

7.1.2   Association Rules

Association rules are generated from frequent itemsets. There are many such rules of type 3 generated from these itemsets. But, only rules which achieve a minimum confidence level are selected as the final set of rules.

Confidence: There are two parts of the rule in Eq. 3. 'X' part is called antecedent and the 'Y' part is called the consequent. Confidence of the rule is calculated using the support of antecedent and consequent combined, and support of antecedent alone.

$$C(X \implies Y) = \frac{S(X \cup Y)}{S(X)} \tag{5}$$

Algorithm 1: Invariant Generation using Association Rule Mining

---

1: Place the data collection infrastructure to capture network packets.
2: Decode network packets for state information generated by sensors and actuators.
3: Save the state information in the historian.
4: Apply feature engineering and feature selection techniques on the data collected from the historian.
5: Generate frequent itemsets using the reduced dataset from step 4.
6: Generate association rules using frequent itemsets.
7: Validate invariants (association rules) generated in step 6 using plant design and component specifications.

7.2   Feature Engineering and Challenges to Generate Invariants

There is a total of 51 attributes in the SWaT dataset. It includes binary, ternary, and real-valued attributes. ARM works only on binary-valued attributes. Therefore, to apply ARM, ternary and real-valued attributes were transformed into binary-valued attributes. Doing so requires special care, otherwise, the generated invariants would not be accurate. Therefore, they can lead to high false alarms. Only 15 attributes were selected for the current study which is described in Table 2. It includes state information of flow meters, motorized valves, and pumps. Flow meters and pumps are giving the state information related to Process 1, 2, 3, and 6. While motorized valves are giving the state information related to Process 1, 2, and 3. There was no attribute selected from Process 4 and 5. Because most of the attributes after transformation into binary-valued attributes did not provide any useful information for ARM. Like, they were having only a single value throughout the dataset. Therefore, they are useless for rule generation. They would produce a large number of rules which do not have any importance.

Table 2 Attributes selected for invariants generation

Attributes	Description
Flow meters	
FIT101	It measures the flow rate into T101
FIT201	It measures the flow rate from Process 1 to Process 2
FIT301, 601	It measures the flow rate in UF Process, and UF-backwash respectively
Motorized valves	
MV101, 201	It controls the flow in T101, and T301
MV301, 303	It controls the UF-backwash
MV302, 304	It controls the flow to the de-chlorination unit, and UF-backwash drain respectively
Pumps	
P101	It pumps raw water to Process 2
P203, 205	They works as dosing pumps for HCl, and NaOCl respectively
P302, 602	They pumps water from T301 to T401, and from T602 to UF unit

7.2.1   Transformation of Attributes

The control strategy of SWaT reveals that actuator mostly remains in stable state
i.e. either open or close. For a small duration of time, they get into a third state
i.e. transition state. The transition is from either open to close or vice versa. This
transition lasts for less than 10 s. This transition makes them the ternary valued
attribute. To convert these ternary valued attributes into binary-valued attribute two
factors were considered. This includes the transition direction and corresponding
FIT. If the transition direction is headed from close to open and the state variable of
the corresponding FIT is greater than 0.5, then the actuator was considered as open.
Similarly, if the transition is headed from open to close and the state variable of
the corresponding FIT is less than or equal to 0.5, then the actuator was considered
closed. A different strategy was used for FIT, which is a real-valued attribute. If the
state variable of FIT is greater than 0.5 then flow was assumed, otherwise, no flow
was assumed.

7.2.2   Large Set of Rules

As discussed earlier, support and confidence are the two important features to generate association rules. It is very important to define an optimal value for support. If the
support value is small, then there would be a very large set of rules. Similarly, if the
support is set to a higher value, then many important rules might not be generated.
Some itemsets have a very low number of transactions in the dataset. Therefore, the
rules associated with these itemsets would not be generated. For example, there are

only 3164 transactions where Pump P602 is in the ON state. While the total number of transactions in the dataset are 410,400. This means that P602=ON has only 3164/410,400 i.e. 0.77% support in the dataset.

7.3 Challenges Solved

ML-based anomaly detection approaches normally suffer from zero-day attacks and high false alarms. While applying supervised learning techniques, a lack of attack data creates a bottleneck. Likewise, unsupervised learning approaches suffer from high false alarms. The study reported in this case study has solved both the problems. This is an unsupervised learning approach, so there is no requirement of attack data. Secondly, the proposed approach is capable of detecting a zero-day attack. Because here invariants are generated using benign data of an operational plant. Later these invariants were placed as monitors for anomaly detection. There were no false alarms observed during the operation of the plant. Further, the invariants generated were having antecedent size = 1–7. This makes the current approach quite effective against distributed attack detection. A sample of invariants generated using the data-centric approach is described in Table 3. There are 7 types of invariants depending on the size of the antecedent. If antecedent size is 1 then it checks pairwise consistency between different actuators or sensors. While if antecedent size is more than 1 then all the sensors and actuators present in the antecedent must be true to reach the conclusion or consequent. The complete list of invariants is available at [39].

Table 3 A sample of invariants generated using data-centric approach

Size	Invariant	
	Antecedent[a]	Consequent
1	MV301=ON	MV101=ON
2	P602=ON, MV101=ON	P302=OFF
3	MV302=ON, MV303=OFF, P602=OFF	MV301=OFF
4	P602=ON, MV301=ON, FIT101>0.5, MV101=ON	MV304=OFF
5	P602=ON, MV301=ON, FIT301<0.5, MV101=ON	MV302=OFF
6	P602=ON, FIT301<0.5, MV301=ON, MV302=OFF, MV304=OFF, P302=OFF	FIT101>0.5
7	FIT601>0.5, FIT301<0.5, MV302=OFF, MV303=ON, MV304=OFF, P302=OFF, FIT101>0.5	MV101=ON

[a] Here comma (,) is representing a Boolean conjunction

8 Case Study-2: System Model Based Attack Detection and Isolation

Attacks on sensor measurements can take the system to an unwanted state. With this technique, we propose an attack detection and isolation method using the process dynamics. The disadvantage of using a system model-based approach for attack detection is that it could not isolate which sensor was under attack. For example, if one of two sensors that are physically coupled is under attack, the attack would reflect in both. To this end, this work proposes an attack isolation method using multiple system models for the same process. On top of modeling the system using system identification techniques, ML algorithms are used to detect and isolate an attack.

Attack Isolation Problem: The attack isolation problem also known as determining the source of an attack, is an important problem in the context of CPS. Anomaly detection research suffers from this issue, especially methods rooted in ML [40]. Using ML methods with the available data might be able to raise an alarm but are not able to find the source of the anomaly. In the context of CPS, if a model is created for the whole process it is not clear where does an anomaly is coming from? In Fig. 8 an example of such a problem is shown from a real water treatment process. In Fig. 8 example of stage 1 of the SWaT testbed [30] is shown. The example depicts two sensors namely a flow meter at the inlet of the raw water storage tank labeled as FIT-101 and a level sensor on top of the tank labeled as LIT-101. A joint physical system model for the stage 1 is created using a Kalman filter (more details on this in the following section). Such a system model captures the dynamics of the physical process. In our case, the physical process is an example of a water storage tank, which collects a limited amount of water to be used by the subsequent stages of the water treatment testbed. It is intuitive to understand that there is a physical relationship between the physical quantities, for example, consider that when water flows in the tank through the inlet pipe then the level of the water should rise in the tank. Hence, water level sensor LIT-101 and inlet flow sensor FIT-101 are physically coupled with each other. In the example attack, an attacker spoofs the flow sensor FIT-101 by spoofing the real sensor measurements of zero flow to 4 m^3/h volumetric flow level.

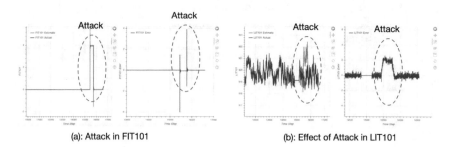

(a): Attack in FIT101   (b): Effect of Attack in LIT101

Fig. 8 This figure shows the attack isolation problem, i.e., due to the physical coupling of the sensors it is hard to isolate the attacks

In the left-hand part (a) of the figure, it can be seen that the attack would be detected using a model-based detector [24, 41] in FIT-101 but if we look at part (b) on the right-hand side, it can be seen that the same attack is detected using the detector for the LIT-101 sensor. For the figure (b) it could be seen that using the system model the estimate for the level tends to increase, for the reason that if there is inflow the level should be increased, but since there is an attack going on, it could be seen that the estimate deviates from the real sensor measurements. The model-based detectors defined for both level sensor and flow sensor would raise an alarm. It is not possible to figure out where is an attack unless manually checked. The problem of attack isolation is important considering the scale and complexity of a CPS.

8.1 Attack Isolation Algorithm

A well known idea in fault isolation literature is to use multiple observers [42, 43]. Consider a dynamic system with p outputs,

$$y_k = [y_k^1, y_k^2, \ldots, y_k^p]^T = C x_k. \tag{6}$$

where y_k is the matrix of p sensor measurements, C is the measurement matrix and x_k is the internal state of the system. For the case of an attack on one sensor i, attack vector $\delta_k^i \neq 0$ and $y_k^i = C_i \hat{x}_k + \delta_k^i$. Again consider the example of two sensors in the water tank example we have considered earlier. To use the idea of bank of observers we would drop one sensor at first and design an observer just using the first sensor, i.e., the flow sensor FIT-101 and then we will design another observer by using the second sensor, i.e., the level sensor LIT-101. Let's consider both the cases one by one.

Case 1: First observer using sensor measurements for FIT-101 gives following state space model:

$$\hat{x}_{k+1} = A \hat{x}_k + B u_k + L_i (y_k^i - C_i \hat{x}_k), \tag{7}$$

$$r_k = C \hat{x}_k - y_k. \tag{8}$$

where $\hat{x}_k$ is the estimate of the system state, u_k is the control input, L_i is the observer/estimator gain for ith output, r_k is the residual vector and A, B, C are the state space matrices. Firstly, the observer is designed using FIT-101 measurements as,

$$\begin{bmatrix} \hat{y}_{k+1}^1 \\ \hat{y}_{k+1}^2 \end{bmatrix} = C \left(\begin{bmatrix} a_{11} & a_{12} \\ a_{21} & a_{22} \end{bmatrix} \begin{bmatrix} \hat{x}_k^1 \\ \hat{x}_k^2 \end{bmatrix} + \begin{bmatrix} b_{11} \\ b_{21} \end{bmatrix} U + \begin{bmatrix} l_{11} \\ l_{21} \end{bmatrix} \begin{bmatrix} e(y_k^1) + \delta_k^1 \\ e(y_k^1) + \delta_k^1 \end{bmatrix} \right) \tag{9}$$

Case 2: Using the second observer designed for LIT-101 gives the output as,

$$\begin{bmatrix} \hat{y}_{k+1}^1 \\ \hat{y}_{k+1}^2 \end{bmatrix} = C \left(\begin{bmatrix} a_{11} & a_{12} \\ a_{21} & a_{22} \end{bmatrix} \begin{bmatrix} \hat{x}_k^1 \\ \hat{x}_k^2 \end{bmatrix} + \begin{bmatrix} b_{11} \\ b_{21} \end{bmatrix} U + \begin{bmatrix} l_{21} \\ l_{22} \end{bmatrix} \begin{bmatrix} e(y_k^2) + \delta_k^2 \\ e(y_k^2) + \delta_k^2 \end{bmatrix} \right) \qquad (10)$$

where δ_k^1 and δ_k^2 are the attack vectors in sensor 1 and sensor 2 respectively. To isolate the attack using a bank of observers, following conditions are considered for p sensors,

> *Condition 1*: if $r_k^j \neq 0$ for one $j \in \{1, 2, \ldots, i-1, i+1, \ldots, p\}$, then sensor j is under attack, while sensor i is the one used to design an observer.
>
> *Condition 2*: if $r_k^j \neq 0$ for all $j \in \{1, 2, \ldots, i-1, i+1, \ldots, p\}$ then sensor i is under attack while sensor i is used to design the observer.

For a simple example, let's consider two observers as designed in Eqs. (9) and (10). In the first case we had used FIT-101 sensor measurements to design an observer and also keep in mind that FIT-101 was free of any attacks. This means according to the condition 1 above FIT-101 residual mean should go to zero but for LIT-101, it does not. Figure 9 shows the results for the case 1. It can be seen that the sensor 1 (FIT-101) residual does not deviate from the normal residual, while the sensor 2 (LIT-101) residual deviates from the normal operation, hence detecting and isolating the source of attack. For the case 2, the observer is designed using the sensor 2 (LIT-101) and also remember that the attack is also present in the LIT-101. Figure 10 shows the results for this case. This case satisfies the condition 2 as stated above and then we see that the attack is present in both the sensors as the observer used is the one which has the attack. This means δ_k^1 was 0 and δ_k^2 was not zero in Eqs. (9) and (10) respectively.

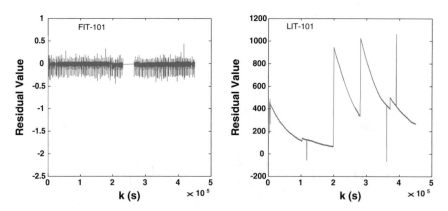

Fig. 9 Sensor 1 FIT-101 is used for observer design but the attack was in sensor 2 LIT-101. Therefore attack can be isolated in residual of LIT-101

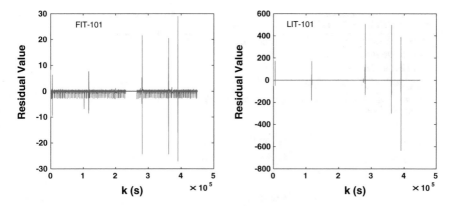

Fig. 10 Sensor 2 LIT-101 used for observer design and the attack was also in sensor 2 LIT-101. Therefore, both the sensor residuals deviate from the normal pattern

However from the results above it could be noticed that the sensor attacks could be isolated using the idea of bank of observers but it would not detect the case when the attack is in multiple sensors at the same time, e.g., multi-point single-stage attacks in a CPS [37]. Towards this end we are proposing the idea of using a Bank of Models (BoM) to isolate and detect attacks on multiple sensors at the same time in a CPS.

Bank of Models (BoM): The idea is to create multiple models of the physical process rather than the multiple observers. For example if you have two sensors that are physically coupled as in the case of FIT-101 and LIT-101, then we will create three models, (1) with both the sensors as output, (2) with FIT-101 only as the output and (3) with LIT-101 only as the output. We call the first method as *Joint* model and the rest two models as *BoM*. We can use these models in conjunction with each other to isolate the attacks and call that model as *Ensemble* of models. By having a separate model the sensors are no longer coupled to each other. These separate models could be used to detect attacks but the accuracy of detection might be low as we will see in the results. Therefore, we propose a method called ensemble of models combining the joint and separate models to make an attack detection decision as well as isolate the attack.

8.2 Empirical Evaluation

To visually present the idea Fig. 11 shows two example attacks and the coupling effects. Attack 1 is carried out on the flow meter FIT-101 by spoofing the flow value to 4 m³/h as shown in figure b and this attack can be observed in the residual value on the right-hand side. However, attack 1 could be seen in figure a in the level sensor LIT-101 as well. The Attack 2 is carried out on the level sensor by spoofing the water level value as shown in Fig. 11a. This attack could be seen in the residual of the level

Algorithm 2: Attack Isolation Method

Result: Output the sensor ID under attack
initialization;
θ_s: {Set of Sensors} ;
$r^i_{joint} = 0, r^i_{BoM} = 0, i \in \theta_s$;
$sensor^i_m.Attack = False$ #Flag i^{th} sensor Attack;
while *Sensor Signal* **do**
 for *i in θ_s* **do**
  $r^i_{joint} = y^i_{joint} - \hat{y}^i_{joint}, r^i_{BoM} = y^i_{BoM} - \hat{y}^i_{BoM}$;
  if $r^i_{BoM}.Attack == True$ && $r^i_{joint}.Attack == True$ **then**
   | $Sensor^i.Attack = True$;
  else
   | $Sensor^i.Attack = False$;
  end
 end
end

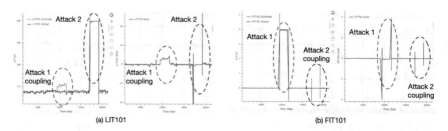

(a) LIT101   (b) FIT101

Fig. 11 This shows how two different attacks on two different sensors are reflected in residuals of both the sensors due to the physical coupling

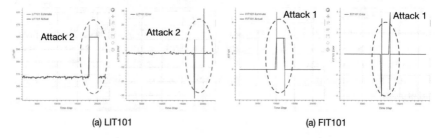

(a) LIT101   (a) FIT101

Fig. 12 In this figure both the attacks as shown in Fig. 11 are shown but for the case when we have two separate models for each sensor. It can be seen that the attacks are isolated to the particular sensor under attack

sensor LIT-101 and also on the right-hand side in the flow sensor FIT-101 residual. In Fig. 12 it can be seen that separate system models for both the sensors were able to isolate both the attacks. Attack 1 only appears in the residual of FIT-101 and Attack 2 is detected only by LIT-101.

8.3 Challenges Solved

The proposed technique has shown how models can be localized for each stage as well as for each device in the CPS. It leads to an attack isolation solution for a set of sensors that are physically coupled. Since the proposed solution is embedded in software it is easy to scale for any number of devices. Lastly, the study has shown the application of supervised learning that is a one-class SVM. The advantage of the one-class SVM is that the knowledge regarding the anomalies/attacks is not required at the training phase.

9 Related Studies

In the field of cyber security, the CIA triad which stands for Confidentiality, Integrity and Availability represents principles that should be guaranteed in any system. For the CPS environment, Availability and Integrity are given the most priority instead of Confidentiality. Attacks on processes may affect the availability and integrity of the data. These attacks on processes may give rise to process anomalies. As such, there are many studies done on detecting process anomalies using ML. One such example is by using convolution neural networks [44]. The researchers did a study on using a variety of deep neural network architecture namely convolutional and recurrent neural networks on the SWaT dataset. The process anomalies were measured based on the changes in the predicted value based on the actual value. The researchers found that the 1D convolution network performs the best despite being less complex.

Apart from process anomalies, researches have also shown promising results on IDS for network anomaly in the CPS network to secure industrial networks. One such research was done using a multi-layer perceptron with a binary classification technique [45]. In the training phase, the industrial protocol, Modbus, was used. The features that represent normal behavior were extracted. The ML algorithm is trained using the network packets and labeled those as normal or malicious. In detection, the neural network predicts the binary decision as either normal or malicious. For testing the anomaly in the network data, the researchers had tested with the network packets that were from unknown IP addresses, ports, functions and combinations of values.

To assess the effectiveness of defense mechanisms for CPS, researchers have used automated smart fuzzing [46]. Fuzzing is an automated testing technique to identify flaws in the system. This ML-based smart fuzzing finds attacks based on networks of CPS to improve the test-suites or benchmark of attacks by only knowing the normal operating range of the senors in the systems. Typically, these test-suites are manually created. First, a model (LSTM/SVR) is created to predict the sensor values based on the current sensor values and the actuator configuration. Second, using the predicated model, a search algorithm is used to cover the configuration of actuators to drive the system to an unsafe state. Using a fitness function to select the fittest configuration, the configuration is then applied to fuzzing networking. After a certain fixed interval, if the system is in an unsafe state, a new attack is found. This black-box technique automates the creation of test suites for CPS on network attacks.

ML brings numerous benefits to CPS. A possible downside to this solution is attacking can be done due to vulnerabilities in ML applied in CPS. Researchers proposed a model, Constrained Adversarial Machine Learning (ConAML) [47], where the model comes up with adversarial examples that are used as ML model inputs that meet with the constraint of the physical system. These vulnerabilities were evaluated in an electric power grid and water treatment systems and found that despite constraints on a physical system, the ConAML can generate adversarial examples that decrease the detection accuracy of the ML defense model.

10 Conclusions and Recommendations for Future Work

In this chapter, we have described two case studies targeted towards solving different practical challenges to the ML-based intrusion detection systems. Challenges highlighted in Sect. 4 led to case studies detailed above. However, there are still some open research areas, and the following are a few recommendations for future work based on these challenges.

Define the scope of the IDS: It is concluded that it is important to define the scope of the ML-based technique design. Sommer and Paxson [40] recommended to define the scope of the IDS for the legacy IT systems. In the realm of CPS, this recommendation becomes even more relevant as these are complex systems composed of both cyber and physical components. An IDS for CPS should have a clearly defined scope. It would be challenging to come up with an IDS which could detect cyber, i.e., in the CPS communications network, as well as physical anomalies.

Distinguish between fault and attack: Most of the studies reported using the SWaT testbed have used the process data. It is a challenge to determine whether the reported anomaly is due to a fault or an attack [48]. It is recommended to design a detector that could distinguish between an anomaly due to a fault with that due to an attack [49]. For example, if a sensor reports a measurement that is not expected by the ML model, can we determine whether this anomalous measurement is due to a cyber-attack or a fault in the sensor?

References

1. Ghassemi M, Naumann T, Schulam P, Beam AL, Ranganath R (2018) Opportunities in machine learning for healthcare. arXiv preprint arXiv:1806.00388
2. Bojarski M, Del Testa D, Dworakowski D, Firner B, Flepp B, Goyal P, Jackel LD, Monfort M, Muller U, Zhang J et al (2016) End to end learning for self-driving cars. arXiv preprint arXiv:1604.07316

3. Junejo KN, Yau DK (2016) Data driven physical modelling for intrusion detection in cyber physical systems. In: Proceedings of the Singapore Cyber-Security Conference (SG-CRC). IOS Press, Tokyo, Japan, pp 43–57

4. Abrams M, Weiss J (2008) Malicious control system cyber security attack case study— Maroochy Water Services, Australia. Tech. Rep., The Mitre Corporation, McLean, VA. http://csrc.nist.gov/groups/SMA/fisma/ics/documents/Maroochy-Water-Services-Case-Study_briefing.pdf

5. Lipovsky R (2016) New wave of cyber attacks against Ukrainian power industry. http://www.welivesecurity.com/2016/01/11

6. Langner R (2011) Stuxnet: dissecting a cyberwarfare weapon. IEEE Secur Privacy 9(3):49–51

7. Junejo KN (2020) Predictive safety assessment for storage tanks of water cyber physical systems using machine learning. Sādhanā 45(1):1–16

8. Nahmias D, Cohen A, Nissim N, Elovici Y (2019) Trustsign: trusted malware signature generation in private clouds using deep feature transfer learning. In: 2019 international joint conference on neural networks (IJCNN). IEEE, pp 1–8

9. Ahmed CM, Zhou J, Mathur AP (2018) Noise matters: using sensor and process noise fingerprint to detect stealthy cyber attacks and authenticate sensors in CPS. In: Proceedings of the 34th annual computer security applications conference, ACSAC 2018, San Juan, PR, USA, 03–07 Dec 2018, pp 566–581

10. Li W, Meng W, Su C, Kwok LF (2018) Towards false alarm reduction using fuzzy if-then rules for medical cyber physical systems. IEEE Access 6:6530–6539

11. Ezeme OM, Mahmoud QH, Azim A (2019) Dream: deep recursive attentive model for anomaly detection in kernel events. IEEE Access 7:18860–18870

12. Weinberger S (2011) Computer security: is this the start of cyberwarfare? Nature 174:142–145

13. Cobb P (2015) German steel mill meltdown: rising stakes in the internet of things. https://securityintelligence.com/german-steel-mill-meltdown-rising-stakes-in-the-internet-of-things/

14. Adepu S, Mathur A (2018) Distributed attack detection in a water treatment plant: method and case study. IEEE Trans Depend Secure Comput 1–8

15. Umer MA, Mathur A, Junejo KN, Adepu S (2017) Integrating design and data centric approaches to generate invariants for distributed attack detection. In: Proceedings of the 2017 workshop on cyber-physical systems security and privacy, pp 131–136

16. Umer MA, Mathur A, Junejo KN, Adepu S (2020) Generating invariants using design and data-centric approaches for distributed attack detection. Int J Crit Infrastruct Prot 28:100341

17. Overschee PV, Moor BD (1996) Subspace identification for linear systems: theory, implementation, applications. Kluwer Academic Publications, Boston

18. Inoue J, Yamagata Y, Chen Y, Poskitt CM, Sun J (2017) Anomaly detection for a water treatment system using unsupervised machine learning. In: 2017 IEEE international conference on data mining workshops (ICDMW). IEEE, pp 1058–1065

19. Tahsini A, Dunstatter N, Guirguis M, Ahmed CM (2020) Deep tactics: a framework for securing cps through deep reinforcement learning on stochastic games. In: 8th IEEE conference on communications and network security (CNS 2020), pp 1–7

20. Ahmed CM, Iyer GRM, Mathur A (2020) Challenges in machine learning based approaches for real-time anomaly detection in industrial control systems

21. Zhang F, Kodituwakku HADE, Hines W, Coble JB (2019) Multi-layer data-driven cyber-attack detection system for industrial control systems based on network, system and process data. IEEE Trans Ind Inform

22. Mitchell R, Chen I-R (2014) A survey of intrusion detection techniques for cyber-physical systems. ACM Comput Surv (CSUR) 46(4):55

23. Shalyga D, Filonov P, Lavrentyev A (2018) Anomaly detection for water treatment system based on neural network with automatic architecture optimization. In: ICML workshop for deep learning for safety-critical in engineering systems, pp 1–9

24. Ahmed CM, Murguia C, Ruths J (2017) Model-based attack detection scheme for smart water distribution networks. In: Proceedings of the 2017 ACM on Asia conference on computer

and communications security, ser. ASIA CCS '17. ACM, New York, NY, USA, pp 101–113 [online]. Available at: http://doi.acm.org/10.1145/3052973.3053011

25. Ahmed CM, Zhou J (2020) Challenges and opportunities in cps security: a physics-based perspective
26. Beaver JM, Borges-Hink RC, Buckner MA (2013) An evaluation of machine learning methods to detect malicious SCADA communications. In: 2013 12th international conference on machine learning and applications, vol 2. IEEE, pp 54–59
27. Hink RCB, Beaver JM, Buckner MA, Morris T, Adhikari U, Pan S (2014) Machine learning for power system disturbance and cyber-attack discrimination. In: 7th international symposium on resilient control systems (ISRCS). IEEE, pp 1–8
28. Priyanga S, Gauthama Raman M, Jagtap SS, Aswin N, Kirthivasan K, Shankar Sriram V (2019) An improved rough set theory based feature selection approach for intrusion detection in SCADA systems. J Intell Fuzzy Syst 36:1–11
29. Kravchik M, Shabtai A (2018) Detecting cyber attacks in industrial control systems using convolutional neural networks. In: Proceedings of the 2018 workshop on cyber-physical systems security and privacy. ACM, pp 72–83
30. Mathur AP, Tippenhauer NO (2016) Swat: a water treatment testbed for research and training on ICS security. In: 2016 international workshop on cyber-physical systems for smart water networks (CySWater), pp 31–36
31. Ahmed CM, Ochoa M, Zhou J, Mathur A, Qadeer R, Murguia C, Ruths J (2018) Noiseprint: attack detection using sensor and process noise fingerprint in cyber physical systems. In: Proceedings of the 2018 ACM on Asia conference on computer and communications security, ser. ASIA CCS '18. ACM
32. Mujeeb Ahmed C, Mathur A, Ochoa M (2017) NoiSense: detecting data integrity attacks on sensor measurements using hardware based fingerprints. ArXiv e-prints
33. Feng C, Li T, Chana D (2017) Multi-level anomaly detection in industrial control systems via package signatures and LSTM networks. In: 2017 47th annual IEEE/IFIP international conference on dependable systems and networks (DSN). IEEE, pp 261–272
34. Ahmed CM, Prakash J, Qadeer R, Agrawal A, Zhou J (2020) Process skew: fingerprinting the process for anomaly detection in industrial control systems. In: 13th ACM conference on security and privacy in wireless and mobile networks (WiSec). ACM
35. Mathur AP, Tippenhauer NO (2016) SWaT: a water treatment testbed for research and training on ICS security. International workshop on cyber-physical systems for smart water networks (CySWater). IEEE, USA, pp 31–36
36. iTrust. iTrust Datasets. https://itrust.sutd.edu.sg/itrust-labs_datasets/
37. Adepu S, Mathur A (2016) Generalized attacker and attack models for cyber physical systems. In: 2016 IEEE 40th annual computer software and applications conference (COMPSAC), vol 1, pp 283–292
38. Agrawal R, Imieliński T, Swami A (1993) Mining association rules between sets of items in large databases. In: Proceedings of the ACM SIGMOD international conference on management of data, vol 22. ACM, New York, NY, USA, pp 207–216
39. iTrust. Dataset and models. Singapore University of Technology and Design. https://itrust.sutd.edu.sg/itrust-labs_datasets/
40. Sommer R, Paxson V (2010) Outside the closed world: on using machine learning for network intrusion detection. In: IEEE symposium on security and privacy. IEEE, pp 305–316
41. Ahmed CM, Adepu S, Mathur A (2016) Limitations of state estimation based cyber attack detection schemes in industrial control systems. In: 2016 smart city security and privacy workshop (SCSP-W), pp 1–5
42. Wei X, Verhaegen M, van Engelen T (2010) Sensor fault detection and isolation for wind turbines based on subspace identification and Kalman filter techniques. Int J Adapt Control Signal Process 24(8):687–707 (online). Available at: http://dx.doi.org/10.1002/acs.1162
43. Esfahani PM, Vrakopoulou M, Andersson G, Lygeros J (2012) A tractable nonlinear fault detection and isolation technique with application to the cyber-physical security of power systems. In: Proceedings of the 51st IEEE conference on decision and control, pp 3433–3438

44. Kravchik M, Shabtai A (2018) Detecting cyberattacks in industrial control systems using convolutional neural networks. In: ACM proceedings of the 2018 workshop on cyber-physical systems security and privacy. ACM, pp 72–83
45. Hijazi A, El Safadi A, Flaus J-M (2018) A deep learning approach for intrusion detection system in industry network. In: BDCSIntell, pp 55–62
46. Chen Y, Poskitt CM, Sun J, Adepu S, Zhang F (2019) Learning-guided network fuzzing for testing cyber-physical system defences. In: 2019 34th IEEE/ACM international conference on automated software engineering (ASE). IEEE, pp 962–973
47. Li J, Lee JY, Yang Y, Sun JS, Tomsovic K (2020) Conaml: constrained adversarial machine learning for cyber-physical systems. arXiv preprint arXiv:2003.05631
48. Ahmed CM, Prakash J, Zhou J (2020) Revisiting anomaly detection in ICS: aimed at segregation of attacks and faults
49. Denning DE (1987) An intrusion-detection model. IEEE Trans Softw Eng SE-13(2):222–232

Applied Machine Learning to Vehicle Security

Guillermo A. Francia III⬤ **and Eman El-Sheikh**

Abstract The innovations in the interconnectivity of vehicles enable both expediency and insecurity. Surely, the convenience of gathering real-time information on traffic and weather conditions, the vehicle maintenance status, and the prevailing condition of the transport system at a macro level for infrastructure planning purposes is a boon to society. However, this newly found conveniences present unintended consequences. Specifically, the advancements on automation and connectivity are outpacing the developments in security and safety. We simply cannot afford to make the same mistakes similar to those that are prevalent in our critical infrastructures. Starting at the lowest level, numerous vulnerabilities have been identified in the internal communication network of vehicles. This study is a contribution towards the broad effort of securing the communication network of vehicles through the use of Machine Learning.

Keywords Controller Area Network · Electronic Control Unit · Machine
Learning · Neural network · Vehicle network · Vehicle security ·
Vehicle-to-everything technology

1 Introduction

Today's automobiles have over 100 Electronic Control Units (ECUs), which are embedded devices that controls the actuators to ensure optimal vehicle performance. These vehicles have multiple wireless entry points, some connected on the Internet, that enable access convenience and online services [1].

G. A. Francia III (✉) · E. El-Sheikh
Center for Cybersecurity, University of West Florida, Pensacola, FL, USA
e-mail: gfranciaiii@uwf.edu

E. El-Sheikh
e-mail: eelsheikh@uwf.edu

© The Editor(s) (if applicable) and The Author(s), under exclusive license to Springer
Nature Switzerland AG 2021
Y. Maleh et al. (eds.), *Machine Intelligence and Big Data Analytics for Cybersecurity
Applications*, Studies in Computational Intelligence 919,
https://doi.org/10.1007/978-3-030-57024-8_19

The proliferation of electronic devices and the rapid advancement of communication technologies ushered the steady progression of vehicular communication from an in-vehicle form to the far-reaching external variety. These advancements introduce unintended consequences towards the security of connected vehicles. Nevertheless, the reality of autonomous vehicles imposes additional pressure on manufacturers to shorten the deployment schedule for the "Vehicle-to-everything" (V2X) technology [2].

As pointed out in [3], there are key challenges with connected vehicle security. These include, but are not limited to:

- the legacy software security issues that are prevalent in millions of vehicles that are currently on the road;
- the need for real-time system update processes for connected cars;
- the unprecedented pace of the design, manufacture, and distribution of modern vehicles;
- the ineffective testing of embedded firmware for base vehicle development;
- the scarcity of research testbeds for connected vehicle security; and
- the lack of connected vehicle security curriculum modules in the information assurance/cyber security academic programs.

These challenges present a rich field for research activities and rightfully, for the benefit of society, need to be addressed with a great sense of urgency.

To this end, we present our contribution to the broad effort of securing the vehicle communication systems, mainly on the local interconnection between the ECUs and sensors. We illustrate the application of Artificial Intelligence (AI), specifically Machine Learning (ML) technologies, on the classification of anomalous network packets and the identification of vehicle models. The importance of this contribution is threefold: first, the pattern recognition of anomalous packets could significantly enhance the design of vehicle intrusion detection systems; secondly, the identification of vehicle models could provide a better understanding of the internal network traffic signatures which are, oftentimes, proprietary and not publicly available; finally, the exploration of various ML technologies could pave the way to advancing the science of AI in areas that are essential to societal needs.

The remainder of the chapter is organized as follows. The next section provides a review of related works on the subject followed by an overview of the Controller Area Network protocol and machine learning methods, including training algorithms. The chapter then presents our vehicle security study, including the dataset, classification results and analysis. It focuses on applying machine learning techniques for the classification of Controller Area Network (CAN) network packets to identify the specific model of certain vehicle models. In addition, the study examines pattern recognition of various types of vehicle network anomalies, which include various types of attack: flooding, fuzzy and malfunction. The chapter ends with some conclusions and directions for future research.

2 Related Works

Machine learning has been increasingly applied to cybersecurity applications. A few studies have been reported on its use for vehicle security. De La Torre et al. surveyed security methodologies developed to secure sensing, positioning, vision, and network technologies in driverless-vehicles, and highlighted how these technologies could benefit from machine learning models [4]. Some research has focused more broadly on using multi-class learning methods to identify attacks at run-time [5].

The majority of existing research on machine learning for vehicle security has focused on detecting anomalies and cyberattacks in the CAN bus, which serves as a protocol for in-vehicle network communication in electric vehicles, using various machine learning methods [6]. The CAN bus is vulnerable to various cyberattacks due to the lack of a message authentication mechanism. Avatefipour et al. utilized an anomaly detection model based on a modified one-class support vector machine in the CAN traffic [6]. The model uses the modified bat algorithm to find the most accurate structure in offline training. Experimental results indicated that the model achieved the highest True Positive Rate (TPR) and lowest False Positive Rate (FPR) for anomaly detection compared to other CAN bus anomaly detection algorithms such as Isolation Forest and classical one-class support vector machine [6].

One study used a three-pronged approach to detect anomalies in the Controller Area Network [7]. This involved cross correlating and validating sensor values across multiple sensors to improve the data integrity of CAN bus message, detecting anomalies using the order of messages from the Electronic Control Unit (ECU) can be used to detect anomalies and using a timing based detector to observe and detect changes in the timing behavior through deterministic and statistical techniques. The results demonstrated that attack detection is possible with good accuracy and low false positive rates but at the cost of longer detection latency. Zhou, Li and Shen used a deep neural network (DNN) method to detect anomalies of CAN bus messages for autonomous vehicles [8]. The system imports three CAN bus data packets, represented as independent feature vectors, and is composed of a deep network and triplet loss network, which are trainable in an end-to-end fashion. The results demonstrated that the proposed DNN architecture can make real-time responses to anomalies and attacks to the CAN bus and significantly improve the detection ratio.

Lokman et al. conducted a thorough review of Intrusion Detection Systems (IDS) for automotive CAN bus system based on techniques of attack, strategies for deployment, approaches to detection and technical challenges [9]. The study categorized anomaly-based IDS into four methods, namely, machine learning-based, statistical-based, frequency-based and hybrid-based. The machine learning-based methods surveyed mostly used supervised or semi-supervised anomaly detection techniques [9]. Although these techniques achieved high accuracy [9], they require completely labeled data, which is impractical especially for a real-time CAN. As machine learning approaches for vehicle security evolve, particularly the need for more unsupervised anomaly detection models, the training efficacy can be improved using dataset pre-processing techniques for the CAN bus system [9].

Kang & Kang used a semi-supervised deep neural network (DNN) method and off-line training to reduce processing time [10]. The model was validated using spoofed tire pressure monitoring system (TPMS) packets [9] to display incorrect TPMS values on the dashboard. Although the system had a 99% anomaly detection ratio, the computational complexity, training and testing time continued to increase as the amount layers were added. Taylor et al. utilized a supervised one-class support vector machine (OCSVM) to classify the CAN traffic flows [9, 11]. The system detected a very small number of packet injections and reduced the false alarm ratio. The authors developed a supervised long short-term memory to train the received CAN input. Although the anomalies could be detected with the lowest rate of false alarm, it worked only for a single CAN ID and did not support online learning [9].

Wasicek and Weimerskirch used a semi-supervised chip tuning-based method to detect attacks that were trying to modify parameters or reflash memories within the ECUs and integrate new hardware to make the CAN network traffic behave abnormally [12]. Although they were able to get higher true-positive detection against false-positive rate, however, the diagonal Receiver Operating Characteristics (ROC) curve is inclined toward no discrimination. Jaynes et al. attempted to automate the process of correlating CAN bus messages with specific Electronic Control Unit (ECU) functions in a new vehicle by developing a machine learning classifier that has been trained on a dataset of multiple vehicles from different manufacturers [13]. The results demonstrated some accurate classification, and that some ECUs with similar vehicle dynamics broadcast similar CAN messages.

Kumar et al. focused on jamming signal centric security issues for Internet of Vehicles (IoV) [14]. They proposed a machine learning-based protocol that focuses on jamming vehicle's discriminated signal detection and filtration for revealing precise location of jamming effected vehicles. The system uses an open-source ML algorithm, CatBoost, to predict the locations of jamming vehicle. The results demonstrate the resistive characteristics of the anti-jammer method considering precision, recall, F1 score and delivery accuracy. Overall, research on machine learning applications for vehicle security continues to expand.

2.1 Controller Area Network (CAN)

The Controller Area Network (CAN) communication protocol works on a two wired half duplex high speed serial network bus topology using the Carrier Sense Multiple Access (CSMA)/Collision Detection (CD) protocol. It implements most of the functions of the lower two layers of the International Standards Organization (ISO) Reference Model. In the CAN protocol, a non-destructive bitwise arbitration method is used during collision. This non-destructive notion implies that messages remain intact even in the presence of collision. It is a message-based protocol which is different from the address-based protocols such as the Medium Access Control (MAC) protocol that uses a physical address to deliver a network frame. Thus, a message is delivered to all nodes attached to the bus. The intended recipient will

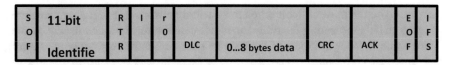

S O F	11-bit Identifie	R T R	I	r 0	DLC	0...8 bytes data	CRC	ACK	E O F	I F S

Fig. 1 CAN 2.0A frame standard format

accept, process, and acknowledge the properly received message; all others will simply discard it. A standard CAN frame, CAN 2.0A, is depicted in Fig. 1.

The fields in a standard CAN 2.0A Frame are described in the following [15]:

- SOF-a 1-bit start of frame field indicating the start of the message.
- ID-an 11-bit identifier that establishes the priority of the message. The lower the value, the higher the priority.
- RTR-a 1-bit remote transmission request indicating data when the bit is dominant. If the bit is recessive, then the message is a remote frame request.
- IDE-a dominant single identifier bit means that standard CAN identifier with no extension is being transmitted.
- r0-a reserved bit.
- DLC-a 4-bit code indicating the number of bytes being transmitted.
- Data-a payload of up to 64 bits of application data can be transmitted.
- CRC-a 16-bit cyclic redundancy check checksum value.
- ACK-a 2-bit acknowledgement field. One bit for acknowledgement; the other as a delimiter.
- EOF-a 7-bit end of frame marker.
- IFS-a 7-bit interframe space contains the time required by the controller to move a correctly received frame to its proper position in the message buffer area.

Inherently, the CAN bus has errors that frequently occur. This is partly due to bus contention. Thus, a device writing a frame onto the CAN bus is also responsible for checking the actual value on the wires. If the value read at a certain time corresponds to the original expected value, everything proceeds. If there is a mismatch from the expected value, the device immediately writes an error message onto the CAN bus in order to recall the previous frame and to notify the other devices to ignore it [16].

3 Machine Learning

A neural network approach to Machine Learning (ML) is a computational system that mimics the human brain's nervous system [17]. It is composed of a large number of highly interconnected processing elements called neurons. Neural networks have been used in numerous applications and continue to provide solutions to problems in areas such as speech and image pattern recognitions, semantic parsing, information extraction, linear and nonlinear regressions, and data classifications [17]. Network training, the process of finding the best value for weights and biases of neurons, remains to be the most difficult problem in neural networks [18]. Training a neural network entails measuring the difference (also called error) between the computed outputs and the target outputs of the training data. The most commonly used error measurement is the mean squared error (MSE), which is the sum of squares of the difference between of two sets of outputs [17]. However, research results by De Boer et al. [19] and McCaffrey [20] postulate that the Cross-Entropy (CE) error measure performs better in problems requiring combinatorial optimization and event simulations. Thus, neural networks utilizing the CE error is gaining more interests in many applications requiring optimal solutions to problems [18].

3.1 Neural Network Training Algorithms

One of the most important steps in designing a neural network is that on determining the most appropriate training algorithm for minimizing the chosen error function. The MATLAB Neural Network Toolbox [18] offers the implementation of the following training algorithms: Levenberg–Marquardt [21], BFGS Quasi-Newton [22], Resilient Backpropagation [23], Scaled Conjugate Gradient [24, 25] and Gradient Descent with Momentum [18, 25]. We limit our discussions to the two training algorithms we used for our study of vehicle security. For detailed discussions on various training algorithms, the astute reader is referred to Kim and Francia [18].

3.1.1 Conjugate Gradient Method

The conjugate gradient method uses the following steps for determining the optimal value (minimum) of a performance index $E(w)$ [18]. Given a starting point (w_0), it selects a direction (p_0). Next, it moves along an optimal direction that it finds through a linear search as illustrated by the following [18]:

$$w_1 = w_0 + \alpha_0 p_0 \tag{1}$$

The next search direction is determined so that it is orthogonal to the difference of gradients [18].

$$\Delta g_1^T p_1 = (g_1 - g_0)^T p_1 = (\nabla E(w_1) - \nabla E(w_0))^T p_1 = 0 \tag{2}$$

Repeating the two steps, we have the algorithm of conjugate gradient method [18]:

$$w_{k+1} = w_k + \alpha_k p_k \tag{3}$$

$$\Delta g_k^T p_k = (\nabla E(w_k) - \nabla E(w_{k-1}))^T p_k = 0 \tag{4}$$

The most common first search direction (p_0) is the negative of the gradient [18]:

$$p_0 = -g_0 = -\nabla E(w_0) \tag{5}$$

A set of vectors $\{p_k\}$ is called mutually conjugate with respect to a positive definite Hessian matrix H [18] if

$$p_k^T H p_j = 0 \quad \text{for } k \neq j \tag{6}$$

It can be shown that the set of search direction vectors $\{p_k\}$ obtained from Eq. (6) without the use of the Hessian matrix is mutually conjugate [18]. The general procedure for determining the new search direction is to combine the new steepest descent direction with the previous search direction [18]:

$$p_k = -g_k + \beta_k p_{k-1} \tag{7}$$

The scalars $\{\beta_k\}$ can be chosen by several different methods. The most common choices are [18]

$$\beta_k = \frac{\Delta g_{k-1}^T g_k}{\Delta g_{k-1}^T p_{k-1}}, \tag{8}$$

which is due to Hestenes and Stiefel [18, 26], and

$$\beta_k = \frac{\Delta g_{k-1}^T g_k}{\Delta g_{k-1}^T g_{k-1}}, \tag{9}$$

which is due to Fletcher and Reeves [18, 26].

While the conjugate gradient algorithms use linear search for linear optimization methods, the scaled conjugate gradient method does not use linear search when updating the error vector [18].

3.1.2 Levenberg–Marquardt Method

The Newton's method, one of the fastest training algorithms [18], performs an update according to the following:

$$w_{k+1} = w_k - H^{-1} g_k \tag{10}$$

It requires the computation of the Hessian matrix (H), which can become very costly if the number of attributes (or variables) is large [18, 27]. The Levenberg–Marquardt algorithm [21] is a variation of the Newton's method and works very well with neural network training where the performance index is MSE [18].

Without having to compute the Hessian matrix, the Levenberg–Marquardt algorithm is designed to approach second-order training speed [18]. When the error function is in form of a sum of squares such as MSE, the Hessian matrix can be approximated by

$$H = J^T J \tag{11}$$

and the gradient can be computed as

$$g_k = \nabla E(w_k) = J^T(w_k)e(w_k) \tag{12}$$

where J is the Jacobian matrix that contains first derivatives of the network errors with respect to the weights and biases, and e is a vector of network errors. If we substitute Eqs. (11) and (12) into Eq. (10), we obtain the following algorithm, known as Gauss–Newton method [18].

$$w_{k+1} = w_k - [J^T(w_k)J(w_k)]^{-1}J^T(w_k)e(w_k) \tag{13}$$

One problem with the Gauss–Newton is that the matrix $H = J^T J$ may not be invertible. This problem can be resolved by using the following modification:

$$G = H + \mu I \tag{14}$$

Since the eigenvalues of G are translation of the eigenvalues of H by μ, G can be made positive definite by finding a value of μ so that all eigenvalues of G are positive. This leads to the Levenberg–Marquardt algorithm [18].

$$\mathbf{w_{k+1}} = \mathbf{w_k} - [J^T(\mathbf{w_k})J(\mathbf{w_k}) + \mu_k I]^{-1}J^T(\mathbf{w_k})e(\mathbf{w_k}) \tag{15}$$

The Levenberg–Marquardt algorithm is known to work fast and stable for various forms of neural network problems with MSE performance index [18].

4   Vehicle Security Study

This research study involves the application of Machine Learning towards the classification of CAN network packets to identify the specific model of certain vehicle models. Further, the study looks into the pattern recognition of various types of vehicle network anomalies.

5 Dataset

The dataset, which is attributed to Han et al. [28], is delineated by vehicle model, Hyundai Sonata, KIA Soul, and Chevrolet Spark, according to the following attack types:

- *Flooding Attack.* In this type of attack, an Electronic Control Unit (ECU) device maintains a dominant status on the CAN bus by utilizing the lowest CAN ID value, 0x000.
- *Fuzzy Attack.* Random values ranging from 0x000 to 0x7FF are generated and injected into the CAN ID and Data fields to form random CAN packets and injected into the CAN bus.
- *Malfunction Attack.* This attack utilizes extracted CAN IDs, 0x316 for Sonata, 0x153 for Soul, and 0x18E for Spark, augmented by random values for the Data field. When these CAN packets are injected into the bus, the vehicles responded abnormally.
- *Attack Free.* These are CAN packets captured during the normal operation of the vehicles.

A summary of the dataset is shown on Table 1.
Each packet on the dataset is assembled as follows:

Timestamp, CAN ID, DLC, Data[0], Data[1], Data[2], Data[3], Data[4], Data[5], Data[6], Data[7], Flag

where.

- Timestamp—operating time
- CAN ID—identifier of CAN message in hexadecimal
- DLC—data length
- Data[0]–[7]—data values in byte
- Flag—T for injected message; R for normal message.

Table 1 Dataset summary [28]

Attack type	Number of packets		
	Hyundai Sonata	KIA Soul	Chevrolet Spark
Flooding	149,547	181,901	120,570
Fuzzy	135,670	249,990	65,665
Malfunction	132,651	173,436	79,787
Free	117,173	192,516	136,934

5.1 Classification of Vehicle Models

CAN packet information may vary from one car manufacturer. The fact that CAN
traffic information is regarded as proprietary by the manufacturers, it is difficult to
decipher the packets for security analysis. A similar work by Crow et al. [29] uses
Multilayer Perceptron (MLP) and a deep Convolutional Neural Network (CNN) to
classify vehicles based on CAN samples. The results of their study reveal an accuracy
of 73.03% for MLP and 99.79% for CNN. Our study utilizes a different set of data
and faster training algorithms. The results are shown below.

Figures 2 and 3 depict the results of our comparative study on the behavior of
two training algorithms in classifying vehicle models. The two training algorithms
are the Scaled Conjugate Gradient and the Levenberg–Marquardt algorithms. The
performance (error-checking) methods used by the training algorithms are the CE
and the MSE methods, respectively. The rationale behind this choice of ML training
algorithms is based primarily on a prior comparative study [18] made by one of the

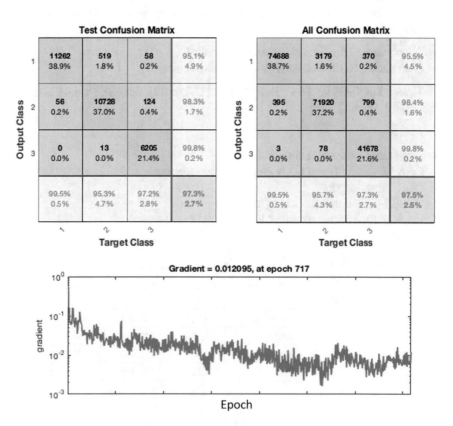

Fig. 2 Vehicle classification with scaled conjugate gradient

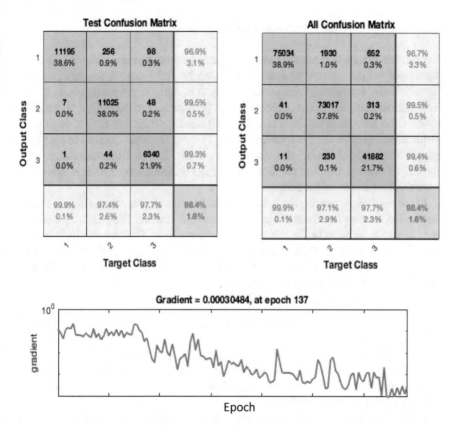

Fig. 3 Vehicle classification with Levenberg–MarquardtThis caption is OK but the graphics for Figures 2 and 3 are incorrectly swapped.

authors on the application of various ML techniques on the pattern recognition of operational data obtained from an industrial control system.

5.1.1 Analysis of the Gradient Plot

A scrutiny of the gradient plots in Figs. 2 and 3 would reveal the major difference between the Scaled Conjugate Gradient (SCG) and Levenberg–Marquardt (LM) training algorithms. Firstly, gradient descent in the LM appears more stable than that in the SCG and secondly, the speed with which convergence is achieved in terms of number of iterations is quite a significant advantage of the LM, with 137 epochs, over the SCG, with 717 epochs.

5.1.2 Analysis of the Confusion Matrix

The confusion matrix provides a visual depiction of the performance metrics of the supervised neural network system. The Matlab Neural Network toolbox produces four confusion matrices: Training, Validation, Test, and All. As the name implies for each confusion matrix, it is a depiction of the performance metrics for that particular stage of the ML process. For example, the Training confusion matrix depicts the performance metrics for the training phase. The All confusion matrix is simply an aggregation of the Training, the Validation, and the Test performance metrics. For the sake of brevity, we omit the Training and the Validation confusion matrices and show the Test and the All confusion matrices.

In the confusion matrix, the first 3 columns represent the actual (target) classes and the first 3 rows represent the predicted (output) classes. The classes are labeled as **1, 2, and 3** for **Sonata, Spark** and **Soul,** respectively. The last row indicates the *Recall* metrics for each of the classes; the last column indicates the *Precision* metrics for each of the classes. A formal definition of the performance metrics is in order and presented in the following discussions.

First, we need to define the following basic terms:

- **True Positives (TP).** These are cases which the system correctly predicted that it belongs to the class.
- **False Positives (FP).** These are cases which the system predicted that it belongs to the class but, in fact, it does not. These are also known as *Type I errors*.
- **True Negatives (TN).** These are cases which the system correctly predicted that it does not belong to the class.
- **False Negatives (FN).** These are cases which the system predicted that it does not belong to the class but, in fact, it does. These are also known as *Type II errors*.

Given those basic definitions, we define *Recall* as the proportion of actual positives that are correctly classified. Formally, it is calculated as

$$Recall = \frac{TP}{TP + FN} \tag{16}$$

Precision is defined as the proportion of positive predictions as truly positive. Formally, it is calculated as

$$Precision = \frac{TP}{TP + FP} \tag{17}$$

With high recall and low precision, there are few class samples that are classified as false negative while, at the same time, there are more class samples classified as false positive. With low recall and high precision, there are more class samples that classified as false negative and, at the same time, there are less class samples that are classified as false positive.

Table 2 Performance metrics

	Class	Sonata	Soul	Spark	
Scaled conjugate gradient	Recall	99.5	95.7	97.3	
	Precision	95.5	98.4	99.8	
	F-measure	97.4	97	98.5	
	Accuracy				97.5
Levenberg Marquardt	Recall	99.9	97.4	97.7	
	Precision	96.7	99.5	99.4	
	F-measure	98.3	98.4	98.5	
	Accuracy				98.4

Accuracy is defined as the proportion of positive and negative predictions that are correctly classified. In short, it measures the ratio of the correctly labeled vehicles to the whole pool of vehicles. Formally, it is calculated as.

$$Accuracy = \frac{TP + TN}{TP + TN + FP + FN} \tag{18}$$

F-measure is defined as the harmonic mean of the precision and recall.

$$F\text{-}measure = \frac{2*Recall*Precision}{Recall + Precision} \tag{19}$$

The F-measure is a representative of the Recall and Precision measures and uses the harmonic mean instead of the arithmetic mean. This implies that the F-measure is biased to the lower value of either the Precision or Recall. The Performance Metrics, which are gleaned from each of the *All Confusion Matrices* in Figs. 2 and 3, are summarized in Table 2.

5.2 Vehicle Network Anomaly Detection

Our study on vehicle network traffic anomaly detection separately examines each type of attacks: Flooding, Fuzzy, and Malfunction. Each dataset, as previously described, contains random records of specific attack and normal CAN network packet. The astute reader is referred to Han et al. [28] for a detailed description of the dataset. We narrowed our focus on the Hyundai Sonata dataset and applied the Levenberg–Marquardt training algorithm and the MSE performance index for pattern recognition.

5.2.1 Flooding Attack Detection

In the CAN network protocol, bus access is event-driven and takes place randomly. A simultaneous access to the bus by one or more nodes is implemented with a nondestructive, bit-wise arbitration. Nondestructive means that the node.

winning arbitration just continues on with the message, without the message being destroyed or corrupted by another node. The allocation of priority is based on the message ID, i.e. the lower it is the higher is its priority. An identifier consisting entirely of zeros is the highest priority message on a network because it holds the bus dominant the longest. Therefore, if two nodes begin to transmit simultaneously, the node that sends a last identifier bit as a zero (dominant) while the other nodes send a one (recessive) and goes on to complete its message. A dominant bit always overwrites a recessive bit on a CAN bus [15]. This arbitration scheme facilitates a Denial of Service or Flooding attack by maintaining a dominant status on the CAN bus using the lowest CAN ID value, 0x000. The realization of this vulnerability is fulfilled in the Flooding Attack dataset.

In this research, we stripped the original dataset out of the Timeline and Flag attributes to produce the ten-attribute input dataset and changed the Flags T and R to binary values 1 and 0, respectively to produce the two-attribute output dataset. These two datasets, which are randomly divided into 70% for training, 15% for validation, and 15% for testing, are fed into the ML system for pattern recognition. A snapshot of the results is depicted in Fig. 4.

5.2.2 Fuzzy Attack Detection

Fuzzy attack involves the injection of random or arbitrary values as input into a system. This type of attack dates all the way back in the 1950s when fuzzy testing was applied to computer programs [30]. For the Fuzzy Attack dataset, random values ranging from 0x000 to 0x7FF are generated and injected into the CAN ID and Data fields to form random CAN packets and injected into the CAN bus [28].

We used the same procedure as that in the Flooding Attack dataset, i.e., we stripped the original dataset out of the Timeline and Flag attributes to produce the ten-attribute input dataset and changed the Flags T and R to binary values 1 and 0, respectively to produce the two-attribute output dataset. We also used the same dataset partitioning scheme–randomly dividing the dataset into 70% for training, 15% for validation, and 15% for testing. A snapshot of the results is depicted in Fig. 5.

5.2.3 Malfunction Attack Detection

The Hyundai Sonata uses the CAN ID, 0x316, to gather information about the vehicle engine's Revolution per Minute (RPM). This CAN ID is augmented with random data payload to complete a CAN frame. These CAN frames, which could make the vehicle behave improperly, are interspersed with normal CAN frames to create

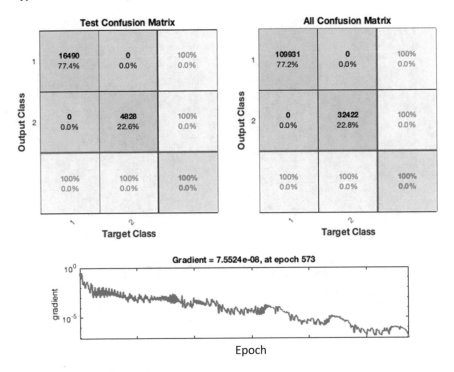

Fig. 4 Flooding attack detection

the entire dataset. Next, we used the same dataset partitioning scheme–randomly dividing the dataset into 70% for training, 15% for validation, and 15% for testing. A snapshot of the results is depicted in Fig. 6. As depicted by the smooth gradient plot on Fig. 6, the training process converged more rapidly, in 38 epochs, compared to that in the Flooding and Fuzzy attacks.

5.2.4 Multiclass (Combined) Attack Detection

The last dataset is comprised of a random selection of all type of attacks and the normal CAN frames. The classes are labeled as **1, 2, 3** and **4 for Flooding, Fuzzy, Malfunction** and **Normal**, respectively. A snapshot of the results is depicted in Fig. 7. The results further validate the efficacy of the ML system using the Levenberg–Marquardt training algorithm for multiclass classification.

A summary of all the results is compiled in Table 3.

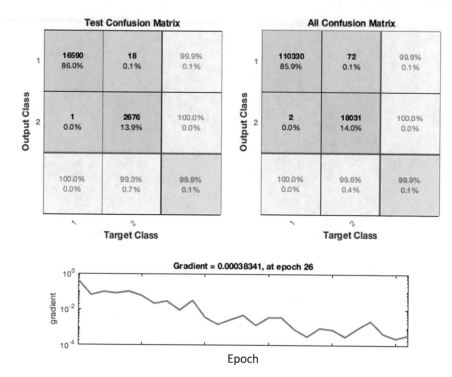

Fig. 5 Fuzzy attack detection

6 Conclusions and Future Directions

In this chapter, we present various applications of ML in vehicle security. We start with pattern recognition of the CAN network packets from a multiclass of vehicle models. Most CAN network traffic packets are very difficult to decipher because their descriptions are considered proprietary by car manufacturers [31]. This is tantamount to "Security by Obscurity." Through a back-propagation neural network ML system, we demonstrate that a given set of CAN network traffic packets can be effectively classified by car manufacturer and model. We then proceed with the detection of three different types of attack on the CAN bus protocol: Flooding (aka Denial of Service or DoS) attack, Fuzzy, and Malfunction (aka Anomaly) attacks. We utilize the datasets gathered and published by vehicle security researchers. The results of this study validate the efficacy of a ML system based on the Levenberg–Marquardt training algorithm. Further, we believe that the results could be used as the foundation for the implementation of an operational intrusion detection system for vehicle security.

While it is tempting to compare the results with those found by Han, et al. [28], we decided not to do so. The reasons for such decision are based on the following:

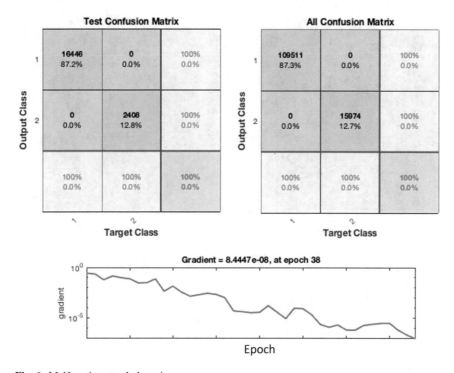

Fig. 6 Malfunction attack detection

- the study conducted by Han, et al. is incongruent with this study with regard to the treatment of the dataset; and
- this study examines the perceived uniqueness of the CAN data due, in part, to the proprietary design information which is often kept as trade secrets by manufacturers. This aspect has not been examined by the Han study and thus, we do not have any result with which to compare our results.

Recognizing the richness and urgency of this research area, we offer the following research directions:

adoption of deep neural networks to optimize the CAN network packet classification system;

introduction of Principal Component Analysis (PCA) to preprocess the dataset attributes for feature reduction; and

investigation of the speed of convergence of various training algorithms on more disparate datasets.

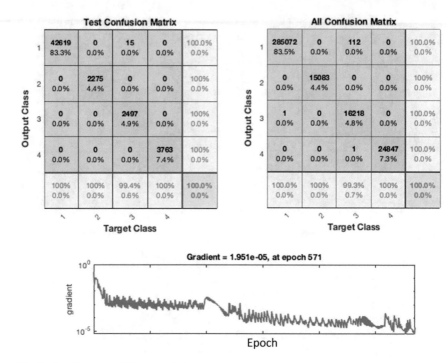

Fig. 7 Combined (multiclass) attack detection

Table 3 Summary of attack classification

	Class	Flooding	Fuzzy	Malfunction	
Single attack versus normal classification	Recall	100	100	100	
	Precision	100	99.9	100	
	F-measure	100	99.9	100	
	Accuracy	100	99.9	100	
Multiclass classification	Recall	100	100	99.3	
	Precision	100	100	100	
	F-measure	100	100	99.6	
	Accuracy				99.9

Acknowledgements This work is partially supported by the Florida Center for Cybersecurity, under grant # 3901-1009-00-A (2019 Collaborative SEED Program) and the National Security Agency under Grant Number H98230-19-1-0333. The United States Government is authorized to reproduce and distribute reprints notwithstanding any copyright notation herein.

References

1. Karahasanovic A (2016) Automotive cyber security. Chalmers University of Technology, Gotehnburg, Sweden
2. Gemalto (2018) Securing vehicle to everything [Online]. Available: https://www.gemalto.com/brochures-site/download-site/Documents/auto-V2X.pdf. Accessed 13 April 2020
3. Francia GA (2020) Connected Vehicle Security. In: 15th international conference on cyber warfare and security (ICCWS 2020), Norfolk
4. Torre GD, Rad P, Choo KR (2020) Driverless vehicle security: challenges and future research opportunities. Future Gener Comput Syst 108:1092–1111
5. Devir N (2019) Applying machine learning for identifying attacks at run-time [Online]. Available: https://www.cs.technion.ac.il/users/wwwb/cgi-bin/tr-get.cgi/2019/MSC/MSC-2019-06.pdf. Accessed 13 April 2020
6. Avatefipour O, Al-Sumaiti AS, El-Sherbeeny AM, Awwad EM, Elmeligy MA, Mohamed MA, Malik H (2019) An intelligent secured framework for cyberattack detection in electric vehicles' CAN bus using machine learning. IEEE Access 7:127580–127592. https://doi.org/10.1109/ACCESS.2019.2937576
7. Vasistha DK (August 2017) Detecting anomalies in controller area network (CAN) for automobiles [Online]. Available: https://cesg.tamu.edu/wp-content/uploads/2012/01/VASISTHA-THESIS-2017.pdf. Accessed 13 April 2020
8. Zhou A, Li Z, Shen Y (2019) Anomaly detection of CAN bus messages using a deep neural network for autonomous vehicles. Appl Sci 9:3174
9. Lokman S, Othman AT, Abu-Bakar M (2019) Intrusion detection system for automotive controller area network (CAN) bus system: a review. J Wireless Com Network 184 https://doi.org/10.1186/s13638-019-1484-3
10. Kang MJ, Kang JW (2016) Intrusion detection system using deep neural network for in-vehicle network security. PLoS One 11(6)
11. Taylor A, LeBlanc S, Japkowiz N (2016) Anomaly detection in automobile control network data with long short-term memory networks. In: 2016 international conference on data science and advanced analytics (DSAA), Montreal
12. Wasicek A, Weimerskirch (2015) Recognizing manipulated electronic control units. SAE
13. Jaynes M, Dantu R, Varriale R, Evans N (2016) Automating ECU identification for vehicle security. In: 2016 15th IEEE international conference on machine learning and applications (ICMLA), Anaheim, CA
14. Kumar S, Singh K, Kumar S, Kaiwartya O, Cao Y, Zhao H (2019) Delimitated anti jammer scheme for internet of vehicle: machine learning based security approach. IEEE Access 7:113311–113323
15. Corrigan S (2016) Introduction to the controller area network (CAN). Texas Instruments, Dallas, TX
16. Maggi F (2017) A vulnerability in modern automotive standards and how we exploited it. Trend Micro
17. Bishop CM (2007) Patern recognition and machine learning. Springer, Belrin
18. Kim J, Francia G (2018) A comparative study of neural network training algorithms for the intelligent security monitoring of industrial control systems. In: Computer and network security essentials. Springer International Publishing AG, pp 521–538
19. De Boer P, Kroese DK, Mannor S, Rubinstein RY (2005) A tutorial on the cross-entropy method. Ann Oper Res 134:19–67
20. McCaffrey J (2014) Neural network cross entropy error. Vis Studio Mag 04:11
21. Marquardt D (1963) An algorithm for least-squares estimation of nonlinear parameters. SIAM J Appl Math 11(2):431–441
22. Dennis JE, Schnabel RB (1983) Numerical methods for unconstrained optimization and nonlinear equations. Prentice-Hall, Englewoods Cliffs, NJ
23. Riedmiller M, Braun H (1993) A direct adaptive method for faster backpropagation learning: the RPROP algorithm. In: Proceedings of the IEEE international conference on neural networks

24. Moller M (1993) A scaled conjugate gradient algorithm for fast supervised learning. Neural Netw 6:525–533
25. Hagan MT, Demuth HB, Beale MH (1996) Neural network design. PWS Publishing, Boston
26. Scales L (1985) Introduction to non-linear optimization. Springer-Verlag, New York
27. Magnus JR, Neudecker H (1999) Matrix differential calculus. John Wiley & Sons Ltd., Chichester
28. Han ML, Kwak BI, Kim HK (2018) Anomaly intrusion detection method for vehicular networks based on survival analysis. Veh Commun 14:52–63
29. Crow D, Graham S, Borghetti B (2020) Fingerprinting vehicles with CAN Bus data samples. In: Proceedings of the 15th international conference on cyber warfare and security (ICCWS 2020), Norfolk, VA
30. Weinberg GM (5 Feb 2017) Fuzz testing and fuzz history [Online]. Available: https://secretsof consulting.blogspot.com/2017/02/fuzz-testing-and-fuzz-history.html. Accessed 6 April 2020
31. Stone B, Graham S, Mullins B, Kabban C (2018) Enabling auditing and intrusion detection for proprietary controller area networks. Ph.D. Dissertation, Air Force Institute of Technology, Dayton, OH

Mobile Application Security Using Static and Dynamic Analysis

Hossain Shahriar, Chi Zhang, Md Arabin Talukder, and Saiful Islam

Abstract The mobile applications have overtaken web applications in the rapid growing of the mobile app market. As mobile application development environment is open source, it attracts new inexperienced developers to gain hands-on experience with application development. However, the data security and vulnerable coding practice are two major issues. Among all mobile operating systems including iOS (by Apple), Android (by Google) and Blackberry (RIM), Android remains the dominant OS on a global scale. The majority of malicious mobile attacks take advantage of vulnerabilities in mobile applications, such as sensitive data leakage via the inadvertent or side channel, unsecured sensitive data storage, data transition and many others. Most of these vulnerabilities can be detected during mobile application analysis phase. In this chapter, we explored some existing vulnerability detection tools available for static and dynamic analysis and hands-on exploration of using them to detect vulnerabilities. We suggest that there is a need of new tools within the development environment for security analysis in the process of application development.

Keywords Android security · Static analysis · Dynamic analysis · iOS

H. Shahriar (✉) · C. Zhang · M. A. Talukder · S. Islam
Department of Information Technology, Kennesaw State University, Kennesaw, USA
e-mail: Hshahria@kennesaw.edu

C. Zhang
e-mail: czhang4@kennesaw.edu

M. A. Talukder
e-mail: mtalukd1@students.kennesaw.edu

S. Islam
e-mail: sislam16@students.kennesaw.edu

© The Editor(s) (if applicable) and The Author(s), under exclusive license to Springer
Nature Switzerland AG 2021
Y. Maleh et al. (eds.), *Machine Intelligence and Big Data Analytics for Cybersecurity Applications*, Studies in Computational Intelligence 919,
https://doi.org/10.1007/978-3-030-57024-8_20

443

1 Introduction

In 2018, a report by Statcounter Globalstats[1] reported that Android dominates the smartphone OS market with 76.6% share and over two billion active users monthly. Consequently, this gigantic market of potential victims has been noticed by cyber-criminals and online malicious users. The attackers exploit typical human nature, by advertising the applications as free to use, which often results in large number of downloads of these malicious applications. Malicious software is generally packed with several forms of malware payloads including but not limited to, trojans, botnets, and spyware. These applications can easily help theft of valuable personal information, such as usernames, passwords, social security numbers, health history, location history and much more.

With this explosive rise in the number of Android malware applications, analysis, detection and prevention against such applications have become a critical research topic. Several techniques, frameworks, tools have been proposed to prevent and detect such applications [1–5]. Consequently, the research in Android security and analysis has transitioned into a wide domain in both academic and enterprise communities. Despite many tools are available, there are limited resources in the literature emphasizing hands-on application of the tools for analyzing malware. To contribute to this area, in this chapter, we explore several popular tools employing both static and dynamic analysis to analyze the source code and behavior of android applications.

The rest of this chapter is structured and organized as follows: Sect. 2 provides an overview of the major related works. We highlight examples of Android static and dynamic analysis tools, their usage, strengths and weaknesses. Section 3 presents hands-on analysis with the tools. Finally, Sect. 4 concludes this chapter.

2 Related Works

In the fast growing mobile application market and mobile app development, data security that protects the privacy and integrity of users data in the mobile application should be a top priority [6–8]. We reviewed a number of the security tools that are known for detecting security problems in the Android Applications. Interested readers can see the extensive survey by Kong et al. [9] for other related work.

2.1 CuckooDroid

CuckooDroid is a premier malware analyzing software. It is capable of methodically examining multiple variants of malware through the use of virtual machines that monitor the behavior in a protected and isolated atmosphere [4]. It is written in

---

[1] http://gs.statcounter.com/os-market-share/mobile/worldwide.

the python programming language and facilitates its analysis in both the static and dynamic dimensions [3]. Cuckoo is an open source malware analysis system. It is used to run and investigate files and gather inclusive analysis results of what the malware is and does even though they are running inside a secluded Windows OS [4]. It can generate the following results:

- Files created, deleted, and downloaded by the malware
- Memory dumps of the malware processes.
- Traces of win32 API calls accomplished by all processes produced by the malware.
- Network traffic trace in PCAP format.
- Full memory dumps of the machines.

The sandbox of Cuckoo started in 2010, as a "Google Summer of Code" project within the Honeynet project [3]. The first beta release of Cuckoo was released in Feb 2011 and then in March 2011 when it was selected again as a supported project. After many versions of the sandbox, Cuckoo sandbox was released in April 2014.

It comprises of a host that is responsible for the sample execution and the analysis in which the guests run. When the host has to launch a new analysis, it chooses the guests and uploads that sample as well as the other components that are required by the guest to function [10]. Cuckoo initializes modules when it first starts up and when there is a new task sent to Cuckoo, it identifies the file by using the machinery modules which is used to interrelate with the diverse possible virtualization systems, and configuration, it installs what is known as the analyzer inside one of the available virtual machines.

Once the analysis has completed, the analyzer refers the results of the analysis to the ResultServer, which in turn will implement whichever processing modules are configured (the modules used to populate the product of the analysis, the report) and produce the report [4]. The analysis took place in the virtualized machine, which has the monitoring system components. The proxy is a Python script that intervals listening to a port in the guest machine. When a new inquiry is launched in the machine, the host sends the equivalent analyzer (the component in charge of managing the analysis inside the machine) and the package module used to accomplish the sample sent, which depends on the type of sample [4].

For instance, the array used to implement an exe file will be altered when it is used to open a PDF sample, or a ZIP file. When the mockup completes its implementation, or a break is reached, the analyzer stops the analysis, collects the results from the monitor, and sends them back to the result server.

2.2 FlowDroid

With more common data breaches, data privacy and security protections become increasingly important. Most privacy leaks are due to flaws in the code that could have been prevented, if the flaws are detected and fixed in a timely manner. FlowDroid is an open-source Java based tool that can analyze Android applications for potential data leakage. FlowDroid is the first full context, object sensitive, field, flow, and static taint analysis tool that specifically models the full Android lifecycle with high precision and recall [1]. While it is not meant to analyze malware [11], the tool can detect and analyze data flows, specifically an Android application's byte code, and configuration files, to find any possible privacy vulnerabilities, also known as data leakage [1]. FlowDroid does different types of taint analysis: objective sensitive, flow, context, field, and lifecycle aware [12]. In regard to reflective calls, FlowDroid can only fix the reflective calls that have constant strings as parameters [13].

This tool can be built using Maven, a build automation tool for Java projects, or Eclipse, an IDE used for software development. The data tracker can be used from the command line [2]. This tool cannot be used in Android Studio since it is not an Android application. These additional tools need to be downloaded in order for FlowDroid to run properly. The tools include Jasmin, Soot, Heros, and GitHub repositories soot-infoflow and soot-infoflow-android [2]. Jasmin is an open-source tool that can convert Java classes ACII descriptions into binary Java class files that can be loaded into a Java interpreter [14]. Soot is an open-source Java optimization framework that has four different types of representation of analyzing and changing Java bytecode [2]. Heros is a general implementation of an IFDS and IDE framework solution that can be integrated into an existing Java program analysis framework [15].

Based on Reaves et al.'s experience with setting up FlowDroid, it took them 1.45 h to fully and properly set up the tool [12]. Reaves et al. had to download missing SDK files that were necessary to run DroidBench, as well as Flow Droid .jar files. Two minutes were spent changing configuration settings of the analyzed mobile applications [12].

FlowDroid uses an analysis technique based on an analysis framework that does not rely on every program path [1]. This means that every program path does not need to be analyzed. Android applications do not contain a main method in their code. Instead, they have methods that are indirectly invoked upon by the Android framework [1]. This leads to a problem where Android analyzing tools cannot start the analyzing process by evaluating the main method of the program [1]. FlowDroid solves this problem by generating and analyzing a fake main method where there is every possible life cycle arrangement of separate application components and callbacks [1]. It is not necessary to go through all the possible paths, because the technique previously stated solves this problem, and it would also be expensive to implement [1].

FlowDroid uses a call graph technique to accurately map components to callbacks, which leads to minimal false positives and lowered taint analysis running time [1]. The tool generates one call graph for each application component [1]. The call graph is used in the process of scanning calls to Android system methods that has a popular callback interface as a parameter [1]. The call graph gets extended until all callbacks are found [1]. Yanick et al. researched detection of logic bombs in Android applications using analysis tools, including FlowDroid. FlowDroid was tested for malware analysis along with three other analysis tools. The other three analysis tools were Kirin, TriggerScope, and DroidAPIMiner. Among the other analysis tools, FlowDroid had the highest false positive percentage, and second lowest false negative percentage [11]. This tool is not great at malware detection, specifically logic bombs, because that was not the intended purpose of this tool. A logic bomb is an unauthorized software that changes the output of the Android application or does applications actions that are not intended [11]. Yanick et al. found that FlowDroid had high false positive percentage of 69.23% and a low false negative percentage of 22.22% [11]. Most Android applications have sensitive data flow, so most of the time when an Android application contained sensitive data flows, it triggers a false positive [11]. This accounts for the high false positive percentage [11]. Not all detected sensitive data flows triggered a false positive because FlowDroid did not relate every sensitive data flows to logic bomb detection [11], which is the main reason for the false negatives [11].

The first limitation of FlowDroid is that it is not very good at multi-threading. The tool executes threads in a consecutive order [1]. The second limitation is that the tool only fixes self-called callbacks if the parameters are constant string [1]. The third limitation is hardware resource. Reaves et al. [12] experienced obstacles when trying to run analysis on real mobile applications. First, FlowDroid was analyzed on a computer system with dual quad core Xeon processors and 48 gigabytes of RAM [12]. The computer with these specifications lacked the acceptable memory space to run the necessary analysis, so Reaves et al. [12] tried an alternative approach, to use an Amazon EC2 Ubuntu virtual machine with 12 virtual CPUs and 64 gigabytes of RAM. Unfortunately, this alternative had the same result. Only one mobile application was able to be analyzed when Reaves et al. used Amazon EC2 virtual machine with 64 gigabytes of RAM and 16 virtual CPUs. The fourth limitation is that leaks are not traced if it was caused by multi-threading and implicit data flows [16]. This can cause false negatives. The fifth limitation is that this tool does not analyze tainted data flows that involve file accesses [16]. This tool is not able to do a complete analysis of all of the Android applications even with 64 gigabytes of RAM and 16 virtual CPUs. It is uncertain if the hardware resource or number of Android applications to analyze could be the problem. More research with different list size of Android applications and hardware resource in the same test are needed to have a more definitive answer to this issue. Moreover, multi-threading needs to be included in a future update. It would help analyze the Android applications quicker. Currently, the tool only fixes self-called callbacks if the parameters are string constants. The type of parameters need to be expanded, so the tool can fix other types of callbacks. Leaks caused by multi-threading and implicit data flows need to be recognized, as well as tainted data flows involving file accesses, so FlowDroid can do a complete and reliable analysis for data leaks.

2.3 DroidBox

DroidBox is a dynamic malware analysis tool for Android applications. DroidBox v4.1.1 is a framework for analyzing automatically Android applications. It uses a modified version of the Android emulator 4.1.1_rc6 enabling to track activities of

Android applications, i.e., tainted data leaked out, SMS sent, network communications, etc. It is composed of two folds: one fold on the guest machine (Android emulator) that tracks the Android application's activity and sends the corresponding DroidBox logs through ADB to the host machine whereas the other fold on the host machine that parses the ADB log to extract the DroidBox log [17]. The release has only been tested on Linux and Mac OS. The Android SDK can be downloaded from http://developer.android.com/sdk/index.html. The following libraries are required: pylab and matplotlib to provide visualization of the analysis result.

Export the path for the SDK tools.

$ export PATH = $PATH:/path/to/android-sdk/tools/
$ export PATH = $PATH:/path/to/android-sdk/platform-tools/

Download necessary files and decompress it anywhere.

wget
https://github.com/pjlantz/droidbox/releases/download/v4.1.1/DroidBox411RC.tar.gz

Setup a new AVD targeting Android 4.1.2 and choose Nexus 4 as device as well as ARM as CPU type by running:
Start the emulator with the new AVD:

$ ./startemu.sh < AVD name>

When emulator has booted up, start analyzing samples (please use the absolute path to the apk):

$ ./droidbox.sh < file.apk >< duration in secs (optional)>

The analysis is not automated currently except for installing and starting packages. Ending the analysis is simply done by pressing Ctrl-C.

DroidBox analyzes incoming and outgoing network activity in an application. It also records and analyzes all file read and/or write activity of an application. All initialized services and loaded classes are recorded and analyzed through DexClass-Loader component. It also reports on any information leakages either through network activity, file operations, or SMS. Furthermore, it monitors security permission protocols and returns warnings if a protocol is circumvented. If there is any broadcast activity, DroidBox monitors and lists all the receivers and listeners. As default, DroidBox monitors all calls and SMS activities, analyzes each one and returns results. Additionally, two graphs are generated visualizing the behavior of the package—one showing the temporal order of the operations and the other one being a treemap that can be used to check similarity between the analyzed packages [17]. Figure 1 shows the temporal order of application behavior. DroidBox maps each activity to a specific timestamp which provides an oveview of the linear behavior of the application based on Android system events. These events could be sending SMS, making a call, reading/writing or some other internal system activities.

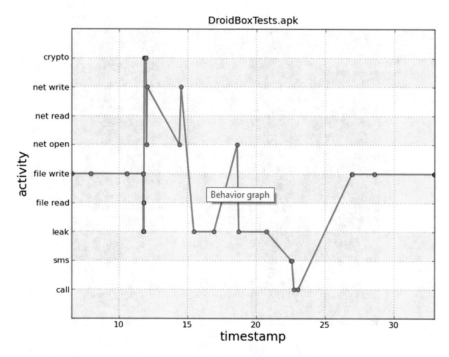

Fig. 1 Apk test result by DroidBox

Figure 2 presents a simple image that shows the similarity between packages and related operations carried out on them. This enables a user to easily locate related packages that were affected as part of a specific operation.

Unlike MobSF that provides output targeting the specific source of a behavior, DroidBox provides an overview of the related packages and type of behavior such as SMS, network leakages, etc. that occurred. Although DroidBox indicates the general behavior and possible suspect packages, it does not pinpoint the source code. Other tools such as MobSF indicates what sources are related to the behavior and captures timeline of behavior. However, DroidBox provides relevant information regarding multiple behavior of packages and application. DroidBox analysis may be used to have an overview of malware behavior and can provide suggestions as to which specific packages and resources in which security testing could begin with. DroidBox is a good analysis tool that can be used in the early states of dynamic malware analysis in Android applications. While it provides some helpful output to identify application behavior and localize affected packages, it does not give detailed information about what the underlying code is. Moreover, it does not provide any output to what specific modules in the code that is responsible for the output behavior and what steps must be taken to remove and or quarantine the affected packages. Droidbox is a purely dynamic analysis tool. It does not perform any static analyses. As a result, for a user who wants to do both dynamic and static analyses, another tool needs to be installed to perform static analysis.

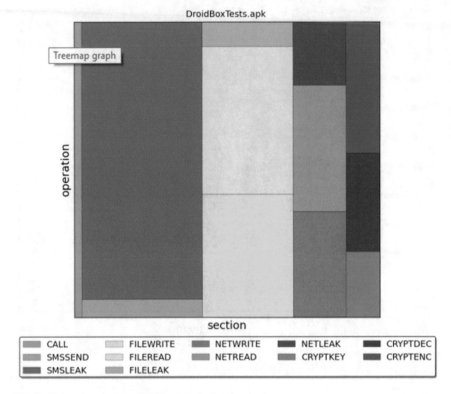

Fig. 2 Comparison between packages and related operation

3   Hands-on Analysis

In this section, we provide hands-on static and dynamic analysis of Android applications. Static analysis approach analyzes the source apk file of android application
and identifies possible security risks without actually running on a mobile device or
emulator. On the other hand, a dynamic analysis approach runs an apk file into a
virtual emulator and analyzes activities performed to identify malicious activities.

We analyzed samples obtained from contagiodump [18]. We first test malware
samples in Mobile Security Framework (MobSF) [19] and Flowdroid [2]. However,
there are more analysis tools available such as DroidBox, CuckooDroid, and Droid-
Safe. Table 1 shows the list of tools and their coverage in terms of static and dynamic
analysis of apk files.

Table 1 Analysis coverage of example tools

Tool	Static analysis	Dynamic analysis
Cuckoodroid	No	Yes
Flowdroid	Yes	No
Droidbox	No	Yes
MobiSF	Yes	Yes

3.1 Static Analysis by MobiSF

Android is a combined form of java and xml. It uses Linux operating system which is developed for embedded systems and mobile devices. The upper layers of Android language is written by itself. For Graphical User Interface (GUI) android uses XML layout files, it also uses event-based library. Static analysis refers to decompiling the APK of an application to its corresponding Java and XML files. To perform static analysis a static analyzer must have features to examine XML with correctness and precision. Java files can be extracted by DEXtoJar decompiler [20].

A Static analysis tool de-compiles the code of an APK to human readable format. So that an analyzer can read the code and identify the vulnerability that an application could have. "Mobile Security Framework (MobSF)" is a combined tool which does the static and dynamic analysis of an APK. MobSF is developed based on Python 3.7. To generate the report MobSF uses html. It runs the local server via command line in the host computer. All the existing static and dynamic analysis tool do the analysis using DroidMon-Dalvik Monitoring Framework [21] and Xposed Module Repository [22]. Xposed is a framework that can change the behavior of the system and apps without touching the APK [23]. MobSF works fine in Windows 7 and 10, macOs(El capitan, high Sierra), Linux(Kali, Ubuntu 16.04). The tools is available in gitHub repository [2]. We did the analysis in Unix operating system. We used macOS High Sierra to perform the analysis. To start the development server the following command has to be executed on the terminal.

Figure 3 shows the terminal command for Unix operating system to start development server. The command will initiate the development server at "". In this server, we can upload the APK file for analysis. The dynamic environment is also available in the development server. See Fig. 4.

Figure 4 shows the development server running at "http://127.0.0.1:8000/". We have several APK files using MobSF. A vulnerable APK has been tested by MobSF that includes SQL injection vulnerability. The application finds data from the database based on user input. It performs a raw query without sanitizing the user input. Which

Fig. 3 Terminal command to start server MobSF

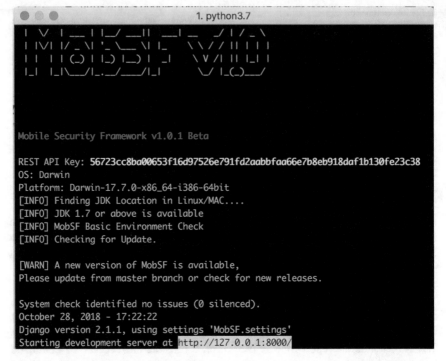

Fig. 4 Development server running by MobSF

dumps all the content from the database. Later, it tries to send the data to a static number that is hardcoded in the application. Figure 5, shows SMS API permission used in the manifest.

MobSF finds the Android API call in an application that also finds the corresponding java file. For example, in this application SMS API has been called in the MainActivity.java file. See Fig. 6.

≡ Android Permissions

PERMISSION	STATUS	INFO	DESCRIPTION
android.permission.SEND_SMS	dangerous	send SMS messages	Allows application

Fig. 5 Permission analysis

⊙ Android API

API	FILES
Inter Process Communication	sqlinjection/sqliexample/sqlinjection0717/MainActivity.java
Send SMS	sqlinjection/sqliexample/sqlinjection0717/MainActivity.java
Java Reflection	android/arch/lifecycle/Lifecycling.java android/arch/lifecycle/ReflectiveGenericLifecycleObserver.java
Crypto	com/h3xstream/findsecbugs/crypto/InsufficientKeySizeBlowfishDetector.java com/h3xstream/findsecbugs/crypto/StaticIvDetector.java
Message Digest	com/h3xstream/findsecbugs/crypto/CustomMessageDigestDetector.java com/h3xstream/findsecbugs/crypto/BadHexadecimalConversionDetector.java
URL Connection supports file,http,https,ftp and jar	com/h3xstream/findsecbugs/injection/custom/CustomInjectionDetector.java
Content Provider	com/h3xstream/findsecbugs/injection/sql/AndroidSqlInjectionDetector.java

Fig. 6 API call

3.2 Dynamic Analysis Using MobiSF

Dynamic analysis refers to analyzing the functionality of an application in an isolated device or emulator. In terms of emulator, it should be launched on a virtual machine in the host computer that keep our host computer safe from being affected. We did dynamic analysis using MobSF's pre-configured Android virtual machine [19]. We used a sample malware apk, which apparently look like a movie player. To start analysis virtual environment has to be created first. See the image below to create a virtual environment in a virtual box.

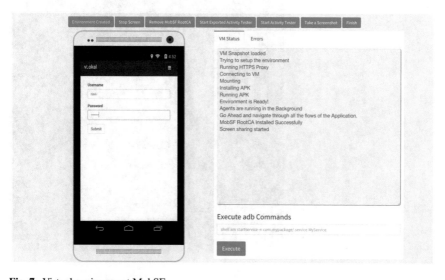

Fig. 7 Virtual environment MobSF

Figure 7 shows the virtual environment created to perform the dynamic analysis of an application. Here we found a couple of options for analysis such as Activity Tester, Screenshot of each screen, and Adb command execution field. After successful installation of the apk, we got a popup window that show a notification for sending a text message to a static number. See Fig. 8.

MobSF also generates the dynamic analysis report. After finishing the analysis process it produces an analysis report in HTML format, as shown below.

Figures 9 and 10 show the dynamic reports which highlight the application sending text message to hard coded numbers. In Fig. 10, we also find that a database journal has been created in the device through this application.

Fig. 8 Dynamic analysis of android application in MobSF

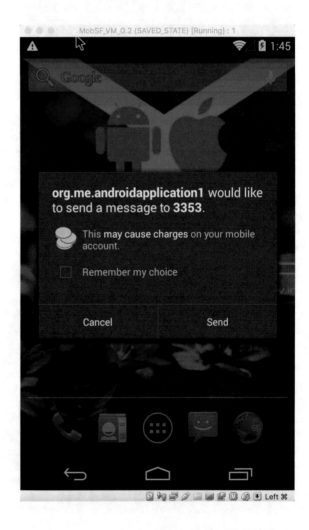

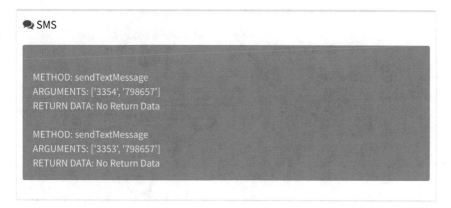

Fig. 9 Dynamic analysis report—Part 1

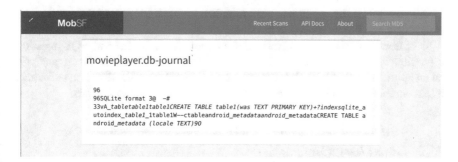

Fig. 10 Dynamic analysis report—Part 2

3.3 Tainted Data Flow Analysis

Android has multiple data sources as well as sinks. Most of the sinks are built in API's, for example, SMS API, CALL API, etc. An application gets data through these sources for processing. Soot and Heroes are FlowAnalysis tool built to analyze java application [15, 24]. Soot uses worklist algorithm to do static data flow analysis. It has FlowAnalyis classes for fixed-point computation of static data flow analysis. There are classes as well for ForwardAnalysis and BackwardAnalysis [24]. Heros is an extension of soot, it requires its implementation as a class that extends Scene-Transformer [15]. FlowDroid is a data flow analysis tool which is built over soot and heros to do static data flow analysis from source to sink of an Android Application [25]. FlowDroid does context-, flow-, field- and object-sensitive analysis. To increase recall it creates a complete model of Android's app lifecycle [26]. Figure 11 shows a model of source to sink data flow design that used in FlowDroid.

In Fig. 11, in main function a source is encountered. The value of the source is attached with the object that has been passed as a formal parameter to the function.

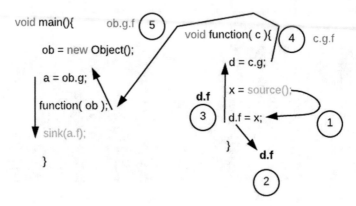

Fig. 11 Data flow from source to sink

In this case when the object has been initialized in main, it ultimately receives the source data and finally passed to the sink.

Data flow analysis can be performed using FlowDroid. In Android, SQL injection is possible because Android has the feature of SQLite Database. We did our analysis on an application that fetches username and password from the database and sends it to a number via SMS API. The following command starts the analysis.

java -jar soot-infoflow-cmd/target/soot-infoflow-cmd-jar-with-

*dependencies.jar *

*-a <APK File> *

*-p <Android JAR folder> *

-s <SourcesSinks file>

java -jar soot-infoflow-cmd/target/soot-infoflow-cmd-jar-with-

*dependencies.jar *

*-a <APK File> *

*-p <Android JAR folder> *

*-s <SourcesSinks file> *

-o <Xml file for output>

Here, '-a' is the path of the APK file, '-p' is the path of the platforms folder that could be found under sdk folder of the android directory, '-s' is a text file where we can declare the possible sources and sinks for the analysis, and '-o' is an additional parameter for the analysis that is used to produce an output file as a XML file.

```
● ● ●                          📄 SourcesAndSinks (1).txt
<android.database.Cursor: java.lang.String getString(int)> -> _SOURCE_

<android.telephony.SmsManager: void
sendTextMessage(java.lang.String,java.lang.String,java.lang.String,android.app.PendingInte
nt,android.app.PendingIntent)> android.permission.SEND_SMS -> _SINK_
```

Fig. 12 SourcesSinks.txt file

```
[main] INFO soot.jimple.infoflow.android.SetupApplication$InPlaceInfoflow - The sink virtualin
voke $r5.<android.telephony.SmsManager: void sendTextMessage(java.lang.String,java.lang.String
,java.lang.String,android.app.PendingIntent,android.app.PendingIntent)>("123456", null, $r1, n
ull, null) in method <sqlinjection.sqliexample.sqlinjection0717.MainActivity: void showResult(
java.lang.String)> was called with values from the following sources:
[main] INFO soot.jimple.infoflow.android.SetupApplication$InPlaceInfoflow - - $r1 = interfacei
nvoke $r4.<android.database.Cursor: java.lang.String getString(int)>(1) in method <sqlinjectio
n.sqliexample.sqlinjection0717.MainActivity: void showResult(java.lang.String)>
[main] INFO soot.jimple.infoflow.android.SetupApplication$InPlaceInfoflow - - $r1 = interfacei
nvoke $r4.<android.database.Cursor: java.lang.String getString(int)>(2) in method <sqlinjectio
n.sqliexample.sqlinjection0717.MainActivity: void showResult(java.lang.String)>
[main] INFO soot.jimple.infoflow.android.SetupApplication$InPlaceInfoflow - Data flow solver t
ook 0.338522 seconds. Maximum memory consumption: 104.322368 MB
[main] INFO soot.jimple.infoflow.android.SetupApplication - Found 1 leaks
Arabins-MacBook-Pro:FlowDroid-2.6.1 arabin$ ▌
```

Fig. 13 Data flow analysis by FlowDroid

Figure 12, is sample SourceSinks.txt file that has all the possible sources and sinks declared in it. Figure 13 shows only one data leak in this application.

We find one sink inside the *onCreate()* method of the MainActivity.java file which is *sendTextMessage()* method in the SMS manager API. The possible sources for this sink are *getString()* methods which has been found in *showResult()* method in the *MainActivity.java* class [27].

4 Conclusion

This paper presents an analysis report of existing mobile security tools available in the market. We find that Flowdroid and MobiSF are suitable tools for detecting malicious Android mobile applications. In addition, we provided a review of other notable tools and the use of the tools, and demonstrated hands-on analyses.

We find a gap in the current tool support on security analysis for the Android application development environments, i.e., Android Studio, the most widely used development environment. Currently it does not have a comprehensive built-in security analysis module. Hence, an important research direction is to develop a security analysis plugin tool to improve the security testing ability. This will enable app developers to test their developed applications before deploying them on the Google Store market for the general public to download and use. The security analysis module as an integral component in the Android app development environment will be able to reduce the risk of privacy invasion and data leakage as well as mitigate malwares and spywares in the Android applications.

References

1. Arzt S et al (2013) FlowDroid: precise context, flow, field, object-sensitive and lifecycle-aware taint analysis for Android apps. In: Proceedings of the 35th ACM SIGPLAN conference on programming language design and implementation - PLDI '14, Edinburgh, United Kingdom, 2013, pp 259–269. https://doi.org/10.1145/2594291.2594299
2. Arzt S, Dann A, Bodden E, Benz M, Amin A (2020) Sable/soot - FlowDroid. Secure software engineering group at Paderborn University and Fraunhofer IEM
3. CuckooDROiD (2004) Installation — CuckooDroid v1.0 Book. https://cuckoo-droid.readth edocs.io/en/latest/installation/. Accessed 24 May 2020
4. CuckooDROiD (2014) What is Cuckoo? — CuckooDroid v1.0 Book. https://cuckoo-droid.rea dthedocs.io/en/latest/introduction/what/. Accessed 24 May 2020
5. Lerch J, Arzt S, Laverdière MA, Benz M, jtoman (2020) Sable/heros. GitHub. https://github. com/Sable/heros. Accessed 24 May 2020
6. 3 Reasons mobile app security should be a top priority. Zimperium Mobile Security Blog (14 April 2020). https://blog.zimperium.com/3-reasons-mobile-app-security-should-be-a-top-priority/. Accessed 23 May 23
7. Alzubaidi A, Roy S, Kalita J (2019) A data reduction scheme for active authentication of legitimate smartphone owner using informative apps ranking. Digit Commun Networks 5(4):205–213. https://doi.org/10.1016/j.dcan.2018.09.001
8. Atkinson JS, Mitchell JE, Rio M, Matich G (2018) Your WiFi is leaking: what do your mobile apps gossip about you? Future Gener Comput Syst 80:546–557. https://doi.org/10.1016/j.fut ure.2016.05.030
9. Kong P, Li L, Gao J, Liu K, Bissyandé TF, Klein J (2019) Automated testing of android apps: a systematic literature review. IEEE Trans Reliab 68(1):45–66. https://doi.org/10.1109/TR.2018. 2865733
10. Li L, Bissyandé TF, Octeau D, Klein J (2016) Reflection-aware static analysis of Android apps. In: 2016 31st IEEE/ACM international conference on automated software engineering (ASE), pp 756–761
11. Fratantonio Y, Bianchi A, Robertson W, Kirda E, Kruegel C, Vigna G (2016) TriggerScope: towards detecting logic bombs in android applications. In: 2016 IEEE Symposium on Security and Privacy (SP), May 2016, pp 377–396. https://doi.org/10.1109/sp.2016.30
12. Reaves B et al (Oct 2016) *droid: assessment and evaluation of android application analysis tools. ACM Comput Surv 49(3):55:1–55:30. https://doi.org/10.1145/2996358
13. Qiu L, Wang Y, Rubin J (2018) Analyzing the analyzers: FlowDroid/IccTA, AmanDroid, and DroidSafe. In: Proceedings of the 27th ACM SIGSOFT international symposium on software testing and analysis, Amsterdam, Netherlands, Jul 2018, pp 176–186. https://doi.org/10.1145/ 3213846.3213873
14. Lhoták O, Bartel A, Arzt S, Benz M (2020) Sable/jasmin. Sable Research Group
15. Bodden E (14 Jan 2020) Example: using heros with soot. GitHub. https://github.com/Sable/ heros. Accessed 24 May 2020
16. Bhosale AS (2014) Precise static analysis of taint flow for android application sets. Carnegie Mellon University
17. Lantz P (2015) Droidbox 4.1.1. GitHub. https://github.com/pjlantz/droidbox. Accessed 24 May 2020
18. Mila (19 Apr 2020) KPOT info stealer samples. Contagio. http://contagiodump.blogspot.com/ 2020/04/kpot-info-stealer-samples.html. Accessed 24 May 2020

19. Abraham A, Schlecht D, Ma G, Dobrushin M, Nadal V (2020) Mobile security framework (MobSF). Mobile Security Framework
20. Ashour SA, Stotz J, Donlon (2020) Dex to Java decompiler
21. CuckooDROiD (2020) Dalvik monitoring framework for CuckooDroid
22. rovo89 Xposed Installer | xposed module repository. https://repo.xposed.info/module/de.robv. android.xposed.installer. Accessed 24 May 2020
23. Spreitzenbarth M, Schreck T, Echtler F, Arp D, Hoffmann J (2015) Mobile-sandbox: combining static and dynamic analysis with machine-learning techniques. Int J Inf Secur 14(2):141–153. https://doi.org/10.1007/s10207-014-0250-0
24. Einarsson A, Nielsen JD (17 Jul 2008) A survivor's guide to java program analysis with Soot. https://www.brics.dk/SootGuide/. Accessed 24 May 2020
25. Talukder M, Shahriar H, Haddad H (2019) Point-of-sale device attacks and mitigation approaches for cyber-physical systems. In: Cybersecurity and privacy in cyber physical systems, CRC Press, pp 368–383
26. Arzt S (2016) Static data flow analysis for android applications. Technische Universitat Darmstadt
27. Talukder MAI et al (Jul 2009) DroidPatrol: a static analysis plugin for secure mobile software development. In: 2019 IEEE 43rd annual computer software and applications conference (COMPSAC), vol 1, pp 565–569. https://doi.org/10.1109/compsac.2019.00087

Mobile and Cloud Computing Security

Fadi Muheidat and Lo'ai Tawalbeh

Abstract The technological wonders that people have been reading about the various literature relating to science-fiction are today coming into reality. Drawing from 3D-holograms, cell phones, robots, artificial intelligence among other novelties devices of modern engineering, human beings are today enabled to perform many tasks including those which were deemed to be impossible about twenty years ago. Mobile and cloud computing technology has evolved with time and becomes an essential component in the nowadays industry. It will keep growing as more devices are connected and more data is fed into the cloud. Most businesses provide their services through mobile applications and devices. Mobile and Cloud computing is making big data analytics, distributed and real-time artificial intelligence and machine learning, blockchain, cryptocurrency, wearables, internet of things, and cyber-physical systems, possible. The dynamic nature of big data and information flow calls for the need to establish strong protection against threats that could emerge from big data. The complexity and sensitivity of these systems and the secure data analysis will require multilevel and kinds of security measures and standards. In this chapter, we are answering some of the questions, concerns, and challenges mobile devices and cloud computing and big data cybersecurity are facing.

Keywords Mobile computing · Cloud computing · Security · Privacy · Authentication · Data integrity · Big data · Artificial intelligence · Encryption · Virtualization

F. Muheidat
California State University, San Bernardino, San Bernadine, CA 92407, USA
e-mail: Fadi.Muheidat@csusb.edu

L. Tawalbeh (✉)
Texas A&M University, San Antonio, TX 78224, USA
e-mail: ltawalbeh@tamusa.edu

© The Editor(s) (if applicable) and The Author(s), under exclusive license to Springer Nature Switzerland AG 2021
Y. Maleh et al. (eds.), *Machine Intelligence and Big Data Analytics for Cybersecurity Applications*, Studies in Computational Intelligence 919,
https://doi.org/10.1007/978-3-030-57024-8_21

1 Introduction

Advancement in technology and hardware has provided a better opportunity for connectivity between all the places and public for better handshaking capabilities, communication abilities, the ability to conduct business regularly while enabling better capabilities of communication and information plethora to be explored through the power of the Internet. The interlinking of the entire world through the ability of the Internet has provided better opportunities for organizations and individuals to have better communication capabilities and faster means of reaching each other for various needs.

It has also enabled the exchange of information in a faster manner from any point to any point in a split second and allow better communications to be established. The ability of the Internet to provide seamless connectivity from one place to the other place has allowed better remote-control capabilities to be established and maintained from a centralized location. With all the capabilities the Internet provides to all the users and businesses to corporates, it has also created concerns of in the way some people use the Internet for unethical activities and behaviors that cause issues to unsuspecting users and create problems in their daily life. The unethical activities of some people have led the vast majority of users to apply security measures on all the infrastructure to safeguard their privacy of information and infrastructure from being hacked are illegally used for other purposes. It has become very important for every organization to have an infrastructure that supports their business purposes to enable better security measures to be incorporated when as a preventive measure to defend from any possible attack through the Internet or from internal attack possibilities. The ability of an organization to provide effective security measures for information privacy and infrastructure security depends on various security measures implemented and enforced regularly when they are also validated frequently for adherence to security and legal requirements.

Cloud computing has emerged as one of the most prominent services for data storage. Cloud computing saves money and time because data are stored in multiple virtual servers. Therefore, users are saved from paying maintenance and license fees, hardware, and software. Cloud computing also provides high agility and scalability since cloud users access to cloud systems directly. Maintaining data integrity in cloud computing, however, is difficult because they do not have control over their outsourced data. According to [1], "Integrity is an extent of confidence that what information is available in the cloud, what is there, and is protected against accidental or intentional alteration without authorization." Cloud users depend on trusted third-parties to protect and maintain their data.

Data integrity is an important part of cloud data storage. Several institutions and organizations today store their data in cloud-based systems. Some of the popular cloud computing systems include "such as SaaS, PaaS, and IaaS and deployment models like Private, Public, and Hybrid" [1]. These cloud systems are vulnerable to hacking and other activities that jeopardize the integrity of data. Cloud data loses its integrity when accessed by unauthorized third parties causing alteration, modification, misuse, and defacement. Digital storage of data requires a service provider to practice utmost honesty and integrity to ensure that data is stored safely. The concept of data integrity also involves reliability, confidentiality, and availability of data during retrieval. Cloud service providers employ different methods including hashing to maintain data integrity. Hashing maintains data integrity by converting key values into a range of indexes of an array. A hash is calculated and sent to a receiver such that if someone alters or modifies the data, the receiver will be able to detect since there will different hash value. Hashing enhances flexibility, integrity, and reduces the latency of data [2]. The most common hashing functions include MD5, CRC, and SHA-1. Hashing function helps in transforming data of arbitrary size onto data of a fixed size.

Mobile digital communication has become an essential and speedily evolving technology because it permits users to transmit knowledge from remote locations to alternative remote or fastened locations [3]. The growth of mobility has changed our lives fundamentally in an unprecedented way. According to Power Research Center [4], Mobile technology has spread rapidly around the globe. Today, it is estimated that more than 5 billion people have mobile devices, and over half of these connections are smartphones. These mobile devices have brought a lot of applications at the palms of people's hands. At the same time, Cloud Computing has emerged as a phenomenon that represents the way by which IT services and functionality are charged for and delivered.

Big data is a vital subject in engineering, medicine, business, health, finance, science, and the entire society. Every day there is a creation of more than 2.5 quintillion bytes of information. About 90 percent of information is already existent with twitter feeds, data shared on the internet, the YouTube videotapes, among another social media information [5]. The creation of a lot of information and the continuous doubling year after year has created room for big data where such information is collected, integrated, analyzed, and used by institutions, individuals in research, analytics, and many more aspects. The process brings more data together every day and such information serve in creating security, determining human behavior, and marketing or generating products that suit them. According to [6] Big data is an abbreviated form of technological advance learning that could pave the way into new strategies of understanding the world and enhancing computing processes in corporations that amass digital data. The dynamic nature of big data and information flow calls for the need to establish strong protection against threats that could emerge from big data. It is necessary to have good security controls for big data through leveraging tools that operate in it. Purports that Role-based access control (RBAC) controls never give effective protection for authentication and authorization processes come with laxities and inefficiencies. Hackers and cybercriminals are consistently researching working out on how to bring down companies, individuals, systems, and use data for selfish reasons. According to [7], all people who engage in big data must consider securing such data through compliance and technological investment in tools that protect Big Data.

Cyber and socio-engineering attacks are the commonest on Big Data meaning layering of IT and business security is of the essence to Big Data. The possession of data transfer workflows among repositories who gather data at high velocity and volumes is important. At the same time, multiple repositories need to actively attack the risks. The Big data Infrastructure that is distributed creates difficulty in defending the environment but standardized physical controls across accessible locations are vital [8]. When scientists pursue access to information, perimeter safeguards are vital.

On the other hand, changes in commercial computer systems are rapid and continuous. New systems are presented every day. Many factors contributed to the spread of mobile devices and to become ubiquitous such as increased computing power, Internet accessibility and energy-efficiency, advances in human–computer interfaces, and low cost of hardware. Further, devices such as phones and personal digital devices (PDAs) have turned into general purpose devices such as smart phones and tablets.

Mobile computing is a technology that allows transmission of data of different forms wirelessly. Mobile computing consists of mobile communication (infrastructure to ensure seamless and reliable communications), mobile hardware (a device with mobility access) ranges from smartphones, tablets, PDAs, printable laptops…etc. Mobile software; the engine of mobile devices; which is the program that runs on mobile devices. Mobile computing has many advantages beside communication and entertainment, they provide a streamline of business process through secured connection; meetings, webinars, and video conferencing that reduced the travel expense. Mobile Computing is an umbrella term used to describe technologies that enable people to access services anytime and anywhere [9].

Mobile computing can be categorized into seven major categories of focus [10];

- *Portability*: The focus was to reduce the size of hardware to enable the creation of computers that could be physically moved around relatively easily.
- *Miniaturization*: Creating new and significantly smaller mobile form factors that allowed the use of personal mobile devices while on the move.
- *Connectivity*: developing devices and applications that allowed users to be online and communicate via wireless data networks while on the move.
- *Convergence*: Integrating emerging types of digital mobile devices, such as Personal Digital Assistants (PDAs), mobile phones, music players, cameras, games, etc., into hybrid devices
- *Divergence*: Opposite approach to interaction design by promoting information appliances with specialized functionality rather than generalized ones
- *Applications*: The latest wave of applications (apps) is about developing matter and substance for use and consumption on mobile devices, and making access to this fun or functional interactive application content easy and enjoyable
- *Digital Ecosystems*: The emerging wave of *digital ecosystems* is about the larger wholes of pervasive and interrelated technologies that interactive mobile systems are increasingly becoming a part of.

The mobile user expects to be able to retrieve data and do computing at any given moment and any given time. And this is precisely why the support for a variety of platforms with a variety of user interfaces is critical for a mobile application. Mobile users expect to start a transaction and leave it unfinished on one device at a given place and time and finish the same transaction later on a different device and at a different place and time [11]. This motivates the move into Mobile Cloud Computing (MCC). Mobile computing enables the mobile devices to offload operations that were infeasible by the limited resources (battery, storage, network, processing power) of the mobile devices. It also, added long term storage, accessibility, reliability (backup), and data sharing. Mobile cloud applications move the computing power and data storage away from mobile devices and into powerful and centralized computing platforms located in clouds, which are then accessed over the wireless connection.

In the following sections, we will study Mobile and Cloud computing in more detail; definitions, characteristics, security measures, and challenges. We will wrap the chapter with real-life applications and conclusions.

2   Cloud Computing and Service Models

Many attempts for defining Cloud Computing; [12] defined it as "*A cloud is a type of parallel and distributed system consisting of a collection of interconnected and virtualized computers that are dynamically provisioned and presented as one or more unified computing resources based on service-level agreements established through negotiation between the service provider and consumers.*" Others defined it as a "*large pool of easily usable and accessible virtualized resources*" [13], and as "*hardware-based services offering compute, network and storage capacity, where hardware is abstracted*" [14]. The United States' National Institute of Standards and Technology (NIST) was the first standards organization to define cloud computing and identify its main characteristics, deployment, and service models. According to the definition published in NIST Special Publication (SP) 800-145, "*cloud computing is a model for enabling ubiquitous, convenient, on-demand network access to a shared pool of configurable computing resources (e.g., networks, servers, storage, applications, and services) that can be rapidly provisioned and released with minimal management effort or service provider interaction*" [15]. Figure 1 shows general view of cloud computing architecture.

Cloud computing is capable of delivering diverse IT services on demand. The internet cloud services can be classified into three service models based on the layers of virtualization: network as a service (Naas), infrastructure as a service (IaaS), platform as a service (PaaS), and software as a service (SaaS).

Other "as a service" models exist such as "Database as a service (DaaS), Application as a service (AaaS), Network as a service (NaaS)"…etc. Figure 2 provides a view of the cloud computing basic services model.

Each layer provides a different service to users. IaaS solutions are sought by users who want to leverage cloud computing from building dynamically scalable computing systems requiring a specific software stack. PaaS solutions provide scalable programming platforms for developing applications. SaaS solutions target

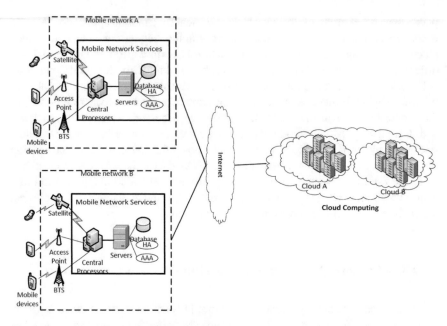

Fig. 1 General view of cloud computing architecture

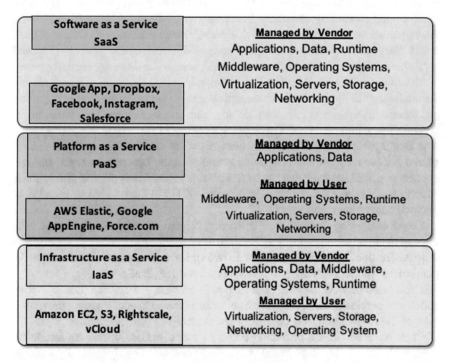

Fig. 2 Cloud computing basic services model

mostly end users who want to benefit from the elastic scalability of the cloud without doing any software development, installation, configuration, and maintenance.

2.1 Infrastructure-as-a-Service (IaaS)

At this layer, IaaS delivers infrastructure on-demand and pay-as-you-go for services such as storage, networking, and virtualization. Virtual machines, created on the provider's infrastructure at the user request, are the compute on demand hardware. The user does not manage or control the underlying cloud infrastructure but has control over operating systems, storage, and deployed applications; and possibly limited control of select networking components. The pricing model is usually defined in terms of dollars per hour, where the hourly cost is influenced by the characteristics of the virtual hardware. Some examples; AWS EC2, Rackspace, Google Compute Engine (GCE).

2.2 Platform-as-a-Service (PaaS)

At this layer, PaaS delivers scalable and elastic runtime environments on-demand and hosts the execution of applications. It encloses everything for the whole software engineering lifecycle, from programming to deployment. Scalability and fault tolerance management is the service provider task, while users are requested to focus on the logic of the application developed by leveraging the provider's APIs and libraries. Some examples; AWS Elastic Beanstalk, Google AppEngine, Windows Azure, Force.com.

2.3 Software-as-a-Service (SaaS)

At the top-level layer, SaaS provides applications and services on demand. Most of the common functionalities of the applications are replicated on the provider's infrastructure and made more scalable and accessible through a browser on demand. The user does not manage or control the underlying cloud infrastructure with the possible exception of limited user-specific application configuration settings. Some examples; BigCommerce, Google Apps, Salesforce, Dropbox, MailChimp, DocuSign, Slack.

2.4 Mobile Cloud Services Model

The concept of mobile cloud computing is categorized in different service models. Similar to the basic cloud services/layers; we can define more services with respect to the "Mobile" context; Mobile Infrastructure-as-a-Service (MIaaS) similar to SaaS and includes services like Dropbox, iCloud, OneDrive, GoogleDrive. Mobile Network-as-a-Service (MNaaS); Mobile Data-as-a-Service (MDaaS); Mobile Multimedia-as-a-Service (MMaaS); Mobile App-as-a-Service (MAaaS); Mobile Backend-as-a-Service (MBaaS). Other classifications can be based on the roles of the commutation entities; service broker, service provider, and service consumer.

- *Mobile as Service Consumer*: Mobile devices use cloud services mainly for computation power.
- *Mobile as Service Provider*: Mobile devices provide sensing services and sending sensing data to the cloud either for further processing or to transfer it to other mobile devices.
- *Mobile as Service Broker:* Mobile devices act as an edge or proxy for wireless sensors or other mobile devices with limited capabilities.

2.5 Cloud Deployment Models

Cloud data sharing among multiple entities brings security threats and integrity concerns. However, the data can be in different states; residing on the servers, or transit; flowing to and from the cloud. Since the data can be stored in the public cloud, or private or both, the NIST Cloud Computing Reference Architecture (NIST SP 500-292) and NIST Cloud Computing Security Reference Architecture (Draft NIST SP 500-299) documents introduce and discuss these deployment models, as shown in Fig. 3:

- *Private*: The cloud's infrastructure is operated for the exclusive use of a single owner. The cloud instance could be managed by the owning organization or run by a third party. The private cloud can be on- or off-premises.
- *Public*: The cloud's infrastructure is available for public use alternatively for a large industry group and is owned by an organization selling cloud services."
- *Community*: Provides a cloud instance that has been organized to serve a common purpose or function.
- *Hybrid*: Provides for the integration of multiple cloud models (private, public, community) where those cloud tenants retain uniqueness while forming a single unit. Common ubiquitous protocols are provided to access data for presentation [15].

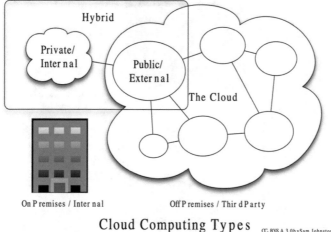

Cloud Computing Types CC- BYS A 3.0b ySam Johnston

Fig. 3   Cloud deployment models

3   Mobile and Cloud Computing Security

Mobile security is crucial for mobile cloud computing. Mobile devices can run both personal applications and companies' applications for supporting business functions. Hence, security and privacy are major concerns for both the cloud and mobile devices. Mobile devices have their own security issues; physical security, untrusted Wi-Fi, or cellular services. Even though these devices use the cloud for storage and computing services, yet some essential data still stored in the local device. Application installed on mobile devices can be a security threat and reason for data leakage and privacy threat. Hence, mobile devices must be scanned for any malicious software and viruses. However, maintaining device security is a resource-intensive task, which requires delegating these tasks to cloud computing resources. Security measures in both the cloud and mobile devices were meant to protect our privacy. When we have our data online, ensuring our privacy comes by securing access to our data and hide it from hackers and intruders. One solution is to build our own personal cloud but that will add extra cost and complexity, another solution is to trust and have faith in the cloud service provides. Cloud providers are thriving to design and implement security measures and guidelines to provide security, privacy, and trust with their users. Security as a Service is another comprehensive view of the cloud and mobile service model that encompasses device, network, and cloud security in one abstract view as shown in Fig. 4.

According to [16], seven security risks for users to consider and raise in the cloud computing areas:

- *Privileged user access*: Sensitive data outsourced in the cloud would bring an inherent level of risk due to the loss of direct physical, logical, and personnel

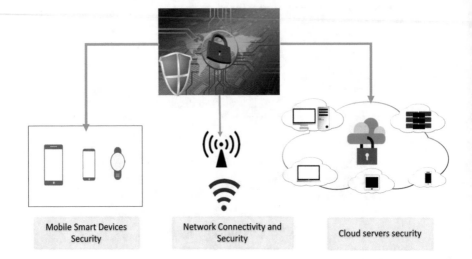

Fig. 4 Mobile and cloud computing security as a service model

control over the data. Customers need to get much information about the personnel who have access and manage their data.

- *Regulatory compliance*: Cloud service providers should be willing to undergo external audits and security certifications. Customers are responsible for the security and integrity of their own data, despite being held and managed by the service provider.
- *Data location*: The data stored in the cloud is not known where it is hosted, which servers, or even geographical location, which may lead to confusion on specific jurisdictions and commitments on local privacy requirements.
- *Data segregation*: Since cloud data is usually stored in a shared space, it is important that each user's data is separated from others with efficient encryption schemes, but encryption is not a cure-all. Encryption accidents can make data totally unusable.
- *Recovery*: Cloud providers should provide proper recovery mechanisms for data and services in case of technological failure or disaster. They need to provide a replica of both data and infrastructure to ensure backup and restore of the data.
- *Investigative support*: It is impossible to investigate illegal or inappropriate activity in cloud computing, because data for multiple customers may be co-located and spread across the virtualized servers, hosts, and data centers.
- *Long-term viability*: Assurance that users' data would be safe and accessible even if the cloud company itself goes out of business or acquired by a larger company. The provider should provide an import tool to download and format the data.

In the light of these seven security risks, privacy violation, illegal access to sensitive information, vulnerability and security threats will add more management burden and responsibilities when providing cloud services.

3.1 Mobile Computing Security

The security risks and challenges are concerned with devising secure and trustable systems from different perspectives and levels. In the technical level, proxy-based security protocols in the networked mobile devices provide an alternative security measure which is cost-effective and efficient in processing. The ability to process all the smart applications and other actions using a ***proxy computer*** with a higher capacity depending on the inputs and outputs to the source and destinations allow security measures to be adequately incorporated and the proxy level and the carrier secured through cryptographic services, as shown in Fig. 5. The efficiency of security measures taken through the proxy services enables process-intensive security measures to be implemented effectively on the proxies and allow the low efficient mobile devices to also enjoy the security measures to secure the communication and applicability of all the aspects [17].

One of the easiest ways to achieving the security measure required is by implementing security measures on the mobile devices that are connecting to the official networks to provide ways of applying the working nature from any location that carries information over the Internet. The ability of the mobile device to retain all the functional capabilities based on the security measures implemented might restrict some of the basic functions from exhibiting properly into the limitations of the security measures which includes SD card memories which can be seen as additional memory, the ability of peripheral components being used in a device that may provide opportunities for attackers to use the components effectively and gain access to the device, the ability of the security measure incorporated into the mobile device may severely hamper the performance of the device altogether and obtaining consent from each user on the security measures to be incorporated me apply additional restrictions and prevent some of the basic functionalities from executing [18].

The ability of mobile technology has provided a specific *custom-built application* to be developed and installed on mobile devices to anyone to support business functionality from any remote location to be established clearly and the activities are

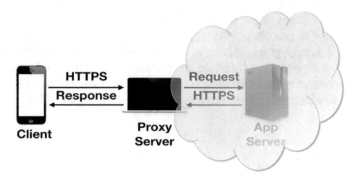

Fig. 5 Proxy based-security

monitored regularly for better outputs. A very important aspect of security measures on mobile devices is staying updated with the security protocols and identification of opportunities to use the security protocols to benefit them while network capabilities to establish security models on a regular basis and derived and validation process through audit verifications [19].

Antivirus and anti-malware programs such as AVG, Norton are used to detecting security threats and protect mobile devices. Data refactoring by simply breaking the data into packets and transfer hem into different routes and paths which distract the hackers and intruders to benefits from partial data packets. Denial of service (DOS) can be prevented and reduced by using virtualization (see below). Single Sign-On in a cloud computing environment, customers log in to multiple applications and services by single username and password. Shifting the security measures from the device into the cloud by implementing strong authentication.

3.2 Mobile Cloud Computing Security

Data Encryption is the key player in cloud computing security and data privacy. Without encryption hackers and sniffers can access the data flowing from and to the cloud [20]. Many communication channels encryption techniques existed such as encrypted Wi-Fi, Virtual Private Network (VPN), Secure Sockets Layer (SSL) and Transport Layer Security (TLS). Cloud data security solutions exist such as authentication, access control, data integrity check, data masking, one-time password authentication, and more. Access control mechanisms are one of the primary sources of validating the user and providing adequate access to conduct their daily job responsibilities and restrict them from having access to sensitive information and infrastructure. The ability of access control mechanisms to deliver security measures through the authentication mechanism and accepted services catalog that allows validation of every user, authentication process to be followed on a consistent basis, the development of various aspects of the security measures including access controls for reading and writing, modification rights, view of sensitive information rights, and different measures of the user authentication process, linking of every possible service that can be provided to the user based on the roles in the organization can be developed through proper process documentation, understanding of the process, understanding the critical use of all the application hosted on the cloud environment by the user. Access control mechanisms are the need to ensure insider attack as well as outsider attacks possibilities to be reduced to manageable levels [21].

3.3 Data Security

Data integrity and security in cloud computing depend upon the cloud service and deployment model. Data integrity can be defined as maintenance of intactness of any data during transactions like transfer, retrieval, or storage by ensuring that the data is unaltered, correct, and consistent. The data may change if and only if an authorized operation is valid on the data [22]. Data concerns that are common in the cloud include; risk of data theft, physical security, mishandling of encryption keys, auditing issues due to virtualization. Cloud service providers employ different methods including hashing to maintain data integrity. Hashing maintains data integrity by converting key values into a range of indexes of an array. A hash is calculated and sent to a receiver such that if someone alters or modifies the data, the receiver will be able to detect since there will different hash value. Hashing enhances flexibility, integrity, and reduces the latency of data [23]. Cloud service providers should implement better infrastructure by providing software and hardware-based security solutions such as proxy servers, firewall data consistency, and recovery tools. In addition to encryption and hashing, in [24] the authors discussed common cloud data integrity and security techniques:

- *Access Control*: restrictive permissions applied by the cloud service provider to allow data owners or their authorized users to access the data.
- *Integrity Check*: Outsourced data can be access and modified by authorized users only. While data is resting in the cloud storage, multiple integrity checks can be applied such as Proof of Retrievability (POR), and Provable Data Possession (PDP).
- *One Time Password (OTP)*: It is a type of *strong authentication*, where a password is generated through a random key and can be used once. It prevents identity theft and access to private data, sometimes. Similar techniques like Two-Factor Authentication (2FA) or Multi-Factor Authentication (MFA) are used nowadays.
- *Data Masking*: a process to hide data where it might be noticeable to someone without owners' permission. It creates a copy of the data that matches the original one. When requesting the data, it is replaced with dummy data and then a mask is applied to it. It can be done statically, *Static Data Masking* (SDM) or dynamically, *Dynamic Data Masking* (DDM), where it can transform data on the fly based on the user role [25].
- *Intrusion Detection System (IDS)*: Monitors network and system traffic for any suspicious activity. It sends alert notifications when threats identified. An effective solution should be able to discover any threats before they fully infiltrate the system. IDS comes in two types; Network-Based Intrusion Detection System (NIDS) that present in devices or computer connected segment of an organization's network and monitor network traffic and keep eyes on ongoing attacks, Host-Based Intrusion Detection System (HIDS) is installed on a server connected to the network and monitor illegal activities on that system.

4   Virtualization Security in Cloud Computing

One of the primary aspects of cloud computing is the virtualization of all the physical servers and hardware that can cater to the maximum efficiency of the available services that can be provided to all the clients. Virtualization allows the creation of a secure, customizable, and isolated execution environment for running applications. The basis of this technology is the ability of a computer program to emulate an executing environment separate from the one that hosts such programs. Virtualization is a software component that provides the capability of providing a sliced capacity of a huge hardware component to deliver operational resources and dynamically configured the needs of the infrastructure to be managed in the runtime.

Virtualization allows sharing resources by creating separate computing environments within the same host, hence better resource utilization. It also provides emulation capabilities that the user can run different programs that are not available in the user devices. Virtualization provides isolation through an abstract layer that provides access to the physical hardware of the host without interference between different users. Different types and levels of virtualization exist. Even we believe it is important to talk about, but the literature is rich with these details. Our focus is on the use of visualization in cloud computing from the security point of view. Of particular interest is the virtual computing environment and execution models; mainly computation programming languages and runtimes and hardware techniques. With all the good and advantages of virtualization, it still sources of security threats such as phishing; where hackers or intruders try to deploy malicious software and steal other guest's sensitive data. They can be embedded with the operating system boot process, and present themselves as a virtual machine. A modified version of programming language runtime can capture user's data and track their program executions.

The ability of virtual servers to provide all the capabilities of the physical server to incorporate security measures allows better security controls to be established on the same physical hardware through virtualized software services. Virtualized servers can provide isolated security measures through the creation of *software-controlled access control* mechanisms on the resources available which enables an additional layer of security measures to be implemented on the soft operating systems through virtualized capabilities. Incorporation of virtualized services allows the creation of a different level of security measures at different layers to develop *isolation* methods and cater to the specific needs of the clients who share the same physical hardware configuration to some extent [26].

Customers run an application on the cloud on one or more virtual machines. These machines run on serves on the datacenters and have access to the underlying memory and CPU resources. One server can host multiple instances of virtual machines managed by a Hypervisor, software that allows creating, running, and controlling virtual machines and its other virtual subsystems as shown in Fig. 6. It provides logical separation among the virtual machines and manages the running machines' shared hardware resources.

With all the virtualization advantages, there exist some security vulnerabilities and challenges such as challenges within the virtual machines; unauthorized access, attacks, communication security within the virtualized environment, security challenges within the hypervisors, and confidentiality of data. Below we will discuss some virtualization security changes and their countermeasures.

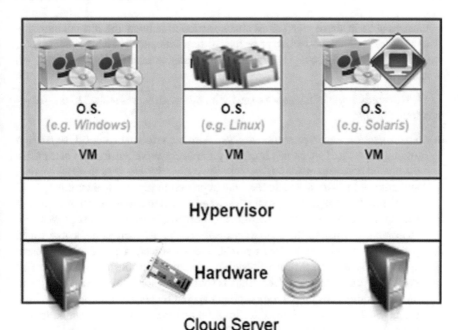

Fig. 6 Virtualizing architecture, single host multiple virtual machines

4.1 Virtualization Security Challenges

Vulnerabilities in the virtual machine and the attacks on the hypervisor have high impacts because of the control the attacker or intruder can have over the virtual machine; data and shared resources. The type and source of intrusion or attacks vary between trusted cloud users, a third party with good intention user, cloud admin user, or external attacks utilizing vulnerabilities in one of the abovementioned sources. In the work by [27] levels of attacks are:

- *Application-level* (guest virtual machine's user-space): these are attacks against user applications, such as through injection of malicious code inside an application to divert its control flow and execute the attacker's code.
- *Kernel-level* (guest virtual machine's kernel space): these attacks target the operating system (OS), such as kernel rootkits which allow an attacker to fully control the system.
- *Virtualization layer:* these attacks exploit the virtualization features in many ways, such as to attack virtual machines residing on the same host.
- *Hypervisor* (to and from the hypervisor): these attacks try to exploit vulnerabilities at the hypervisor to gain control of it and of all the virtual machines on top of it; other ways require escaping from a virtual machine to attack the hypervisor.

- *lower levels*: in these attacks, an attacker tries to subvert the levels below the hypervisor such as the hardware or the System Management Mode (SMM), for example, to directly access the memory to modify or read the hypervisor virtual space.

We will study some of the attacks and their countermeasures related to the above-mentioned levels:

(a) **Virtualization layer channel attacks**: Although there is logical separation among the virtual machines hosted by the same server. Attackers can utilize the shared hardware resources such as the cash to have access to sensitive data and control the virtual machines. One approach to protect against this type of attack is to prevent packets probing between virtual machines and hence prevent data leakage. Some countermeasures to implement include *Hard Isolation*; a technique to separate the shared hardware resources such as dedicated cache per virtual machine or one virtual machine per server, or a white list of users approved by the client. Cash Flushing: a technique to flush the shared cache every time the allocation of the cache is switched among virtual machines, with the downside of losing the temporal and spatial locality of the cache. *Limiting Cache Switching*: ta technique to limit the amount of data that can be leaked across virtual machines by limiting how often the cache is switched from a virtual machine to another. This makes it hard for another virtual machine to attain benefits from the data the previous virtual machine has accessed when probing the cache.

(b) **Hypervisor level malicious attacks**: a malicious misbehave of the virtual machine makes the hypervisor assigns to it more resources than what it is supposed to obtain. This extra allocation of resources for the malicious virtual machine comes at the expense of the other virtual machines that share the same server as the malicious virtual machine, where these victim virtual machines get allocated less share of resources than what they should actually obtain, which in turn degrades their performance (unfair use of resources) [28]. The hypervisor should implement a fair sharing resource sharing mechanism by checking the virtual machine utilizing the underlying hardware such as the CPU cores and memory every millisecond. Then priorities these machines by the usage of the resources. This way the misbehaving virtual machine will yield the resources to other virtual machine sand hence pre the service degradation.

(c) **Cloud Service provider inside attacks**: this occurs when we assume service providers administrators and mangers of the as trusted entities [29]. Application's sensitivity level determines the level of the concern, as the cloud data center administrators will have the ability to access and modify the collected data. Some countermeasures to protect the data from these attacks includes; *Homomorphic encryption*: is a form of encryption that allows specific types of computations to be executed on ciphertexts and obtain an encrypted result that is the ciphertext of the result of operations performed on the plain text [30]. This encrypted result when decrypted matches the result of performing the computational operation on the unencrypted input data. Another solution is to chop the data collected by the smart object into multiple chunks and then to use a secret key to perform certain permutations on those chunks before sending the data to the cloud servers. This allows storing the data on the cloud servers in an uninterpretable form for the cloud administrators. Only authorized entities that have the secret key can return the stored data to an interpretable form by performing the correct permutations [31].

5 Implementation and Real-Life Applications

One of the efficient ways of implementing security measures is expecting the unexpected and preparing for future possibilities through effective monitoring and controlling process being established in a centralized location. Establishing additional security measures for specific events, relations, capability enhancement, success stories to require some security to be provided in time where all the details that are bumped from the facility can be verified for its authenticity and fit to the purpose. An effective way of identifying the future possible incidents is through anomaly detection from advanced detection applications and methods incorporated in all the applications that are used on a regular basis.

5.1 Big Data, Cloud and Cybersecurity in Healthcare

Big data is an emerging technology that draws attention among many industries like a cloud. Cloud Analytics is used across industries in different innovative ways so that any individual or company can enhance their services or products depending on the need of the hour. The healthcare industry uses cloud analytics to improve healthcare services by minimizing costs and improving the quality provided. To achieve this feat, large data needs to be analyzed to answer new challenges. The scenario is similar to governments as well. They produce large amounts of data every single day. Analysis of such huge data helps governments in decision making as well as understand patterns using advanced technologies like machine learning [32].

In this subsection we discuss Big Data Security and privacy issues in Healthcare and how data analysis can help solving this challenge. There are many ways of applying big data analysis from diagnosis to treatment to population health management. Data governance is more and more important in healthcare as the healthcare industry creates large volumes of data. The main motivation for big data analytics in healthcare is to prevent adverse health events occurring like chronic diseases diabetes. Imperatively it's tough to collect, link and analyze the patient data, thus a logical method called patient-centric model would be helpful to measure by considering patient health by including elements like clinical, physical, psychological, social and environmental.

In the United States the base healthcare-related privacy compliance called HIPAA (The Health Insurance Portability and Accountability Act) and it mandatory for all the healthcare parties. While implementing Big Data Analytics for US base healthcare parties we have to make sure that the data centers are HIPAA certified. The patient-centric model is more analytical and logical in collecting and linking the patient data it also strengthens the compliance policies with HIPAA certified data centers [33]. The reason HIPAA focuses on ensures security policies and procedures rather implementing the data models for Big Data Analytics related activities.

More instances and copies of cloud base data storage are actually gets created by cloud providers like Google App Engine and Amazon EC2. These services providers create and manage several copies of data at different locations so that the data retrieval process can be optimized with less time of the cycle. The same time and process optimization technical architecture can be a compliance threat for healthcare parties because different healthcare parties like payers, third party administrators, patients, providers operate healthcare-related data with authentication obligations and with HIPAA security compliances. The multiple instances and copies of data can be an opportunity to get access to data [34].

Many organizations are approaching cloud storage like AWS, Google cloud. As a solution all data of the consumers is been stored in the cloud which is in the outside of the organization. Where there are more chances of losing the date due to less security. For instance, Data breach has happened in Facebook, capital one. They have introduced a new act called CCNA and CCPA.

The motive of this act is to remove the data of the customers if they are not using the accounts.

Big Data analytics is major advantage for the health care domain in which they are using extensively for creating the lab reports all the information related to the patients is been stored.

Like lab reports, diseases which makes them to understand the patients and price has been estimated. Data Analytics is also a big success in commercial sectors such as entertainment where they are storing the information of the customers and improve their productivity.

5.2 Healthcare: Wearables Applications

Apple watch is another device with a health application that has an interface for human use. Apple Watch is among the smartwatches designed and developed with health-oriented capabilities. The device has an application that can be used to monitor and track the heartbeat rate, count steps taken, burned of calories in the human body, physical activities such as yoga and swimming, and reminds an individual when to take a break from work. For healthcare centers to have dependable, usable and well interactive devices, they need to have an implementation of awareness in human–computer interaction designed to monitor patients who require regular checking [35]. Primarily, there has been digitalization application in much-advanced healthcare as

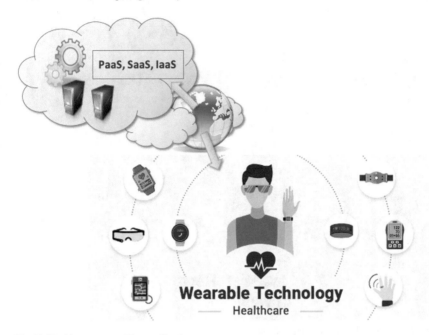

Fig. 7 Healthcare: wearables applications

shown in Fig. 7, with societal processes of changes as a result of new ICT solutions that have to bring in different and entirely new methods of handling things within the industry of healthcare [36].

Based on human–computer interaction, there has been a discipline as a result of wearable computing devices in the designing, evaluation, and implementation of systems concerning the health status for patients in the healthcare centers that have taken this initiative. As a result, it has helped to improve the functioning of healthcare to patients making monitored services to be more active and efficient. Both of these devices' applications used to help human beings in their daily activities as well as making life to be simpler. Most of the health technology applications are used for the purposes of healthcare while those of gesture applications are mainly used for securities purposes.

5.3   Healthcare: ECG Cloud Application

This section is adapted from [37]. An important healthcare application is the use of cloud technologies to support doctors in providing more effective diagnostic processes, of interest is the electrocardiogram (ECG) data analysis on the cloud. An electrocardiogram is a test that measures the electrical activity of the heartbeat. With each beat, an electrical impulse travel through the heart. This wave causes the

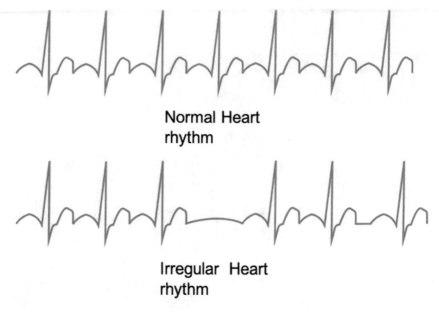

Normal Heart rhythm

Irregular Heart rhythm

Fig. 8 Heart rhythm regular and irregular

muscle to squeeze and pump blood from the heart. Figure 8 shows normal and irregular heartbeat. The analysis of the shape of the ECG waveform is used to identify arrhythmias (the heart beat is not having a steady rhythm) to detect heart disease.

With its remote computation and monitoring, cloud computing provides tools to monitor and analyze patient's heartbeat data in minimal time with immediate real-time time alert if attention is required. This way a patient at risk can be constantly monitored without going to a hospital for ECG analysis, and doctors can instantly be notified of cases that require their attention.

Figure 9 shows the ECG remote monitoring system infrastructure; (1) wearable computing devices equipped with ECG sensors (with embedded Bluetooth, Wi-Fi or ZigBee enabled data communication and processor module) constantly monitor the patient's heartbeat. (2) the information is transmitted to the patient's mobile device which is (3) then forwarded to the nearest cloud computing service through wireless or cellular networks. (4) the ECG data analysis will start as a service in SaaS layer and then (5) will be stored in the Amazon S3 service and issue a processing request to the scalable cloud platform. End user (6) will review and make health decisions.

The runtime platform is composed of a dynamically sizable number of instances running the *workflow engine*, a middleware for the execution, composition, management, and monitoring of workflows across heterogeneous systems and *Aneka, a middleware* for cloud application development and deployment. Each of the ECG processing jobs consists of a set of operations involving the extraction of the waveform from the heartbeat data and the comparison of the waveform with a reference

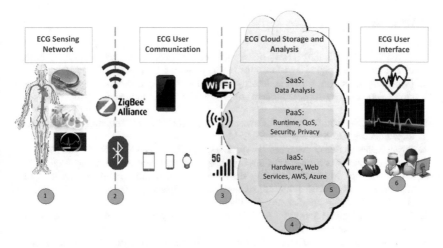

Fig. 9 ECG cloud application

waveform to detect anomalies. Doctors and first-aid personnel can be notified if anomalies are found.

6 Summary

Cloud computing is the developing paradigm of distributing IT services to consumers as a utility service over the Internet. The cloud offers resources to multiple users dynamically, in real-time, and according to their needs, with pay as you use. The key to privacy protection in the mobile cloud environment is the strict separation of sensitive data from non-sensitive data followed by the encryption of sensitive elements. It is important to have a security solution to meet the needs of mobile and cloud customers. In this chapter, we studied mobile and cloud computing technology from a security and privacy perspective. An introduction to the topics provided, followed by details of the mobile and cloud computing definitions and the seven categories of focus; portability, Miniaturization, Connectivity: Convergence, Divergence, Applications, and Digital Ecosystems. Cloud basic deployment and service models and the extended "Anything as a Service" models discussed. We described different types of security and countermeasures for mobile devices, mobile cloud computing services, and infrastructure, and described how the different attacks at each domain work and what defensive countermeasures can be applied to prevent, detect, or mitigate those attacks. We wrapped up the chapter with two application utilizing the mobile and cloud computing services; wearables, and ECG cloud analysis.

Acknowledgements The authors would like to thank the Chancellor of the Texas A&M University system for supporting this research through the Chancellor Research Initiative (CRI) grant.

References

1. Kumar NG, Rao KK (2014) Hash-based approach for providing privacy and integrity in cloud data storage using digital signatures. Int J Comput Sci Inf Technol 5(6):8074–8078
2. Harfoushi O, Obiedat R (2018) Security in cloud computing using hash algorithm: a neural cloud data security model. Modern Appl Sci 12(6):143. https://doi.org/10.5539/mas.v12 n6p143
3. Kulkarni G, Solanke V, Shyam G, Pawan K (2014) Mobile cloud computing: security threats. https://doi.org/10.1109/ECS.2014.6892511
4. Power Research Center. Accessed 18 Apr 2020. https://www.pewresearch.org/global/2019/02/05/smartphone-ownership-is-growing-rapidly-around-the-world-but-not-always-equally/
5. Cloud Security Alliance (2013) Big data analytics for security intelligence. Cloud security alliance. Retrieved from https://downloads.cloudsecurityalliance.org/initiatives/bdwg/Big_Data_Analytics_for_Security_Intelligence.pdf
6. Geeta V, SivaJyothi P, Rao TVN (2015) Big data analytics for detection of frauds in matrimonial websites. Databases 6:8
7. Big data analytics for cybersecurity. Retrieved from https://education.dellemc.com/content/dam/dell-emc/documents/en-us/2015KS_Krishnappa-Big_Data_Analytics_for_Cyber_Security.pdf
8. AlMadahkah AM (2016) Big data in computer cyber security systems. Int J Comput Sci Netw Secur IJCSNS 16(4):56
9. Pattnaik PK, Mall R (2015) Fundamentals of Mobile Computing. PHI Learning Pvt, Ltd
10. Kjeldskov J (2014) Mobile interactions in context: a designerly way toward digital ecology. Synthesis Lect Human-Centered Inform 7(1):1–119
11. B'far, R. (2004). Mobile computing principles: designing and developing mobile applications with UML and XML. Cambridge University Press
12. Buyya R, Yeo CS, Venugopal S, Broberg J, Brandic I (2009) Cloud computing and emerging IT platforms: vision, hype, and reality for delivering computing as the 5th utility. Future Gener Comput Syst 25(6):599–616
13. Vaquero LM, Rodero-Merino L, Caceres J, Lindner M (2008) A break in the clouds: towards a cloud definition
14. McKinsey & Co. (2009) Clearing the air on cloud computing. Technical Report
15. NIST SP 800-145 (2011) The NIST definition of cloud computing, Sept 2011. Available at https://csrc.nist.gov/publications/PubsSPs.html#800-145
16. Brodkin J (2008) Gartner: seven cloud-computing security risks. https://www.infoworld.com/d/security-central/gartner-seven-cloudcomputing-security-risks-853
17. Burnside M, Clarke D, Mills T, Maywah A, Devadas S, Rivest R (2002) Proxy-based security protocols in networked mobile devices. In: Proceedings of the 2002 ACM symposium on applied computing (SAC '02). Association for Computing Machinery, New York, NY, USA, pp 265–272. https://doi.org/https://doi.org/10.1145/508791.508845
18. Tawalbeh H, Hashish S, Tawalbeh L, Aldairi A (2017) Security in wireless sensor networks using lightweight cryptography. J Inf Assur Secur 12(4)
19. Balapour A, Reychav I, Sabherwal R, Azuri J (2019) Mobile technology identity and self-efficacy: implications for the adoption of clinically supported mobile health apps. Int J Inf Manage 49:58–68
20. Tawalbeh L, Jararweh Y, Mohammad A (2012) An integrated radix-4 modular divider/multiplier hardware architecture for cryptographic applications. Int Arab J Inf Technol IAJIT 9(3)
21. Al-Ruithe M, Benkhelifa E, Hameed K (2019) A systematic literature review of data governance and cloud data governance. Pers Ubiquit Comput 23(5–6):839–859
22. Chalse R, Selokar A, Katara A (2013) A new technique of data integrity for analysis of the cloud computing security. CICN, pp 469–472

23. Jararweh Y, Al-Ayyoub M, Al-Quraan M et al (2017) Delay-aware power optimization model for mobile edge computing systems. Pers Ubiquit Comput 21:1067–1077. https://doi.org/10.1007/s00779-017-1032-2
24. Kumar J (2019) Cloud computing security issues and its challenges: a comprehensive research 8:10–14
25. Ravikumar GK (2011) Design of data masking architecture and analysis of data masking techniques for testing. Int J Eng Sci Technol 3(6):5150–5159
26. Tim Mather SK (2009) Cloud security and privacy. O'Reilly Media
27. Sgandurra D, Lupu E (2016) Evolution of attacks, threat models, and solutions for virtualized systems. ACM Comput Surv (CSUR) 48:46
28. Jararweh Y, Ababneh H, Alhammouri M, Tawalbeh Lo'ai (2015) Energy efficient multi-level network resources management in cloud computing data centers. J Netw 10(5):273
29. Tawalbeh LA, Ababneh F, Jararweh Y, AlDosari F (2017) Trust delegation-based secure mobile cloud computing framework. Int J Inf Comput Secur 9(1–2):36–48
30. Jararweh Y, Tawalbeh L, Tawalbeh H, Moh'd A (2013) 28 nanometers FPGAs support for high throughput and low power cryptographic applications. J Adv Inf Technol 4(2):84–90
31. Jararweh Y, Al-Sharqawi O, Abdulla N, Tawalbeh Lo'ai, Alhammouri M (2014) High-throughput encryption for Cloud computing storage system. Int J Cloud Appl Comput (IJCAC) 4(2):1–14
32. Rashed AH, Karakaya Z, Yazici A (2018) Big data on cloud for government agencies: benefits, challenges, and solutions. In: Proceedings of the 19th annual international conference on digital government research: governance in the data age, pp 1–9
33. Rosenbloom ST, Smith JRL, Bowen R, Burns J, Riplinger L, Payne TH (2019) Updating HIPAA for the electronic medical record era. J Am Med Inform Assoc 26(10):1115–1119
34. Patil HK, Seshadri R (2014) Big data security and privacy issues in healthcare. In: 2014 IEEE international congress on big data. IEEE, pp 762–765
35. Hughes R, Muheidat F, Lee M, Tawalbeh Lo'ai A (2019) Floor based sensors walk identification system using dynamic time warping with cloudlet support. In: 2019 IEEE 13th international conference on semantic computing (ICSC). IEEE, pp 440–444
36. Muheidat F, Tyrer HW (2016) Can we make a carpet smart enough to detect falls? In: 2016 38th annual international conference of the IEEE engineering in medicine and biology society (EMBC). IEEE, pp 5356–5359
37. Buyya R, Vecchiola C, Selvi T (2013) Mastering cloud computing. Morgan Kaufmann, Burlington Massachusetts, USA. ISBN: 978-0-12-411454-8

Robust Cryptographical Applications for a Secure Wireless Network Protocol

Younes Asimi, Ahmed Asimi, and Azidine Guezzaz

Abstract In this chapter, we aim to discuss three dynamics systems to enhance the data confidentiality and integrity on wireless network security. We start it by a quick, dynamic and random synchronous generator of the unpredictable binary sequences. We combine a large theory concept to produce a lightweight and random stream cipher. This generator combines a linear feedback shift registers LFSRs, the arithmetic of quadratic fields and Boolean functions. Encryption and decryption are done by XRO'ing the output pseudorandom number generator with the plaintext and ciphertext, respectively. It will be used as a symmetric key cipher to avoid serious security problems. To underline the data integrity, we evaluate our system by a dynamic integrity check code DCRCn(M,G) that calculates a checksum of primitive polynomials generator. It proves her robustness by her aptitude to regenerated primitive signals of any generator polynomial. The core of this integrity check process is founded on a dynamic primitive polynomials generator. We aim to prove the data integrity of any transmitted information without broadcast any information about the used generator polynomial. To close, we highlight our cryptographic applications.

Y. Asimi (✉) · A. Asimi
Information Systems and Vision Laboratory, Team: Security, Cryptology, Access Control and Modeling, Agadir, Morocco
e-mail: asimi.younes@gmail.com

A. Asimi
e-mail: asimiahmed2008@gmail.com

Y. Asimi
Technology High School Guelmim, IbnZohr University, Agadir, Morocco

A. Asimi
Department of Mathematics, Faculty of Sciences, Ibn Zohr University, Agadir, Morocco

A. Guezzaz
Department of M2SC Team, Technology High School Essaouira, Cadi Ayyad University, Marrakech, Morocco
e-mail: a.guzzaz@gmail.com

© The Editor(s) (if applicable) and The Author(s), under exclusive license to Springer Nature Switzerland AG 2021
Y. Maleh et al. (eds.), *Machine Intelligence and Big Data Analytics for Cybersecurity Applications*, Studies in Computational Intelligence 919,
https://doi.org/10.1007/978-3-030-57024-8_22

485

Keywords Wireless network security · Linear feedback shift registers · Boolean functions · Data confidentiality and integrity · Dynamic CRC · Synchronous stream cipher generator

1   Introduction

If we do a little analysis of the cryptographic applications proposed so far, we see that the researchers, in most cases, have played on two evolutionary channels: length of the output and the internal complexity. They innovated relevant solutions capable of ensuring the uncorrelation between the input information (shared key, password, ...) and the calculated outputs. In the context of a protocol or a cryptographic primitive (hash function or encryption algorithm), the security level is calculated based on the lowest million. Hence, the construction of these applications had to be seen as an interlinked chain of sensitive elements. Thus, we cannot separate the internal states of the external states of a given primitive. The proposed solution also had to have a balance between internal performance and output length to even the probabilistic and the structural attacks [6–10, 13–15, 37–39].

A stream cipher is an encryption algorithm that encrypts one bit or byte of plaintext at a time. It uses an infinite stream of pseudorandom bits as the key. It defines the core of data confidentiality in the wireless network. The sensitive role requires robust primitives able to ensure the best security. At the same time, it should able to defend a weakly protected environment without infecting the performance. This property makes wireless networks more exposed to sophisticated attacks (Fig. 1).

In digital life, the critical issues are proved by the ability, of a given cryptographic primitive, to ensure high performances without affecting its resources or security level even with intensive computations. The flexibility of its internal processes can be viewed as an extra wizard solution. For this reason, we have subdivided

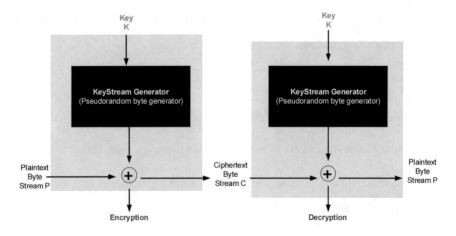

Fig. 1   Stream cipher

our synchronous stream cipher into quasi-independent tasks. The main object is to remove as much as possible the traceability and correlation between the output of a given process and his successor. A stream cipher is a required field of symmetric-key encryption algorithms that encrypts the plaintext by using a keystream generator. The latter can be a synchronous or an asynchronous stream cipher. The most proved attacks on the wireless networks are founded on deterministic nature of proposed stream cipher [9, 10, 13–15, 24, 26, 38, 39, 44]:

- Core of a stream cipher founded on a predictable generator;
- Use a classical initial permutation to construct a key K (like RC4 and AES);
- Usually prepend initialization vector (IV) to the Key;
- Key of a fixed length;
- Redundant key;
- Determinist internal behaviour;
- Use a classical balancing function to build the initial vectors of the same length;
- Weak statistical properties;
- Key must be as long as plaintext:

$$\text{Ciphertext}(\text{Key}, M) = M \oplus \text{Key}$$

- Insecure if keys are reused:

$$(M1 \oplus \text{PRNG}(\text{Pwd})) \oplus (M2 \oplus \text{PRNG}(\text{Pwd})) = M1 \oplus M2$$

In cryptography, the result of applying a cryptographic system over data provides cryptographic outputs of fixed length [6–10, 37–39]. The space of primitive signals regenerated by a given application (reversible or irreversible function) reflects its capacity to withstand correlation attacks. It also adds its internal and external ability to resist collisions. With the high computing capacity of quantum computers, a robust cryptographic system should be like a chain of internal and external primitives interlinked to fulfil the following recommended properties:

- His core should be founded on a strong generator that products primitive signals uncorrelated, unpredictable and independents under the minimal perturbations;
- It ensures the cryptographic quality of internals and externals states in order to avoid correlation attacks;
- It provides strong and dynamics internals and externals states under the minimal perturbations to withstand structural attacks;
- It improves the uncorrelation between the output's length and the input's length.

The most proven cryptanalysis has found its analytical study on the deterministic nature of the proposed cryptographic systems. The statistical dependence makes these cryptographical primitives more exposed to the must intelligent attacks [9, 10, 13–15, 17, 24–26, 37, 38, 40]. To resist these attacks, the solution should be able to break any correlation between the external elements and the internal behaviour of a proposed system. To have good statistical properties, we highlight our solution by the

Table 1 Notations

DA	Destination MAC address
SA	Sender MAC address
$\oplus$	XOR operation
$\parallel$	Concatenation operation
DICV	Dynamic integrity check value
$\mathbb{N}$	Set of natural numbers
FCS	Frame check sequence
CRC	Cyclic redundancy checksum
TSC	TKIP sequence counter
ICV	Integrity check value
RC4	Rivest cipher 4
AES	Advanced encryption standard
DES	Data encryption standard
TKIP	Temporal key integrity protocol
DCRCn(M,G)	Dynamic cyclic redundancy control
IV	Initial vector
WEP	Wired equivalent privacy
WPA2	Wi-Fi protected access
CCMP	Counter-mode/CBC-MAC protocol
GCMP	Galois counter mode protocol
MAC	Medium access control
Keystream	Output secret-key

robust cryptographic primitives, idem, the recommendations require a system whose internal behaviour is unpredictable. We covet by this work to improve the data confidentiality and integrity to enhance the wireless network security. To prevent the most proved attacks, all proposed primitives in this work, fill the recommended properties by unbreakable systems [4, 5, 7, 13, 16, 19, 23]. We aim to evolve wireless network security by a dynamic CRC that calculates a checksum of primitive polynomials generator and a symmetric key system founded on quadratic fields [1, 22, 37] (Table 1).

2 Related Works

Wireless network security requires both robust and lightweight cryptographic systems, which include a set of objectives meeting of the specifications of this network. A secure protocol should be able to withstand possible attacks without affecting network performances. The WEP protocol is the first encryption solution standardized by IEEE whose objective is to prove a security level equivalent to wired networks. It relies on a shared secret key between all the devices on the network. Hence, if a wireless user has been compromised, the encryption of exchanged data becomes useless; also, the opponent takes a global control of the entire network thus simplifying the replay attack [35, 38]. There are also security vulnerabilities linked to the use of breakable cryptographic primitives (RC4, AES and CRC 32) and the lack of physical protection [2–4, 40, 45]. The deterministic nature of these applications defines the most exploited vector by the proven attacks [9, 10, 13–15, 17, 24–26, 37, 38, 40]. For minimal security, security experts have thought of changing the length of the used secret keys. In the same sense, in the event of perceived risks, they require the periodic updating of these keys in secure communication channels.

For this reason, we find that the WEP protocol combines filtering by MAC addresses and a key rotation mechanism to withstand key reinstallation attacks [38]. The main goal is to secure the change and activation of a new WEP key. Besides, this standard adds individual WEP keys allowing personal identification of user stations at an access point. Indeed, this mechanism adds a load of management and security of the keys in a large company; instead, it presents a vital solution reinforcing the adjacent cryptography between the wireless users of the same access point.

The EAP protocol is another security alternative based on the 802.1x architecture that was introduced by an overview designed to manage multiple authentication methods (password, smart card, etc.) [35, 36]. The IETF organization standardized it. In this architecture, wireless users' identification is centralized in an authentication server (named RADIUS) [33]. Here, access points act as an intermediary between network supplicants and the authentication server. The authentication method is negotiated between wireless users and the authentication server at the time of synchronization. Idem; The 802.1x architecture defined EAPoL protocol to secure the communications between supplicants and access points in a context of the local WiFi network; knowing that communications between access points and servers authentication are encapsulated in RADIUS requests. As this architecture does not require such an authentication method, this has given rise to a multitude of authentication proposals aimed to strengthen the security level of the EAP protocol (OTP, GTC, SIM, TLS, PEAP, etc.). The security of this protocol is strongly linked to the used authentication method. In general, methods built on authentication tunnels (PEAP, TTLS) remain the utmost security. Instead, they are limited by a weak ability, of internal encryption mechanisms, to resist against session theft and replay attacks. Likewise, for robust security, you must configure your system by criteria for mutual verification of digital certificates to prohibit connections using false certificates.

To improve the security of WiFi networks, using robust encryption, the WiFi Alliance announced the WPA protocol as a transitive phase towards the 802.11i standard (WPA2). The WPA protocol is based on TKIP encryption (based itself on RC4 encryption) and only manages infrastructure type networks. For robust security, in 2004, the 802.11i working group announced the WPA2 protocol as a powerful solution that manages the two-encryption methods (AES and TKIP) and both types of networks (Infrastructure and Ad Hoc). These protocols coexist these authentication modes: a shared secret key (Personal) and the centralized 802.1x architecture (Enterprise). The WPA2 protocol appears as a real solution for robust security. Rather, it is limited by the complexity of setting up in an Ad Hoc network such as the WSN network. It also adds to concern related to the use of unbreakable cryptographic applications [37, 38]. To highlight, an authentication system that combines an encryption key generating method (EAP/TLS, TTLS, PEAP, FAST), an 802.1x architecture and a robust encryption algorithm seems the most secure than WEP, WPA-PSK or WPA2-PSK.

TKIP encryption was introduced as a successor to RC4 encryption following numerous proven security vulnerabilities on the WEP protocol. To avoid re-use of RC4 keys, TKIP encryption uses an extended (48-bit) initialization vector (IV) which changes it with each packet to protect against replay attacks. In each packet sent, it is clearly sent to reconstruct the RC4 keys. Therefore, it gives more opportunity for opponents to break this encryption mode [13–15, 23, 24, 45]. It is also based on the Michael protocol, which calculates a global integrity control code (MIC) (MSDU level) using a hash function [18, 19]. This protocol is seen as a robust solution strengthening the CRC linear integrity control. Unfortunately, it was broken in 2008 [11, 12, 18]. For robust WiFi security, following the security and performance of AES encryption, the 802.11i standard announced, in 2004, the CCMP protocol as a robust alternative solution defined using AES encryption and the CBC integrity control algorithm [20, 21, 35, 37, 38]. This protocol has been introduced to explain how AES encryption and CBC code could be combined in order to prove strong encryption in WiFi networks. To avoid replay attacks, this encryption combines counter mode (CM) and nonce to avoid two encrypted messages by identical keys giving the same output. Besides, the CBC code is calculated on the whole message concatenated by the CCMP and MAC header in order to defend against attacks aimed at changing MAC addresses.

Admittedly, CCMP encryption (AES and CBC) remains the most robust so far for the security of WiFi networks [20, 34, 35]. Rather, his application space is limited by the computing power requirements and the proven failures on the AES [4–8, 38–40, 43]. In general, AES key length defines the most attractive target helping cryptanalysis to find key recovery, key reinstallation or preimage attacks on AES [4–8, 38]. Key reinstallation attack (KRACK) targets the 4-way handshake used by a given supplicant to negotiate a new encryption key. This attack forges handshake messages of WPA2 when any used cipher (TKIP, AES-CCMP, and GCMP) [38, 39]. This novel attack technique works against personal and enterprise networks. It affects all device supports WiFi networks. To even the critical reinstallation attack, we must invest by a zero-knowledge protocol that negotiates a new authentication and encryption parameters by session without disclosure of any information about its sensitive data (like password, shared encryption key or a packet number). In this chapter, we highlight the wireless network security by two robust cryptographic applications: random stream cipher algorithm and dynamic integrity check code DCRCn(M,G).

3   Synchronous Stream Cipher Generator

Most cryptographic applications use the theory of polynomials over finite fields to build their algebraic structure [1, 22, 27–29]. We find that this theory plays a primary role in the most robust cryptographic systems. Rijndael encryption algorithm (AES) exploits the proven aptitude by the primitive polynomial of degree eight over a binary field to build the S-box [1, 28, 30, 31]. Idem, we use it to construct the linear Feedback Shift Register (LFSR) [1, 32, 41, 42]. To propose a robust and dynamic stream cipher, we have combined all these cryptographic primitives with building our synchronous generator. We covet to have an ideal solution that achieves the unpredictable behaviour of the internal and external states. It improves her sturdiness by her ability to take as input the keys of arbitrary length. This primitive allows us to map a binary string of arbitrary length to a primitive signal of arbitrary length. Here, we don't need to use a balancing function to meet input length requirements [4, 5, 7, 11, 13, 15, 16]. In terms of security, this proposed solution exceeds probabilistic attacks, which are based on the terminal nature of the inputs and outputs of a given function. Besides, the number and complexity of initial vectors do not depend only on the number of the input key's component elements, but rather reflect the quadratic nature of all its component elements. We covet to resist against any attempt, reaction or control implemented by actives eavesdropping on the transmitted packets over the network. This synchronous stream cipher generator starts with a secret-key KS = (Z1, ..., ZN) where Zi are positive integers, is based on the:

- Quadratic field arithmetic to generate three positive integers, and according to the quadratic nature of each the input-key component element;
- Federation the linear feedback shift registers LFSRs and the dynamic boolean functions to construct the binary sequences;
- Filtration of linear feedback shift registers LFSRs with a primitive polynomial of length eight;
- Use of a random balancing function to build the initial vectors of the same length;
- Modular congruence;
- Build an encryption function mapping without affecting its cryptographic quality.

3.1   Process of Generating the Initial Vectors

The most used encryption algorithms, like AES, RC4 or DES, start their encryption process by an input key of a fixed length [4, 5, 7, 11, 13, 15, 16, 40]. They carry out under the action of a balancing function. This characteristic weakens the robustness of cryptographic applications. To bypass this weakness, we dynamite the input space, which is necessary to perform a robust stream cipher without affecting the user behaviour. In our proposal, the length of the input key does not define a primitive element; quite, his dynamism which events our solution against the must intelligent probabilistic attacks like pre-computation attacks. We covet to prevent and to meet of the most intelligent recommendations by the proved cryptosystems to attain good statistical properties. This process generates the initial vectors $IV = \{X_1, \ldots, X_N\}$ from a secret-key $K_S = (z_i, \ldots, z_N)$ where z_i are positive integers for all $i \in \{1, \ldots, N\}$. The quadratic nature of these positive numbers allows us three random number named d_i, n_i and m_i. Then, we concatenate their binary representation to construct the initial vector. The steps of this process are articulated as follows:

- The user will have to choose an input key of arbitrary length (like password or secret key);
- The ASCII code of each character constituting an input key give us a positive integer z_i;
- We use this positive number to compute the positive integers d_i with $d_i = z_i$ mod 2^m with $m \in \mathbb{N}$ and $m \geq 8$ for all $i \in \{1, \ldots, N\}$.
- Based on quadratic nature of each d_i, we assign only two positive integers n_i and m_i:

  - Assume that $d_i = s_i^2 r_i$ where $r_i = 1$ or r_i is a square free integer, we then get $n_i = r_i$ and $m_i = s_i^2$.
  - Assume that d_i is a square free integer, we assign only one fundamental unit ε_i of the quadratic field $\mathbb{Q}\left(\sqrt{d_i}\right)$ together with:

$$\varepsilon_i = \begin{cases} n_i + m_i\sqrt{d_i} & \text{if } d \equiv 2 \text{ or } 3 \mod 4 \\ \frac{n_i + m_i\sqrt{d_i}}{2} & \text{if } d \equiv 1 \mod 4 \end{cases}$$

This process runs as follows (Fig. 2).

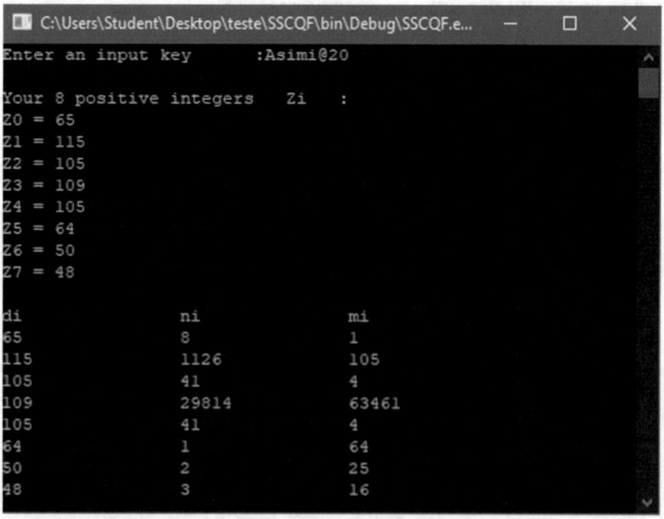

Fig. 2 Generation of three positive integers d_i, n_i and m_i

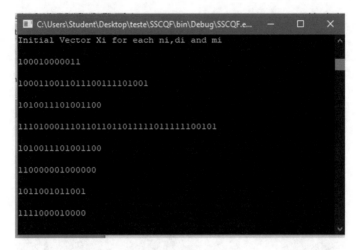

Fig. 3 Generation of the initial vectors

- For each couple of these three numbers, we associate an initial vector as follows:
  For all $i \in \{1, \ldots, N\}$, $X_i = \overline{n_i}^2 || \overline{d_i}^2 || \overline{m_i}^2$.

  The unbalancing result illustrates as follows (Fig. 3).
  We constate that the number of initial vectors X_i depends on the number of characters constituting an input key. Instead, the complexity of each initial vector is strongly linked to the quadratic nature of each positive number d_i.

3.2　Balancing Process of the Initial Vectors

The vectors X_i for all $i \in \{1, \ldots, N\}$ are not necessarily of the same length. The goal of this process is to balance these vectors to a length multiple to the length of the used primitive polynomial. Here, we use a primitive polynomial of degree eight. As a result, shifting the length of this primitive polynomial completely changes the internal behavior of our generator. For balancing these initial vectors, we then choose a vector of a maximal length, for example X_k of length $l_k = L'$, and we proceed as follows:

- For each vector $X_i = (x_{i1}, \ldots, x_{il_i})$ one assigns the only vector $Y_i = (y_{i1}, \ldots, y_{iL})$ defined as follows:

  - If $L \equiv 0 \bmod 8$, $L = L'$, we get:

$$
\begin{cases}
y_{ij} = x_{ij} & \text{for all } 0 \le j \le l_i \\
y_{i(l_i+t)} = x_{i(t \bmod li)} \oplus x_{kt} & \text{for all } 0 \le t \le L - l_i
\end{cases}
\tag{1}
$$

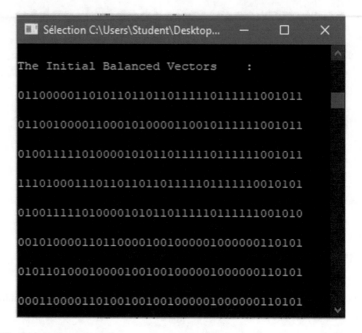

Fig. 4 The initial balanced vectors

- Otherwise, $L = L' + (8 - L' \mod 8)$, we get:

$$
\begin{cases}
y_{ij} = x_{ij} & \text{for all } 0 \leq j \leq l_i \\
y_{i(l_i+t)} = x_{i(t \mod li)} \oplus x_{kt} & \text{for all } 0 \leq t \leq L' - l_i \\
y_{i(L'+s)} = \sum_{t=0}^{s} x_{it} \oplus x_{ks} & \text{for all } 0 \leq s \leq 8 - \left(L' \mod 8\right)
\end{cases}
\tag{2}
$$

This figure demonstrates the results of this balancing function (Fig. 4).

From these results, we show that this execution grants us eight initial vectors Y_i of length eight. In addition, each initial vector gives us five linear feedback shift registers of length eight.

3.3 Keystream Generation Process

These generated initial vectors Y_i for all $i \in \{1, \ldots, N\}$ have the same length L divisible by eight. We subdivide it into $L/8$ binary sequences of length eight. In this case, each initial vector gives us five binary sequences. We use these last to initialize the linear feedback shift registers filtered by the primitive polynomial of degree eight [1, 28, 29, 41, 42]. If we change the degree of the used primitive polynomial, we will get a new internal behavior of the same input key. This internal flexibility of our solution makes it more robust against intelligent attacks. It suffices to note here that all disturbances on the inputs can completely change the number of created registers for a given execution process. At this time, we then obtain, for each Y_i, five linear feedback shift registers, namely $LFSR_{i1}, \ldots, LFSR_{iL/8}$. The output binary sequence of all $LFSR_{ij}$ will be combined with five Boolean functions $R_1, \ldots, R_{L/8}: \{0, 1\}^N \rightarrow \{0, 1\}$ defined as follows: For each $j \in \{1, \ldots, L/8\}$, the Boolean function R_j combines the output bits of $LFSR_{ij}$ for all $i \in \{1, \ldots, N\}$, together with $R_j(x_1, \ldots, x_N) = R(x_1, \ldots, x_{j-1}, 1, x_{j+1}, \ldots, x_N)$ and $R(x_1, \ldots, x_N) = \sum_{i=1}^{i=N} x_i + \sum_{i<j=1}^{N} x_i x_j$ mod 2. The concatenation of the output binary sequences of all Boolean functions gives us a Keystream. To encrypt (or decrypt) a given data, we apply XOR operation between the keystream bits and the plaintext bits (or ciphertext bits). This figure presents just a part of a regenerated Keystream (Fig. 5).

4 Dynamic Primitive Polynomials Generator

In the wireless network, the data integrity solicits a shared generator polynomial between communicating parties to start a given session. It uses the CRC 32 code to generate a cyclic checksum of 32-bit. It is easy to practice intelligent attacks (like fixed points or falsification attack packages) over a determinist system that provides the fixed outputs length [35, 37–39]. To prove the best protection level, TKIP protocol has used Michael protocol that defines a set of security mechanisms strengthening data integrity against attacks. Instead, these countermeasures make networks more vulnerable to DDoS attacks [35]. In our purpose, we enhance the data integrity by dynamic generator polynomials per session. We use primitive polynomials generator over a binary field of a fixed degree n [1, 28, 29, 35, 37]. This process builds on a shared session encryption key between an access point and a wireless user to calculate the degree of a generator polynomial (Fig. 6).

As you know, an adversary can easily sniff or store all the traffic in the wireless network. It can achieve total control over the network infecting then all adjacent wireless users. To prove the cryptography adjacent between all wireless users, the wireless access point should be able to ensure the zero-knowledge parameters concept. The interest is to protect the connection links between an access point and wireless users against eavesdropping or falsification attacks that a user (legal or illegal) can execute on all sessions opened by the same network. Our concept aims to strengthen the privacy of interns in the same network (Fig. 7).

This dynamic cyclic redundancy control is significant because it provides an additional sub-layer to guard the integrity of the transmitted data in the network. It implements the concept of generator polynomial per session. In the wireless network, a shared session key between the communicating entities allows us to determine the degree $n \in [32,64]$ of a primitive polynomial. This number will be used by a generator of the primitive polynomials, in each session, to generate a primitive polynomial of

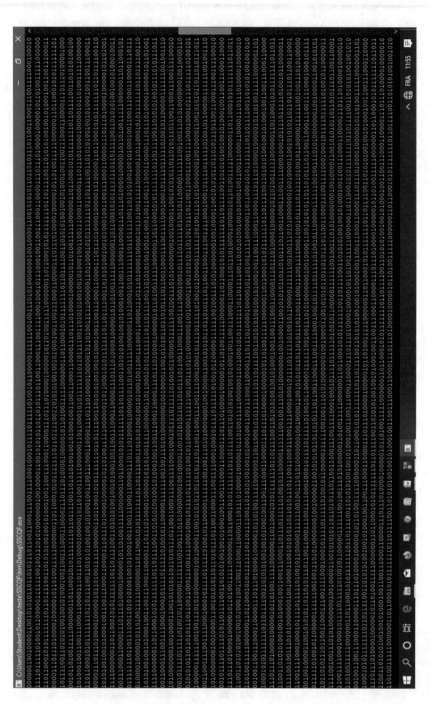

Fig. 5 A part of a given keystream

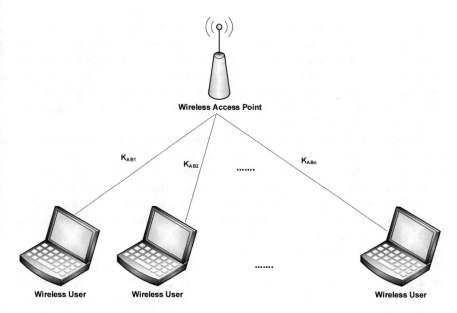

Fig. 6 Shared sessions encryption keys between an access point and all wireless users

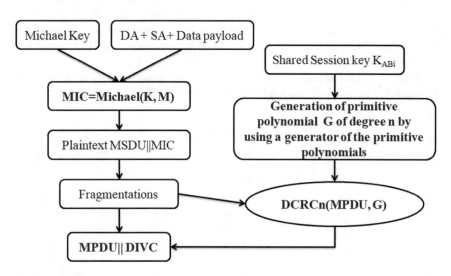

Fig. 7 Michael protocol with dynamic cyclic redundancy control

degree n in both sides [37]. The proposed protection mechanism gives us tow practical sub-layers to defend various vulnerabilities. It is founded on these four functions:

$$DCRC_n(M, G) \; with \, n \in \mathbb{N}*$$

Function Clear_Zeros(): It removes the zeros to the left of a binary sequence. Practically, to calculate the checksum R of a bit sequence S, we apply the XOR operation between this binary sequence and a generator polynomial G. We start with the most significant bit, repeating this operation until we have a suite of a smaller length strictly at the degree of G. For having carried out all these iterations, we will have to delete after each iteration the zeros positioned on the left of the sequence by using the *Clear_Zeros()* function.

Function Null(): In our proposition, the degree of the generator polynomial is not predefined. It is dynamic and related to the complexity of the encryption-shared key during a given session. This function regenerates, for each case, the number of zeros needed to execute the dynamic CRC principle. In general, it adds k zeros to the right of a binary sequence. It corresponds to the multiplication of $M(x)$ by x^k.

Function Calculate_CRC(): It allows us to calculate an integrity check sum $R(x)$ specific to a given message $M(x)$ and a generator polynomial $G(x)$ of any degree. It relies on the two functions defined above.

Function DCRCn(): It is used to add an integrity checksum $R(x)$ to a given message $M(x)$. The result is the safe message $M_0(x)$. To check the integrity of $M'(x)$, just recall the *Calculate_CRC()* function, if the result is zero, then the processed message $M'(x)$ is safe.

In this paper [1], the author has invested by two fast interesting programs for building the irreducible and primitive polynomials over a binary field. We interest to primitive polynomials generator as a core of our cryptographic primitives. To highlight its strong characteristics, we illustrate the evolution number of regenerated primitive polynomials according to a given degree n (Fig. 8).

Idem, we present same regenerated primitive polynomials of degree at least than 12. For more details, you can see this paper [1] (Table 2).

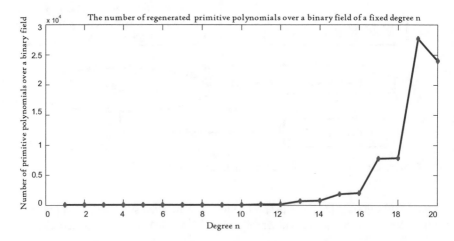

Fig. 8 The number of regenerated primitive polynomials over a binary field of a given degree n

Table 2 Primitive polynomials of degree n over a binary field

Degree n	Primitive polynomials of degree n
3	$x^3 + x^2 + 1$ $x^3 + x + 1$
4	$x^4 + x^3 + 1$ $x^4 + x + 1$
5	$x^5 + x^4 + x^3 + x + 1$ $x^5 + x^4 + x^3 + x^2 + 1$
6	$x^6 + x^5 + 1$ $x^6 + x^5 + x^3 + x^2 + 1$
7	$x^7 + x^6 + x^4 + x^2 + 1$ $x^7 + x^6 + x^3 + x + 1$
8	$x^8 + x^5 + x^3 + x + 1$ $x^8 + x^7 + x^6 + x^3 + x^2 + x + 1$
9	$x^9 + x^5 + 1$ $x^9 + x^7 + x^5 + x^3 + x^2 + x + 1$
10	$x^{10} + x^9 + x^8 + x^6 + x^2 + x + 1$ $x^{10} + x^9 + x^8 + x^7 + x^6 + x^4 + x^3 + x + 1$
11	$x^{11} + x^{10} + x^9 + x^6 + x^5 + x^4 + x^3 + x^2 + 1$ $x^{11} + x^{10} + x^9 + x^8 + x^5 + x^4 + x^2 + x + 1$
12	$x^{12} + x^{11} + x^{10} + x^7 + x^6 + x^3 + x^2 + x + 1$ $x^{12} + x^{10} + x^9 + x^7 + x^5 + x^4 + x^3 + x^2 + 1$

As you see, the evolution of the regenerated primitive polynomials number of given degree judges the importance of integrating this program as a core of our cryptographic applications. We use it to have unpredictable behaviour system to protect the sensitive data against the most intelligent attacks like key reinstallation attacks [6, 8, 11, 14, 15, 19, 23, 37, 38]. It will allow us to build a zero-knowledge authentication and encryption protocol.

5 Security Issues

A secure crypto graphical application doesn't mean that has currently proposed, but, that has a high-security level. It sees as a robust contribution that resists against the most intelligent attacks like attacks of replay, suppression and interception of messages. Matching by classical solutions, that produce the outputs of fixed length, we enhanced this field by tow unpredictable cryptographic applications. We aim to energize their internal and external spaces to resist against probabilistic and structural attacks. In this section, we present a security study that expresses our expectations of performance and security terms:

Synchronous Stream Cipher Generator: As Symmetric Block cipher, AES algorithm is most efficient compared to DES, 3DES and RC4. However, the CPU process time and memory define tow important factors that infect her performance [9, 10,

Table 3 Comparison between RC4, AES and our solution

Encryption algorithm	RC4	AES	Our proposal
Difference	Stream cipher	Block cipher	Stream cipher
Key space	Limited	Limited	Dynamic key
Behavior	Determinist	Determinist	Dynamic
Computational time	Good	Very poor	Good
Attack proved	Exist	Exist	Not Exist
Errors propagation	Not	Yes (CBC)	Not

13–15, 24, 26, 38, 40, 44, 45]. Idem, it adds the determinist internal and external behavior. This nature gives more the chance of attackers to prove the tractability between its chaining variables. In order to resist against the more intelligent attacks, our stream cipher uses an arbitrary input secret key. This nature gives more the chance of attackers to prove the tractability between its chaining variables. To resist against the more intelligent attacks, our stream cipher uses an arbitrary input secret key. This nature enriches her robustness against probabilistic attacks. To prove unpredictable behaviour, the minimal perturbations on the input secret key can change his internal behaviour. Because the length of each initial vectors is founded on the quadratic nature of each element constructed a given input secret key. These initial vectors define the core of the regenerated linear feedback shift registers LFSRs. Here, we can change just the length of these regenerated registers to obtain another unpredictable behaviour. For resilient security, it inspires her robustness by her ability to respond to any internal or external minimal perturbations. The interest is to ensure the cryptographic adjacent in the wireless network (Table 3).

Dynamic Primitive Polynomials Generator: To emphasize the integrity adjacent, even the same wireless network of the exchanged data, a dynamic errors detection system DCRCn(M,G) is proposed. We highlight it by a primitive polynomials generator that regenerates primitive polynomials of a given degree. Here, the communicating parties need just to generate a secure shared integer number to execute this mechanism that adapts to any generator polynomial. It is seen as a main solution that allows us to enhance **Michael protocol** against attacks.

6 Highlights and Future Work

The advancement of information technologies could be going with a strong evolution in terms of cryptographic applications. These defensive mechanisms should be able to protect network and computer systems against intelligent attacks. They should meet all security requirements to defend against weakness in the design, configuration, implementation, or management of a network or system. Indeed, network security requires both robust and lightweight cryptographic solutions that include a set of objectives in light of the new network's specifications. Idem, we must take into account the limited resources that characterize wireless networks like low computational power, limited memory and low energy power to which is added the absence of physical protection. For this reason, in our model, we have defined some specific goals to achieve:

A limited inputs and outputs space to resist against collision, replay and key recovery attacks;

Balancing between the performance and security level to address the recommendations required by the wireless network;

Avoid the proved weakness on the existing cryptographic applications.

The un-correlation of primitive signals affords a vital property that encryption algorithms are expected to possess to preserve their collision resistance. In this work, we innovate tow quick, dynamic and robust cryptographic applications. The first allows us to ensure the confidentiality of transmitted data on the network. These synchronous generator products the unpredictable primitive signals under minimal internal or external perturbations. After that, we have evolved the data integrity by a dynamic protective sub-layer. It grants a robust and efficient integrity control without broadcast any information about the used generator polynomial:

Dynamic, uncorrelated and unpredictable primitive signals of the same input secret-keys lengths.

Each minimal perturbation on input key or used polynomial primitive length changes the internal behaviour of our synchronous stream cipher generator completely;

A random balancing function has used to keep the complexity of generated initial vectors. We also strengthen the integrity of the exchanged data by a dynamic cyclic redundancy control.

As you know, the robustness of each security protocol builds on her aptitude to resist against the intelligent attacks. Sure, it takes several dimensions like the used cryptographic applications, her architecture and physical protection. It suffices to remember here that almost all the vulnerabilities announced on the WPA2 protocol take advantage of the deterministic nature of its cryptographic primitives or the computational limitations of the wireless environment [4, 6–8, 37, 39]. These weaknesses encourage us to invest in this field by a robust protocol whose interest is to fill the vulnerabilities security proven on WPA2.

7 Conclusion

Indeed, the confidentially and integrity of transmitted or stored data represent the most attractive vectors for attackers. They play a crucial role in the communications chain between wireless users. Their privacy is strongly linked to the ability of a computer system to protect them from sophisticated attacks. In this work, we highlight the wireless network security by tow efficient cryptographic applications.

The targeted vectors aim to strengthen the confidentiality and integrity of transmitted data in the wireless networks. The first checks the data integrity of given information without broadcast any information about the used generator polynomial. Then, a synchronous stream cipher algorithm founded on quadratic fields confirms its highest to ensure the data confidentiality by the unpredictable primitive signals' generator.

References

1. Asimi A (2016) Determination of irreducible and primitive polynomials over a binary finite field. Int J Pure Eng Math (IJPEM) 4(1): 45–59. ISSN 2348-3881
2. IEEE 802.11 (1999) Part 11: wireless LAN Meduim access control (MAC) and physical layer (PHY) specifications: higher speed physical layer extension in the 2.4 GHZ band. Institute of Electrical and Electronic Engineers
3. IEEE Std 802.16e (2005) IEEE standard for local and metropolitan area networks, amendment for physical and medium access control layers for combined fixed and mobile operation in licensed bands
4. Biryukov A, Khovratovich D (2009) Related-key cryptanalysis of the full AES-192 and AES-256. Cryptology ePrint Archive, Report 2009/317
5. Schneier S (2007) AES news, crypto-gram newsletter, 15 Sept 2002. Archived from the original on 7 July 2007. Retrieved 07-2007
6. Nikolić I (2009) Distinguisher and related-key attack on the full AES-256. Advances in cryptology—CRYPTO 2009. Springer Berlin/Heidelberg, pp 231–249
7. Biryukov A, Dunkelman O, Keller N, Khovratovich D, Shamir A (2009-08-19) Key recovery attacks of practical complexity on AES variants with up to 10 rounds. Archived from the original on 28 January 2010. Retrieved 2010-03-11
8. Ashokkumar C, Giri RP, Menezes B (2016) Highly efficient algorithms for AES key retrieval in cache access attacks. In: 2016 IEEE European symposium on security and privacy (EuroS&P), 21–24 Mar 2016
9. Sen Gupta S, Maitra S, Paul G, Sarkar S (2013) (Non-) random sequences from (non-) random permutations—analysis of RC4 stream cipher. J Cryptol Appear
10. Stigge M, Plötz H, Müller W, Redlich J-P (May 2006) Reversing CRC—theory and practice. Humboldt University Berlin, Berlin 17. Retrieved 4 Feb 2011
11. Beck M, Tews E (2008) Practical attacks against WEP and WPA
12. Beck M (2010) Enhanced TKIP Michael attacks
13. Isobe T, Ohigashi T (10–13 Mar 2013) Security of RC4 stream cipher. Hiroshima University. Retrieved 2014-10-27
14. Sepehrdad JP, Vaudenay S, Vuagnoux M (2011) Discovery and exploitation of new biases in RC4. Lecture Notes Computer Science 6544: 74–91.https://doi.org/10.1007/978-3-642-195 74-7_5
15. Jump Green M (2013) Attack of the week: RC4 is kind of broken in TLS. In: Cryptography engineering. Retrieved 12 Mar 2013
16. AlFardan JN, Bernstein D, Paterson K, Poettering B, Schuldt J (2013) On the security of RC4 in TLS. Royal Holloway University of London. Retrieved 13 Mar 2013
17. Fluhrer IM, Shamir A (2001) Weaknesses in the key scheduling algorithm of RC4. In: Proceedings of the 8th annual international workshop on selected areas in cryptography, Toronto, Canada
18. Wool A (2004) A note on the fragility of the "Michael" message integrity code. IEEE Trans Wirel Commun
19. Huang J, Seberry J, Susilo W, Bunder M (2005) Security analysis of Michael: The IEEE 802.11i message integrity code. In: Second international symposium on ubiquitous intelligence and smart worlds (UISW2005), Lecture Notes in Computer Science 3823, Springer-Verlag, Berlin

20. Whiting D, Housley R, Ferguson N (2003) Counter with CBC-MAC (CCM). RFC 3610, Sept 2003
21. National Institute of Standards and Technology. FIPS Pub 197: Advanced Encryption Standard (AES). November 26, 2001.
22. Asimi Y, Asimi A (2015) A synchronous stream cipher generator based on quadratic fields (SSCQF). Int J Adv Comput Sci Appl (IJACSA) 6(12):151–160
23. Gilis MCJ-L (2010) Attacks against the WiFi protocols WEP and WPA. October–December 2010
24. Fluhrer S, Mantin I, Shamir A (2001) Weaknesses in the key scheduling algorithm of RC4
25. Attack K (2004) https://www.netstumbler.org/f49/need-security-pointers-11869/
26. KoreK, Chopchop Attack (2004) https://www.netstumbler.org/f50/chopchop-experimental-wep-attacks-12489/
27. Rosen KH (1996) Applied cryptography, The CRC Press series on discretemathematics and its applications
28. Lidl R, Niederreiter H (1967) Finite fields, encyclopedia of math, and its Appl, 20, Addison-Wesley Publ. Co, Reading, Mass (1983), Reprint: Cambridge University Press, Cambridge (1967)
29. Lidl R, Niederreiter H (1994) Introduction to finite fields and their applications, 2nd edn. Cambridge University Press, Cambridge
30. Daemen J, Rijmen V AES proposal : Rijndael. https://csrc.gov/encryption/aes/Rijndaelamme nded.pdf
31. Arjen Lenstra K, Citibank NA Computional methods in public key cryptology, 1 North gate road, Mendham, NJ 0794-3104, U.S.A. https://www.arjen.lenstra@citigroup.com
32. Golomb SW (1967) Shift register sequences. Holden-Day, San Francisco
33. Rigney C, Willens S, Rubens A, Simpson W (2000) Remote authentication dial in user service (RADIUS). RFC 2865, June 2000
34. National Institute of Standards and Technology (2001) FIPS Pub 197: advanced encryption standard (AES), 26 Nov 2001
35. He C, Mitchell J (2005) Security Analysis and Improvements for IEEE 802.11i
36. Blunk L, Vollbrecht J, Aboba B, Carlson J, Levkowetz H (2003) Extensible authentication protocol (EAP). Internet draft. draft-ietf-eap-rfc2284bis-06.txt, 29 Sept 2003
37. Asimi Y, Asimi A et al (2018) Unpredictable cryptographic primitives for the robust wireless network security. Proc Comput Sci 134:316–321
38. Vanhoef M, Piessens F (2017) Key reinstallation attacks: forcing nonce reuse in WPA2. In: Proceedings of the 2017 ACM SIGSAC conference on computer and communications security ser. CCS '17, pp 1313–1328
39. Moen V, Raddum H, Hole KJ (2004) Weakness in the temporal key hash of WPA. ACM SIGMOBILE Mobile Comput Commun Rev 8(2):76–83
40. Singhal N, Raina JPS (2011) Comparative Analysis of AES and RC4 algorithms for better utilization. Int J Comput Trends Technol:177–181
41. Cerda JC, Martinez CD, Comer JM, Hoe DHK (2012) An efficient FPGA random number generator using LFSRs and cellular automata. In: IEEE 44th southeaster symposium on system theory, Mar 2012
42. Arnault F, Berger T, Minier M, Pousse B (2011) Revisiting LFSRs for cryptographic applications
43. Abd Elminaam DS, Kader HMA, Hadhoud MM (2010) Evaluating the performance of symmetric encryption algorithms. Int J Netw Secur 10(3):213–219
44. Charbathia S, Sharma SA (2014) Comparative study of rivest cipher algorithms. Int J Inf Comput Technol 4(17):1831–1838
45. Saarinen M-JO (2012) The BlueJay ultra-lightweight hybrid cryptosystem. In: IEEE symposium on security and privacy workshops (SPW), pp 27–32

A Machine Learning Based Secure Change Management

Mounia Zaydi and Bouchaib Nassereddine

Abstract IT change management process is a fundamental part of information security operations service and strategy. An inoperative change management process leads to serious outcomes, including information loss, reputation damage, exposure to risks and network failures as well as a negative impact on revenues. IT incidents, especially security incidents, may generate via various causes; component failure, errors by users or through implementation IT changes trouble. The change control is the key factor to minimize incidents, it ensures non-vulnerabilities in the presence of changes and proper cybersecurity and it offers a measure of trust to involve stakeholders. To control this change considered as an intelligent and secure step to deal with the impact factors before incidents occur. Therefore, in this study, a new model proposed. It based on Machine Learning algorithms and CRISP-DM methodology to predict the relevant named changes causing incidents, in order to allow strategic decisions at the time and to expose the shortcomings on which change managers should focus to improve them and thus ensure the right level of cybersecurity.

Keywords ITSM · IT change management · ML · Prediction · Classification · Security · KNN

1 Introduction

IT Service Management (ITSM) is a discipline for managing IT operations and services [1]. There are several ITSM frameworks available to help organizations manage their IT services while conforming to their specific business requirements and needs. The challenge of these ITSM frameworks is to improve the business performance of organizations through enhancing IT service delivery. Thus, ITSM focuses on methodologies and tools to deliver efficient and high-quality service [2].

M. Zaydi (✉) · B. Nassereddine
IT, Networks, Mobility and Modelling Laboratory (IN2M), Faculty of Sciences and Techniques, Hassan 1st University, Settat, Morocco
e-mail: m.zaydi@uhp.ac.ma

© The Editor(s) (if applicable) and The Author(s), under exclusive license to Springer
Nature Switzerland AG 2021
Y. Maleh et al. (eds.), *Machine Intelligence and Big Data Analytics for Cybersecurity Applications*, Studies in Computational Intelligence 919,
https://doi.org/10.1007/978-3-030-57024-8_23

In that sense, there is a wide range of tools that can help optimize IT services and business operations, increase employee productivity or customer satisfaction costs which is the important value for any organization in any industry [3]. It is an IT approach characterized by IT services, customers, service level agreements (SLA) and daily activities management of an IT function through processes [4]. As a result, the use of advanced IT infrastructures becomes an essential need in IT organizations. Medium and large organizations are taking advantage of technology to provide better services and improve their internal management. However, the use of advanced IT infrastructures can increase management complexity, especially about the deployment of change, which results in higher costs to maintain a healthy IT environment [5, 6]. In this way, the Information Technology Infrastructure Library (ITIL) [7] presents a set of best practices to help organizations to maintain their IT infrastructures properly. ITIL provides a set of processes [7], of which Incident Management (IM) is one of the most popular. The purpose of this process is to identify and resolve IT problems to restore normal service operations as quickly as possible and to minimize the impact on business operations [8], to guarantee and maintain the best level of service in terms of quality and availability. The service's usual operation is defined through an SLA in a contractual document describing the performance criteria that a provider promises to meet when delivering a service [8]. Thus it represents the corrective measures and possible penalties that will take effect if performance is less than the promised level. Practitioners in this field define SLA as the agreed time for incident-resolution. IM is a key factor in ensuring that the SLA is well respected, resulting in financial penalties and unsatisfied customers for the service provider. However, the complexity of the underlying problem, may take several days before an incident is resolved, which makes it crucial for service providers to have effective and proactive risk management in place to prevent incidents. To provide sustainable, high-availability services, IT service providers need to make frequent IT changes to their customer's infrastructure to continue to enable smooth and efficient operations.

The issue at this level is the probability changes failure while causing incidents affecting the quality of service. According to Gartner report [9], the failure of change is one of the biggest problems facing IT service providers nowadays. Incidents have different causes. It can come from a defect in a computer component, a wrong manipulation by a user or produced by IT changes. This cost of repairing these incidents is pretty high, which makes it challenging to manage their impact on the business. The importance of change management lies in modifying the infrastructure using an organized and standardized method to move the environment to a new, consistent and functional desired state. The change management process aims to minimize the potential risks that can affect the IT environment by ensuring that clear and standardized procedures are in place to enable all changes to be handled effectively [10]. The keys to ensuring this are: identifying relevant changes (that cause incident affecting client business), taking proactive measures to manage and reduce risks, preparing prompt corrective action. The analysis conducted to identify the parameters that affect the occurrence of incidents must be in-depth, well-conducted and intelligent. If this study is done incorrectly, the changes will lead to incidents. Such predictive modelling could be done through artificial intelligence, in particular,

Machine Learning (ML). The ML has proven its effectiveness in various economic and scientific fields. However, research applying the ML in the field of ITSM remains insufficient, whereas integrating intelligence in this field can bring several benefits in different areas. The life cycle of ML models is the same in all domains; it is the data and their types that change. This cycle includes data pre-processing (1) to transform the raw data into features, learning (training) a model with the features (2) and providing the model to respond to predictive queries (3) [11]. Although ML is applied to ITSM, it is not yet explored comprehensively in change management. Whereas integrating intelligence in this field can bring several benefits in different areas. However, research in this area remains insufficient.

This chapter focuses on the effect changes had on the occurrence of incidents, to detect the causes that produce them. A predictive model is proposed to analyze the changes that cause incidents, with the aim Is to fully understand every aspect of the changes in-depth and testing them properly before their implementation to minimize the probability of incidents. This model can be integrated as an extension in an ITSM tool, which will analyze the incidents and can then learn the relevant changes to conduct to adjust the parameters of the changes before their implementation. With preventing changes leading to incidents comes Reducing significantly of risks factors and implementation of a more effective risk mitigation measures at the organizational level through improvements on processes, people and/or tools.

This chapter is organized into three main sections. The first presents current work in this field the second section emphasizes the dependency between changes and incidents and the multiple causes of the latter, to propose a model that will predict the incidents caused by the changes, based on the CRISP-DM (cross-industry standard for data mining) methodology [12]. This methodology starts first with data preparation and cleaning; second, it focuses on modelling to select the ML algorithms used to create the predictive model, and model evaluation. In this part, a case study is conducted using a dataset of incidents and changes from a large service company operating in the energy and gas sector in France. Finally, the study moves to the deployment phase. The last section summarizes the research contribution and presents future guidelines.

2 Literature Review

In this field, we distinguish between various work in different areas, including analysis, data treatment and process automation, which contributes to reducing operating costs. In-Text Mining [13], Sarawagi presents the automatic extraction and integration of information [3]. In the field of ITSM, Gupta et al. Proposes an information extraction technique for the incident management process [14], which can be extended to other processes, such as events, problems or changes. Change management remains the most complicated process in ITSM, and its successful management has a direct influence on other critical processes such as incident and problem management. As the resolution of incidents impacts the Service Level Agreement

(SLA), which describes the performance criteria that the service provider should meet when providing a service [15], and the cause of these incidents can be related to changes, so predicting the relevant changes (those that cause the incidents) early on is necessary. In this context, various works are being carried out either for data extraction, automation or the use of learning machines to set up prediction, recommendation or prevention models, Sarawagi presents a method for automatic information extraction and integration [16]. Another method of information extraction in the field of incident management is proposed by Gupta et al. [2]. Keller et al. present a workflow-based system for automatic generation of change plans called FIELDS [3]. Thus, to reduce incident resolution time, a model based on ML is proposed [8]. Dees and End proposed a model to predict the duration of the change and its global impact [17]. The objective is to predict the workload of the Service Desk according to the interactions with the affected configuration items. Incident analysis is conducted to identify unusual trends and patterns in the operation of incident management [18], as well as to find the causality between incidents to predict which ones occur or are repeated [19]. A decision-making model is introduced, which can automate the totality of the incident management process and improve the efficiency of the responses to incidents provided [20]. Thus, ML has been applied in the field of change management to explore the effectiveness of automation in relation to cost [6]. We have identified many works performed in ITSM based on machine learning techniques. To automate a process task such as [21] proposed a model to automate the process of changes based on change previously processed through the CMDB to find similar ones to handle them quickly. The study [22] focused on the relationship between incident management and problem management in detecting, recording and correcting errors that are occurring and impacting the IT infrastructure, and how human intervention reduces productivity. In this context, the authors of this study have identified the importance of automation to increase productivity and reduce the costs of incident resolution. Therefore, they have proposed a model of incident ticket correlation to promote the automation of this process and minimise human involvement. Another study [23] is being carried out in the same direction to automate incident management based on a correlation model of critical words of Cis from the CMDB. However, we regret the lack of work based on ML techniques to improve IT security. We are exploiting this lead by studying the change management process and its relationship with IT incidents, designing a model to predict the changes that cause these incidents.

Generally, research in this area often aims to automate some activities within the service operation processes to make the Service Desk more efficient [24]. While recognizing that it has vital processes in ITSM, including incident and change management that can positively or negatively impact ITSM stability and availability, at this level, it seems necessary to understand the dependency between incidents and changes as described in the 3rd section and predict changes leading to incidents.

3 IT Change Management

Change management is generally considered as a back-office function minimizing risks of failure or IT service outages. Taking the time to evaluate changes before they are implemented enables to discover and anticipate incidents before they occur.

A high number of incidents resulting from changes is a good indicator that the changes are not being properly tested. Major incidents resulting from changes is one of the most effective measures of collection, as it shows the impact of implemented changes on the level of service. It is not a measure of system failures but service failures. It makes teams accountable for the impact they have on the business and measures the business interruption caused by IT itself.

The change data encompasses a set of attributes it describes, including ID, type, closing date, ..., and the closing code that signifies the status of the change after implementation. We distinguish in this sense between those successful (implemented as planned), failed or in progress (processing, validation or implementation), the value of this field seems essential to determine the incidents that are produced as a result of the changes.

A failed change causes several incidents of different criticality and urgency levels on the IT infrastructure, so it can be concluded that the failed changes are the ones that cause the IT incidents. However, this hypothesis is not always verified because we have observed that even successful changes cause incidents.

The failure or success of the changes certainly has an impact on the financial strategy of the company and can be the leading cause of a financial penalty for one of the customers. Success can retain the customer and thus have satisfied customers, and even attract the attention of new customers. However, in this research we are not interested in the business aspect but rather in the technical issue, analyzing the leading cause of incidents that occurred as a result of a change without taking into consideration its closing code.

To do so, we will analyze all changes and similar incidents to determine the leading cause of the incidents for:

- Anticipate relevant changes: and facilitate root cause analysis upstream and downstream of the change;
- Find and analyze the systemic causes of the changes that lead to the incidents at the level of people, processes and technologies;
- Continuously monitor existing risk factors and obtain early warning indicators for new risk factors; the risk factors that lead to incidents are continually evolving at the same pace as IT processes, people and technology. As a result, it will be necessary to monitor change data to detect emerging risk factors continuously.

4 Methodology

To help managers and those involved in the change management process to predict the change that will cause the incidents, we are going to build a predictive model. The proposed model can predict changes that will cause incidents, and thus adjust the responsible parameters to affect change or to minimize the impact of resulting incidents. The model will clarify the dependency between changes and incidents in the IT infrastructure. To perform this predictive modelling, we're using machine learning algorithms and the proposed following architecture, as shown in Fig. 1, inspired by CRISP-DM methodology.

We will use an appropriate classification model, trained based on incidents and changes history log, to predict if the change triggered the incident. In this case, the target attribute will describe whether or not the change will result in an incident. The model will be tested and evaluated using predefined criteria.

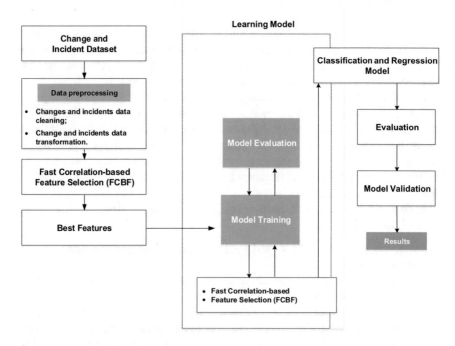

Fig. 1 The proposed model archtecture

4.1 Business Understanding

The first phase of CRISP-DM methodology aims to understand objectives and requirements from a business perspective. For the sake of this paper, this phase is aimed at defining the business objectives of the case study. The clients of the IT service provider, who presents our case study, report many incidents related to networks, applications and hardware. The increasing number of these incidents presents a severe violation of the SLA, which has a significant impact on the firm's financial and budgetary aspect, this is due to the penalties related to the degradation of service quality caused by the said violation. This has motivated the actions initiated by the top management of the service provider with the primary objective of eradicating the causes of these violations and increasing the support of global IT management, thus ensuring the proper functioning and efficient operation of the infrastructure of their clients. It is at this level that this study took place, which consists of an in-depth analysis of the causes of the incident. The goal is to detect IT changes from an incident history log.

4.2 Preparing Data

The dataset used in this project provided by an IT services company operating in the energy and gas sector in France, due to confidentiality issues, cannot be mentioned. The dataset consists of two files, the first one contains the change records, and the second one contains the incident records. These two files included detailed descriptions of the incidents and changes that occurred, associated configuration items, opening and closing hours, related incidents, related changes and other attributes shown in Tables 1 and 2.

Incidents can be caused by a dysfunction of an IT component (printer, server, cable, storage bay), a user's wrong manipulation and/or caused by IT changes. IT change has a significant cost in IT management; that's why in this paper, we explore change-related incidents. Thus for incident and change managers, it is interesting to know about the likelihood that a change may lead to incidents, as well as their severity and scale, to anticipate solutions to minimize their impact and especially to respect the SLA. The objective is to design a model able to predict whether or not a change will cause the incidents.

The main idea was to retain the most relevant attributes in the construction of the model to allow prediction. Attributes are transformed, and new ones are created, we also removed several attributes that did not have a significant impact on the classification and obtained a final set of predictors. Among the most significant attributes in this model, we distinguish *Change_ID*, *originated_from*, *CI-Name*, *CI-Type*, *Risk-assessment* and *Related_change*. Next, we had to define and create the target attributes for our model. The target variables, in this case, were not explicitly specified in the dataset but could be transformed from certain attributes. We created

Table 1 Incidents dataset attributes

Attribute	Representation	Information attribute	Description
CI name (Aff)	CI name (Aff)	Alphanumeric	Name of affected configuration items by the change
CI type (Aff)	CI type (Aff)	Alphanumeric	Type of affected configuration item
Change ID	Change ID	Integer	Change identifier, a unique number, automatically incremented via the ITSM tool
Change type	Change type	Alphanumeric	Change type can be applicative, software, hardware…
Risk assessment	Risk assessment	Alphanumeric	specifies level (critical, high, medium, minor) of impact on the business
Emergency change	Emergency change	Alphanumeric	Indicate if the change is urgent
CAB-approval needed	CAB-approval	Alphanumeric	indicates if the change needs approval of the Change Advisory Committee
Planned start	Planned start	Date	Date and time of start implementation of the change
Planned end	Planned end	Date	Date and time of end implementation of the change
Scheduled downtime start	Scheduled downtime start	Date	Date and time of the scheduled service interruption during change implementation
Scheduled downtime end	Scheduled downtime end	Date	date and time of the scheduled recovery after change implementation
Actual start	Actual start	Date	Effective date and time of change implementation
Actual end	Actual end	Date	Effective date and time of service recovery
Requested end date	Requested end date	Date	Date and time of recovery of the requested service after change implementation

(continued)

Table 1 (continued)

Attribute	Representation	Information attribute	Description
Change record open time	Change record open time	Date	Open date and time of the change record
Change record close time	Change record close time	Date	Date and time of closure of the change record
Related incidents	Related incidents	Alphanumeric	Number of incidents caused by the change

the attribute *relevent_change*, its value was derived from the values of the ***Change.ID (change data set)*** and ***Related.Change (incident data set),*** indicating if the change was really responsible for the incident or not. We compared the values of these two attributes, in case they were equal, the value of the *relevent_change* was set to YES, and in case they were different, we set the attribute to NO.

4.3 Feature Selection

Initially, the algorithm selects characteristics correlated above a threshold given by SU with the class variable (relevent change). After this initial filtering, it detects the predominant correlations of the characteristics with the class. The definition is that, for a predominant characteristic "X", no other characteristic is more correlated to "X" than "X" is to the class. The IT changes and incidents dataset has 34 attributes. Since a large number of irrelevant and redundant attributes are involved in these expression data, the Relevent_Change classification task is made more com.

This phase consisted of model learning, which the type used for our data is classification. We used Weka platform [25], which provides a full collection of machine learning algorithms and pre-processing tools for cleaning and transforming data, as well as for the machine learning process. It includes implementations of the most popular machine learning algorithms.

We used the algorithms Random Forest (RF), Naive Bayes classifier (NBC), K Nearest Neighbors (KNN.) that are commonly used to process data for binary decision problems in machine learning.

Performance of supervised ML methods, especially for classification, usually depends on learning samples, allowing a non-optimistic classification error estimate. The higher number of learning samples used in testing (use a learning base and a test base), the more reliable the classification rules will be. At the same time, it is necessary to keep a significant number of test samples for the evaluation of these performances to be meaningful. The technique of cross-validation is frequently used to meet these two needs: it consists of dividing the initial set into several subsets of equal size, with each sub-set being used as a test frame. In contrast, the union of all the other sub-sets is used as a learning frame [26].

Table 2 Changes dataset attributes

Attribute	Representation	Information attribute	Description
CI name (aff)	CI name (aff)	Alphanumeric	Configuration Item responsible for the incident
CI type (aff)	CI type (aff)	Alphanumeric	Type of the configuration Item
Service comp WBS (aff)	Service comp WBS (aff)	Alphanumeric	Managers of CIs registered in CMDB under the form of CIs-service-responsible
Incident ID	Incident ID	Integer	incident unique identifier (ID)
Status	Status	Alphanumeric	Incident status ()
Priority	Priority	Integer	Combines impact (impact of service disruption) and urgency (degree of urgency in resolving the incident)
Category	Category	Alphanumeric	Used to categorize incidents according to their similarity to other incidents
KM number	KM number	Integer	Knowledge Document number, is referring to the knowledge base
Open time	Open time	Date	Opening time of the record in the IT service management tool
Reopen time	Reopen time	Date	when the incident is closed and re-opened for a given reason (same incident occurs again)
Resolved time	Resolved time	Date	Date and time of incident resolution
Closed time	Closed time	Date	date and time of closing recording incident
Handle time	Handle time	Integer	Time required to resolve the incident
Closure code	Closure code	Alphanumeric	Code describing service interruption type
Alert status	Alert status	Alphanumeric	Alert status based on service level agreed (if it is respected or not)
Related incidents	Related incidents	Integer	Number of related incidents, in the case of a major incident

(continued)

Table 2 (continued)

Attribute	Representation	Information attribute	Description
Related changes	Related changes	Integer	If a change is related, it is recorded here (field takes multiple)

To avoid unstable operation results and test the reliability of the prediction model, the tenfold cross-validation algorithm was used. It consists of dividing the initial set of 30,000 change records into 10 blocks. Then, 10 evaluation learning phases are repeated where a hypothesis h is obtained by learning on 9 blocks of data and tested on the remaining block.

5 Performance Evaluation

Assessing classification algorithms is a key part of any data mining process to measure its efficiency and determine how well these algorithms have responded to the original issue.

For that, we compare the accuracy measures based on precision, recall, TP rate and FP rate values for K-NN, BN and RF. The results are shown in Table 3. We can see that the best results are those generated by KNN classifier.

Besides we have used a confusion matrix, focusing on the most important criteria identified in this area [27], namely, error rate, that indicates the proportion of misclassified changes, and accuracy, that defines the probability of classifying the change accurately or not.

According to the graphs (Fig. 2), we notice that the KNN had the lowest error rate and the highest precision value of 99.3% (Table 3), followed by Naïve Bayes

Table 3 Comparative table of precision measurements used classifiers

Classifiers	Evaluation criteria					Class
	TP rate	FP rate	Precision	Recall	F-measure	
Naive Bayes	0.905	0.122	0.998	0.905	0.949	Yes
	0.878	0.095	0.105	0.878	0.187	No
	0.905	0.121	0.987	0.905	0.94	
K Nearest Neighbors	0.995	0.259	0.997	0.995	0.996	Yes
	0.741	0.005	0.675	0.741	0.706	No
	0.992	0.256	0.993	0.992	0.992	
Random Forest	0.854	0.09	0.925	0.854	0.92	Yes
	0.91	0.146	0.627	0.91	0.135	No
	0.854	0.091	0.776	0.854	0.911	

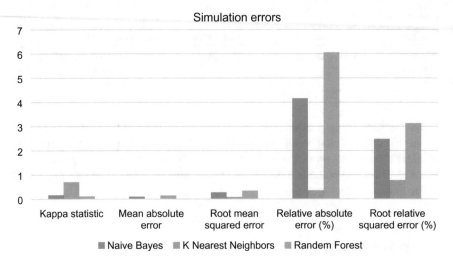

Fig. 2 Comparative error graph of used classifiers

with 98.7% and the Random Forest, which ranked third with 77.6%. However, these values are very close, thereby validating that changes cause incidents.

Concerning precision metric and efficiency of classifiers in terms of time to build the model, correctly classified instances, incorrectly classified instances and accuracy, Table 4 shows the accuracy performance rate. The KNN has a correct accuracy of 98.629%, followed by Naïve Bayes with 89.516% and 84.406% for Random Forest. However, the KNN is performing better at the classification level in this dataset. The two other algorithms provide prediction results close to KNN results in terms of relevant compensation changes (which effectively caused the incidents).

As a measure of performance of the classification and prediction system, the confusion matrix shown in the table (Table 5) was used. Each column of the matrix represents the number of occurrences in the estimated class, while each row represents the number of occurrences in the actual (or reference) class. One of the advantages of the confusion matrix it quickly shows whether the system can classify correctly.

The NO class, representing the fact that a change did not cause the incident, was the class with a relatively high error rate.

Table 4 Classifiers performance

Evaluation criteria	Naive Bayes	K Nearest Neighbors	Random Forest
Time taken to build model (s)	0.09	0.02	0.36
Correctly classified instances	27,387	30,042	25,864
Incorrectly classified instances	2888	233	4411
Accuracy (%)	89.516	98.629	84.406
Error	0.105	0.014	0.156

Table 5 Confusion matrix of changes

	Yes	No	
Naive Bayes	27,055	2842	Yes
	46	332	No
K Nearest Neighbors	29,762	135	Yes
	98	280	No
Random Forest	25,520	4377	Yes
	34	344	No

NB identified 27,055 positive records, 29,762 by KNN and 25,520 identified by RF. Compared to 46, 98 and 43 respectively at NB, KNN and RF for the negative records.

The most important task is to confirm if the change causes the incident or not. This classification was more precise, and the three models confirmed this.

6 Conclusion

The main purpose of this work is to affirm the possibility that change can cause incidents. To do this, we developed a model predicting changes leading to incidents. We used all the data from the change and incident files and specified the main area that we tried to explore. It deals with the dependency and relationship between changes and incidents. Therefore we built a predictive model to solve this issue, using the Random Forest (RF), Naive Bayes classifier (NBC) and K Nearest Neighbors (KNN), are the models likely to perform best in the dataset used in this study. Although the accuracy and error rate was marginally different, all the three models confirmed that the most likely cause of incidents occurred during the implementation of IT changes. The final step of the CRISP-DM methodology is the deployment of this model through, among other things, its integration into the organization's ITSM tool, aiming to predict any change request once it arrives in the backlog, for a more refined and early analysis before its implementation, this last phase will be the subject of future work.

References

1. Galup SD, Dattero R, Quan JJ, Conger S (2009) An overview of IT service management. Commun ACM 52(5):124
2. Marcu P, Grabarnik G, Luan L, Rosu D, Shwartz L, Ward C (2009) Towards an optimized model of incident ticket correlation. In: 2009 IFIP/IEEE international symposium

3. Hochstein A, Zarnekow R, Brenner W (2005) ITIL as common practice reference model for IT service management: formal assessment and implications for practice. In: The 2005 IEEE international conference on e-technology, e-commerce and e-service, 2005. EEE'05. Proceedings. IEEE, pp 704–710

4. Iden J, Eikebrokk TR (2013) Implementing IT service management: a systematic literature review. Int J Inform Manag 33:512–523

5. David JS, Schuff D, Louis RS (2002) Managing your total IT cost of ownership. Commun ACM 45(1):101–106

6. Brown AB, Hellerstein JL (2005) Reducing the cost of IT operations: is automation always the answer? In: HOTOS'05: Proceedings of the 10th conference on hot topics in operating systems. USENIX Association, Berkeley, CA, USA, pp 12–12

7. ITIL (2008) Information technology infrastructure library (ITIL). Office of Government Commerce (OGC). [Online]. Available http://www.itil.co.uk/

8. Zuev D, Kalistratov A, Zuev A (2018) Machine learning in IT service management. Procedia Comput Sci 145:675–679

9. Gartner IT Security Summit (2005) Best practices for continuous application availability

10. Pozgaj Z, Strahonja J (2008) It service management based on ITIL methodology. In: An enterprise Odyssey. International conference proceedings. University of Zagreb, Faculty of Economics and Business, p 965

11. Popp M (2019) Comprehensive support of the lifecycle of machine learning models in model management systems. Master's Thesis

12. da Rocha BC, de Sousa Junior RT (2010) Identifying bank frauds using CRISP-DM and decision trees. Int J Comput Sci Inform Technol 2(5):162–169

13. Weiss SM, Indurkhya N, Zhang T, Damerau (2005) Text mining: predictive methods for analyzing unstructured information. Springer

14. Gupta R, Prasad KH, Mohania M (2008). Information integration techniques to automate incident management. In: NOMS 2008–2008 IEEE network operations and management symposium. IEEE, pp 979–982

15. Malega P (2014) Escalation management as the necessary form of incident management process. J Emerg Trends Comput Inf Sci 5(6):641–646

16. Sarawagi (2002) Automation in information extraction and integration (tutorial). In: VLDB

17. Dees M., van den End F (2014) A predictive model for the impact of changes on the workload of Rabobank group ICT's service desk and IT operations. Rabobank Group ICT's Service

18. Li TH, Liu R, Sukaviriya N, Li Y, Yang J, Sandin M, Lee J (2014) Incident ticket analytics for IT application management services. In: 2014 IEEE international conference on services computing. IEEE, pp 568–574

19. Liu R, Lee J (2012) IT incident management by analyzing incident relations. Presented at the November 12, 2012

20. Yun M, Lan Y, Han T (2017) Automate incident management by decision-making model. In: 2017 IEEE 2nd international conference on big data analysis (ICBDA). IEEE, pp 217–222

21. Li H, Zhan Z (2012, April) Bussiness-driven automatic IT change management based on machine learning. In: 2012 IEEE network operations and management symposium. IEEE, pp 1374–1377

22. Marcu P, Grabarnik G, Luan L, Rosu D, Shwartz L, Ward C (2009, June) Towards an optimized model of incident ticket correlation. In: 2009 IFIP/IEEE international symposium on integrated network management. IEEE, pp 569–576

23. Gupta R, Prasad K, Mohania M (2008) Automating ITSM incident management process. In: Proceedings of the 5th IEEE international conference on autonomic computing, Chicago, June 2008, pp 141–150

24. Andrews AA, Beaver P, Lucente J (2016) Towards better help desk planning: predicting incidents and required effort. J Syst Softw 117:426–449

25. Moghimipour I, Ebrahimpour M (2014) Comparing decision tree method over three data mining software. Int J Stat Prob 3(3):147

26. Alaoui A (2012) Application des techniques des métaheuristiques pour l'optimisation de la tâche de la classification de la fouille de données. Doctoral dissertation, USTO
27. Fawcett T (2006) An introduction to ROC analysis. Pattern Recogn Lett 27(8):861–874

Intermediary Technical Interoperability Component TIC Connecting Heterogeneous Federation Systems

Hasnae L'Amrani, Younes El Bouzekri El Idrissi, and Rachida Ajhoun

Abstract The spread of digital identity raises many new opportunities and challenges concerning identity. A set of identity management systems has been developed to handle such identities. The aim is both to enhance the end-user experience and to provide secure access for users. Nowadays, we have a large number of heterogeneous identity management initiatives. Proof of its eligibility for identity management is provided under the federation system. The strength of security domains within federated systems is a trusted agreement between communicating entities. However, Federated systems are challenged by the interoperability issue across those federated heterogeneous systems. This work aims to provide a technical interoperability approach for the different federations. The researchers are offering a technical interoperability component TIC, as a midway tool that will enable identity data to be interchanged between heterogeneous federations in total transparency.

Keywords Digital identity · Federated system · Interoperability · Security · Cross-Domain · Technical interoperability

1 Introduction

It is often said that humans couldn't feel alive without identity. Identity story starts with the Stone Age. However, real identities registration and normalization started just after the First World War. The first identification document was a passport introduced in England. In this day and age, a new revolution in identity concept was introduced. The cyber identity or digital identity is the reflection of citizen identity

H. L'Amrani (✉) · R. Ajhoun
Smart Systems Laboratory (SSL), National School of Computer Science and Systems Analysis (ENSIAS), Mohamed V University, Rabat, Morocco
e-mail: hasnae90lamranii@gmail.com

Y. El Bouzekri El Idrissi
Systems Engineering Laboratory, National School of Applied Science (ENSAK), Ibn Tofail University, Kenitra, Morocco

© The Editor(s) (if applicable) and The Author(s), under exclusive license to Springer Nature Switzerland AG 2021
Y. Maleh et al. (eds.), *Machine Intelligence and Big Data Analytics for Cybersecurity Applications*, Studies in Computational Intelligence 919,
https://doi.org/10.1007/978-3-030-57024-8_24

in the cyber world. Facebook, Whatsapp, YouTube, Facebook Messenger, Instagram, Snapchat, Twiter, Pinterest, etc. are all social platforms that require digital identity to benefit from their services. In January 2020, Facebook reached 2, 5 billion active users monthly, YouTube reached 2 billion users, Whatsapp 1600 million [1] etc. These are massive numbers of users' identities to deal with. The digital identity is indeed a critical subject facing users' privacy.

While talking about these huge identities, many questions are asked. How these identities are managed? Should users retain and remember all identities used for different services? Identity federation systems as one of identity management systems, could it be a solution to manage digital identity? Could heterogeneous federation technologies interoperate easily? Those are the main research question that the researchers intend to deal with in this paper.

Actually, these identities should undoubtedly be managed by an efficient identity management system. The researchers have found that there are many problems while using numerous identities by one user. They found that it's hard to remember all user identities to have access to all requested services separately. Identity federation technologies are systems that provide trust for users, services providers, and identity providers to spread identities. Based on trust concept, users could use only one identity to guarantee access to all requested authorized services. Single sign-on with federation technologies proved it benefices against identity theft and data privacy.

However, federation technologies front issues in communication interoperability. Since federated systems technologies do not use the same methods to ensure user's authentication, authorization, and attributes exchange, the interoperability issue is the challenge that heterogeneous federation technologies deal with. Therefore the researchers provide an interoperability prototype to ensure interoperability among different communicating federation technologies. This paper presents how we carry out this study and the achieved results. In the definition of terms section, researchers define the most important concepts related to this work. Furthermore, the related works section introduces the main works that treat interoperability issues. After that, the materials and methods section presents the problem statement and the proposed prototype. Besides, the results section contains the main obtained results. Finally the conclusion and future works.

2   Definitions of Terms

There's a whole variety of identities in the real world, even in the virtual world. Until now, it's more than more sophisticated to unassociate all these real and virtual identities. Such a need to manage the identity of internet users clearly shows besides being essential, keeping in mind the increased development of web users.

In this paper, we define terms from the Recommendation of the International Telecommunication Union ITU-T [2] to use standardized definitions in this work.

The term *Digital Identity* means "a digital representation of information known about a specific individual, group or organization".

According to the same reference, the term *Entity* is "something that has separate and distinct existence and that can be identified in context".

The term *Context* means "an environment with defined boundary conditions in which entities exist and interact".

The term *Federation* indicates "An association of users, service providers, and identity providers".

As stated in the same reference, an *Identity provider (IdP)* is "an entity that verifies, maintains, manages and may create and assign identity information of other entities".

The term *Service Provider (SP)* or *Relying Party (RP)* involves " an entity that relies on an identity representation or claim by a requesting/asserting entity within some request context".

A trust connection between federations is named the *Circle Of Trust (CoT)*. The term Circle Of Trust is represented as "a federation within; the trusted Service Provider (SP) and Identity Provider (IdP) that have business relationships, where the IdP outputs are allowed by all other IdPs of the same CoT" [3].

After the definition of technical terms connected directly to identity federation, now, the researchers will present terms about the communication process between federated entities. In keeping with the standardized definitions of ITU-T, the term *Identification* means "the process of recognizing an entity within a context".

The *Authentication* term presents "a process used to achieve sufficient confidence in the binding between the entity and the presented identity".

Despite the term *Authorization* implies " the gathering of rights and based on these rights, the guarantee of access".

Among different federated entities, Attributes and Pseudonyms are communicated. The term *Attributes* refers to "information bound to an entity that specifies a characteristic of the entity".

The *Pseudonym* term symbolizes "An identifier who's binding to an entity is not known or is known to only a limited extent, within the context in which it is used".

The last term to define is the *Interoperability* term, in consonance with the Institute of Electrical and Electronics Engineers (IEEE), interoperability is "the ability of two or more systems or components to exchange information and to use the information that has been exchanged".

3   Related Works

The main purpose of this research is to make interoperable a set of identity federation technologies, which are distinguished according to their specifications and properties [3]. Besides that, the main function of both the federation and how identities are managed, simply stay the same [4].

Initially, the researchers started the study from the Service Oriented Architecture (SOA) paradigm, then researchers found that the concept of identity federation could be adapted to all evolution of data management and storage. For comparison purposes, the case of the cloud computing revolution, there is a mutual mutation from cloud computing to cloud federation and the cloud federation is the future of the Cloud [5]. In this work, researchers' target federation systems, they choose to treat this subject away from SOA, Cloud, etc.

There are extensive researches just on the topic of interoperability within identity federation systems. The Oxford Computer Group [6] established interoperability among two federation technologies to web-based applications, and solutions adopted for the study were Shibboleth V1.3 and ADFS V1 that is the implementation of WS-Federation V1.0. Furthermore, this cited paper has demonstrated that ADFS and Shibboleth interoperate handling a module named ADFS extensions concerning Shibboleth, established in the SP or IdP which require interoperability in order to maintain communication established with other technologies. Hence, this proposal does not hold a generic case of interoperability within all identity management existent technologies, it currently regards Shibboleth and WS-Federation. This extension solution exposes further limits in terms of embedding interoperability treatment in the providers [7].

An impressive continuing of the previously presented work target description of the interoperability within couple federation architectures, SAML and WS-Federation [8]. This approach is primarily highlighted by describing a technical comparison between the two SAML 2 and WS-Federation 1.1B technologies, which subsequently represent a review of the exchanged queries and the structure of federation data in the passive client situation. However, they describe the issue without contributing a solution.

4   Materials and Methods

Identity federation systems have introduced an impressive concept of data management and identity security, hence several federated identity technologies are used to manage identities. While these technologies are numerous, the difference between technologies brings many benefits for entities using federation. Nevertheless, this variation of federal technology triggers federated systems technology interoperability issues.

Researchers highlight issues of interoperability in federation technologies in this section. We demonstrated the situation of the communication interrupting in different federation technologies. Researchers would then present the prototype proposed to tackle this issue.

4.1 Federations' Technologies and Interoperability Challenges

Identity federation systems provide many solutions to manage identity propagation among domains. SAML1.0, SAML2.0, WS-Federation, OpenID, OpenID Connect, OAuth 1.0, OAuth 2.0, etc. are all federation technologies used to ensure identity federation among entities inside and outside federation domains. The researchers generate a comparative study within federation technologies to gain evidence of the dissimilarities between federated technologies.

Table 1 presents the concurrent federated technologies. At this point we use the following criteria to define the difference between those technologies' performances: authentication, authorization, single sign-on, attributes exchange and pseudonyms existence. The collect of information about those technologies gives us a global view about the capabilities of each technology [9, 10].

Researchers used authentication and authorization as comparison criteria to show the need for both authentication/authorization to ensure federation. Here we insist on the existence of authentication criteria in comparing federation technologies, however, authorization could be added with a supplementary authorization standard that could change from one federation technology to another. Attribute exchange with pseudonyms are required also to guarantee the efficiency of communication between federations' technologies [11].

Table 1 Existing federation technologies

Technology features	WS-Federation	SAML	OAuth	OpenID	OpenID connect
Federation version	WS-Federation 1.2	SAML2.0	OAuth 2.0	OpenID 2.0	OpenID Connect 1.0
Authentication	Enabled	Enabled	Not enabled	Enabled	Enabled
Authorization	Enabled	Enabled/User roles (attributes)	Enabled	With OAuth 1.0	With OAuth 2.0
Token	SAML 1.1/2.0 assertions	SAML2.0 assertions	SAML2.0 assertions	id_token	id_token
Token type	XML document	XML document	JSON web token (JWT)	JSON web token (JWT)	JSON web token (JWT)
Attributes exchange	Enabled	Enabled	OpenID connect based	Enabled	Enabled
Pseudonym	Enabled	Enabled	OpenID connect based	Pairwise Pseudonymous	Pairwise Pseudonymous

Researchers have found that authentication is a basic level to ensure while communicating. According to the comparative study shown in Table 1, most of the federation technologies support authentication. However, authorization is ensured based on the discussed federation technology or based on a combination with other federation technology.

Tokens are required for guarantee the exchange of assertions that carry authorizations attributes, user attributes... Attributes are necessarily required to the use and re-use of data among different federations. Pseudonyms strength is the guarantee of certain levels of security while communicating within heterogeneous domains.

Taking everything into consideration we deduced that federation technologies support the basic levels of federated systems, despite, they use different ways to ensure these required processes. This difference generates interoperability issues among the cited federation technologies.

4.2   Problem Statement

It will be perfect if we can define interoperability in a unique way, except the diversity of contexts where interoperability is required prevents us, for this reason, concept of interoperability has various definitions while switching domains and contexts. To describe the interoperability issues among Federated Identity Management systems (FIM), we assume a User U(A) belonging to federation A Fed(A), who request access to a service S(B) deployed in a service provider B SP(B), which is an element of a federation B Fed(B): SP(B) ∈ Fed(B). Researchers assume too, that Fed (A) has also an IdP and SP that are elements of Fed (A): SP(A) ∈ Fed(A) and IdP(A) ∈ Fed(A).

When a user from federation Fed(A) aims to gain a service in federation Fed(B) which is part of the same trust circle. At this point, researchers consider that the trust issue is solved, hence the federations can exchange the identity information safely. We begin the communication processes by an access request to the resources. Therefore, the user from the Fed(A) has to be authenticated in the Fed(B) identity provider IdP(B) which is an element of F(B): IdP(B) ∈ Fed(B). The issue observed is that, the request is not understandable by the IdP(B). Here we found that both Fed(A) and Fed(B) can't speak the same language. This is due to the difference in federations technologies.

Figure 1 presents, the case of massive number of communicating entities, where several service providers establish federations with multiple identity providers. All entities knowing and trusting each other. This is the most complicated case regarding topology.

However, this is the clearest case at the level of explaining the possibilities in terms of issues that communication among the entities of the federation can address. This model presents the plurality of service providers and the diversity of identity providers that provide user authentication. Above all, the level of complexity in processing communication between federation entities increases with the complexity

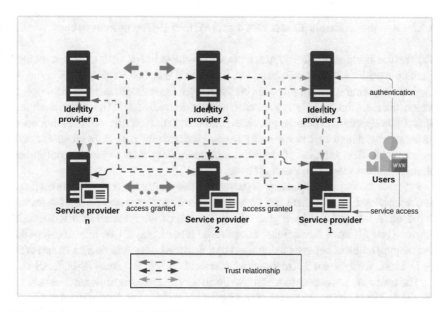

Fig. 1 Case of multiple entities federation

of the model used. However, the problem increases enormously when all these entities are adopting different technologies for the implementation of their federations.
Actually, we can say that there is nonexistence of interoperability between different
federations' technologies.

The goal of this work is creating a common infrastructure i.e. interoperability platform, where the exchanged requests among heterogeneous federation technologies
could be translated into other federations' requests [12]. There are many levels of
interoperability to deal with, however the researcher target the technical level. Since
there are technical problems detected in the communication among different federations [13]. Technical interoperability within identity federation systems refers to two
or more federation systems that may operate and exchange identity data between
each other, with the system communication interface as the endpoint responsible for
these exchanges, it must be identified, known and accessible. The interface that is
represented by the edge layer of the two systems enables exchange and interaction.

In the context of exchange between identity federation systems that are characterized by their heterogeneity, ensuring communication between these heterogeneous
systems requires analyzing the aspect of interoperability. In the previous studies,
researchers treated the identities portability issue among federated models. They
proposed to solve this problem by an interoperable approach [3]. Systems that want
to communicate and exchange identity information needs to be able to interoperate
without particular concern by using common structures and types of data. These
communication systems can evolve independently without the risk of breaking this
interoperability.

4.2.1   Communication Restriction Level Within Different Federations

To meet the interoperability requirements, we conducted a simulation for the communication workflow with two federations, both based on a heterogeneous technology. Under this example, communication between these federations has been interrupted, which indicates that there is a problem in this communication. In this work, the interoperability approach is being implemented to address the problem of communication discontinuity, due to the heterogeneity of these federations' protocols, standards, and technologies. The solution of technical interoperability aims to ensure continuous communication among different systems.

Figure 2 shows the architecture deployed to achieve a real exchange between these two federations. We note here that the user is already a trusted user for the federation based on WS-Federation. We give an explanation of the communication scenario between two different federations. The purpose is to display all the relevant steps for exchanging requests between federations based on the two standards SAML and WS-Federation. We have set up the previously selected SAML2.0 and WS-Federation.

The user requests access to a protected web service from his browser. The service provider (SP) redirects user to the identity provider (IdP) and transports the SAML request. This request is sent to IdP. An error occurred when redirecting to the Security Token Service (STS) identity provider.

The purpose of the previous scenario is to allow a SAML service provider to exchange data about a subject with a WS-Federation identity provider and vice versa, in a transparent and technology independent manner. At this level, it can be seen that the WS-Federation's identity provider can no longer analyse the received SAML request.

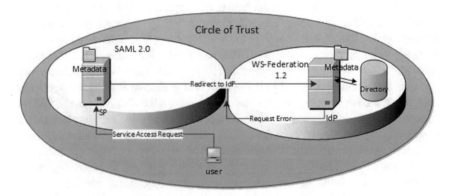

Fig. 2 Communication interruption while sending authentication request from SAML2.0 to WS-Federation

4.3 Problem Discussion

Following the implementation of identity federation technologies, it is noted that each of these technologies has its specific way to negotiate and propagate identity.

After the observation of the parameters used by each of the SAML and WS-Federation (Table 2), Researchers noticed that there is homogeneity in the exchanged flows and heterogeneity in the structure and parameters of the requests. We have mentioned in the following table the characteristics of some parameters and their correspondence SAML and WS-Federation [14].

Each federation relies on its own properties, so the exchange between the two entities has produced results that need to be taken into account. In summary, SAML and WS-Federation reach the same result except with different treatments [15].

In the first place, the user requests the protected service through an HTTP request, to the SAML configured SP, once the request is received, the SP redirects the user to the SSO (Single Sign-On) service of the WS-Federation configured STS IdP. The request consists of the URL of STS SSO service in charge of authenticating users based on Active Directory and takes in parameters the usual SAML values, specifically the SAMLRequest request transferred in GET to this service [16].

The STS receives the SAMLRequest and RelayState parameters [14]. Under these conditions, the identity provider returns an access error to the server with a reference number, which is mentioned bellow:

196d480a-0c97-4ae0-b95a-7fae8e0a1f4e

This code generated by the ADFS service of the STS identity provider shows a request error and indicate the revision of the structure. In normal cases, the STS receives a request that contains the parameters Wa = wsignin1.0, Wtrealm, Wctx [14]. When receiving the parameters for the SAML specification, the STS is confused because it cannot interpret the request by providing the appropriate response. For

Table 2 Comparative analysis between SAML and WS-Federation parameters

Parameters	Function	SAML	WS-federation
Connection request	Request to the authentication server	SAMLRequest	Wa = wsignin1.0
	Inform the ADFS server about the SP	RelayState	Wtrealm
	Status of the SP before communicating the ADFS	RelayState	Wctx
Authentication protocol	Authentication protocol used	Form based	Form based microsoft ntlm protocol
Token	Type of token used	SAML2.0	SAML 1.1

this reason, the ADFS returns the error code that demands requests understandable by the STS [17].

4.4 Prototype Proposal

Interoperability is the faculty of two entities to interact, communicate and exchange information bringing into account protocols and policies use of each entity. To ensure interaction with no constraints. Interoperability of federated systems in a federation context refers to the potential to manage and exchange identity information with two or more federated systems [18, 19] (Fig. 3).

Each federation system has its own way of negotiating and propagating requests that include all identity information. The discussion of the problem of structural heterogeneity, of existing federation technologies, leads to the study of the appropriate solution for ensuring technical interoperability between these federations. The solution must ensure the exchange between heterogeneous federations in a flexible and transparent way. In addition, the diversity of requests and responses used by existing federation technologies requires the creation of a solution that supports the matching of all parameters of inter-federation exchanged requests.

The researchers found there are many positions to implement the interoperability third party. They discussed the existent scenarios, which support federated systems to achieve this capability [3]. To achieve technical interoperability between different federation technologies, it is important to study the interoperability supported scenarios. For this purpose, we consider the possibilities to take charge for interoperability according to the following cases:

After analyzing the feasibility of supporting interoperability, authors concluded that every deployment scheme examined has specific strengths and weaknesses. In relation with the requirement to implement interoperability, federated systems must be easily embedded in communication while having independent handling. Scenario 3 is a benefit in terms of integration, independent handling and centralized usability. Based on this finding, the establishment of a third party entity that will take care of the interoperability treatment was the solution adopted. However, in the case of other scenarios, the implementation still always depended on the technology where

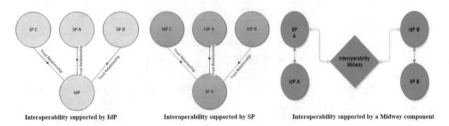

Fig. 3 Interoperability scenarios

implemented. Because of that, the authors proposed an approach based on a third party entity to ensure the interoperability task.

4.4.1 Proposal to Solve Interoperability Issue

We propose a model for transparent communication between federated systems with a guarantee of interoperability among the heterogeneous federated systems technologies. The requirements form this proposed midway are:

- Bilateral Interoperability between the federated systems.
- Authentication and authorization supported.
- SSO multi-domains.
- Trust between federated systems.

Equivalent to what is exposed in Fig. 4, an outline of the functional process regarding the suggested proposal, since the source, when the interoperability midway acquires the request, and as long as the spread of identities among heterogeneous federations is going on. The researchers explain the main steps to convert a request from source federation to destination federation:

- Detection of the source and destination federation.
- After the phase of detecting the federation destination; this information is extracted from the request that is the input to this midway (Input). In this step, we encounter two situations, one is simple, and the other is sophisticated.
- The first situation: in the case of a federation with a homogeneous destination, i.e. the detected technology of the source federation is the same as that of the destination detected in the second step. The communication process continues normally and the request is transmitted to the target.
- The second situation: In the case of the federation of a different destination from that of the source (heterogeneous federation). Many steps to be followed before sending the request to the detected destination.
- First, we should detect the type of request. Is that a request or response? Is it an authentication request or attribute request? Is it a Get, Post or Redirect request..., we extract the main characteristics of the detected request. In the area "Conversion" conversion to standard form, it is possible to recognize the technology of federated entities requiring interoperability, which is registered in an internal database, which includes a list of entities and the technology associated with each entity. The next step is the mapping of federated requests, based on the identification of the type of request, and then involving the required processing.
- The task of converting the query is handled by the mapping box between the query parameters, and their equivalents in the standard form table, and then reconstructing its parameters according to the structure supported by the destination federation technology. The output of the conversion box in standard form is a request corresponding to the form of the destination federation. After all, the new reformulated request is transferred to the destination federation.

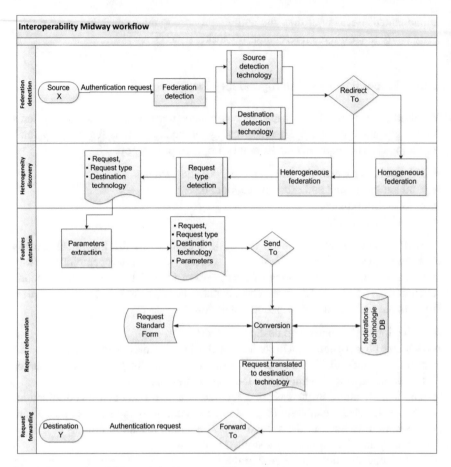

Fig. 4 Proposed model for technical interoperability approach

5 Results

From the material and method section, researchers assume that the proposal should be implemented in a practical environment. In this result section, researchers present the technical environment used to implement the proposed prototype. Researchers' results are demonstrated in a graphical user interface. Thereafter, TIC main elements are presented separately. The main results subsection will show the result after translating a request from one federation technology to another federation technology.

5.1 Implementation

Researchers discuss the deployment of the Technical interoperability component TIC in this subsection. Technical Interoperability Component is the midway component in our proposed solution to ensure technical interoperability between the different federations.

First, the researchers deployed a virtualization infrastructure based on Oracle solutions. Figure 5 describes the architecture used to implement the chosen solution. Researchers used two machines, the first one will be used to install and configure the VM manager (Virtual Machine Manager) and the second one will be assigned to the server. The virtualization technology used is Oracle VM Server/Manager.

The two machines are connected via an Ethernet cable to establish a connection between them, which is an internal wireless network between both machines. The machine or server will be installed and then added in a pole. Then, Virtual machines are imported (if exist) or installed in these poles, this will be managed via the VM manager. Those virtual machines are the hosts where we have installed different federation technologies (Example: SAML, WS-Federation, Shibboleth, Open-ID) with their identities and servers providers.

The virtual infrastructure is built to install different federation technologies. Due to the multiplicity of federation technologies. Moreover, one of those virtual machines used to implement the proposed interoperability component which has to be an intermediary tool to receive exchanges requests between different federations and then translate them based on the request source, destination, type, and parameters.

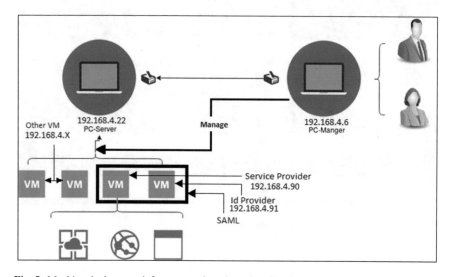

Fig. 5 Machine deployment infrastructure based on virtualization solution

5.1.1   TIC Graphical User Interface (GUI)

Figure 6 present Graphical User Interface of TIC components which makes able to translate received request to the appropriate federation.

The core components of technical interoperability Component TIC, are implemented in Java (J2EE) and MySQL database has been used for the storage of parameters and attributes. The graphical interface (Fig. 6) of the part responsible for converting requests from on technology to another.

The researchers present the results obtained in this research work. If the user gives as input to the TIC midway a request then the TIC can detect the federation technology. Thereafter, TIC will process the request to know if it is a **get**, **post** or **redirect** request. Then, based on the request type and the destination federation parameter extracted from the request, the TIC will translate the request to the appropriate type of request appropriately with the destination federation technology.

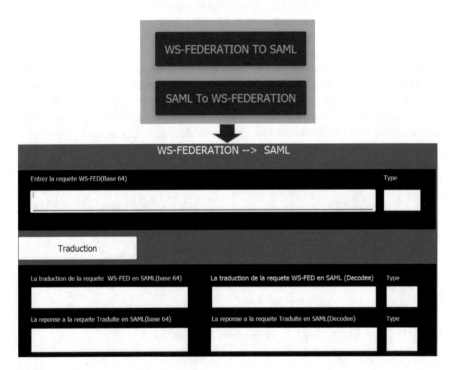

Fig. 6 The TIC component graphical interface

5.1.2   TIC Component Main Elements

The TIC components gives as result the translated request from SAML2.0 to WS-Federation, and vice versa. If the request is URL, Base64 decoded or encoded, TIC can inverses the operation. Then the tool translate the request to the appropriate form.

Figure 7 present the conception followed to implement the TIC component. The researchers used 4 classes to ensure the conception of this translation tool: decodebase64, SAML2, Translate, WSFED.

To translate the received requests, the following steps are followed:

1. The system will receive the request encoded in base 64, decode it and then detect and retrieve if it is SAML2.0 or WS-Federation.
2. If it is a SAML2.0 request, it detects its type (**GET, POST, REDIRECT**), translates it into a WS-Federation request of the same type (for example if the request is SAML2.0 of type **GET** after translation the system must return a WS-Federation request of type **GET**).
3. After translation comes the response phase, the system sends the response of the translated request as follows:

   - If the translated request is of type **GET** the system sends a request of type **REDIRECT.**
   - Otherwise if it is of the **REDIRECT** type, the system sends a request of the **POST** type.

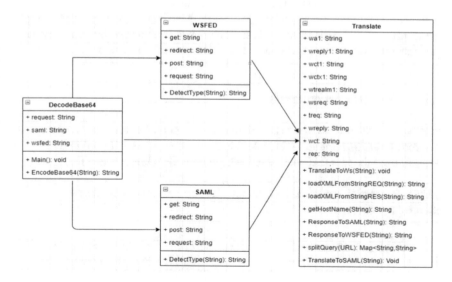

Fig. 7 TIC component implementation class diagram

5.2 *Main Results*

The researchers were able to implement the first translation direction of the TIC component which is the translation of SAML2.0 requests to WSFED requests. As well as the other direction from WSFED to SAML2.0. This is the first translation process within two different federation technologies. The TIC interoperability component is designed to be standard tool to interoperate all federation technologies

In the following we will specify the purpose of each of these classes (Fig. 7): decodebase64, SAML2, Translate, WSFED to successfully translate requests exchanged by federations.

The role of the decodebase64 class encodes or decodes received requests. If we consider the example of a SAML2 technology request. After the decoding, we move on to the processing, i.e. translation and sending of the response according to the types of the request (GET, POST or REDIRECT). The function saml.Detecttype() is imported from the SAML2 class.

If the request is WSFED, the same processing will be done, i.e. we will have a translation and send a response according to the types of requests (GET, POST or REDIRECT). The first case is the translation of a SAML request to a WS-Federation request. We declare the parameters and then translate the requests to WSFED.

LoadXMLFromSting() function provides a DOM parser as a basis. Its purpose is to browse the xml string which composes the SAML2 authentication request (SAMLRequest) request and then extract the parameters required to translate into the WS-Federation request. Then we search in the XML block for the parameters IssueInstant and saml: Issuer and we give their values (content) to the wct and wreply (parameters of WS-Federation request). The query is translated by assigning each parameter the value that corresponds to it in WS-Federation request.

5.2.1 Final Translation Result

Finally, we will present the result of translating an authentication query request. The result obtained implicitly includes all the previous steps followed to translate a request, from the moment when it is received until the moment when it is sent.

SAML Request Encoded in base 64 and Request after decodebase64
Request encodebase64

aHR0cHM6Ly9pZHAub3BlbmNsYXNzcm9vbXMuY29tL2NvdXJzZXL1NTTz
9TQU 1MUmVxdWVzdD1yZXF1ZXN0UmJlbGF5U3RhdGU9dG9rZW4=

Request decodeBase64

https://idp.openclassrooms.com/courses/SSO?SAMLRequest=requestRelayState
=token

Same thing for the inverse process, when TIC receives request from WS-Federation, it will decode it and translate it from WS-Federation form to SAML2.0 form.

WS-Federation Request Encoded in base 64 and Request after decodebase64 Request encodebase64

aHR0cHM6Ly9pZHAub3BlbmNsYXNzcm9vbXMuY29tIC9jb3Vyc2VzL1NTTz
9TQU 1MUmVxdWVzdD1yZXF1ZXN0JlJlbGF5U3RhdGU9dG9rZW4=

Request decodeBase64

https://sts.openclassrooms.com/sts?wa=wsignin1.0wreply=https://www.openclassr
ooms.com/courseswct=2019-03-17T19:06:21Z

Translation from WS-Federation Request to SAML2.0 Result The next figure present the total results obtained while trying to translate request from WS-Federation to SAML2.0 (Fig. 8):

Result after translation

https://idp.idp.openclassrooms.com/SSO?SAMLRequest =
<samlp:AuthnRequest xmlns:samlp = "urn:oasis:names:tc:SAML:2.0:protocol"
xmlns:saml = "urn:oasis:names:tc:SAML:2.0:assertion"ID = "aaf23196-1773-2113-474a-fe114412ab72" Version = "2.0" IssueInstant = null AssertionConsumerServiceIndex = "0" AttributeConsumingServiceIndex = "0" > < saml:Issuer > null </saml:Issuer > < samlp:NameIDPolicy AllowCreate = "true" Format = "urn:oasis:names:tc:SAML:2.0:nameid-format:transient"/> </samlp:AuthnRequest > &RelayState = token

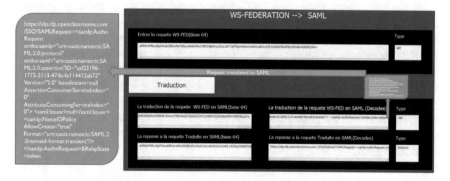

Fig. 8 Translation from WS-Federation request to SAML2.0 result

6 Conclusion and Future Works

The identity federation solution is very suitable to solve this problem of domain change. Several other problems show up while the process of the changing domain. Researchers aimed to achieve at the same time a transparent transition of private data and interoperability between different domains, all with an advanced level of security. The results demonstrate two things. First, there is a way to translate requests from different federation technologies. Second, the translation could be based on this standard framework among all existing federation technologies. The researchers detailed the process of interoperability between homogeneous federated systems, also heterogeneous federated systems. Therefore, they have designed a mechanism that ensures interoperability between mixed federated systems.

The proposed mechanism is an intermediate midway named TIC that is part of the circle of trust of federations. These trusted federations constitute a meta-system called the federation of the federation. This proposal midway solution receives requests from federations, detects the technology of the source federation, the destination federation, translates exchanged requests, and then takes charge of matching requests between the communicating federations.

At this point, we succeeded in tackling the first problem that blocks communication among different federations, which is the nonexistence of technical interoperability among existing federation technologies in order that the first level to deal with in this work is the technical interoperability layer is guaranteed.

Interoperability targets several layers. In this paper, we have worked on the technical layer. Due to the issues involved in the heterogeneity of federation technologies, as well as the multiplicity of the meaning of the exchanged attributes. Several layers of interoperability, other than the technical layer, have to be addressed. For these reasons, the second layer that the researchers target is the resolution of the absence of semantic interoperability among federation systems. The purpose of semantic interoperability is to guarantee the use of information exchanged between federations, whatever its form or semantics.

In the next works, researches deal with semantic interoperability as an achievement for the improvement of technical interoperability. We will investigate the encountered challenges, potential semantic technologies to solve these challenges and the specification of further detailed modules of the semantic approach addressed. The researchers' results cast a new light on a techno-semantic level of interoperability that will enhance communication among heterogeneous federation technologies.

Both technical and semantic interoperability layers are evidently created a unique mechanism to solve communication issues between federation systems. Now that we guarantee techno-sematic interoperability, we will achieve the Federation of Federation goal.

References

1. DataReportal (2020) Digital 2020 Zimbabwe (January 2020) v01. Jan-2020. [Online]. Available https://www.slideshare.net/DataReportal/digital-2020-zimbabwe-January-2020-v01. Accessed 15-Mar-2020

2. Itu T, Itu TSSO (2016) Series X: data networks, open system communications and security cyberspace security—identity management Baseline identity management terms and definitions

3. L'Amrani H, Berroukech BE, El Bouzekri El Idrissi Y, Ajhoun R (2017) Toward interoperability approach between federated systems. In: ACM international conference proceeding series, vol. Part F1294

4. Beer Mohamed MI, Hassan MF, Safdar S, Saleem MQ (2019) Adaptive security architectural model for protecting identity federation in service oriented computing. J King Saud Univ Comput Inf Sci (xxxx)

5. Kanwal A, Masood R, Shibli MA (2014) Evaluation and establishment of trust in cloud federation. In: Proceedings of the 8th international conference on ubiquitous information management and communication, ICUIMC 2014

6. Group OC (2007) Achieving interoperability between active directory federation services and shibboleth

7. France M (2012, June) Using AD FS 2. 0 for interoperable SAML 2. 0-based federated web single sign-on

8. Ates M, Gravier C, Lardon J, Fayolle J, Sauviac B (2007) Interoperability between heterogeneous federation architectures: illustration with SAML and WS-Federation. In: Proceedings of international conference signal image technology internet based system SITIS 2007, pp 1063–1070

9. Baldoni R (2012) Federated identity management systems in e-government: the case of Italy. Electron Govern

10. Damien C (2016) SP vs. IdP initiated SSO I Damien Carru's Blog: it's a federated world. [Online]. Available https://blogs.oracle.com/dcarru/sp-vs-idp-initiated-sso. Accessed 19-Nov-2019

11. Pérez-Méndez A, Pereñíguez-García F, Marín-López R, López-Millán G, Howlett J (2014) Identity federations beyond the web: a survey. IEEE Commun Surv Tutorials 16(4):2125–2141

12. EL Haddouti S, Dafir Ech-Cherif EL Kettani M (2019) A hybrid scheme for an interoperable identity federation system based on attribute aggregation method. Computers 8(3):51

13. Type P (2013, August) Federated identity management for research collaborations the need for federated identity management

14. David Gregory M (2014) ADFS Deep-Dive: comparing WS-Fed, SAML, and OAuth-Microsoft Tech Community-257584. [Online]. Available https://techcommunity.microsoft.com/t5/Core-Infrastructure-and-Security/ADFS-Deep-Dive-Comparing-WS-Fed-SAML-and-OAuth/ba-p/257584. Accessed 19-Nov-2019

15. Groß T, Pfitzmann B (2015) Proving a WS-Federation passive requestor profile. In: Proceedings of 2004workshop on secure web services, SWS 2004, pp 77–86

16. onelogin Saml Developer Tools (2015) SAML Attribute and NameID Extractor I SAML-Tool.com

17. Svidergol B, Meloski V, Wright B, Martinez S, Bassett D (2018) Active directory federation services. Mastering Wind. Server® 2016, pp 423–455

18. Pierre-dit-mery L (2015) Référentiel Général d'Interopérabilité Standardiser, s'aligner et se focaliser pour échanger efficacement Direction Interministérielle du Numérique et du Système d'Information et de Communication de l'Etat

19. Oh SR, Kim YG (2019) Interoperable OAuth 2.0 Framework. In: 2019 International conference on platform technology and service—PlatCon 2019—proceedings, pp 2–6

Printed in the United States
by Baker & Taylor Publisher Services